T0227741

Algebraic
and Stochastic
Coding Theory

Dave K. Kythe | Prem K. Kythe

CRC Press
Taylor & Francis Group
Boca Raton London New York

CRC Press is an imprint of the
Taylor & Francis Group, an **Informa** business

CRC Press
Taylor & Francis Group
6000 Broken Sound Parkway NW, Suite 300
Boca Raton, FL 33487-2742

International Standard Book Number: 978-1-4398-8181-1 (Hardback)

Library of Congress Cataloging-in-Publication Data

Kythe, Dave K.
  Algebraic and stochastic coding theory / Dave K. Kythe, Prem K. Kythe.
    p. cm.
  Includes bibliographical references and index.
  ISBN 978-1-4398-8181-1 (hardback : acid-free paper)
    1. Coding theory. 2. Stochastic analysis. I. Kythe, Prem K. II. Title.

QA268.K98 2012
003'.54--dc23
2011050830

Visit the Taylor & Francis Web site at
http://www.taylorandfrancis.com

and the CRC Press Web site at
http://www.crcpress.com

To Kiran, Bharti, Jay, Keshav, and Ida,
so near and dear to us.

Contents

Preface

The purpose of writing this book is primarily to make the subject of coding theory easy for the senior or graduate students who possess a thorough understanding of digital arithmetic, Boolean and modern algebra, and probability theory. In fact, any reader with these basic prerequisites will find this book interesting and useful. The contents cover the technical topics as well as material of general interest. This book explains many important technical features that have permeated the scientific and social culture of our time.

The book does not stop there. It takes advanced readers, who have a good understanding of probability theory, stochastic analysis, and distributions, to recent theoretical developments in coding theory and shows them the limitations that the current research on the subject still imposes. For example, there still exist the following unanswered questions: What kind of overhead factor can we expect for large and small values of coding blocks? Are the different types of recent codes equivalent or do they perform differently? How do the published distributions fare in producing good codes for a finite number of blocks? Is there a significant random variation in code generation for the given probability distributions? In addition, there are patent issues that deal with more efficient hardware and the expectation of the current advancement on the Internet and data storage.

As one can see from the above description and the Table of Contents, the book does not deal with any aspect of cryptography as this subject warrants a separate book, although certain rules used in this book are still useful in cryptography. What are the key benefits of this book for readers? Why should they read this book? The answer lies, among many factors, in the following factual observations:

1. During the early 1960s when the space race began as a result of the Sputnik crisis, the computers on-board unmanned spacecraft corrected any errors that occurred, and later the Hamming and Golay codes were used for better transmission of data, especially during the 1969 Moon landing and the 1977 NASA Deep Space Mission of Voyager 1 and 2.

2. Subsequent NASA missions have used the Reed–Solomon codes and the above codes for data transmission from the SkyLab and International Space Station and all other space missions to Mars (2001) and the Hubble Space Telescope.

3. These codes provided error-free and high-speed data transmission with low overhead on the Internet and data storage.

4. Low-density parity-check codes, combined with different erasure codes including Gallager, symmetric, IRA, Fountain, and Tornado codes, opened up a brand new approach to improve wide-area network data transmission.

5. Current advancements in probability theory and stochastic analysis of belief propagation, along with the development of robust soliton distribution used in the Luby transform (LT) and Raptor codes, produce a new direction in high-speed transmission of very large packets of messages as text and digital media over noisy transmission channels, very worthwhile in an age when people can watch movies on their smartphones.

During the past fifteen years, research in coding theory has been devoted to error correction as well as the recovery of missing (lost) bits. Distributions and belief propagation play a significant role in this "search-and-rescue mission." The entire outlook in coding theory seems to be geared toward stochastic processes, and this book takes a bold measured step in this direction.

This book provides useful material not only to students but also to industry professionals, researchers, and academics in areas such as coding theory, signal processing, image processing, and electromagnetics. Despite the general books on coding theory available in the market, for example, those cited in the Bibliography, which are all good in different aspects, this book can be distinguished by two basic characteristics: simplicity of analytic and computational approach, and a sufficiently thorough discussion of the modern inroads into the stochastic process of belief propagation and the associated decoder that is used in the universal family of Fountain codes.

Overview

Our aim is to provide a book that satisfies the following criteria:

- Is simple enough for students to understand and grasp the beauty and power of the subject.

- Has enough mathematical rigor to withstand criticism from the experts.

- Has a large number of examples (208 to be exact) that thoroughly cover every topic in each chapter.

- Provides motivation to appeal to students, teachers, and professionals alike, by providing sufficient theoretical basis and computational techniques and resources.

- Possesses a robust self-contained text, full of explanations on different problems, graphical representations where necessary, and a few appendices for certain background material.

- Subject material of this textbook is arranged in a manner that tries to eliminate two detrimental effects on readers, namely, to prevent mathematically inclined readers from feeling left out in a book that consists of dull manipulations; and, for technically oriented readers, that the subject of coding theory need not be treated simply by stray examples without providing logical justification for the methodology.

Outline

The approach adopted in this book is more direct. To provide a solid footing in the subject, the first task is to provide background material in Boolean algebra with related logic and operations. This is necessary not only to build a strong basis for the development of the subject but also to provide many readers a chance to review these important topics. Chapter 1, being of a historical nature, describes the codes preceding the Hamming codes and those leading to the ASCII-67 code via its development from the original Morse code to Boudot's, Murry's, Bacon's, and Gauss and Weber's telegraph codes. The binary-coded decimal codes are also included because of their importance in storing high-precision floating point numbers on some calculators. Chapter 2 is a review of certain useful material from (i) the number system, including the least common multiple, greatest common divisor, and the modulo computation; (ii) Boolean and bitwise operations including the XOR swap algorithm and XOR linked lists, bit shifts together with arithmetic, logical and circular shifts, and arithmetic of one's and two's complement; (iii) Fletcher's and cyclic redundancy checksums; (iv) residues, residue classes, and congruences; and (v) lexicographic order.

Linear codes begin in Chapter 3, which contains the following topics: vector spaces over finite fields; communications channels and Shannon's theorems on channel capacity; definitions of Hamming and Lee distance, Hamming cubes, and Hamming weight; residual codes; tree algorithm, and some definitions needed for the structure of linear codes; parity bits and their applications; and descriptions of six important decoding methods. The chapter ends with the definition of wedge and bar products. Chapter 4 deals with the single-error correcting binary linear Hamming (7,4) and (11,7) codes and all sixteen possible Hamming (7,4) codes. Hamming's original algorithm is described because of its historical significance, and equivalence codes and q-ary Hamming codes are included.

Chapter 5 describes the extended Hamming codes that are single-error-correction double-error-detection (SEC-DED) binary linear codes. This chapter includes a detailed account encoding of Hamming (8,4), (13,8), (32,26), and (72,64) codes and the modified Hamming (72,64) code. Hsiao code, as an extended Hamming code, with its encoder and decoder, occupies a special place in this description. Chapter 6 explains all known bounds in coding theory, which include sphere-packing bound, Johnson bound, Gilbert–Varshamov bound, Hamming bound, Plotkin bound, Griesmer bound, Zyablov bound, Reiger bound, linear programming bound, and stochastic bounds for SEC-DED codes. The Krawtchouk polynomials are defined with certain useful results that are needed in linear programming bound.

Linear binary and ternary Golay codes are discussed in Chapter 7, along with the extended Golay binary and ternary codes. The notion of perfect codes and their examples are presented. It is interesting to note that Hamming codes were introduced by Golay and Hamming. The first phase of coding theory constructed by using the Boolean and bitwise operations and the notion of the parity-check and generator matrices conclude this chapter.

The next phase in the development of coding theory is based on the Galois field theory and the associated arithmetic, which are the subject matter of Chapter 8. The Hadamard matrix and the Hadamard code with its encoding and decoding procedures are discussed in Chapter 9. The graphical design called the Hadamard tiling pattern is introduced. Other topics included in this chapter are the Hadamard transform, lexicodes, octacode, simplex codes, and block codes. Based on the Galois theory, cyclic codes are discussed in Chapter 10, BCH codes in Chapter 11, Reed–Muller codes in Chapter 12, and Reed–Solomon codes in Chapter 13. The encoding and decoding algorithms for these codes are explained, and the problem of burst errors and erasures achieved by the Reed–Solomon codes is thoroughly investigated. The concept of concatenated codes and their types are discussed in Chapter 13.

The third phase in the development of coding theory starts in Chapter 14 where the concept of belief propagation and the related stochastic process are presented. This chapter deals with the belief propagation algorithm, the notion of stopping time, sum-product algorithm, and log-likelihood ratios (LLRs) with the associated algebra.

A new class of linear codes that was invented by Gallager [1962, 1963], the low-density parity-check (LDPC) codes, is discussed in Chapter 15. The use of Tanner graphs for encoding such codes is thoroughly explained. Two types of communication channels, namely, the binary erasure channel (BEC) and the binary symmetric channel (BSC) are introduced, and the hard-decision and soft-decision algorithms are presented as offshoots of the belief propagation for decoding of LDPC codes. Special LDPC codes, namely, the Gallager codes, irregular-repeat-accumulate (IRA) codes, systematic and nonsystematic codes, are examined in Chapter 16. The encoding and decoding of Turbo codes by the log-likelihood ratio method are presented, and the BP decoder algorithms are provided together with the probabilistic shuffled and probabilistic check-shuffled decoding, which are useful not only for the LDPC codes but also for the classes of modern codes discussed in subsequent chapters.

Discrete distributions, especially degree distribution and robust soliton distributions, are the cornerstones of modern coding theory. They are presented in Chapter 17, along with some other distributions, namely, the Gaussian distribution, Poisson distribution, and continuous and discrete probability distributions. Polynomial interpolation and the Chernoff bound are also discussed in this chapter.

Erasure codes, forward error correcting codes, Tornado codes, rateless codes, online codes, and Fountain codes are discussed in Chapter 18. These codes, and especially the Tornado codes and the concept of a pre-code, provided the groundwork for the development of modern coding theory.

The universal Fountain code family is composed of two classes: one consisting of the Tornado codes and the Fountain code, and the other consisting of the Luby transform (LT) codes and the Raptor codes. These two classes of codes, known as the modern codes, are presented in detail in Chapters 19 and 20, respectively. This presentation also includes the erasure codes, rateless codes, and online codes. The encoding and decoding of modern codes and their performance are both discussed. The asymptotically good Raptor codes, finite-length Raptor codes, and systematic Raptor codes are described in detail. The strategy for decoding these codes involves the introduction of a pre-code that detects and corrects erasures, usually by using an LDPC or a Reed-Solomon code, and then using the BP decoder for decoding the entire message at a rate close to Shannon's capacity with a very high probability of success.

These chapters are followed by five appendices, detailing the ASCII table, and some useful tables in finite and Galaois fields, plus a short essay on some useful groups and on the discrete Fourier transform, and freely available software resources on the Internet. This is followed by the Bibliography of literature used and cited in the book, along with the Index.

Instead of splitting more than two hundred examples into some examples and some exercises, we decided to keep them as they are, to make the book more reader-friendly. There are over one hundred illustrations in the book, mostly graphs and diagrams about various aspects encountered during the development of the subject.

Layout of the Material

The material of this book ranges from average to more challenging sections, which can be adapted for students and readers depending on their background. The basic prerequisite, however, is a thorough knowledge of Boolean algebra, especially the bitwise operations, Galois field arithmetic, and probability theory. This background assumes that the readers are at least at the senior undergraduate level. Readers with skills in Boolean algebra and finite fields can easily go through the first seven chapters. Readers with additional knowledge of Galois fields can steer through the thirteenth chapter, and those with knowledge of probability theory and stochastic processes will read through the entire book. Depending on their interest, readers can decide how far ahead they would like to go. Some portions of the book, if found too demanding, can be omitted at the first reading until more skills are acquired. Books dealing with a broad subject like this one cater to the needs of students and readers from different branches of science, engineering, and technology. The simplest rule to get full use of such a book is to decide on the requirements and choose the relevant topics.

Acknowledgments

The authors take this opportunity to thank Nora Kanopka, Executive Editor for Engineering and Environmental Sciences, Taylor & Francis/CRC Press, for encouragement to complete this book. We also thank the Project Editor Karen Simon for doing a thorough editing job, and the artist Shayna Murry for the creative cover design. Thanks are due to at least six referees who took the time to review the book proposal and communicate some valuable suggestions, and finally, to our wives for their patience and cooperation during the preparation of this book, and to our friend Michael R. Schäferkotter for help and advice freely given whenever needed.

Dave K. Kythe
Prem K. Kythe

Notations and Abbreviations

A list of the notations, definitions, acronyms, and abbreviations used in this book is given below.

$\{a_{ij}\}$, elements of an $n \times n$ matrix A

A, an alphabet, e.g., binary $A = \{0,1\}$, ternary $A = \{0,1,2\}$, and q-ary $A = \{0,1,2,q-1\}$; also, a matrix

$|A|$, length of A; number of elements in A

AES, American Encryption Standard

AND, Boolean logical operator denoted by & or $\wedge$

AND, bitwise operator (& or $\wedge$)

ASCII, American Standard Code for Information Interchange

AWGN, additive white Gaussian noise

A^T, transpose of a matrix A

$A_q(n,d) = \max_{C} |C|$, largest possible size of C for which there exists an (n,k,d) code over the alphabet A

b, base or radix of a number system

b_i, $i = 1, \ldots, n$, received source files (treat like c-nodes)

$b|n$, b divides n with remainder zero; the residue of n modulo b (Gaussian notation for the remainder r such that $n = bq + r$, $0 \leq r < b$)

bpu, bits per (channel) use

$B = \{\mathbf{v}_1, \ldots, \mathbf{v}_k\} \in V$, basis of a vector space V, also, $B = \{\mathbf{b}_1, \ldots, \mathbf{b}_k\} \in V$, basis of a finite field

$B_t(\mathbf{c})$, ball of Hamming radius t about a codeword $\mathbf{c}$

BCD, binary coded decimal

BEC, binary erasure channel

BEP, bit error probability

BER, bit error rate

BPA, belief propagation algorithm

BPSK, binary phase shift keying

BSC, binary symmetric channel

c-node (circle), represents a codeword in a Tanner graph

$\mathbf{c}$, codeword ($\mathbf{c} \in C$), $\mathbf{c} = \{c_i\}$

c_i, $i = 1, \dots, n$, codeword $\mathbf{c}$ in an (n, k) code

$c_{i \backslash j}$, all c_i except c_j

C, linear code

$\widetilde{C}$, extended code

$C^{\perp}$, dual code of C (orthogonal complement of C)

$|C|$, size of a code C; number of codewords in C; same as M in an (n, M, d) code

$\mathcal{C}$, pre-code

$C_1(n, d)$, set of all q-ary codes of length n and minimum distance d

$C_1|C_2$, bar product of two linear codes C_1 and C_2

$\mathcal{C}(x) = \sum\limits_{i=0}^{n} \nu_i x^n$, weight enumerator of a code C

$\mathfrak{C}$, Shannon capacity of a transmission channel

$\mathbb{C}$, set of complex numbers

CRC, cyclic redundancy check

CW, continuous wave

d, Hamming distance, or simply, distance; also, degree of a distribution

$d(C)$, Hamming distance of a code C

$d = \min\limits_{\mathbf{x}, \mathbf{y} \in C, \mathbf{x} \neq \mathbf{y}} d(\mathbf{x}, \mathbf{y})$, minimum distance of C

$d_{\min} = 2t + 1$, minimum distance of a code, where t is the number of correctable errors

d_κ, $\kappa = 1, \dots, k$, data bits in a codeword $\mathbf{c} = \{c_i\}$, $i = 1, \dots, n$

$\mathbf{d} = \{d_1, d_2, \dots, d_k\}$, data bits vector or dataword d_κ

$d(\mathbf{x}, \mathbf{y})$, Hamming distance between two strings $\mathbf{x}$ and $\mathbf{y}$ of length n

$\dim(C)$, dimension of a code

$\dim(V)$, dimension of V

DB, data bits

DEC-TED, double-error-correction triple-error-detection

DED, double-bit error detection

DIMM, dual in-line memory module

DNS, domain name system

DRAM, dynamic random access memory

e, erasure in BEC

e_i, error values

$\mathbf{e}_i$, i-th unit vector (a zero vector with a 1 in the i-th place, counting from the first)

e.g., for example (Latin *exempli gratia*)

Eq(s), equation(s) (when followed by an equation number)

ECC, error correction code

EAN, European Article Number

F_0, field of integers mod p

F_p, prime field of modulus p

F_{2^r}, binary field of order 2^r

F_q^n, vector space

F_q, finite field, subspace of F_q^n

F_q^*, nonzero elements of a finite field F_q

$F_2[x]$, binary Galois field of polynomials

FPGA, field-programmable gate array

gcd, greatest common divisor of two integers

glog, logarithm in $GF2^m$

galog, anti-logarithm in $GF2^m$

$GF(q)$, Galois field of q elements

$\mathbf{G}$, generator matrix

$\mathcal{G}_{11}$, ternary Golay (11, 6, 5) code

$\mathcal{G}_{12}$, extended ternary Golay (12, 6, 6) code

$\mathcal{G}_{23}$, binary Golay (23, 12, 7) code

$\mathcal{G}_{24}$, extended binary Golay (24, 12, 8) Code

H_n, harmonic sum $= \sum_{k=1}^{n} \frac{1}{k} = \ln k + \gamma + \varepsilon_n$, where $\gamma \approx 0.577215665$ is the
  Euler's constant and $\varepsilon_n \sim \frac{1}{2n} \to 0$ as $n \to \infty$

$\mathbf{H}$, parity-check matrix

$\mathfrak{H}$, Hadamard matrix

i.e., that is (Latin *id est*)

iff, if and only if

IBM, International Business Machines

IF-THEN, Boolean logical operator (conditional)

ITA, International Telegraphy Alphabet

I_n, $n \times n$ unit matrix

k, number of data bits in a codeword

lcm, least common multiple of a finite number of integers

LR, likelihood ratio

LLR, log-likelihood ratio

LSB (lsb), least significant bit, or right-most bit

LUT, syndrome lookup table

$m = n - k$, number of parity-check bits in a code

min wt(C), smallest of the Hamming weights of nonzero codewords of C

mod p, modulo p

MSB (msb), most significant bit, or left-most bit

MBMS, Multimedia Broadcast/Multicast Services

MD5, message digest 5 (hash functions)

MDS code, minimum distance separable code

MPA, message passing algorithm

MRF, Markov random field

M_n, Mersenne numbers of the form $2^n - 1$

M_{23}, M_{24}, Mathieu groups

n, length of a string; length of a code C; total number of bits in a codeword

$n \equiv m \pmod{b}$, n and m are congruent mod b

(n, k) code, a code with k data bits, $n - k = m$ parity-check bits, and n bits in a codeword

(n, k, d), an (n, k) code with minimum Hamming distance d

$(n, k, d)_q$, an (n, k) q-ary code with minimum Hamming distance d

(n, M, d) code, a code of length n, size M and distance d, where n is the number of bits in each codeword, and $M = |C|$ the size of the code

$n! = 1 \cdot 2 \cdot \ldots (n-1)n$, $0! = 1! = (-1)! = 1$

NOT, bitwise operator ($\neg$)

$\mathbb{N} = \mathbb{Z}^+$, set of natural numbers

op. cit., in the work cited (Latin opere citado)

OR, bitwise operator ($|$)

OR, Boolean logical operator

Ordo (Order-o), Big O: $f(x) = O(x^\alpha)$ as $x \to 0$ means: $f(x) = x^\alpha g(x)$, where $g(x)$ is bounded in the neighborhood of $x = 0$. Little o: $f(x) = o(x^\alpha)$ means: $f(x)/x^\alpha \to 0$ as $x \to 0$; in particular, $f(x) = o(1)$ means that $f(x) \to 0$ as $x \to \infty$

p, prime number

p-node (square), represents a parity bit in a Tanner graph

$p(x)$, polynomial in $F_2[x]$

$p^*(x)$, reciprocal polynomial to $p(x)$ in $F_2[x]$

p_j, $j = 1, \ldots, m$, parity bits in an (n, k) code, $m = n - k$

$p_1, p_2, p_4, \ldots, p_{2^j}$, $j = 0, 1, 2, \ldots$, parity or parity check bits (index notation)

$p_0, p_1, p_2, \ldots, p_j$, $j = 0, 1, 2, \ldots$, parity or parity check bits (exponent notation)

$p_{j \backslash i}$, all p_j except p_i

P_j, $j = 1, \ldots, K$, packets (treat like p_j or v_j)

$\mathfrak{p}$, probability for the transmitted symbol to be received with error; $1 - \mathfrak{p} = \mathfrak{q}$ is the probability for it to be received with no error

$\mathfrak{p}_i$, probability at the i-th order check, $i = 0, 1, \ldots$

$\mathfrak{P}(x|y)$, conditional probability of occurrence of x such that y occurs

$\mathfrak{p}(r)$, probability function

$\mathfrak{p}_n$, probability of a miscorrection or undetected n-tuple errors

pdf, probability density function

PR, probability ratio

q, number of elements in a q-ary alphabet

QWERTY, modern keyboard based on a layout by Christopher Latham Sholes in 1873. The name comes from the first five letter keys below the number keys on a keyboard

RAM, random access memory

$\mathrm{Res}(C, \mathbf{c})$, residual code of C

REF, row echelon form

RREF, reduced row echelon form

$\mathbf{R}^T$, same as the parity-check matrix $\mathbf{H}$

$\mathcal{R}$, relation (between elements of a set)

$\mathbb{R}$, set of real numbers

$\mathrm{sgn}(x)$, sign ($\pm$) of a number x

$\mathrm{supp}(\mathbf{c})$, set of coordinates at which a codeword $\mathbf{c}$ is nonzero

$S(k)$, Sylvester matrix

$\mathbf{s}$, symbol (sequence of one or more bits)

SDA, standard decoding array

SEC, single-bit error correction

SEC-DED, single-bit error correction double-bit error detection

S4ED, single 4 error detection

SPA, sum-product Algorithm

SLT or SLUT, syndrome lookup table (also, LUT, lookup table)

SRAM, static random access memory

$\langle S \rangle$, span of a subspace $S \subset V$; $\langle S \rangle = \{\mathbf{0}\}$ for $S = \emptyset$

$S(5, 8, 24)$, Steiner system

$S^\perp$, orthogonal complement of S

$t = \lfloor (d-1)/2 \rfloor$, number of correctable errors

t_j, $j = 1, \ldots$, fraction of symbols for information at the j-th iteration

u_κ, $\kappa = 1, \ldots, k$, received data bits in a received word $\mathbf{w}$

v_j, $j = 1, \ldots, m$, received parity bits in $\mathbf{w}$

$\mathbf{v} \cdot \mathbf{w}$, scalar (dot) product of $\mathbf{v}.\mathbf{w} \in F_q^n$

$\mathbf{v} \perp \mathbf{w}$, orthogonal vectors $\mathbf{v}$ and $\mathbf{w}$

$\mathbf{v} \wedge \mathbf{w}$, wedge product of vectors $\mathbf{v}$ and $\mathbf{w}$

$\mathbf{v} * \mathbf{w}$, multiplication of vectors $\mathbf{v}$ and $\mathbf{w}$

V, vector space over a finite field F_q

w, number of nonzero entries

w_i, $i = 1, \ldots, n$, the received word $\mathbf{w}$

$\mathrm{wt}(\mathbf{x})$, Hamming weight of a word $\mathbf{x}$

$\mathbf{w}$, the received word $\{w_i\} = \{u_\kappa\}\{v_j\}$

XOR, Boolean logical operator (Exclusive OR)

xor, bitwise operator ($\oplus$)

$[x]$, integral part of a real number x

(x), decimal (or fractional) part of a real number

$(x)_b$, a number x in base b

$\lceil x \rceil$, ceiling of x (smallest integer not exceeded by x)

$\lfloor x \rfloor$, floor of x (largest integer not exceeding x)

x_κ, $\kappa = 1, \ldots, k$, input (data) symbols

$\mathbf{x}$, a string of length n in an alphabet A; $\mathbf{x} = \{x_1, \ldots, x_n\}$; also, input symbols
$\mathbf{x} = \{x_i\}$ composed of finitely many sequences of codewords $\{c_i\}$

XCHG, exchange command

Z, ring

$\mathbb{Z}$, set of integers

$\mathbb{Z}^+$, set of positive integers, same as $\mathbb{N}$

$\varepsilon > 0$, a small positive quantity; a quantity that gives the trade-off between the loss recovery property of the code and the encoding and decoding complexity

ν_i, error-location numbers

π, permutation

$\pi(x)$, number of primes which are less than or equal to a given value of x

$\chi(x)$, quadratic characteristic function of $GF(q)$

$\mathbf{0} = [0, 0, \ldots, 0]$, zero vector, null vector, or origin in $\mathbb{R}^n$

$\mathbf{1} = [1, 1, \ldots, 1]$, unit vector in $\mathbb{R}^n$

$\binom{n}{m} = \dfrac{n!}{m!\,(n-m)!}$, $\binom{n}{0} = 1$, binomial coefficients

$\gg$, right arithmetic shift

$\ll$, left arithmetic shift

$\pm$, plus $(+)$ or minus $(-)$

$\backslash$, exclusion symbol, e.g., $c_i \backslash j$ means a set of c_i excluding c_j

$\oplus$, bitwise XOR operation (summation mod 2)

$\otimes$, Kronecker product; $A \otimes B$ is an $mp \times nq$ matrix made up of $p \times q$ blocks, where the (i, j)-block is $a_{ij} B$; also, circular convolution of two DT sequences x_k and y_k, denoted by $x_k \otimes y_k$

$\mathbf{d} \curvearrowright \mathbf{c}$, the dataword $\mathbf{d}$ generates the codeword $\mathbf{c}$ under a generator polynomial

$\star$, linear convolution, e.g., $x_k \star y_k$ is the linear convolution between two DT sequences x_k and y_k

$\diamond$, concatenation of codes

$\boxplus$, LLR addition

$\sim$, asymptotically equal to, e.g., $a_n \sim b_n$ implies that $\lim_{n \to \infty} a_n / b_n = 1$

SP, $\sqcup$, space (bar)

#, number of

■, black square for bit 1 in Hadamard graph (tiling) diagram

∎, end of an example or of a proof

Note that acronyms can be found at *http://www.acronymfinder.com*.

1

Historical Background

We first provide a brief historical sketch leading up to the Hamming codes. These codes became completely operational in 1950 when Hamming published his ground-breaking research on single-bit error correction for an erroneous bit during transmission over a noisy channel (Hamming [1950]). Before this development, Semaphore flag signaling and telegraphy were two commonly used systems of sending messages; however, they were always repeated to ensure error-free delivery. Details of the Semaphore system can be found at the International Scouts Association website: http://inter.scoutnet.org/semaphore/, and development in telegraphy is discussed below.

1.1 Codes Predating Hamming

There were some simple error-detecting codes predating Hamming codes, but they were not effective. The concept of parity was known at that time to the extent that it adds a single bit that indicates whether the number of 1 bits in a given data are even or odd. If a single bit changes during the transmission, the received message will change parity and the error can be easily detected. Note that the bit that changed can be the parity bit itself. The convention has been as follows: A parity value of 1 indicates that there is an odd number of 1s in the data, and a parity value of 0 indicates that there is an even number of 1s in the data. It means that the number of data plus parity bits should contain an even number of 1s. This parity checking code was not very reliable, because if the number of bits changed is even, the check bit will remain even and no error will be detected. Also, parity does not indicate which bit contained the error, even if it can detect the error. As a result, the entire data received must be discarded and retransmitted again. On a noisy transmission channel, this procedure would take a very long time for a successful transmission or may never succeed. Since parity checking uses only one bit at a time, it can never correct the erroneous bit.

Another method for simple error detection was the two-out-of-five code, also called the $\binom{5}{2}$ code, which was initially a binary-coded decimal code (see §1.3). The purpose of this code was to ensure that every block of 5 bits (known as a 5-*block*) had exactly two 1s. The computer could detect an error if the input did not contain exactly two 1s. This code would definitely detect single-bit errors most of the time, but if one bit flipped to 1 and another to 0 in the same 5-block, the code would still remain true and the errors would go undetected.

Another code in use at that time was known as a *3-repetition code*. It repeated every data bit several times to ensure that it was transmitted successfully. For example, if the data bit to be transmitted was 1, this code would send '111'. If the three bits received were not identical, an error has occurred. In practice, if the transmission channel is not very noisy, only one bit would mostly change in each triplet. Thus, 001, 010, and 100 each would mean that the transmitted bit was 0, while 110, 101, and 011 would mean that the bit transmitted was 1. Other cases would amount to an erroneous transmission that would be retransmitted until the correct bit was obtained. The code was, in its own right, known as an error-correcting code, although such a code cannot correct all errors. For example, if the channel flipped two bits and the receiver got 001, the code would detect the error and conclude, wrongly, that the transmitted bit was 0. If the number of bits is increased to four, the corresponding repetition code of this type would detect all 2-bit errors but fail to correct them; but at five, the code would correct all 2-bit errors and fail to correct all 3-bit errors. It was concluded that this kind of repetition code was highly ineffective, since the efficiency decreased as the number of times each bit was duplicated increased.

1.2 Codes Leading to ASCII

We now present a brief history of the different codes that led to the American Standard Code for Information Interchange (ASCII-1) that was developed during the period 1958 to 1965. The immediate predecessors of this code were mainly the following three telegraphy codes: one each by Murray, Baudot, and Morse. All these codes were character codes used to compress information whether in print or in serial transmission in electrical and electronic communications. These codes reduced the transmissions to discrete symbols that were initially not digital.

A few definitions are in order. By a *code* we mean one thing that stands for another. Naturally a code must be smaller than the thing it stands for, and it is the case except for the Morse code. A code consists of *characters* which are symbols either machine-printed or handwritten. Finally, there is a *function code* that causes the machine to do something, such as ring a bell, carriage return, and so on, and is given a mnemonic name to distinguish it from the characters used in the code.

Telegraphy started in 1837 and used a Morse code, said to be invented by Samuel Morse. The original Morse code looked like this:

1.2.1 Original Morse Code:

```
1 ·    E –    T    2 ··   I ·–   A –·   N – –   M     3 ···   S ··
  –    U · – ·    R · – –.   W – ··   D – · –   K – – ·   G – – –
O    4 ····   H ··· –   V ·· – ·   F · – ··   L – ···   B · – – ·   P – ·· –
X – · – ·   C – – ··   Z – · – –   Y · – – –   J – – · –   Q    5 ·····
4 – ····    6 ··· – –    3 – – ···    7 · – · – ·    + – ··· –    = – ·· – ·
/ ·· – – –    2 – · – – ·    _ – – – ··    8 · – – – –    1 – – – – ·
9 – – – – –    0    6 ·· – – ··    ? · – · – ·    · – ···· –    – – · – – · –
```

Notice that no code was assigned to M, O, Q, and 0.

Modern International Morse code is given in Table 1.2.1. This code was standardized in 1865 at the International Telegraphy Congress in Paris and later designated the standard International Morse Code by the International Telecommunications Union (ITU). In the United States it is largely known as the American Morse Code or the 'railroad code'. Now it is rarely used in telegraphy, but it is very popular among amateur radio operators, commonly known as CW (continuous wave) telegraphy.

Table 1.2.1 Morse Code

A · –	N – ·	0 – – – – –
B – · · ·	O – – –	1 · – – – – –
C – · – ·	P · – – ·	2 · · – – –
D – · ·	Q – – · –	3 · · · – –
E ·	R · – ·	4 · · · · –
F · · – ·	S · · ·	5 · · · · ·
G – ·	T –	6 – · · · ·
H · · · ·	U · · –	7 – – · · ·
I · ·	V · · · –	8 – – – · ·
J · – – –	W · – –	9 – – – – ·
K – · –	X – · · –	· – · · – · –
L · – · ·	Y – · – –	, – – · · – –
M – –	Z – – · ·	? · · – – · ·

Morse's original scheme did not involve any transmission of codes for characters, but instead it transmitted something like a numeric code, which was encoded at the transmitter's side and decoded at the receiver's end using a huge 'dictionary' of words with each word numbered. But this tedious practice stopped in 1844 when Alfred Vail sent the famous message: WHAT HATH GOD WROUGHT, which was sent from the Supreme Court in Washington, D.C., to

Baltimore.[1] This message was composed from a set of finite symbols of the Roman alphabet. This character code was also a recording system, which recorded the signals received on a narrow strip of paper with a pen, in the form of small wiggles depending on the changes in the voltage on the wire, which were decoded later. What this system used was a simple alphabetic code, now known as the Morse code, although Morse never invented it, but he got credit because of the hardware he created. This hardware had the following four different states of the voltage on the wire: voltage-on long (dash or 'dah'), voltage-on short (dot or 'dit'), voltage-off long (to mark space between characters and words), and voltage-off short (space between dashes and dots). However, in practice there were only dashes, dots, and spaces. The number of these symbols assigned to letters, spaces, and punctuation marks were designated generally on the basis of the frequency of their occurrence.

Emile Baudot, in cooperation with Johann Gauss and Wilhelm Weber, designed his own system for 'printing telegraph' in France in 1874. Unlike the Morse code, all symbols in Baudot's code were of equal length: five symbols, which made encoding and decoding much easier. This system involved a synchronous multiple-wire complex, where the job of human operators was to do the 'time slicing' to generate codes on a device with five piano-type keys operated by two fingers on the left hand and three from the right: the right index and two adjacent fingers and the left index and middle fingers. Both encoding and decoding were mechanical and depended only on two stages of the wire: voltage-on and voltage-off, and there was an automatic printing of encoding and decoding. By virtue of being a fixed-length code, this system, though crude, was a pioneer in the field. Baudot's code follows.

1.2.2 Baudot's Code:

0	1	2	3	4	5	6	7	8	9	10	11	12	13	14	15
LTRS															
undef	A	E	E'	I	O	U	Y	FIGS	J	G	H	B	C	F	D
LTRS	C	X	Z	S	T	W	V	(note)	K	M	L	R	Q	Z	P
FIGS															
undef	1	2	&	3	4	O	5	FIGS	6	7	H	8	9	F	0
LTRS	.	,	:	;	!	?	'	(note)	(	)	=	−	/	No	%

---

[1] This phrase is part of the following well-known poem: "The sun has now risen; the stone has been cast;/ Our enemies are the enemies of Thought;/ So prescient indeed was the first question asked;/ for truly, What hath God wrought?" The following original statement in Morse's handwriting is kept at the Smithsonian Archives in Washington, D.C.: "This sentence was written in Washington, DC, by me at the Baltimore Terminus at 8h. 45m. in A.M. on Friday May 24th 1844, being the first ever transmitted from Washington to Baltimore by Telegraph and was indicted by my much beloved friend Annie G. Ellsworth. (signed) Sam F. B. Morse. Superintendent of Elec. Mag. Telegraphs." It is the first public Morse Code message of 22 characters including spaces sent in 1844 from Baltimore to Washington that changed the history.

Although the above code lists only 32 symbols (characters), actually 64 symbols can be produced using the typewriter's SHIFT key which moves the paper and platen up or down to achieve two different kinds of cases or rows of characters. Thus, by using the same number of print hammers, the shift key doubles the number of symbols that are designated as FIGS (figures) and LTRS (letters). For example, to type PAY 50 DOLLARS the encoder would press the following keys:

P A Y [SP] [FIGS-SHIFT] 5 0 [SP] [LTRS-SHIFT] D O L L A R S

where [SP] denotes the space bar (⎵), and [FIGS-SHIFT] and [LTRS-SHIFT] refer to the typewriter's SHIFT key. Note that this code produced only uppercase letters, numbers, and punctuation.

An improvement on Boudot's code was made during 1899–1901 by Donald Murray who developed an automatic telegraphy system with a typewriter-like key encoding system that generated synchronized bit-level codes. The Murray code avoided Baudot's 'piano' key system which required impressing the symbols onto wires with fingers, and instead required the operator (encoder) to press keys with appropriate labels on the typing machine. The simplifying criterion was one-lever one-punch movement on a paper tape, which reduced the cost and wear on the machinery considerably. Murray's code follows:

1.2.3 Murray's Code:

	0	1	2	3	4	5	6	7	8	9	10	11	12	13	14	15
LTRS																
0	BLANK	E	LF	A	LTRS	S	I	U	CR	D	R	J	N	F	C	K
1	T	Z	L	W	H	Y	P	Q	O	B	G	FIGS	M	X	V	DEL
FIGS																
0	BLANK	3	LF	undef	LTRS	'	8	7	CR	2	4	7/	−	1/	(	9/
1	5	.	/	2	5/	6	0	1	9	?	3/	FIGS	,	£	)	DEL

There were two binary-ordered telegraphy codes used during this period. They were known as the Bacon's code and Gauss and Weber's Code. According to *A History of Science and Engineering in the Bell System* these code are defined below.

1.2.4 Bacon's Telegraph Code:

AAAAA	*A*	*AABBA*	*G*	*ABBAA*	*N*	*BAABA*	*T*
AAAAB	*B*	*AABBB*	*H*	*ABBAB*	*O*	*BAABB*	*V(U)*
AAABA	*C*	*ABAAA*	*I(J)*	*ABBBA*	*P*	*BABAA*	*W*
AAABB	*D*	*ABAAB*	*K*	*ABBBB*	*Q*	*BABAB*	*X*
AABAA	*E*	*ABABA*	*L*	*BAAAA*	*R*	*BABBA*	*Y*
AABAB	*F*	*ABABB*	*M*	*BAAAB*	*S*	*BABBB*	*Z*

There was another telegraphy code, known as the Gauss and Weber Telegraph

Code, defined as follows:

1.2.5 Gauss and Weber's Telegraph Code:

$RRRRR$ (A)	$LLRLL$ (I/Y)	$RRRLL$ (R)	$RLRLL$ (3)
$RRRRL$ (B)	$LRRRL$ (K)	$RRLRL$ (S/Z)	$RLLRL$ (4)
$RRRLR$ (C)	$RLRRR$ (L)	$LLRLR$ (T)	$LLLRR$ (5)
$RRLRR$ (D)	$RRLLL$ (M)	$RLLLR$ (U)	$RLLRR$ (6)
$RLRLR$ (E)	LLLLL (N)	$LRRLL$ (V)	$LLLRL$ (7)
$LRRRR$ (F)	$LRLLL$ (O)	$LLLLR$ (W)	$LLRRL$ (8)
$LRLRR$ (G/J)	$LRLRL$ (P)	$RLLLL$ (1)	$LRRLR$ (9)
$RLRRL$ (H)	$LLRRR$ (Q)	$RRLLR$ (2)	$LRLLR$ (0)

The rights to Murray's design were purchased by the Western Union Telegraph Company, which modified it by eliminating fractions and other obsolete characters in the FIGS case and replaced them with their own characters. This improved code was used until 1950s, when improvements were required because the telegraphy networks had grown very large internationally. An improved code, known as the ITA# 2 (International Telegraphy Alphabet) code, was adopted by the International Telegraph and Telephone Consultative Committee (also known as CCITT in French):

1.2.6 ITA2:

0	1	2	3	4	5	6	7	8	9	10	11	12	13	14	15
LTRS															
0	BLANK	E	LF	A	SP	S	I	U	CR	D	R	J	N	F	C	K
1	T	Z	L	W	H	Y	P	Q	O	B	G	FIGS	M	X	V	LTRS
FIGS															
0	BLANK	3	LF	–	SP	$'$	8	7	CR	WRU	4	BEL	,	undef	:	(
1	5	+	)	2	undef	6	0	1	9	?	undef	FIGS	.	/	=	LTRS

1.2.7 US TTY:

0	1	2	3	4	5	6	7	8	9	10	11	12	13	14	15
LTRS															
0	BLANK	E	LF	A	SP	S	I	U	CR	D	R	J	N	F	C	K
1	T	Z	L	W	H	Y	P	Q	O	B	G	FIGS	M	X	V	LTRS
FIGS															
0	BLANK	3	LF	–	SP	BEL	8	7	CR	$	4	$'$	,	!	:	(
1	5	"	)	2	#(f)	6	0	1	9	?	&	FIGS	.	/	;	LTRS

The ITA2 replaced all codes used prior to the 1930s, and all teletype equipment began using this code or its American version, the USTTY code. These codes have retained the five-bit telegraphy scheme initiated by Baudot, but they did not have any automatic controls, except for the WRU function (WHO ARE YOU) and the BEL function that rang the bell. These codes were slow but teletypes and ITA2 codes were used in early computing machinery until the late 1950s, until Alan Turing envisioned computing with numbers. First, ITA2 type coding machines began storing all bit patterns on five-level paper tapes in the reverse order (left-to-right), yet some machines still punched holes from right-to-left. There was no confusion reading these tapes as long as one remembered which machine produced the tapes. The reversible ITA2 code is as follows:

Character	Defined	Reversed	Result
BLANK	00000	00000	Symmetrical
SPACE	00100	00100	Symmetrical
LTRS	11111	11111	Symmetrical
FIGS	11011	11011	Symmetrical
CR	01000	00010	Equals LF
LF	00010	01000	Equals CR

A solution to this situation was to make bitwise symmetrical all the characters that are connected with 'transmission control'. Thus, the codes for FIGS, LTRS, SPACE, BLANK, CR, and LF remain the same when reversed. This produced ASCII-1967 code which is as follows:

1.2.8. ASCII-1967 Code:

FIELDATA

	0	1	2	3	4	5	6	7	8	9	10	11	12	13	14	15
0	IDL	CUC	CLC	CHT	CCR	CSP	a	b	c	d	e	f	g	h	i	j
1	k	l	m	n	o	p	q	r	s	t	u	v	w	x	y	z
2	$D0$	$D1$	$D2$	$D3$	$D4$	$D5$	$D6$	$D7$	$D8$	$D9$	SCB	SBK	undef	undef	undef	undef
3	RTT	RTR	NRR	EBE	EBK	EOF	ECB	ACK	RPT	undef	INS	NIS	CWF	SAC	SPC	DEL
4	MS	UC	LC	HT	CR	sp	A	B	C	D	E	F	G	H	I	J
5	K	L	M	N	O	P	Q	R	S	T	U	V	W	X	Y	Z
6	)	–	+	<	=	>	_	\$	*	(	"	:	?	!	,	ST
7	0	1	2	3	4	5	6	7	8	9	'	;	/	.	SPEC	BS

The FIELDATA character code was part of a military communications system; it remained in use between 1957 and the early mid-1960s. Although it is not standardized and failed to be commercially useful, it affected ASCII development. Later a few variants of this code appeared, namely the FIELDATA Standard Form, FIELDATA variant COMLOGNET, and FIELDATA variant SACCOMNET 465L.

The ASCII (American Standards for Information Interchange) was developed by the committee X3.4 of the American Standards Association (ASA). This committee was composed of persons from the computing and data communications industry, including IBM and its subsidiary Teletype Corporation. AT&T also needed a new efficient character code for its business. The standard X3.4-1963 code was published on June 17, 1963. It had the following characteristics: 7-bits; no case shifting; control codes rearranged and expanded; redundancies and ambiguities eliminated; 4-bit 'detail' grouping of FIELDATA retained; explicit sub-set groupings: 4-bits encompasses BCD 0-9; 5-bits encompasses A-Z; and so on; COBOL graphic characters retained; superset of all in-use character sets; a few non-English alphabets included, like the British Pounds Sterling symbol; and the typewriter keyboard layout improved.

ASCII-1967 was an International Reference Version (IRV), definitely a US version of the ECMA-6 code with lots of changes incorporated, such as dropping of and improving upon some message format control characters; juggling some control characters, notably ACK and the new NAK to increase their Hamming distance (see §3.3.1); adding graphic characters above the lowercase characters; replacing ∧ (carat) by up-arrow, and left-arrow by −. The X3.4-1967 code included the following graphic characters:

$$! @ \# \$ \% \wedge \& * ( ) - + = \{ \} [\ ] - : ; ' , . / ? \& <> "$$

The current ASCII code is given in Table A.1 in Appendix A. In addition to these codes, there are additional ANSII codes that use decimal digits ≥ 128 to represent other international symbols that have become part of the English language. These codes are available on the Internet.

1.3. BCD Codes

We will eventually consider binary linear (n, k) codes, which together with nonlinear (n, M, d) codes have played a very significant role in coding theory. During the earliest period of development of this subject, the *binary-coded decimal* (BCD) codes (or systems, as they were called then) were popular, and some of them have been used even in modern times These codes were based on the premise that, in addition to the number systems with base b, there are other special number systems that are hybrid in nature and are useful in computation, as computer inputs and outputs are mostly in decimal notation.

1.3.1 Four-Bit BCD Codes. These codes are defined as follows.

(a) **8421 code.** A number system in base b requires a set of b distinct symbols for each digit. In computung the decimal ($b = 10$) and the binary ($b = 2$) number systems we need a representation or coding of the decimal digits in terms of binary symbols (called *bits*). This requires at least four bits, and any 10 out of the 16 possible permutations of these four bits represent the decimal digits. A systematic arrangement of these 10 combinations is given in Table 1.3.1, where d denotes the decimal digit.

In the BCD code, the weights of the positions are the same as in the binary number system, so that each decimal digit is assigned a combination of bits, which is the same as the number represented by the four components regarded as the base 2 number. This particular code is also called *direct binary coding*.

The nomenclature 8421 follows from the weights assigned by the leftmost 1 in the successive bits in this representation.

The 8421 code uses four bits to represent each decimal digit. For example, the number 697 is represented by the 12-bit number 0110 1001 0111, which has 3 four-bit decades. Although this number contains only 0s and 1s, it is not a true binary number because it does not follow the rules for the binary number system. In fact, by base conversion rules we have $(697)_{10} = (1010111001)_2$. Thus, it is obvious that arithmetic operations with the 8421 code or any other BCD code would be very involved. However, as we shall soon see, it is quite easy for a computer program to convert to true binary, perform the required computations, and reconvert to the BCD code.

Table 1.3.1 BCD Codes

d	8421	Excess-3	2421	5421	5311	7421
0	0000	0011	0000	0000	0000	0000
1	0001	0100	0001	0001	0001	0111
2	0010	0101	0010	0010	0011	0110
3	0011	0110	0011	0011	0100	0101
4	0100	0111	0100	0100	0101	0100
5	0101	1000	1011	1000	1000	1010
6	0110	1001	1100	1001	1001	1001
7	0111	1010	1101	1010	1011	1000
8	1000	1011	1110	1011	1100	1111
9	1001	1100	1111	1100	1101	1110

A digital computer can be regarded as an assembly of two-state devices as it computes with 0s and 1s of the binary system. On the other hand, we are accustomed to decimal numbers. Therefore, it is desirable to build a decimal computing system with two-state devices. This necessity has been responsible for the development of codes to encode decimal digits with binary bits. A minimum of four bits are needed. The total number of four-bit codes that can be generated is given by the permutations $^{16}p_{10} = \dfrac{16!}{6!} = 29{,}059{,}430{,}400$. As the 8421 code shows, although 10 out of 16 possible permutations of 4 bits are used, all of the above numbers are available. Hence, the choice of a particular code is obviously important. The following features are desirable in the choice of a code: (i) Ease in performing arithmetical operations; (ii) economy in storage space; (iii) economy in gating operations, error detection, and error correction; and (iv) simplicity.

(b) **Excess-3 code.** This code represents a decimal number d in terms of the binary equivalent of the number $d + 3$. It is a self-complementing but not a weighted code, and since it does follow the same number sequence as binary, it can be used with ease in arithmetical operations.

(c) 2421 code. This code is a self-complementing weighted code, commonly used in bit counting systems.

Other weighted codes are: 5421 code, 5311 code, and 7421 code, which are presented in Table 1.3.1.

1.3.2 Addition with 8421 and Excess-3 Codes. Since every four-bit BCD code follows the same number sequence as the binary system, the usual binary methods may be used. But, since in the binary notation there are 16 representations with four bits, while in BCD only 10 of these representations are used, we require some correction factors in order to account for the 6 unused representations.

(a) BCD Addition. A common method is to add two numbers in a decade in the binary manner and, if necessary, add appropriate correction factors. If addition is performed in a decade-by-decade fashion (i.e., serial addition with parallel decades, called the *multiple decade addition*), we can use either 8421 or Excess-3 code. If addition is performed in parallel, then Excess-3 code is better than the 8421 code.

In the 8421 code, the sum will be correct if it does not exceed $(9)_{10} = (1001)_2$. If the sum lies between $(10)_{10} = (1010)_2$ and $(15)_{10} = (1111)_2$, the correction factor $(+6)_{10} = (0110)_2$ must be added and it generates a carry $(1)_{10} = (0001)_2$ to the next decade. If the sum exceeds $(15)_{10} = (1111)_2$, a carry is generated by the initial addition, but the correction factor $(+6)_{10} = (0110)_2$ must still be added to the sum. Thus, we have the following three cases to consider to find the appropriate correction factor:

CASE 1. If sum $\leq (9)_{10}$, no correction is needed. For example,

$$
\begin{array}{r}
0\,1\,0\,0 = (4)_{10} \\
+\ 0\,0\,1\,1 = (3)_{10} \\
\hline
0\,1\,1\,1 = (7)_{10}
\end{array}
$$

CASE 2. If $(10)_{10} \leq$ sum $\leq (15)_{10}$, the initial sum has an illegitimate representation (i.e., has one out of the six unused representations). Add the correction factor of $(+6)_{10}$, which gives the correct sum and a carry. For example,

$$
\begin{array}{rl}
1\,0\,0\,0 = (8)_{10} & \\
+\ 0\,1\,0\,0 = (4)_{10} & \\
\hline
1\,1\,0\,0 = (12)_{10} & \text{illegitimate representation} \\
+\ 0\,1\,1\,0 = (6)_{10} & \text{correction factor} \\
\hline
1 \leftarrow 0\,0\,1\,0 = (2)_{10} & \text{plus a carry}
\end{array}
$$

CASE 3. If $(16)_{10} \leq$ sum $\leq (18)_{10}$, the initial sum gives a carry, but because the initial sum is incorrect, a correction factor $(+6)_{10} = (0110)_2$ is added. For example,

$$
\begin{array}{rl}
1\ 0\ 0\ 1 & = (9)_{10} \\
+\ 1\ 0\ 0\ 0 & = (8)_{10} \\
\hline
1 \leftarrow\ \ 0\ 0\ 0\ 1 & = (1)_{10} \qquad \text{incorrect sum plus a carry} \\
+\ 0\ 1\ 1\ 0 & = (6)_{10} \qquad \text{correction factor} \\
\hline
1 \leftarrow\ \ 0\ 1\ 1\ 1 & = (7)_{10} \qquad \text{plus a carry from initial addition}
\end{array}
$$

Example 1.3.1. (Multiple decade addition) Compute $(547)_{10} + (849)_{10}$ in the 8421code.

$$
\begin{array}{ccc}
0\ 1\ 0\ 1 & 0\ 1\ 0\ 0 & 0\ 1\ 1\ 1 = (547)_{10} \\
+1\ 0\ 0\ 0 & 0\ 1\ 0\ 0 & 1\ 0\ 1\ 1 = (849)_{10} \\
\hline
 & 1 \leftarrow & 0\ 0\ 0\ 0 \\
 & & +\ 0\ 1\ 1\ 0 \quad = (6)_{10}\ \text{correction factor} \\
\hline
1\ 1\ 0\ 1\ \rceil & 1\ 0\ 0\ 1 & 0\ 1\ 1\ 0 \quad = (396)_{10} \\
+\ 0\ 1\ 1\ 0 & & \qquad\qquad \text{plus a carry} \\
\hline
1 \leftarrow\ 0\ 0\ 1\ 1 & \text{case 1} & \text{case 3} \\
\text{case 2}
\end{array}
$$

$\blacksquare$

In the Excess-3 code when two numbers are added, their sum will contain an excess of 6. If the sum $\leq (9)_{10}$, it is necessary to subtract $(3)_{10} = (0011)_2$ in order to return to the Excess-3 code. If sum $> (9)_{10}$, the excess 6 contained in the initial sum cancels the effect to the six illegitimate (i.e., unused) representations, but it is necessary to add $(3)_{10} = (0011)_2$ to return to the Excess-3 code. Thus, the following three steps are needed in carrying out the Excess-3 addition: (i) Add the two BCD numbers in the binary manner, (ii) check each decade for a carry, and (iii) subtract $(3)_{10}$ from each decade in which a carry does not occur, and add $(3)_{10}$ to each decade in which a carry occurs. Hence, in the Excess-3 addition there are only two cases to determine a correction factor of $(+3)_{10}$ or $(-3)_{10}$ depending on whether or not a carry occurs in the initial addition.

CASE 1. If sum $\leq (9)_{10}$, the correction factor is $(-3)_{10} = (-0011)_2$. For example,

$$
\begin{array}{rl}
0\ 1\ 1\ 1 & = (4)_{10} \\
+\ 0\ 1\ 1\ 0 & = (3)_{10} \\
\hline
1\ 1\ 0\ 1 & = \quad \text{uncorrected sum} \\
-\ 0\ 0\ 1\ 1 & = \quad \text{correction factor of } (-3)_{10} \\
\hline
1\ 0\ 1\ 0 & = (7)_{10}
\end{array}
$$

CASE 2. If sum $\geq (10)_{10}$, the correction factor is $(+3)_{10} = (0011)_2$. For example,

$$
\begin{array}{rcl}
1\,0\,1\,1 &=& (8)_{10} \\
+\,0\,1\,1\,1 &=& (4)_{10} \\
\hline
1 \leftarrow 0\,0\,1\,0 &=& \text{uncorrected sum} \\
+\,0\,0\,1\,1 &=& \text{correction factor of } (+3)_{10} \\
\hline
0\,1\,0\,1 &=& (2)_{10}
\end{array}
$$

Note that in these examples the correction factor $(+3)_{10}$ or $(-3)_{10}$ is written in binary, while the number to be added and the sum are in the Excess-3 code. ∎

Example 1.3.2. (Multiple decade addition) Compute $(558)_{10} + (243)_{10}$ in the Excess-3 code.

$$
\begin{array}{ccccl}
1\,0\,0\,0 & 1\,0\,0\,0 & 1\,0\,1\,1 & = & (558)_{10} \\
+\,0\,1\,0\,1 & 0\,1\,1\,1 & 0\,1\,1\,0 & = & (243)_{10} \\
\hline
 & & 1 \leftarrow 0\,0\,0\,1 & & \\
1 \leftarrow 0\,0\,0\,0 & +\,0\,0\,1\,1 & & = (+3)_{10} \text{ correction factor} \\
\hline
1\,1\,1\,0 \quad +\,0\,0\,1\,1 & 0\,1\,0\,0 & & = (801)_{10} \\
-\,0\,0\,1\,1 \quad 0\,0\,1\,1 & & & \\
\hline
1\,0\,1\,1 & & &
\end{array}
$$

∎

(b) BCD Subtraction. To subtract with the 8421 or Excess-3 code, the subtrahend is first complemented and then added to the minuend. (In a self-complementing BCD code, such as the 8421 or the Excess-3 code, the base-minus-one complement of any number is easily obtained by changing all 0s to 1s and all 1s to 0s.) A sign bit 0 is attached to the left-most decade of the minuend and 1 is attached to the leftmost decade of the complement of the subtrahend. During the first step of this addition, the individual bits are added as in the binary system. A carry propagates from each digit decade to the leftmost decade and from the leftmost decade to the sign bit. If the sign bit produces a carry, it is added to the least significant decade as an end-around carry.

The next step in the subtraction process is to apply an appropriate correction factor to each decade. The correction factor follows the same procedure as in BCD addition. In the 8421 code, a correction factor of $(0110)_2 = (6)_{10}$ is added to the decade in which a carry occurs. In the Excess-3 code we must add a correction factor of $(+3)_{10} = (0011)_2$ to each decade in which a carry occurs, and a correction factor of $(-3)_{10}$, which is $(1100)_2$, to complement each decade in which a carry did not occur.

Example 1.3.3. Compute $(51)_{10} - (34)_{10}$ in the Excess-3 code.

sign bit		digit decades		
0		1 0 0 0	0 1 0 0	$= (+51)_{10}$
+ 1		1 0 0 1	1 0 0 0	$= -$ complement of $(34)_{10}$
1	←	0 0 0 1	1 1 0 0	
1 ← 0			1	end around carry
0		0 0 0 1	1 1 0 1	
		+ 0 0 1 1	+ 1 1 0 0	correction factors
		0 1 0 0	1 0 0 1	
			1	end around carry
0		0 1 0 0	1 0 1 0	$= (+17)_{10}$

■

Example 1.3.4. Compute $(51)_{10} - (34)_{10}$ in the 8421 code. By following the method of Example 1.3.3 in the 8421 code the answer is found as $0\,0001\,0111 = (17)_{10}$. ■

1.3.3 Codes Larger than Four Bits. The BCD $\binom{n}{m}$ codes do not, in general, follow the pure binary system as has been seen in previous sections. Sometimes, in order to provide special features like the detection of errors and simplification of decoding, coding systems use more than 4 components. One such example is the $\binom{n}{m}$ codes, where $\binom{n}{m}$ represents the number of distinct ways to choose m items from n. Those $\binom{n}{m}$ codes for which $\binom{n}{m} = 10$ are examples of BCD codes. They are: the $\binom{10}{1}$ code and the $\binom{5}{2}$ code; and as such they are useful because both of them have 10 bits and therefore can be used to represent the decimal digits (0 through 9). For example, the $\binom{5}{2}$ code has been used by the U.S. Post Office. Both are defined as follows:

Decimal Digit	$\binom{5}{2}$ code	$\binom{10}{1}$ code
0	00110	0000000001
1	00011	0000000010
2	00101	0000000100
3	01001	0000001000
4	01010	0000100000
5	01100	0000100000
6	10001	0001000000
7	10010	0010000000
8	10100	0100000000
9	11000	1000000000

The $\binom{n}{m}$ codes are only error detection codes with codewords of length n bits, such that each codeword contains exactly m counts of 1s. A single-bit error will affect the codeword to have either $(m+1)$ or $(m-1)$ 1s. The simplest way to implement an $\binom{n}{m}$ code is to append a string of 1s to the original data until it contains m counts of 1s, then append enough zeros to make a code of length n. For example, a $\binom{7}{3}$ code is created as follows:

Original 3 Data Bits	Appended Bits
0 0 0	1 1 1 1
0 0 1	1 1 1 0
0 1 0	1 1 1 0
0 1 1	1 1 0 0
1 0 0	1 1 1 0
1 0 1	1 1 0 0
1 1 0	1 1 0 0
1 1 1	1 0 0 0

In the $\binom{5}{2}$ code, if one neglects the code assigned to 0, this system can be interpreted as a weighted code with weights $6, 3, 2, 1, 0$. Thus, it is called a *semi-weighted code*.

The $\binom{10}{1}$ code is also known as the *ring counter code* and was widely used in counting operations and punched card machines, which are described below.

1.3.4 Biquinary Code. It is a weighted code constructed as a mixed $\binom{2}{1}$ and $\binom{5}{1}$ code in seven columns, of which the first two columns are for the 'bi' part and the remaining five for the 'quinary' part with weights 5 0 and 4 3 2 1 0, respectively, as shown at the top of the following table.

d	5	0	4	3	2	1	0
0	0	1	0	0	0	0	1
1	0	1	0	0	0	1	0
2	0	1	0	0	1	0	0
3	0	1	0	1	0	0	0
4	0	1	1	0	0	0	0
5	1	0	0	0	0	0	1
6	1	0	0	0	0	1	0
7	1	0	0	0	1	0	0
8	1	0	0	1	0	0	0
9	1	0	1	0	0	0	0

This code is a self-checking code and is used in an error detecting coding system. An old digital computer IBM 650 used a self-checking $\binom{5}{2}$ code for efficient storage and a biquinary code for checking the executions of arithmetical operations.

2

Digital Arithmetic

In this chapter we describe constructive procedures in the form of error detecting, correcting, and decoding codes that are used for encoding messages being transmitted over noisy channels. The goal of such codes is to decode messages with no error rate or the least error rate. Most of these codes involve certain basic iterative procedures for simple error-correcting codes, which are described in detail in the following chapters. During the past half century, coding theory has shown phenomenal growth, with applications in areas such as communication systems, storage technology, compact disc players, and global positioning systems. Before we enter into these developments, we must review some basic digital logic and related rules that are useful for the development of the subject.

2.1 Number Systems

In addition to the decimal number system, we will discuss binary, ternary, octal, duodecimal, and hexadecimal systems.

2.1.1 Decimal Numbers. This system, also known as the base-10 system, uses ten symbols (units) 0 through 9 and positional notation to represent real numbers in a systematic manner. The decimal (from Latin *decimus*, meaning 'tenth') system is also known as *denary* from Latin *denarius* which means the 'unit of ten'. The real numbers are created from the units by assigning different weights to the position of the symbol relative to the left or right of the decimal point, following this simple rule: Each position has a value that is ten times the value of the position to the right. This means that each positional weight is a multiple of ten and is expressible as an integral power of ten. The positional scheme can be expressed as follows:

$$\boxed{10^p} \cdots \boxed{10^3}\boxed{10^2}\boxed{10^1}\boxed{10^0}.\boxed{10^{-1}}\boxed{10^{-2}}\boxed{10^{-3}} \cdots \boxed{10^{-q}}$$

$\uparrow$
Decimal point

Figure 2.1.1 Positional scheme of the decimal number system.

where p and q are non-negative integers. The part of a positive rational number x on the left of the decimal point is called its integral part, denoted by $[x]$, and the one on the right of the decimal point its fractional part, denoted by (x). Thus, $x = [x] + (x)$. As seen in Figure 2.1.1, the integral part $[x]$ is a sequence of $p + 1$ digits, whereas the fractional part (x) is a sequence of q digits. The integral part $[x]$, represented by the sequence of $p + 1$ digits $\{a_p, a_{p-1}, \ldots, a_2, a_1, a_0\}$, where $0 \le a_i < 10$, has the following decimal representation:

$$[x] = a_p \, 10^p + a_{p-1} \, 10^{p-1} + \cdots + a_2 \, 10^2 + a_1 \, 10^1 + a_0 \, 10^0. \qquad (2.1.1)$$

Here, each digit a_i is multiplied by a weight 10^i, $i = 0, 1, \ldots, p$, which determines its position in the sequence. The fractional part (x), being a sequence of q decimal digits $a_{-1}, a_{-2}, \ldots, a_{-q}$, where $0 \le a_{-j} < 10$ for $j = 1, 2, \ldots, q$, has the decimal representation

$$(x) = a_{-1} \, 10^{-1} + a_{-2} \, 10^{-2} + \cdots + a_{-q} \, 10^{-q}. \qquad (2.1.2)$$

Here, each digit a_{-j} is multiplied by a weight 10^{-j}, $j = 1, 2, \ldots, q$, which determines its position in the sequence. The representations (2.1.1) and (2.1.2) justify the name of this system as the base-10 system, and the number 10 is called the *base* or *radix* of this system.

2.1.2 Division Algorithm. This algorithm describes a procedure to determine the decimal representation of a positive integer N as follows: Divide N, represented by (2.1.1), by the base 10. This gives the integral quotient

$$N_1 = a_p \, 10^{p-1} + a_{p-1} \, 10^{p-2} + \cdots + a_2 \, 10 + a_1 \qquad (2.1.3)$$

and the remainder a_0. Then divide N_1, given by (2.1.3), by the base 10. which yields the integral quotient

$$N_2 = a_p \, 10^{p-2} + a_{p-1} \, 10^{p-3} + \cdots + a_2,$$

and the remainder a_1. This process is then applied to N_2, which on repeated application $(p+1)$-times will yield all the digits $a_0, a_1, \ldots, a_p$ of the required decimal representation. This process is called an algorithm because it repeats but terminates after a finite number of steps.

Example 2.1.1. To derive the decimal representation for the integer 958, this algorithm yields $\dfrac{958}{10} = 95 + \dfrac{8}{10} \to a_0 = 8$; $\dfrac{95}{10} = 9 + \dfrac{5}{10} \to a_1 = 5$; $\dfrac{9}{10} = 0 + \dfrac{9}{10} \to a_2 = 9$. Thus, the decimal representation of 958 is $9 \left(10^2\right) + 5 \left(10\right) + 8 \left(1\right)$ ∎.

Since we are so familiar with the decimal system, this example may not appear to be impressive at all, but it explains the power of the algorithm to determine the sequence of digits a_i. Its generalization to base $b > 1$, known as the *radix representation theorem*, is as follows:

Table 2.1.1 Different Number Systems

Decimal	Binary	Octal	Duodecimal	Hexadecimal
0	0	0	0	0
1	01	1	1	1
2	10	2	2	2
3	11	3	3	3
4	100	4	4	4
5	101	5	5	5
6	110	6	6	6
7	111	7	7	7
8	1000	10	8	8
9	1001	11	9	9
10	1010	12	α	A
11	1011	13	β	B
12	1100	14	10	C
13	1101	15	11	D
14	1110	16	12	E
15	1111	17	13	F

Theorem 2.1.1. (Radix Representation Theorem) *For a base $b > 1$ and a number $N \in \mathbb{Z}^+$, there exists a unique sequence of nonnegative integers $a_0, a_1, \ldots$ and a unique nonnegative integer p such that*

(i) $a_p \neq 0$; (ii) $a_i = 0$ if $i > p$; (iii) $0 \leq a_i < b$ for all i; and (iv) $N = \sum_{i=0}^{p} a_i b^i$.

Let the positive integer $b > 1$ denote the base or radix of a number system. Then, by the radix representation theorem, a positive integer N can be written as

$$N = a_p b^p + a_{p-1} b^{p-1} + \cdots + a_2 b^2 + a_1 b^1 + a_0 b^0,$$
$$0 \leq a_i < b, \quad i = 0, 1, \ldots, p. \tag{2.1.5}$$

To determine the sequence of digits a_i, $i = 0, 1, \ldots, p$, we can use the division algorithm where we replace 10 by b. The representation so derived is unique since $0 \leq a_i < b$. Thus, for a mixed number $x = [x] + (x)$, we can write

$$x = a_p b^p + a_{p-1} b^{p-1} + \cdots + a_0 b^0 \text{ radix point } + a_{-1} b^{-1} + \cdots + a_q b^{-q},$$
$$0 \leq a_i < b. \tag{2.1.6}$$

The radix point in the above representation is called a decimal point if $b = 10$, a binary point if $b = 2$, an octal point if $b = 8$, a duodecimal point if $b = 12$, and a hexadecimal point if $b = 16$.

Example 2.1.2. Counting in different number systems, such as binary, octal, duodecimal, and hexadecimal, vis-à-vis decimal, is illustrated in Table 2.1.1. ∎

2.1.3 Base Conversion. For any integer base $b > 1$, the integral and fractional parts have independent representations, except in the case of representations of unity; i.e., $(1)_{10} = (1.000\ldots)_{10}$, or $(1)_{10} = (0.999\ldots)_{10}$, or, in general, we represent $(1)_b = 0.(b-1)(b-1)\ldots$ in a base b system. Thus, the two parts may be treated separately. Since our interest is in integers, we will describe a method to convert a given integer from one base into another, as follows.

2.1.4 Conversion of Integers. The division algorithm (§2.1.2) for a base-b system yields the algorithm

$$\frac{n_k}{b} = n_{k+1} + \frac{a_k}{b}, \quad k = 0, 1, 2, \ldots, p; \quad n_0 = n, \quad 0 \le a_k < b, \qquad (2.1.9)$$

where the quantities n_k are integers. This algorithm is used to convert positive integers from one base to another, where b denotes the new base. A simple method for converting integers between binary and octal is to use the binary equivalents of octal digits as given in the following table:

Octal	Binary	Octal	Binary
0	000	4	100
1	001	5	101
2	010	6	110
3	011	7	111

This table gives binary-coded octals. To find the octal digit equivalent to a binary number, the method consists of partitioning the given binary number into 3-bit binary digits starting from the rightmost digit and then writing their equivalent octal digits from the above table.

Example 2.1.3. The binary number 101111011010 is converted to octal as follows:

Binary	101	111	011	010
	↓	↓	↓	↓
Octal	5	7	3	2

Hence, $(101111011010)_2 = (5732)_8$. Note that the leading zeros are added to the remaining leftmost one or two digits without affecting the binary number in order to complete the leftmost 3-bit binary digit. The above process can easily be reversed. If an octal number is given, say $(1534)_8$, then the equivalent binary number is found as follows:

Octal	1	5	3	4
	$\downarrow$	$\downarrow$	$\downarrow$	$\downarrow$
Binary	001	101	011	100

Hence, $(1534)_8 = (001101011100)_2 = (1101011100)_2$, after discarding the two leading zeros. ∎

2.1.5 lcm and gcd. Let $[a_1, \dots, a_n]$ denote the least common multiple (lcm) of the integers $a_1, a_2, \dots, a_n$. One method of finding lcm is as follows: Prime number factorize $a_1, a_2, \dots, a_n$. Then form the product of these primes raised to the greatest power in which they appear.

Example 2.1.4. To determine the lcm of $18, 24, 30$, note that $18 = 2 \cdot 3^2; 24 = 2^3 \cdot 3; 30 = 2 \cdot 3 \cdot 5$. Then the required lcm $= 2^3 \cdot 3^2 \cdot 5 = 360$. ∎

Let (a, b) denote the greatest common divisor (gcd) of a and b. If $(a, b) = 1$, the numbers a and b are relatively prime. One method, known as Euclid's algorithm, of finding (a, b) is as follows: Assume $a > b$ and dividing a by b yields $a = q_1 b + r_1$, $0 \le r_1 < b$. Dividing b by r_1 gives $b = q_2 r_1 + r_2$, $0 \le r_2 < r_1$. Continuing this process, let r_k be the first remainder, which is zero. Then $(a, b) = r_{k-1}$.

Example 2.1.5. To determine $(112, 42)$, by Euclid's algorithm we have: $112 = 2 \cdot 42 + 28; 42 = 1 \cdot 28 + 14; 28 = 2 \cdot 14 + 0$. Thus $(112, 42) = 14$. ∎

2.1.6 Modulo. For integers m, n and p, the numbers m and n are said to be congruent modulo p, written $m = n \bmod(p)$, if $m - n$ is a multiple of p, i.e., m/p and n/p have equal remainders. The following results hold:

(i) $m_1 = n_1 \bmod (p)$, $m_2 = n_2 \bmod (p) \Rightarrow$ (i) $cm_1 = cn_1 \bmod (p)$;

(ii) $m + 1 \pm m_2 = (n_1 \pm n_2) \bmod (p)$; and (iii) $m_1 m_2 = n_1 n_2 \bmod (p)$.

2.2 Boolean and Bitwise Operations

The distinction between Boolean logical and bitwise operations is important. This section is devoted to these two topics, which play a significant role in the construction of different codes.

2.2.1 Boolean Logical Operations. The truth tables for classical logic with only two values, 'true' and 'false', usually written T and F, or 1 and 0 in the case of the binary alphabet $A = \{0, 1\}$, are given in Table 2.2.1 for most commonly used operators AND, OR, XOR, XNOR, IF-THEN, and THEN-IF. The operator NOT is defined by NOT $0=1$, and NOT $1=0$. The others are:

Table 2.2.1. Boolean Logical Operators

p	q	AND	OR	XOR	XNOR	IF-THEN	THEN-IF
0	0	0	0	0	1	1	1
0	1	0	1	1	0	1	0
1	0	0	1	1	0	0	1
1	1	1	1	0	1	1	1

2.2.2 Bitwise Operations. A bitwise operation is carried out by operators like NOT, AND, OR, and XOR, which operate on binary numbers or one or two bit patterns at the level of their individual bits. These bitwise operators are defined as follows.

NOT ($\neg$). This operator, also known as the *complement*, is a unary operation that performs a logical negation at each bit. Thus, digits that were 0 become 1, and conversely. For example, NOT $0110 = 1001$. In certain programming languages, such as C or C++, the bitwise NOT is denoted by $\sim$ (tilde). Caution is needed not to confuse this bitwise operator with the corresponding logical operator '!' (exclamation point), which treats the entire value as a single Boolean, i.e., it changes a true value to false, and conversely. Remember that the 'logical NOT' is not a bitwise operation.

AND (& or $\wedge$). This bitwise operation takes two binary representations of equal length and operates on each pair of corresponding bits. In each pair, if the first bit is 1 AND the second bit is 1, then the result is 1; otherwise the result is 0. This operator, as in the C programming languages, is denoted by '&' (ampersand), and must not be confused with the Boolean 'logical AND' which is denoted by '&&' (two ampersands). An example is: $0101 \& 0011 = 0001$. The arithmetic operation '+' and bitwise operation '& ' are given side-by-side in Table 2.2.2. In general, the expressions $x + y$ and $x \& y$ will denote the arithmetic and bitwise addition of x and y, respectively.

OR (|). This operation takes two binary representations of equal length and produces another one of the same length by matching the corresponding bit, i.e., by matching the first of each, the second of each, and so on, and performing the logical inclusive OR operation on each pair of corresponding bits. Thus, if in each pair the first bit is 1 or the second bit is 1 or both, then the result is 1; otherwise it is 0. Thus, for example, 0101 OR $0011 = 0111$. In C programming languages the bitwise OR is denoted by | (pipe), and it must not be confused with the logical OR which is denoted by $\vee$ (from Latin *vel*) or by || (two pipes).

XOR ($\oplus$). This bitwise operator, known as the *bitwise exclusive-or*, takes two bit patterns of equal length and performs the logical XOR operation on each pair of the corresponding bits. If two bits are different, then the result is 1; but if they are the same, then the result is 0. Thus, for example, $0101 \oplus 0011 = 0110$. In general, if x, y, z are any items, then (i) $x \oplus x = 0$, (ii) $x \oplus 0 = x$, (iii) $x \oplus y = y \oplus x$, and (iv) $(x \oplus y) \oplus z = x \oplus (y \oplus z)$. In C programming languages, the bitwise XOR is denoted by $\oplus$.

Table 2.2.2 Arithmetic and Bitwise Operations

| Bitwise Operations | | | | | Arithmetic and Bitwise Operations | | | | | |
p	q	AND	OR	XOR	p	q	$+$	AND	OR	XOR
0	0	0	0	0	0	0	0	0	0	0
0	1	0	0	1	0	1	1	0	0	1
1	0	0	0	1	1	0	1	0	0	1
1	1	1	1	0	1	1	10	1	1	0

The bitwise XOR operation is the same as addition mod 2. The XOR function has the following properties, which hold for any bit values (or strings) a, b, and c:

PROPERTY 1. $a \oplus a = 0$; $a \oplus 0 = a$; $a \oplus 1 = \sim a$, where $\sim$ is bit complement; $a \oplus b = b \oplus a$;

$a \oplus (b \oplus c) = (a \oplus b) \oplus c$; $a \oplus a \oplus a = a$, and if $a \oplus b = c$, then $c \oplus b = a$ and $a \oplus a = b$.

PROPERTY 2. As a consequence of Property 1, given $(a \oplus b)$ and a, the value of the bit b is determined by $a \oplus b \oplus a = b$. Similarly, given $(a \oplus b)$ and b, the value of a is determined by $b \oplus a \oplus b = a$. These results extend to finitely many bits, say a, b, c, d, where given $(a \oplus b \oplus c \oplus d)$ and any 3 of the values, the missing value can be determined. In general, for the n bits $a_1, a_2, \ldots, a_n$, given $a_1 \oplus a_2 \oplus \cdots \oplus a_n$ and any $(n - 1)$ of the values, the missing value can be easily determined.

PROPERTY 3. A string **s** of bits is called a *symbol*. A very useful formula is

$$\mathbf{s} \oplus \mathbf{s} = \mathbf{0} \quad \text{for any symbol s.} \qquad (2.2.1)$$

2.2.3 Applications. Some applications involving the above bitwise operations are as follows:

The bitwise AND operator is sometimes used to perform a *bit mask* operation, which is used either to isolate part of a string of bits or to determine whether a particular bit is 1 or 0. For example, let the given bit pattern be 0011; then, to determine if the third bit is 1, a bitwise AND operation is performed on this bit pattern and another bit pattern containing 1 in the third bit. Thus, $0011 \, and \, 0010 = 0010$. Since the result is non-zero, the third bit in the given bit pattern is definitely 1. The name 'bit masking' is analogous to use masking tape to mask or cover the parts that should not be changed.

The bitwise AND operator can be combined with the bitwise NOT to *clear* bits. Thus, consider the bit pattern 0110. In order to clear the second bit, i.e., to set it to 0, we apply the bitwise NOT to a arbitrary bit pattern that has 1 as the second bit, followed by the bitwise AND to the given bit pattern and the result of the bitwise NOT operation. Thus, [NOT 0100] AND 0110 = 1011 AND 0110 = 0010.

The bitwise OR is sometimes applied in situations where a set of bits is used as flags. The bits in a given binary number may each represent a distinct Boolean variable. By applying the bitwise OR to this number, together with a bit pattern containing 1, will yield a new number with that set of bits. As an example, given the binary number 0010, which can be regarded as a set of four flags, where the first, second, and fourth flags are not set (i.e., they each have value 0) while the third flag is the set (i.e., it has value 1), the first flag in this given binary number can be set by applying the bitwise OR to another value with first flag set, say 1000. Thus, 0010 OR 1000 = 1010. This technique is used to conserve memory in programs that deal with a large number of Boolean values.

The bitwise XOR operation is used in assembly language programs as a short-cut to set the value of a register to zero, since operating XOR on a value against itself always yields zero. In many architectures this operation requires fewer CPU clock cycles than the sequence of operations that are needed to load a zero value and save it to the registers. The bitwise XOR is also used to toggle flags in a set of bits. For example, given a bit pattern 0010, the first and the third bits may be toggled simultaneously by a bitwise XOR with another bit pattern with 1 in the first and the third bits, say 1010. Thus, $0010 \oplus 1010 = 1000$.

2.2.4 XOR Swap Algorithm. The standard method of swapping requires the use of a temporary storage variable in computer programming. But the XOR swap algorithm uses the XOR bitwise operation to swap values of variables that are of the same data type, without using a temporary variable. This algorithm and its C program are given below. Note that although it can be proved that this algorithm works, it is not foolproof: The problem is that if X and Y use the same storage location, the value stored in that location will be zeroed out by the first XOR command, and then remain zero; it will not be swapped with itself. In other words, this problem does not arise because both X and Y have the same value, but from the situation that both use the same storage location.

```
X := X XOR Y        |   void xorSwap (int *x, int *y)
Y := X XOR Y        |   {
X := X XOR Y        |     if (x != y) {
                    |       *x ^= *y;
                    |       *y ^= *x;
                    |       *x ^= *y;
                    |     }
                    |   }
```
_____

XOR Swap Algorithm | C Code

Note that the C code does not swap the integers passed immediately. Instead, it first checks if their memory locations are distinct. The rationale behind this step is that the algorithm only works when x and y refer to different integers; otherwise it will set $^*x = {}^*y = 0$.

In practice this swap algorithm with a temporary register is very efficient. Other limited applications include the following situations: (i) on a processor where the portion of the program code permits the XOR swap to be encoded in a smaller number of bytes, and (ii) when there are few hardware registers are available, this swap may allow the register allocator avoid spilling the register. Since these situations are rare, most optimizing compilers do not generate XOR swap code. Modern compilers recognize and optimize a temporary variable-based swap rather than high-language statements that correspond to an XOR swap. Many times this situation is translated in the compiler as simply two variables swapping memory addresses. Often, if supported by architecture, the compiler can use an XCHG (exchange) command to perform the swap in a single operation. The best precaution is to use XCHG to swap registers and not memory. Again, since XOR swap may zero out the contents of some location, it must not be used in a high-level language if aliasing is possible.

2.2.5 XOR Linked Lists. These lists are a data structure used in computer programming. By employing the bitwise XOR operation ($\oplus$), they decrease storage requirements for doubly-linked lists.[1] An XOR linked list compresses the same information into one address field by storing the bitwise XOR of the address for previous and the address for the next in one field:

$$\cdots \quad A \qquad B \qquad C \qquad D \qquad E \quad \cdots$$
$$\longleftrightarrow \quad A \oplus C \quad \longleftrightarrow \quad B \oplus D \quad \longleftrightarrow \quad C \oplus E \quad \longleftrightarrow$$

As one traverses the list from left to right, the address of the previous item is XOR-ed with the value in the link field; e.g., at C the address of B is XOR-ed with D. The list is continued to be traversed in this manner. At any starting point, one needs the addresses of two consecutive items. If the addresses of two consecutive items are reversed, then the list will be traversed in the

_____

[1] A doubly-linked list stores addresses of the previous and next list items in each list node requiring two address fields.

opposite direction. However, given only one item, the addresses of the other elements of the list cannot be immediately obtained. Two XOR operations are required to traverse from one item to the next, under the same instructions in both cases. Consider a list with items $\{\cdots$ B C D $\cdots\}$; let the address of the current list item C be contained in two registers R_1, R_2, and let there be a work register that contains the XOR of the current address C with the previous address D, assuming right to left traversing, i.e., it contains $C \oplus D$. Then, using the instructions

$X \quad R_2$

Link $\quad R_2 \leftarrow C \oplus D \oplus B \oplus D$! ($B \oplus C$, "Link" being the link field in the

current record, containing $B \oplus D$)

$XR \quad R_1, R_2$

$R_1 \leftarrow C \oplus B \oplus D$! (B obtained)

The eol (end of list) is marked by imagining a list item at address 0 that is placed adjacent to the endpoint, as in $\{0$ A B C $\cdots\}$. Then the XOR link field at A is $0 \oplus A$. An eol can be made reflective by taking the link pointer as 0. The zero pointer is called a *mirror*. Note that the XOR of the left/right adjacent addresses to 0 is zero.

2.2.6 Bit Shifts. The bit shifts are generally regarded as bitwise operations since they operate on the binary form of an integer instead of its numerical value. Bit shifts are strictly not bitwise operations because they do not operate on pairs of corresponding bits. In a bit shift operation, the digits are moved (i.e., shifted) to the left or to the right. Since the registers in a computer processor have a fixed number of bits available for storing numbers, some bits are 'shifted out' of the register at one end while the same number of bits are 'shifted in' from the other end. The differences between bit shift operators depend on how the values of the shifted-in bits are computed.

2.2.7 Arithmetic Shifts. In this type of shift, the bits that are shifted out of either end are discarded. In a left arithmetic shift, zeros are shifted in on the right. In a right arithmetic shift, the sign bit is shifted in on the left, thus preserving the sign of the operand. The left and the right arithmetic shifts are denoted by $\ll$ and $\gg$, respectively. Using an 8-bit register, these two bit shift operations by 1 bit to the left and to the right, respectively, are represented as follows:

Left shift: $00010101 \ll 1$ yields 00101010,

Right shift: $00010101 \gg 1$ yields 00001010.

In the case of the left shift, the leftmost digit is shifted past the end of the register, and a new bit 0 is shifted into the rightmost position. In the right shift case, the rightmost bit 1 is shifted out (perhaps into the carry flag) and a new bit 0 is copied into the leftmost position, thus preserving the sign of the number. These shifts in both cases are represented on a register in Figure 2.2.1(a)–(b).

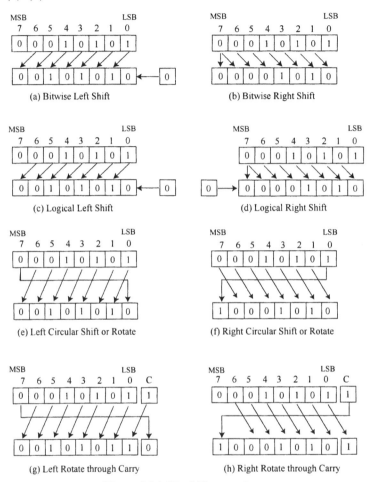

Figure 2.2.1 Bit shift operations.

In Figure 2.2.1, the bits marked as MSB and LSB refer to the most significant byte (or octet) and the least significant byte, respectively.[2]

---

[2] These terms take their meaning from the least significant bit (lsb) and the most significant bit (msb), defined in a field containing more than one bit that has a single value. They are similar to the most (least) significant digit of a decimal integer. The lsb (sometimes called the *right-most bit*, by convention in positional notation) is the bit position in a binary integer giving the units value, i.e., determining whether the number is even or odd. A similar definition is valid for msb, which is sometimes called the leftmost bit. The least significant bits change rapidly if the number changes even slightly. For example,

Multiple shifts are often combined into a single shift by a certain number of bits. Thus, in general, the expressions $x \ll b$ ($x \gg b$) will denote the bitwise shift to left (right) by b bits. Other examples in 16-bit register are given below.

Example 2.2.1. Let 16-bit binary numbers A, B, C and D be given by

$$A = 0110110010111010,$$
$$B = 0101100001100101,$$
$$C = 0010001000110010.$$
$$D = 0101010101010101.$$

Then

$$(A \gg 1) = 0011011001011101,$$
$$(B \gg 2) = 0001011000011001,$$
$$(C \gg 4) = 0000010101010101,$$
$$(A\&D) = 0100010000010000. \quad \blacksquare$$

2.2.8 Logical Shifts. In logical shifts the bits that are shifted out are discarded and zeros are shifted in on either end, as shown in Figure 2.2.1(c)–(d). Notice that the logical and arithmetic left shifts are exactly the same operations. But in the logical right shift, the bit with value 0 is inserted instead of the copy of the sign bit. This means that logical shifts are suitable only for unsigned binary numbers, whereas the arithmetic bitwise shifts are suitable for signed two's complement binary numbers.

2.2.9 Circular Shifts. The circular shifts, or the bit rotation operation, both left and right, are shown in Figure 2.2.1(e)–(f). In this operation, the bits are 'rotated' to the effect that the left and the right ends of the register seem to be joined. In the circular left shift, the value that is shifted in on the right is the same value that was shifted out on the left, and the converse holds for the circular right shift. This operation is used if it is required to retain all the existing bits.

2.2.10 Shift Registers. Let $a, a_0, a_1, \ldots, a_{k-1}$ be given elements of a finite field F_q, where $k > 0$ is an integer. Then a sequence $\{s_0, s_1, \ldots, \}$ of elements of F_q satisfying the relation

$$s_{n+k} = a_{k-1}s_{n+k-1} + a_{k-2}s_{n+k-2} + \cdots + a_0 s_n + a \quad \text{for } n = 0, 1, \ldots, \quad (2.2.2)$$

is called a *k-th order linear recurring sequence* in F_q, and the terms $s_0, s_1, \ldots,$ which uniquely determine the rest of the sequence, are called the *initial values*.

---

adding $1_{10} = (00000001)_2$ to $(3)_{10} = (00000011)_2$ yields $(4)_{10} = (00000100)_2$, where the (rightmost) three of the least significant bits have changed from 011 to 100, whereas the three most significant bits (which are leftmost) remained unchanged (000 to 000).

If $a = 0$, then (2.2.2) is called a *homogeneous* linear recurrence relation; otherwise it is an inhomogeneous recurring relation.

In terms of electric switching circuit configuration, the generation of linear recurring sequences can be implemented on a feedback shift register. There are four types of devices in use: (i) an *adder*, which has two inputs and one output, where the output is the sum in F_q of the two inputs; (ii) a *constant multiplier*, which has one input and yields one output that is the product of the input with a constant element of F_q; (iii) a *constant adder*, which is similar to a constant multiplier, except that it adds a constant element of F_q to the input; and (iv) a *delay element* (a 'flip-flop' type device), which has one input and is regulated by an external synchronous clock so that its input at a given time appears as its output one unit of time later. The feedback shift register that generates a linear recurring sequence satisfying (2.2.2) is shown in Figure 2.2.2.

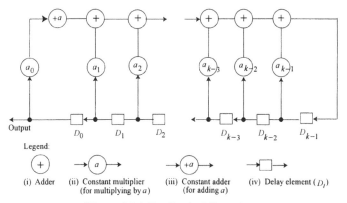

Figure 2.2.2 Feedback shift register.

2.2.11 Rotate through Carry. This operation is similar to the 'Rotate No Carry operation', except that the two ends of the register (on the right end in both left and right-rotate-through-carry operations) are regarded as separated by the carry flag (marked by C over it). The bit that is shifted in (in each case) is the old value of the carry flag, and the bit that is shifted out (on either end) is used as the new value of the carry flag. These operations are presented in Figure 2.2.1(g)–(h). This operation is useful when performing shifts on numbers larger than the processor's word size. If a large number is stored in two registers, the bit that is shifted out at the end of the first register must enter at the other end of the second; this bit is 'saved' in the carry flag during the first shift, and ready to shift in during the second shift without any extra operation.

A single 'rotate through carry' is used to simulate a logical or arithmetic shift of one position by prior setting up the carry flag. Thus, if the carry flag is 0, then X RIGHT-ROTATE-THROUGH-CARRY-BY-ONE defines a *logical* right shift; but if the carry flag contains a copy of the sign bit, then X RIGHT-ROTATE-THROUGH-CARRY-BY-ONE is an *arithmetic* right shift. This feature has led some microcontrollers, such as PICs, to use only *rotate* and *rotate through carry*, instead of the logical or arithmetic shifts.

The left and right shifts operations in the C family of languages are denoted by '$\ll$' and '$\gg$', respectively. The number of bits to be shifted is the same as above, i.e., $x \ll b$ or $x \gg b$. In Java, all integer types are signed, and the operators $\ll$ and $\gg$ perform arithmetic shifts; moreover, the operator $\ggg$ performs the logical right shifts, but there is no operator $\lll$ because the arithmetic and logical left shift operations are identical in Java.

2.2.12 One's and Two's Complement Arithmetic. One's and two's complement (or negative) of a binary number are part of a system of arithmetic that is devised to avoid the addition and subtraction circuitry, and examine the signs of the operands (plus or minus) to determine if the numbers are to be added or subtracted.

The one's complement of a binary number is performed by simply inverting every bit, i.e., by changing the 1s into 0s and 0s into 1s. The two's complement of an N-bit binary number is obtained by subtracting the given number from a larger power 2^N by computing its two's complement. In practice, it is accomplished by first starting at the least significant bit (lsb), copying all zeros (working leftward toward the msb) until the first 1 is reached, then copying that 1, and reversing (flipping) all the remaining bits. The 2^N-bit two's complement system can be used for all binary numbers in the range -2^{N-1} to $+2^{N-1} - 1$.

Example 2.2.2. The two's complement for an 8-bit binary system is presented in Table 2.2.3.

Table 2.2.3. Two's Complement

msb	Binary Number	Decimal	Comment
0	1111111	$= 127$	$2^8 - 1$
0	1111110	$= 126$	
0	0000010	$= 2$	2^1
0	0000001	$= 1$	2^0
0	0000000	$= 0$	
1	1111111	$= -1$	1's complement
1	1111110	$= -2$	2's complement
1	0000001	$= -127$	
1	0000000	$= -128$	

The leftmost bit (most significant bit) represents the sign of the binary number, 0 for plus sign and 1 for the minus sign. ∎

The general rule to find the b's complement of a number N for the radix (base) b number system, with the integer part of m digits and the fractional part of n digits, is given by the formula

$$N^{**} = (b^m - N) \bmod (b^m),$$

where $0 \leq N < b^{m-1} - b^{-n}$. Also, the $(b-1)$'s complement of a number N is given by

$$N^* = b^m - b^{-n} - N.$$

Alternatively, the b's complement of a number N can also be found by adding b^{-n} to the $(b-1)$'s complement of the number, that is,

$$N^{**} = N^* + b^{-n}.$$

Example 2.2.3. For a given positive binary number, first convert it to decimal and then perform the one's or two's complement on it, and then convert it back to the original system. Thus,

(a) Let the given number be $N = (11.11)_2$, which is equal to $(3.75)_{10}$. Then, its two's complement is given by $\left(2^2\right)_{10} - (11.11)_2 = \left(2^2\right)_{10} - (3.75)_{10} = (0.25)_{10} = (0.01)_2$. The one's complement of the given number is simply $(00.00)_2$. Alternatively, by adding the lsb of the one's complement, i.e., adding $\left(2^{-2}\right)_{10} = (0.01)_2$, we obtain the above two's complement of the given number. ∎

(b) Given $N = (11.11)_{10}$, first convert it into its binary form, i.e., $(11.11)_{10} = (1011.0000111)_2$ with $m = 4$ integer bits and $n = 7$ fractional bits. Then the two's complement is $\left(2^4\right)_{10} - (11.11)_{10} = (4.89)_{10} = (100.11100011)_2$. ∎

(c) Given $(153.25)_{10}$, convert it to binary, then take the one's complement, and next, obtain the two's complement by adding .01. Thus,

$$(153.25)_{10} = (10011001.01)_2$$

one's complement $\rightarrow (01100110.10)_2$ (keep the last 0)

$$\underline{+ \qquad\qquad .01 \qquad\qquad}$$

two's complement $\rightarrow (01100110.11)_2$ ∎

(d) Given a four-bit hexadecimal number $N = (117)_{16}$, first convert it into its binary form, and then using the technique in (c) above, find the two's complement and convert it back to hexadecimal. Thus,

$$(117)_{16} = (100010111)_2$$

one's complement $\rightarrow (011101000)_2$

$$\underline{+ \qquad\qquad 1 \qquad\qquad}$$

two's complement $\rightarrow (011101001)_2 = (E9)_{16}$. ∎

It is obvious that the two's complement system represents negative integers obtained by counting backward and wrapping around. The rule about the boundary between the positive and negative numbers is based on the msb (most significant or the leftmost bit). Thus, the most positive 4-bit number is 0111 (7) and the most negative is 1000 (8), because the leftmost bit is the sign bit (0 for positive and 1 for negative). However, the msb represents not only the sign bit but also the magnitude bit; the absolute value of the most negative n-number is too large to represent in n bits alone. For example, in 4 bits, $|-8| = 8$ is too large to represent only in 4 bits; thus, a 4-bit number can represent numbers from -8 to 7. Similarly, an 8-bit number can represent numbers from -128 to $127 = (2^8 - 1)$.

This system is used in simplifying the implementation of arithmetic on computer hardware. It allows addition of negative operands without a subtraction circuit and another circuit to detect the sign of a number, as the addition circuit can perform subtraction by taking the two's complement of the operand, which requires only one additional cycle or its own adder circuit.

This system is based on the following fact: The 2^n possible values on n bits form a ring of equivalence classes, which are the integers mod 2^n. Each class represents a set $\{j + 2^n k : k$ an integer$\}$ for some integer j, $0 \leq j \leq 2^n - 1$. There are 2^n such sets, and addition and multiplication is well defined on every one of them. If we assume that these classes represent the numbers from 0 to $2^n - 1$, and if we ignore overflow, then they are the unsigned integers. But each one of these integers is equivalent to itself minus 2^n. Thus, the classes can represent integers from -2^{n-1} to $2^{n-1} - 1$, by subtracting $\lfloor 2^{n-1}, 2^n - 1 \rfloor$, or 2^n from half of them. For example, for $n = 8$, the unsigned bytes are 0 to $2^8 - 1 = 255$; then subtracting $2^8 = 256$ from the top half (128 to 255) yield the signed bytes -128 to 127, as shown in Table 2.2.3. For more details, see Koren [2002] and Wakerly [2000].

2.3 Checksum

A *checksum* is a function that computes an integer value from a string of bytes. Its purpose is to detect errors in transmission. There are different checksum functions with different speed or robustness, and some of them are discussed below. The most robust checksum functions are cryptographic hash functions that are typically very large (16 bytes or more). MD5 hash functions are typically used to validate Internet downloads.

A weak checksum, known as the Internet checksum, is designed for IPv4, TCP, and UDP headers, and is 16 bits (2 bytes) long. Although it is very fast, it is very weak and results in undetected errors. The cyclic redundancy checks (CRC) are moderately robust functions and are easy to implement. They are described below. The checksum CRC32C (Castagnoli et al. [1993]) is extensively used in communications, and it is implemented on hardware in

Intel CPUs as part of the SSE4.2 extensions. This checksum is easy to use because it uses certain built-in GNU functions with user-friendly instructions.

The checksum function is the negative (complement) of the sum of the original data items; it is transmitted along with all the data items to detect errors during transmission through space (telecommunications) or time (storage). It is a form of redundancy check used to protect the correctness of the data transmitted or stored. It works by adding up the data items and attaching the negative of the sum with the original data prior to transmission or storage. Suppose the data consists of five 4-bit numbers, say (7, 11, 12, 0, 6). In addition to sending these numbers, the sum of these numbers is also sent, i.e., data that is sent is $(7, 11, 12, 0, 6, 36)$, where 36 is the sum of the original five numbers. The receiver adds the five numbers and compares the result. If the two are the same, the receiver assumes no error, accepts the five numbers, and discards the sum. Otherwise, there is an error somewhere, and the data is rejected. However, the task of the receiver becomes easier if, instead of the sum, the negative (complement) of the sum is sent along with the original data, i.e., $(7, 11, 12, 0, 6, -36)$ is transmitted. The receiver can then add all the numbers including the checksum, and if the result is zero, it assumes no error; otherwise, there is an error. The checksum is represented in *1's complement arithmetic*, which is explained by the following example.

Example 2.3.1. To represent 21 in decimal by 1's complement arithmetic. note that $(21)_{10} = (10101)_2$. Since it has 5 bits, the leftmost bit is wrapped and added to the remaining (rightmost) four bits, thus giving $0101 + 1 = 0110 = (6)_1$. Now, to represent -6 in one's complement arithmetic, the negative or complement of a binary number is found by inverting all bits. Thus, $(+6)_{10} = (0110)_2$ yields $(-6)_{10} = (1001)_2 = (9)_{10}$. In other words, the complement of 6 is 9. In general, the complement of a number in 1's complement arithmetic is found by subtracting the number from $2^n - 1$, which in the case of this example is $16 - 1 = 15$. This example can be schematically represented as follows:

Sender		Receiver
7		7
11		11
12		12
0		0
6		6
sum→ 36	transmitted $\boxed{7, 11, 12, 0, 6, 9} \rightarrow$	sum→ 45
wrapped sum→ 6		wrapped sum→ 15
checksum→ 9		checksum→ 0

DETAILS OF WRAPPING AND COMPLEMENTING:

$\underbrace{10}_{\text{wrap}}\,0100 \quad = 36$

$\underline{10}$

$\quad\quad 0110 \;=6$

$\quad\quad \underline{1001 \;=9}$

$\underbrace{10}_{\text{wrap}}\,1101 \quad = 45$

$\underline{10}$

$\quad\quad 1111 \;=15$

$\quad\quad \underline{0000 \;=0}$ ∎

This simple form of checksum, which simply adds up the checksum byte to the data, is not capable of detecting many different types of errors. This defect still remains in this simple method even if the bytes in the transmitted message are reordered, or zero-valued bytes are inserted (or deleted), or by simply relying on the sum to be zero.

There are more sophisticated methods of redundancy checks that are available, some of which are described below.

2.3.1 Fletcher's Checksum. This checksum is an algorithm for computing a position-dependent checksum. It involves dividing the binary dataword so that it can be protected from errors into short 'blocks' of bits, and computes the modular sum of all such blocks.[1]

Example 2.3.3. Let the data be a message to be transmitted containing 140 characters, each stored as an 8-bit byte. This makes the dataword of 1120 bits. For the sake of convenience, we will choose a block size of 8 bits, although it is not necessarily the only choice; similarly, a convenient modulus is 255 although this too is not the only choice. With these choices the simple checksum is computed by adding all the 8-bit bytes of message and getting this sum modulo 255, with a remainder r. The checksum value is transmitted with the message, where its length is now increased to 141 bytes (1128 bits). The receiver at the other end re-computes the checksum and compares it to the value received, and this process determines if the message was altered during the transmission. ∎

This checksum process is not very robust. There are certain specific weaknesses. First, it is insensitive to the order of the blocks in the message; the checksum value remains the same even when the order of blocks is changed. The other weakness is that the range of checksum values is small, being equal to the chosen value of the modulus. However, these weaknesses are mitigated by computing a second checksum value, which is the modular sum of the values taken at the addition of each block of dataword, with the same modular value. This is done as follows.

Each block of the dataword is taken in sequence; the block's value is added

---

[1] Note that in this definition the data in its entirety is called a 'word' and the parts into which it is divided are designated as 'blocks'.

to the first sum; and the new value of the first sum is then added to the second sum, where both sums start with the same zero value, or any other prescribed value. At the end of the dataword, the modulus operator is applied and the two values are combined to form the new checksum value. Note that once a block is added to the first sum, it is then repeatedly added to the second sum together with every block thereafter. For example, if two blocks have been exchanged, the one that was initially the first block will be added to the second sum one fewer times, and the block that was originally the second one will be added to the second sum one more time. The final value of the first sum will remain the same, but the second sum will be different, thus detecting the change in the message.

This algorithm was developed by John G. Fletcher at the Lawrence-Livermore Laboratory in the late 1970s (see Fletcher [1982]). There are two versions of the Fletcher checksum: Fletcher-16 and Fletcher-32. In Fletcher-16 the dataword is divided into 8 bit blocks. The resulting two 8-bit sums are combined into a 16-bit Fletcher checksum. The algorithm is to multiply the second sum by $2^8 = 256$ and add it to the checksum value. This stacks the sums side-by-side in a 16-bit word with a checksum at the least significant end. The modulus 255 is generally implied in this algorithm. In Fletcher-32 the dataword is divided into 16 bit blocks, which results in two 16-bit sums that are combined to form a 32-bit Fletcher checksum. In its algorithm the second sum is multiplied by $2^{16} = 65536$ and added to the checksum, thus stacking the sums side-by-side in a 32-bit word with the checksum at the least significant end. This algorithm uses the modulus 65535. A modified version of Fletcher-32 algorithm is Adler-32, which can be found in Maximo [2006].

2.3.2 Cyclic Redundancy Check (CRC). The CRC check is a type of function with input as a data of any length, and output as a value of certain space. A CRC can be used as a checksum to detect accidental alteration of data during transmission or storage. It is useful because (i) it is simple to implement in binary hardware; (ii) it is easy to analyze mathematically; and (iii) it detects very efficiently common errors caused by noise. The computation method for a CRC resembles a long division operation, in which the quotient is discarded and the remainder becomes the result. The length of the remainder is always less than or equal to that of the divisor.

Example 2.3.4. (Division in CRC encoder) Let the dataword be 1001 and the divisor 1011. Then the long division process proceeds as follows:

```
                 1 0 1 0   Quotient
Divisor    1 0 1 1 │ 1 0 0 1 │0 0 0│ ←— Dividend: Augmented dataword
           ─────────  1 0 1 1
                      ─────────
                       0 1 0 0
Leftmost bit 0:        0 0 0 0
Use 0000 divisor       ─────────
```

```
                          1 0 0 0
                          1 0 1 1
```

Leftmost bit 0:	0 1 1 0	
Use 0000 divisor	0 0 0 0	

```
                           1 1 0          | Remainder |
```

This yields:

Codeword | 1 0 0 1 | 1 1 0 |

Dataword Remainder or check bits ■

Example 2.3.5. Division in the CRC decoder for two cases:

CASE 1.   Codeword | 1 0 0 1 | 1 1 0 |

```
                    1 0 1 0   Quotient
                   _____
Divisor    1 0 1 1 | 1 0 0 1 1 1 0  ←— Codeword
                     1 0 1 1
                     _____
                       0 1 0 1
                       0 0 0 0
                       _____
                         1 0 1 1
                         1 0 1 1
                         _____
                           0 0 0 0
                           0 0 0 0
                           _____
                             | 0 0 0 |
```

Since the remainder is zero, the dataword 1 0 0 1 is accepted.

CASE 2. Suppose that the dataword is changed from 1001 to 1000. Then the codeword is | 1 0 0 0 | 1 1 0 |.

```
                    1 0 1 1   Quotient
                   _____
Divisor    1 0 1 1 | 1 0 0 0 1 1 0  ←— Codeword
                     1 0 1 1
                     _____
                       0 1 1 1
                       0 0 0 0
                       _____
                         1 1 1 1
                         1 0 1 1
                         _____
                           1 0 0 0
                           1 0 1 1
                           _____
                             | 0 1 1 |
```

Since the remainder is not zero, the dataword 1 0 1 1 is discarded. ■

2.4 Ring Counters

A *ring counter* results if the output of a shift counter is fed back to the input. The data pattern contained in the shift register keeps on recirculating as long as clock pulses are applied. For example, the data pattern in Figure 2.4.1 repeats every four clock pulses. In loading a data pattern, all 0s or all 1s do not count. The loading of data in the parallel-in/serial-out shift register is configured as a ring counter, in which any random pattern may be loaded, although the most useful pattern is a single 1.

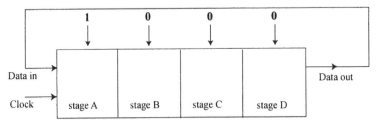

Figure 2.4.1. Parallel-in/serial-out shift register.

After the binary data **1 0 0 0** is loaded into the ring counter, the data pattern for a single stage repeats every four clock pulses in the four stages, where the waveforms for all four stages are similar except for the one clock time delay from one stage to the next. The corresponding circuit for this example is divided into 4 counters, and the ratio between the clock input to any one of the outputs is 4:1. We would require ten stages to recirculate the bit **1** every 10 clock pulses. There is another method of initializing the ring counter to **1 0 0 0**, in which the shift waveforms remain the same as above, but a more reliable 'self correcting' counter, like a synchronous binary counter, is used. The details can be found at http://www.allaboutcircuits.com/vol_4/chpt_12/6.html.

Table 2.4.1 Switch-Tail Ring Counter

d	Switch Tail Ring Counter Code
0	0 0 0 0 0
1	0 0 0 0 1
2	0 0 0 1 1
3	0 0 1 1 1
4	0 1 1 1 1
5	1 1 1 1 1
6	1 1 1 1 0
7	1 1 1 0 0
8	1 1 0 0 0
9	1 0 0 0 0

The *switch-tail ring counter*, also known as the *Johnson counter*, is a shift register that feeds back to itself and overcomes many drawbacks of the ring counter. Given a division ratio it requires only one half of the stages of a ring counter. Details can be found at the above website. It is defined in Table 2.4.1.

2.5 Residues, Residue Classes, and Congruences

For each pair of integers n and b, $b > 0$, there exists a pair of integers q and r such that

$$n = bq + r, \quad 0 \leq r < b.$$

The quantity r is called the *residue of n modulo b* and is denoted (in Gaussian notation) by $b|n$. For example, $5|15 = 0$, $5|17 = 2$. Further, if $n \geq 0$, then $b|n = r$ is the remainder, and q is the quotient when n is divided by b. The quantities q and r are unique (proof of uniqueness follows from the division algorithm).

Consider the class in which a comparison is made of the remainders when each of the two integers n and m are divided by b. If the remainders are the same, then $b|(n-m)$, and we say that the two numbers n and m have the same residue modulo b, so that n and m differ by an integral multiple of b. In this case we say that n and m are *congruent modulo b* and write $n \equiv m \pmod{b}$.

The symbol $\equiv$ is an *equivalence relation* (with respect to a set); that is, it is a relation $\mathcal{R}$ between the elements of a set such that if α and β are arbitrary elements, then either α stands in a relation $\mathcal{R}$ to β (written as $\alpha \mathcal{R} \beta$), or it does not. Moreover, $\mathcal{R}$ has the following properties:

   (i) $\alpha \mathcal{R} \alpha$ (reflexive);
   (ii) if $\alpha \mathcal{R} \beta$, then $\beta \mathcal{R} \alpha$ (symmetric); and
   (iii) if $\alpha \mathcal{R} \beta$ and $\beta \mathcal{R} \gamma$, then $\alpha \mathcal{R} \gamma$ (transitive).

The equality between numbers is an equivalent relation for either $\alpha = \beta$, or $\alpha \neq \beta$; $\alpha = \alpha$; if $\alpha = \beta$, then $\beta = \alpha$; and if $\alpha = \beta$ and $\beta = \gamma$, then $\alpha = \gamma$. Other examples are congruency of triangles, similarity of triangles, parallelism of lines, children having the same mother, or books by the same author. The congruency $n \equiv m \pmod{b}$ possesses the above three properties. In fact, we have

Theorem 2.5.1. *Congruence modulo a fixed number b is an equivalence relation.*

PROOF. There are three cases:

(i) $b|(n - n)$, so that $n \equiv n \pmod{b}$.

(ii) If $b|(n - m)$, then $b(m - n)$; thus, if $n \equiv m \pmod{b}$, then $m \equiv n \pmod{b}$.

(iii) If $b(n - m)$ and $b(m - l)$, then $n - m \equiv kb$, $m - l \equiv jb$, where k and j

are integers. Thus, $n - l = (k + j)b$, i.e., if $n \equiv m$ (mod b) and $m \equiv l$ (mod b), then $n \equiv l$ (mod b). ∎

Lemma 2.5.2. *If* $a|bc$ *and* $(a, b) = 1$, *then* $a|c$.

PROOF. If $(a, b) = 1$, then there exist integers x and y such that $ax + by = 1$. Multiply both sides of the equality by c. Then $acx + bcy = c$, and a divides both ac and bc. Hence a divides c. ∎

Theorem 2.5.3. *The following results are true:*

(i) If $m \equiv n$ *(mod b) and* $u \equiv v$ *mod (b), then the following congruencies hold:*

(a) $m + u \equiv n + v$ *(mod b),*

(b) $m\,u \equiv n\,v$ *(mod b),*

(c) $k\,m \equiv k\,n$ *(mod b) for every integer k;*

(ii) if $k\,m \equiv k\,n$ *(mod b) and* $(k, b) = d$, *then* $m \equiv n \left(\bmod\, \dfrac{b}{d}\right)$, *where* $(k, n) = d$ *means d is the g.c.d. of k and b.*

(iii) If $f(x)$ is a polynomial with integral coefficients and $m \equiv n$ (mod b), then $f(m) \equiv f(n)$ *(mod b).*

PROOF. (i) The results follow from the definition of congruency:

(a) If $m \equiv n$ (mod b), then $b|(m - n)$. If $u \equiv v$ (mod b), then $b|(u - v)$. Hence, $b|(m - n + u - v)$, or $b|\left((m + u) - (n + v)\right)$, which implies that $m + u \equiv n + v$ (mod b).

(b) Similarly, if $b|(m - n)$ and $b|(u - v)$, the $b|\left((m - n)(u - v)\right)$, or $b\left(mu + nv - mv - nu\right)$, or $b|\left(mu - nv + v(m - n) + n(v - u)\right)$. But since $b|(m - n)$ and $b|(v - u)$, we get $b|(mu - nv)$, which means that $mu \equiv nv$ (mod b).

(c) Similarly, $b|k(m - n)$ for every integer k. Thus, $km \equiv kn$ (mod b).

(ii) In view of Lemma 2.5.2, this part can be reworded as follows: If $b|k(m - n)$ and $(k, b) = d$, then there exist integers x and y such that $kx + by = d$, or, on multiplying both sides by $(m - n)$, we get $k(m - n)x + b(m - n)y = d(m - n)$, which, since $b|k(m - n)$ and $b|b(m - n)$, implies that $b|d(m - n)$, or $\dfrac{b}{d}|(m - n)$.

(iii) Let $f(x) = c_0 + c_1 x + \cdots + c_k x^k$. If $m \equiv n$ (mod b), then, by (i) (a) and (b), for every integer $j \geq 0$

$$m^j \equiv n^j \quad (\bmod\ b),$$

$$c_j m^j \equiv c_j n^j \quad (\bmod\ b).$$

Add these last congruences for $j = 0, 1, \ldots, k$. Then

$$c_0 m^0 + c_1 m^1 + \cdots + c_k m^k \equiv c_0 n^0 + c_1 n^1 + \cdots + c_k n^k \quad (\bmod\ b),$$

or

$$c_0 + c_1 m + \cdots + c_k m^k \equiv c_0 + c_1 n + \cdots + c_k n^k \pmod{b},$$

which yields $f(m) \equiv f(n) \pmod{b}$. ∎

Let p be a prime number and let the base $b = p$. Then the integers mod p form a field; its elements are the congruence classes of integers mod p, with addition and multiplication induced from the usual integer operations. For example, the addition and multiplication tables for integers mod 3, with 0, 1, 2 as representatives of the congruent classes, are as follows:

+	0	1	2		·	0	1	2
0	0	1	2		0	0	0	0
1	1	2	0		1	0	1	2
2	2	0	1		2	0	2	1

Figure 2.5.1 Addition and multiplication tables for integers mod 3.

Theorem 2.5.3(iii) is a basic result for applications. Consider the well-known rule in the decimal system that a number is divisible by 9 iff the sum of digits in its decimal representation is divisible by 9. For example, given $n = 3,574,856$, we have $3 + 5 + 7 + 4 + 8 + 5 + 6 = 38$, which is not divisible by 9, so neither is n divisible by 9. Here $n = 3 \times 10^6 + 3 \times 10^5 + 7 \times 10^4 + 4 \times 10^3 + 8 \times 10^3 + 5 \times 10 + 6$, so that $n = f(1)$, where

$$f(x) = 3x^6 + 5x^5 + 7x^4 + 4x^3 + 8x^2 + 5x + 6.$$

On the other hand, $f(1)$ is exactly the sum of digits, $f(1) = 3 + 5 + 7 + 4 + 8 + 5 + 6$. Since $10 \equiv 1 \pmod{9}$, we conclude by Theorem 2.5.3(iii) that $f(10) \equiv f(1) \pmod{9}$, which implies in particular that either $f(10)$ and $f(1)$ are both divisible by 9, or neither is.

The same argument also applies in the general case. The decimal representation of n is always the expression of n as the value of a certain polynomial $f(x)$ for $x = 10$, and invariably $f(1) \equiv f(1) \pmod{9}$. The above rule can be expanded as follows: If $n = f(1)$ and $m = g(10)$, then

$$n + m = f(1) + g(10) \equiv f(1) + g(1) \pmod{9},$$
$$nm = f(10)g(10) \equiv f(1)g(1) \pmod{9}.$$

Let $F(10) = n + m$ and $G(10) = nm$. Then

$$F(10) \equiv F(1) \equiv f(1) + g(1) \pmod{9},$$
$$G(10) \equiv G(1) \equiv f(1)g(1) \pmod{9}.$$

These last two congruences stated in words imply that the sum of the digits in $n + m$ is congruent (mod 9) to the sum of all the digits in n and m, and the sum of the digits in nm is congruent (mod 9) to the product of the sum of the digits in n and the sum of the digits in m.

This statement provides a weak practical check on the correctness of the arithmetic operation called 'casting out nines'. As an example, suppose that if we miscomputed $47 + 94$ as 131, we could recognize the existence of an error by noting that $(4 + 7) + (9 + 4) = 11 + 123 = 24 \equiv 6 \pmod 9$, whereas $1 + 3 + 1 = 5 \equiv 5 \pmod 9$. Similarly, it cannot be that $47 \times 19 = 793$, because $(4 + 7) \times (1 + 9) = 110 = 1 + 1 + 0 = 2 \equiv 2 \pmod 9$, whereas $9 + 9 + 3 = 19 \equiv 1 \pmod 9$. However, this method is not always valid, as it is also true that $47 \times 19 \neq 884$ even though $8 + 8 + 4 = 20 \equiv 2 \pmod 9$. Hence, this method does not provide an absolute check on accuracy.

If two numbers are each congruent (mod b) to a common residue r, they, together with all numbers so congruent, are said to belong to an *equivalence class*. There are b such distinct classes, each labeled by the associated residue.

While dealing with congruence modulo a fixed number b, the set of all integers break down into b classes, called the *residue classes* (mod b), such that any two elements of the same class are congruent and two elements from different classes are incongruent. The residue classes are also called *arithmetic progressions* with common difference b.

In these cases of residue classes, it suffices to consider an arbitrary set of representatives of the various residue classes, i.e., a set consisting of one element from each residue class. Such a set $a_1, a_2, \ldots$ is called a *complete residue system modulo b* and is characterized by the following two properties:

(i) If $i \neq j$, then $a_i \not\equiv a_j$, (mod b);

(ii) If a is an integer, there is an index i with $1 \leq i \leq b$ for which $a \equiv a_i$ (mod b).

Examples of complete residue systems (mod b) are provided by the set of integers $0, 1, 2, \ldots, b - 1$, and the set $1, 2, \ldots, b$. However, the elements of a complete residue system need not be consecutive integers. For example, for $b = 5$, the five residue classes (or arithmetic progressions with common difference 5) are

$$\ldots, -10, -5, 0, 5, 10, 15, \ldots,$$
$$\ldots, -9, -4, 1, 6, 11, 16, \ldots,$$
$$\ldots, -8, -3, 2, 7, 12, 17, \ldots,$$
$$\ldots, -7, -2, 3, 8, 13, 18, \ldots,$$
$$\ldots, -6, -1, 4, 9, 14, 19, \ldots.$$

We could also take $[10, 1, 7, 18, -6]$ as a residue class for $b = 5$; we could choose any one element from each row, as the first row is representative of all integers of the form $5n$, which are divisible by 5 with 0 residue; the second row is representative of all integers of the form $5n + 1$; the third row is representative of all integers of the form $5n + 2$; and so on.

2.6 Integral Approximation

While computing, a number represented in a positional notation must in practice employ only a finite number of digits. Therefore, it is desirable to approximate a number x by an integer. There are two approximations available for this purpose:

(i) *Floor of* x, denoted by $\lfloor x \rfloor$, is the integral part $[x]$ and is defined as the largest integer not exceeding x.

(ii) *Ceiling of* x, denoted by $\lceil x \rceil$, is defined as the smallest integer not exceeded by x.

Example 2.6.1. $\lceil 2.718 \rceil = 3$, $\lfloor 2.318 \rfloor = 2$, $\lfloor -2.318 \rfloor = -3$, $\lceil 2.000 \rceil = 2$, $\lfloor -2.000 \rfloor = -2$. ∎

The following result is obvious.

Theorem 2.6.1. *The following statements are true:*

(a) $\lceil x \rceil = -\lfloor x \rfloor$,

(b) $\lfloor x \rfloor \leq x \leq \lceil x \rceil$,

(c) $n = b \lfloor n \operatorname{div} b \rfloor + b | n$ *for all positive and negative integers* n. *In this result,* $q = \lfloor n \operatorname{div} b \rfloor$, *and* $b > 0$.

Example 2.6.2. The logarithm function $\log_2 n$ of an integer n determines the number of binary bits needed to represent n. Thus, the number of binary bits to represent n is given by $\lceil \log_2 n \rceil$. For example, $\lceil \log_2 10000 \rceil = 14$, so it will take 14 binary bits to represent 10000. In fact, $10000_{10} = 10011100010000_2$. Special cases are the exact powers of 2; e.g., $\log_2 1024 = 10$, but it takes 11 bits to represent 1024 in binary, since $1024_{10} = 10000000000_2$. ∎

2.7 Lexicographic Order

The lexicographic order on words is usually understood simply as the alphabetic order found in a dictionary, where the words consist of strings of the alphabet set in the order $\{a, b, c, \ldots, x, y, z\}$. If the alphabet is binary, consisting of the digits 0 and 1 (called bits), a precise definition of the lexicographic order is needed. In a formal description, we define a word X as a non-empty string of letters as $X = x_1 x_2 \cdots x_r$, where $r > 0$ is an integer and each x_i is an element of the set of the alphabet and r is the length of the word X. The lexicographic order on words X and Y is defined as the relation $X < Y$ if X comes strictly before Y in the dictionary.

Assuming that the alphabet is ordered as $a < b < c < \cdots < x < y < z$, let $X = x_1 \cdots x_r$ and $Y = y_1 \cdots y_s$ be two words. Then $X < Y$ iff there is a non-negative integer t such that (i) $t \leq r$ and $t \leq s$, (ii) $x_i = y_i$ for every positive integer $i \leq t$, and (iii) either $x_{t+1} < y_{t+1}$, or $t = r$ and $t < s$. An example of this relation is: $a < aa < aaa < ab < aba$. Since the lexicographic order is an order relation on words, the following facts show that this order relation is different from all other kinds of mathematical order relations that exist on the sets $\mathbb{N}, \mathbb{Z}, \mathbb{Q}$ and $\mathbb{R}$: (i) There is a first word a, but no last word; (ii) every word has an immediate successor (e.g., the immediate successor of $mawm$ is $mqwma$; (iii) not every word has an immediate predecessor (e.g., $mqwm$ has no immediate predecessor, as for every predecessor of $mqwm$ we can find another word between them, like $mqwl < mqwld < mqwm$).

The idea of least upper bound is important in lexicographic order. Consider a bounded set of words $E = \{$all words beginning with c or $d\}$. One upper bound for E is the word e. It is easy to check that the upper bounds for E are precisely the words beginning with one of the letters $e, f, g, \ldots, z$, but then e is the only element that is the least upper bound of E. In this case, we say that $e = \sup E$. Some other examples are: (i) Let F be the set of all words with no double letters. The set F is bounded, and zz is a least upper bound for F. (ii) Let G be a bounded set of words with no least upper bound, say, $G = \{a, ab, aba, abab, \ldots\}$, which is a set of all words beginning with a and consisting of an alternating string of as and bs. One upper bound for G is b, but ac is the smaller upper bound, and abb is smaller than that, and $abac$ is smaller than that. In fact, the set G of words is bounded, but has no upper bound.

Example 2.7.1. The set of all permutations of 5 elements in a lexicographic order is as follows:

1 2 3 4 5	1 3 2 4 5	1 4 2 3 5	$\cdots$
1 2 3 5 4	1 3 2 5 4	1 4 2 5 3	$\cdots$
1 2 4 3 5	1 3 4 2 5	1 4 3 2 5	$\cdots$
1 2 3 4 5	1 3 2 4 5	1 4 3 5 2	$\cdots$
1 2 5 3 4	1 3 5 2 4	1 4 5 2 3	$\cdots$
1 2 5 4 3	1 3 5 4 2	1 4 5 3 2	$\cdots$

Example 2.7.2. Subsets of 4 elements out of 6 in a lexicographic order:

1 2 3 4	1 3 4 5	2 3 4 6
1 2 3 5	1 3 4 6	2 3 5 6
1 2 3 6	1 3 5 6	2 4 5 6
1 2 4 5	1 4 5 6	3 4 6 6
1 2 4 6	2 3 4 5	

The binary lexicographic order is defined similarly. An example is the binary representation in the ASCII Table A.1, and many other examples are scattered throughout this book.

3

Linear Codes

3.1 Linear Vector Spaces over Finite Fields

We will assume that the reader is familiar with the definition and properties of a vector (linear) space and finite field F_q, which is simply a subspace of the vector space F_q^n. Linear codes C are vector spaces and their algebraic structures follow the rules of a linear space. Some examples of vector spaces over F_q are:

(i) For any q, $C_1 = F_q^n$, and $C_2 = \{\mathbf{0}\}$ = the zero vector $(0, 0, \ldots, 0) \in F_q^n$;

(ii) For any q, $C_3 = \{(\lambda, \ldots, \lambda) : \lambda \in F_q^n\}$;

(iii) For $q = 2$, $C_4 = \{(0, 0, 0, 0,), (1, 0, 1, 0), (0, 1, 0, 1), (1, 1, 1, 1)\}$;

(iv) For $q = 3$, $C_5 = \{(0, 0, 0), (0, 1, 2), (0, 2, 1)\}$.

From these examples, it is easy to see that for any q, $C_2 = \{\mathbf{0}\}$ is a subspace of both C_3 and $C_1 \in F_q^n$, and C_3 is a subspace of $C_1 = F_q^n$; for $q = 2$, C_4 is a subspace of F_2^4; and for $q = 3$, C_5 is a subspace of F_3^3. Thus, a nonempty subspace C of a vector space V over F_q is a subspace iff the following condition is satisfied: If $\mathbf{x}, \mathbf{y} \in C$ and $\lambda, \mu \in F_q$, then $\lambda \mathbf{x} + \mu \mathbf{y} \in C$. Moreover, a linear combination of $\mathbf{v}_1, \ldots, \mathbf{v}_k \in V$ is a vector of the form $\lambda_1 \mathbf{v}_1 + \cdots + \lambda_k \mathbf{v}_k$, where $\lambda_1, \ldots, \lambda_k \in F_q$ are some scalars. A set of vectors $\{\mathbf{v}_1, \ldots, \mathbf{v}_k\} \in V$ is linearly independent if $\lambda_1 \mathbf{v}_1 + \cdots + \lambda_k \mathbf{v}_k = \mathbf{0}$ implies that $\lambda_1 = \cdots = \lambda_k = 0$. For example, any set S that contains $\{\mathbf{0}\}$ is linearly independent; and for any F_q, the subspace $\{(0, 0, 0, 1), (0, 0, 1, 0), (0, 1, 0, 0)\}$ is linearly independent.

Let V be a vector space over F_q and let $S = \{\mathbf{v}_1, \ldots, \mathbf{v}_k\}$ be a nonempty subset of V. Then the (linear) *span* of S, denoted by $\langle S \rangle$, is defined by $\langle S \rangle = \{\lambda_1 \mathbf{v}_1 + \cdots + \lambda_k \mathbf{v}_k : \lambda_i \in F_q\}$. If $S = \emptyset$, then this span is defined by $\langle S \rangle = \{\mathbf{0}\}$. In fact, the span $\langle S \rangle$ is a subspace of V, and is known as the subspace *generated* (or *spanned*) by S. Thus, for a subspace C of V, a subset S of C is called a *generating set* (or *spanning set*) of C if $C = \langle S \rangle$. If S is already a subspace of V, then $\langle S \rangle = S$. Some examples are:

(i) For $q = 2$ and $S = \{0001, 0010, 01000\}$,

$$\langle S \rangle = \{0000, 0001, 0010, 01000, 0011, 0101, 0110, 0111\}.$$

(ii) For $q = 2$ and $S = \{0001, 1000, 1001\}$, $\langle S \rangle = \{0000, 0001, 1000, 1001\}$.

(iii) For $q = 3$ and $S = \{0001, 1000, 1001\}$,

$$\langle S \rangle = \{0000, 0001, 0002, 1000, 2000, 1001, 1002, 2001, 2002\}.$$

Let V be a vector space over F_q. Then a nonempty subset $B = \{\mathbf{v}_1, \dots, \mathbf{v}_k\}$ of V is called a *basis* for V if $V = \langle B \rangle$ and B is linearly independent. A vector space V over a finite field F_q can have many bases, but all bases must contain the same number of elements. This number is called the *dimension* of V over F_q and is denoted by $\dim(V)$. Let $\dim(V) = k$. Then V has exactly q^k elements and $\dfrac{1}{k!} \displaystyle\prod_{i=0}^{k} (q^k - q^i)$ different bases. For example,

Let $q = 2$, $S = \{0001, 0010, 0100\}$ and $V = \langle S \rangle$. Then

$$V = \{0000, 0001, 0010, 0100, 0011, 0101, 0110, 0111\}.$$

Since S is linearly independent, so $\dim(V) = 3$ and $|V| = 2^3 = 8$, and the number of different bases are

$$\frac{1}{k!} \prod_{i=0}^{k} (q^k - q^i) = \frac{1}{3!} \left(2^3 - 1\right) \left(2^3 - 2\right) \left(2^3 - 2^2\right) = 28. \quad \blacksquare$$

Let $\mathbf{v} = \{v_1, \dots, v_n\}$, $\mathbf{w} = \{w_1, \dots, w_n\} \in F_q^n$. Then

(i) The *scalar product* (or *dot product*) of $\mathbf{v}$ and $\mathbf{w}$ is defined as $\mathbf{v} \cdot \mathbf{w} = v_1 w_1 + \cdots + v_n w_n \in F_q$;

(ii) The vectors $\mathbf{v}$ and $\mathbf{w}$ are said to be orthogonal if $\mathbf{v} \cdot \mathbf{w} = 0$ (*orthogonality condition*); and

(iii) If S is a nonempty subset of F_q^n, then the *orthogonal complement* $S^\perp$ of S is defined as $S^\perp = \{\mathbf{v} \in F_q^n : \mathbf{v} \cdot \mathbf{s} = 0 \text{ for all } \mathbf{s} \in S\}$. If $S = \emptyset$, then $S^\perp = F_q^n$. In fact, $S^\perp$ is always a subspace of the vector space F_q^n for any subset S of F_q^n, such that $\langle S \rangle^\perp = S^\perp$. Some examples are as follows:

Example 3.1.1. For $q = 2$ and $n = 4$, if $\mathbf{u} = \{1, 1, 1, 1\}$, $\mathbf{v} = \{1, 1, 1, 0\}$ and $\mathbf{w} = \{1, 0, 0, 1\}$, then $\mathbf{u} \cdot \mathbf{v} = 1 \cdot 1 + 1 \cdot 1 + 1 \cdot 1 + 1 \cdot 0 = 1$, $\mathbf{v} \cdot \mathbf{w} = 1 \cdot 1 + 1 \cdot 0 + 1 \cdot 0 + 0 \cdot 1 = 1$, and $\mathbf{u} \cdot \mathbf{w} = 1 \cdot 1 + 1 \cdot 0 + 1 \cdot 0 + 1 \cdot 1 = 0$ (all calculations in mod 2). Thus, $\mathbf{u} \perp \mathbf{w}$.

Example 3.1.2. For $q = 2$ and $S = \{0100, 0101\}$, let $\mathbf{v} = (v_1, v_2, v_3, v_4) \in S^\perp$. Then $\mathbf{v} \cdot (0, 1, 0, 0) = 0$, which gives $v_2 = 0$, and $\mathbf{v} \cdot (0, 1, 0, 1) = 0$ which in turn gives $v_2 + v_4 = 0$. Thus, $v_2 = v_4 = 0$. Since v_1 and v_3 can be either 0 or 1, we conclude that $S^\perp = \{0000, 0010, 1000, 1010\}$. $\quad \blacksquare$

Finally, note that $\dim(\langle S \rangle) + \dim\left(S^\perp\right) = n$. Thus, we use the data in Example 3.1.2 to get $\langle S \rangle = \{0000, 0100, 0001, 0101\}$. Since S is linearly independent, we have $\dim(\langle S \rangle) = 2$. Using the above value of $S^\perp$, and noting that the basis for $S^\perp$ is $\{0010, 1000\}$, we get $\dim\left(S^\perp\right) = 2$. Thus, $\dim(\langle S \rangle) + \dim\left(S^\perp\right) = 2 + 2 = 4 \, (= n)$.

3.2 Communication Channels

Before we begin to enter the wide field of coding, a few definitions and explanations are in order. Any information in the form of messages is transmitted over a communication channel that may be regarded as a pipe through which the information is sent. There are different kinds of channels depending on the type of information: The channel is visual if the information is sent through smoke signals or semaphore flags; it is wired if the information is audio, as in telephone or telegraph (Morse code); and it is digital if the information is based on discrete (discontinuous) values. Another important concept is that of noise, which is interference that is always present during communications. Noise can be man-made or due to natural phenomena: The smoke or flag signals may get distorted by the presence of wind; several people speaking at once creates noise; atomic action and reaction in matter (above absolute zero) generates noise. Thus, noise is a fundamental fact of nature and it affects the information media (communication systems and data storage) very significantly.

Coding is defined as *source coding* and *channel coding*, where the former involves changing the message source to a suitable code to be transmitted on a channel. An example of source coding is the ASCII code, which converts each character to a byte of 8 bits. Channel coding aims at the construction of encoders and decoders, which result in fast encoding of messages, easy transmission of encoded messages, fast decoding of received messages, maximum rate of transmission of messages per second, and maximum detection and correction of errors in messages created by a noisy channel.

Shannon's theorem states that if a discrete channel has the capacity $\mathfrak{C}$ and a discrete source of entropy H per second, and if $H \leq \mathfrak{C}$, then there exists a coding system such that the output of the source can be transmitted over the channel with an arbitrarily small frequency of errors (Shannon [1948]). In this remarkable result, the word 'discrete' refers to information transmitted in symbols, which can be letters on a teletype, or dots and dashes of Morse code, or digital information. The notion of entropy per second implies that if our information source is transmitting data at a rate less than the communication channel can handle, we can add some extra bits to the data bits and push the error rate to an arbitrary low level, with certain restrictions. These extra bits are called the parity-check bits, which are discussed below in detail. In general, the basic structure of a communication system is described in the following diagram:

$$\boxed{\text{Info source}} \rightarrow \boxed{\text{Encoder}} \rightarrow \boxed{\text{Noisy channel}} \rightarrow \boxed{\text{Decoder}} \rightarrow \boxed{\text{Info sink}}$$

3.2.1 Shannon's Theorems. The *noisy-channel coding theorem*, also known as the *fundamental theorem of information theory*, or just *Shannon's theorem*, affirms the fact that no matter how noisy a communication channel is, it is always possible to communicate information (digital data) almost error-free up to a given maximum rate throughout the channel. This rate is called the *Shannon limit* or *Shannon capacity* (Shannon [1948]). This theorem describes the maximum possible efficiency of error correcting codes as confronted by noise interference and data corruption. However, it does not describe the methods for constructing such codes. This theorem simply states that given a noisy channel with channel capacity $\mathfrak{C}$ and data transmitted at a rate R, then if $R < \mathfrak{C}$ there exist codes that allow the probability of error at the receiver to be arbitrarily small. This means that it is theoretically possible to transmit data error-free at any rate below a limiting rate. If $R > \mathfrak{C}$, an arbitrarily small probability cannot be achieved, and all codes will have a probability of error greater than a certain possible minimum level that will increase as the rate increases. The case $R = \mathfrak{C}$ is not covered by this theorem. The channel capacity $\mathfrak{C}$ is generally calculated from the physical properties of the channel; for a band-limited channel with Gaussian noise, it is calculated using the Shannon-Hartley theorem.

Theorem 3.2.1. (Shannon's Theorem) *(i) For every discrete memory-less channel, the channel capacity $\mathfrak{C} = \max_{P_X} I(X, Y)$ has the following property: For any $\varepsilon > 0$ and $R > \mathfrak{C}$, and for large enough n, there is a code of length n and rate $\geq R$ and a decoding algorithm, such that the maximal probability of block error is $\leq \varepsilon$.*

(ii) If the probability of error $\mathfrak{p}_b$ is acceptable, then rates up to $R(\mathfrak{p}_b)$ are achievable, where $R(\mathfrak{p}_b) = c/(1 - H_2(\mathfrak{p}_b))$, and $H_2(\mathfrak{p}_b)$ is the binary entropy function defined by $H_2(\mathfrak{p}_b) = -[\mathfrak{p}_b \log \mathfrak{p}_b + (1 - \mathfrak{p}_b) \log(1 - \mathfrak{p}_b)]$.

(iii) For any $\mathfrak{p}_b$, rates $> R(\mathfrak{p}_b)$ are not achievable.

PROOF. A proof for this theorem is based on an achievability result and a matching result. These two features serve to bound the set of all possible rates at which a transmission can be carried out successfully on a noisy channel. The matching property shows that these bounds are sharp. For details based on achievability and weak converse for discrete memoryless channels, see MacKay [2003]. ∎

Theorem 3.2.2. (Channel coding theorem for a nonstationary memory-less channel) *Under the assumption that the channel is memoryless, but its transition probabilities are time dependent (i.e., they change with time) in a way known both at the transmitter and receiver, the channel capacity is given* *by*

$$\mathfrak{C} = \liminf_{\substack{\max \\ \mathbf{p}(X_1), \mathbf{p}(X_2), \cdots}} \frac{1}{n} \sum_{i=1}^{n} I(X_i; Y_i), \qquad (3.2.1)$$

such that this maximum value is attained at the capacity-achieving distributions for each respective channel. In other words,

$$\mathfrak{C} = \liminf \frac{1}{n} \sum_{i=1}^{n} \mathfrak{C}_i, \qquad (3.2.2)$$

where $\mathfrak{C}_i$ is the capacity of the ith channel.

PROOF. An outline of the proof is based on the argument that achievability follows from random coding with each symbol chosen randomly from the capacity achieving distribution for that particular channel. However, the limit in (3.2.2) becomes important in cases where the expression $\frac{1}{n} \sum_{i=1}^{n} \mathfrak{C}_i$ fails to converge. ∎

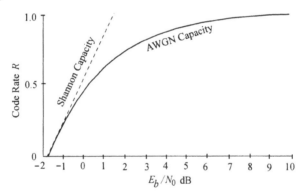

Figure 3.2.1. Capacity of AWGN (one-dimensional signaling).

The channel capacity $\mathfrak{C}$ is measured in bits per channel use (modulated symbol). Consider an additive white Gaussian noise (AWGN) channel with one-dimensional input $y = x + n$, where x is a signal with average energy variance E_s and n is the Gaussian with variance $N_0/2$. Then the capacity of AWGN with unconstrained input is

$$\mathfrak{C} = \max_{p(x)}\{I(X; Y)\} = \frac{1}{2} \ln\left(1 + \frac{2E_s}{N_0}\right) = \frac{1}{2} \ln\left(1 + \frac{2RE_b}{N_0}\right),$$

where E_b is the energy per information bit. This capacity is achieved by a Gaussian input x. However, this is not a practical modulation. If we consider antipodal BPSK modulation, then $X = \pm\sqrt{E_s}$, and the capacity is

$$\mathfrak{C} = \max_{\mathfrak{p}(x)}\{I(X;Y)\} = I(X;Y)\Big|_{\mathfrak{p}(x):\mathfrak{p}=1/2}$$

$$= H(Y) - H(N)$$

$$= \int_{-\infty}^{\infty} \mathfrak{p}(y)\log_2\mathfrak{p}(y)\,dy - \frac{1}{2}\log_2(\pi e N_0), \qquad (3.2.3)$$

since the maximum is attained when two signals are equally likely. The integral term in (3.2.3) is computed numerically using FFT defined by the convolution $\mathfrak{p}_X(y) \star \mathfrak{p}_N(y) = \int_{-\infty}^{\infty} \mathfrak{p}_X(u)\mathfrak{p}_N(y-u)\,du$.

3.3 Some Useful Definitions

These definitions are very useful in understanding the operation and application of linear codes.

3.3.1 Hamming Distance. The Hamming distance between two strings of equal length is equal to the number of positions at which the corresponding symbols are different. It is a metric on the vector space $\mathcal{X}$ of words of equal length, and it measures the minimum number of substitutions that will be needed to change one word into the other; that is, it measures the number of *errors* that transform one string into the other. Thus, if $\mathbf{x}$ and $\mathbf{y}$ are strings of length n in an alphabet A, then the Hamming distance between $\mathbf{x}$ and $\mathbf{y}$, denoted by $d(\mathbf{x}, \mathbf{y})$, is defined to be the number of places at which $\mathbf{x}$ and $\mathbf{y}$ differ. Let $\mathbf{x} = x_1, \cdots, x_n$, and $\mathbf{y} = y_1, \cdots, y_n$. Then

$$d(\mathbf{x}, \mathbf{y}) = d(x_1, y_1) + \cdots + d(x_n, y_n), \qquad (3.3.1)$$

where $d(x_i, y_i) = \begin{cases} 1 & \text{if } x_i \neq y_i, \\ 0 & \text{if } x_i = y_i \end{cases}$.

Examples 3.3.1. Let $\mathbf{x} = 01101$, $\mathbf{y} = 11011$, $\mathbf{z} = 10111$. Then $d(\mathbf{x}, \mathbf{y}) = 3$, $d(\mathbf{y}, \mathbf{z}) = 2$, $d(\mathbf{x}, \mathbf{z}) = 3$.

Examples 3.3.2. Let $\mathbf{x} = 1010111$, $\mathbf{y} = 1000101$. Then $d(\mathbf{x}, \mathbf{y}) = 2$.

Examples 3.3.3. Let $\mathbf{x} = 5078612$, $\mathbf{y} = 5198317$. Then $d(\mathbf{x}, \mathbf{y}) = 4$.

Examples 3.3.4. Let $\mathbf{x} = $ paint, $\mathbf{y} = $ saint. Then $d(\mathbf{x}, \mathbf{y}) = 1$. ∎

Theorem 3.3.1. *Let* $\mathbf{x}, \mathbf{y}, \mathbf{z}$ *be strings of length* n *in the alphabet* A. *Then*

(i) $0 \leq d(\mathbf{x}, \mathbf{y}) \leq n$;

(ii) $d(\mathbf{x}, \mathbf{y}) = 0$ *iff* $\mathbf{x} = \mathbf{y}$;

(iii) $d(\mathbf{x}, \mathbf{y}) = d(\mathbf{y}, \mathbf{x})$; *and*

(iv) $d(\mathbf{x}, \mathbf{z}) \leq d(\mathbf{x}, \mathbf{y}) + d(\mathbf{y}, \mathbf{z})$.

PROOF. Parts (i), (ii) and (iii) follow from the definition of the Hamming distance, i.e., $d(\mathbf{x}, \mathbf{y})$ is non-negative and symmetric. To prove (iv), if $\mathbf{x} = \mathbf{z}$, then it is true since $d(\mathbf{x}, \mathbf{z}) = 0$. If $\mathbf{x} \neq \mathbf{z}$, then either $\mathbf{y} \neq \mathbf{x}$ or $\mathbf{y} \neq \mathbf{z}$, thus showing that (iv) is true. ∎.

3.3.2 Lee Distance. This is a distance between two strings $\{x_1 x_2 \ldots x_n\}$ and $\{y_1 y_2 \ldots y_n\}$ of the same length n over a q-ary alphabet $A = \{0, 1, \ldots, q-1\}$ of size $q \geq 2$. As a metric it is defined as $\sum_{i=1}^{n} \min\left(|x_i - y_i|, q - |x_i - y_i|\right)$. For $q = 2$ or 3, it coincides with the Hamming distance. As an example, let $q = 4$. Then the Lee distance between the strings $\{2011\}$ and $\{1230\}$ is $1 + 2 + 2 + 1 = 6$. This distance, named after Lee [1958], is applied for phase modulation while the Hamming distance is used for orthogonal modulation.

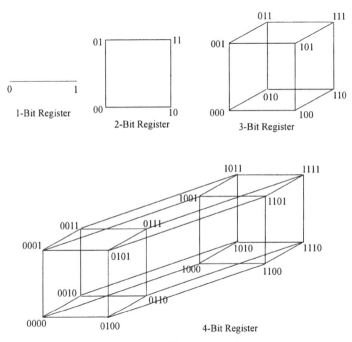

Figure 3.3.1. Hamming cubes.

3.3.3 Hamming Cubes. The metric space of binary strings of length n with the Hamming distance as defined above, is called the *Hamming cube*, which is equivalent to a metric space on the set of distances between the vertices of a hypercube. The vertices of this cube with real coordinates represent the binary strings of a given length. Such cubes for $n = 1, 2, 3, 4$ are shown in Figure 3.3.1. Note that the cubes for $n = 1$ and 2 are, respectively, a line segment of length 1 and a square of side 1; for $n = 3$, it is a geometric cube, and for $n \geq 4$ a hypercube. The vertices are chosen according to the value of

n. Note that the Hamming distance is equivalent to the Manhattan distance[1] between the vertices. Thus, for example,

For $n = 1$: $d(0, 1) = 1 = d(1, 0)$;

For $n = 2$: $d(00, 11) = d(00, 10) + d(10, 11) = 1 + 1 = 2$;

For $n = 3$: $d(100, 011) = 3$; $d(010, 111) = 2$;

For $n = 4$: $d(0100, 1001) = 3$; $d(0110, 1110) = 1$; $d(0000, 1111) = 4$.

The value of these distances can be verified by traversing the minimum path along the sides of the Hamming cubes, starting from the string a and ending in the string b. Each Hamming cube represents an n-bit register.

3.3.4 Hamming Weight. The Hamming weight of a string (or word) is the number of nonzero characters of an alphabet A. Thus, it is equal to the Hamming distance $d(\mathbf{x}, \mathbf{y})$ of all zero strings of the same length. If $\mathbf{x}$ is a word, then the Hamming weight of $\mathbf{x}$, denoted by $\mathrm{wt}(\mathbf{x})$, is defined to be the number of nonzero characters in $\mathbf{x}$, that is,

$$\mathrm{wt}(\mathbf{x}, \mathbf{y}) = d(\mathbf{x}, \mathbf{0}),$$

where $\mathbf{0}$ is the zero word. For every element in $x \in \mathbf{x}$, we have

$$\mathrm{wt}(x) = d(x, 0) = \begin{cases} 1 & \text{if } x \neq 0, \\ 0 & \text{if } x = 0. \end{cases}$$

Thus, if $\mathbf{x} = \{x_1, x_2, \ldots, x_n\}$, then $\mathrm{wt}(\mathbf{x}) = \mathrm{wt}(x_1) + \mathrm{wt}(x_2) + \cdots + \mathrm{wt}(x_n)$. Moreover, if $\mathbf{x}$ and $\mathbf{y}$ are two words (strings) of the same length, then $d(\mathbf{x}, \mathbf{y}) = \mathrm{wt}(\mathbf{x} - \mathbf{y})$. In the case of a binary linear string of bits in $A = \{0, 1\}$, the Hamming weight is equal to the number of 1's in the string. It is also called the *population count* (or *popcount*).

If $\mathbf{x}$ and $\mathbf{y}$ are two strings of the same length in the binary case, then

$$\mathrm{wt}(\mathbf{x} + \mathbf{y}) = \mathrm{wt}(\mathbf{x}) + \mathrm{wt}(\mathbf{y}) - 2\,\mathrm{wt}(\mathbf{x}\,\&\,\mathbf{y}). \tag{3.3.2}$$

If C is a code, not necessarily linear, then the *minimum Hamming weight* of C, denoted by $\min \mathrm{wt}(C)$, is the smallest of the weights of the nonzero codewords of C. If C is a linear code, then $d(C) = \min \mathrm{wt}(C) = \mathrm{wt}(C)$.

Example 3.3.5. Let the alphabets be $A = \{0, 1\}$ and $B = \{a - z, \sqcup\}$. Then, $\mathrm{wt}(1101) = 3$, $\mathrm{wt}(11011001) = 5$, $\mathrm{wt}(00000000) = 0$, wt(hamming weight) = 13, counting the space ($\sqcup$), which is a zeroword. ∎

Example 3.3.6. Let a binary linear code be $C = \{0000, 1000, 0100, 1100\}$. Then $\mathrm{wt}(1000) = 1$, $\mathrm{wt}(0100) = 1$, $\mathrm{wt}(1100) = 2$. Hence, $d(C) = 1$. ∎

---

[1] The Manhattan distance between two vertices is the shortest distance between them along the axes and is equal to the minimum number of axes traversed in reaching from the starting vertex to the end vertex.

Example 3.3.7. Let $\mathbf{x} = 0011$, $\mathbf{y} = 0101$. Then, carrying out the bitwise operations, we have $\mathbf{x}\,\&\,\mathbf{y} = 0001$, $\mathbf{x} + \mathbf{y} = 0110$, $\mathrm{wt}(\mathbf{x} + \mathbf{y}) = 2$; $\mathrm{wt}(\mathbf{x}) = 2$, $\mathrm{wt}(\mathbf{y}) = 2$, $2\,\mathrm{wt}(\mathbf{x}\,\&\,\mathbf{y}) = 2$. Hence, we see that the property (3.3.2) is satisfied. ∎

Another formula to determine the Hamming weight is

$$\mathrm{wt}\,(\mathbf{x}\,\&\,(\mathbf{x} - 1)) = \mathrm{wt}(\mathbf{x}) - 1 \text{ for all } \mathbf{x} \in F_q^n. \qquad (3.3.3)$$

The operation $\mathbf{x}\,\&\,(\mathbf{x} - 1)$ eliminates the lsb 1 of $\mathbf{x}$. Thus, for example, if $\mathbf{x} = 10010100$, then $\mathbf{x} - 1 = 10010011$, and $\mathbf{x}\,\&\,(\mathbf{x} - 1) = 10010000$. This formula is very efficient if the input is sparse or dense.

3.3.5 Residual Code. Let C be a linear binary (n, k, d) code in F_q, and let $\mathbf{c}$ be a codeword in C with $\mathrm{wt}(\mathbf{c}) = w$. The *residual code* of C with respect to $\mathbf{c} \in C$, denoted by $\mathrm{Res}(C, \mathbf{c})$, is a linear $(n - d, k - 1, \lceil d/q \rceil)$ code. The generator matrix of $\mathrm{Res}(C, \mathbf{c})$ is obtained by discarding the first row and the first d columns from the generator matrix of $\mathrm{Res}(C, \mathbf{c})$. Note that $w = |\operatorname{supp}(\mathbf{c})|$. [2] Residual codes are used in the Griesmer bound of error correcting codes (see §6.8). They are also known as *punctured cyclic codes* (Solomon and Stiffler [1965]). The following results, which are part of the above definition of residual codes, are proved in Ling and Xing [2004].

Theorem 3.3.2. *If C is a linear (n, k, d) code over F_q and $\mathbf{c} \in C$ is a codeword of weight d, then $\mathrm{Res}(C, \mathbf{c})$ is an $(n - d, k - 1, d')$ code, where $d' \geq \lceil d/q \rceil$.*

Theorem 3.3.3. $\mathrm{Res}(C, \mathbf{c})$ *has dimension at most $k - 1$.*

3.3.6 Tree Algorithm. In order to determine the Hamming weight, this algorithm performs several operations in parallel on the subwords (which are subsets of a word) as follows.

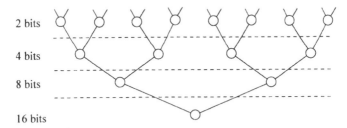

2 bits

4 bits

8 bits

16 bits

Figure 3.3.2. Tree algorithm for a 16-bit word.

First, it computes the Hamming weight of all the subwords of 2 bits; next, it computes the Hamming weight of 4-bit words, then successively that of 8-bit words, and so on. The successive computations of a 16-bit word are

---

[2] The support of $\mathbf{c}$, denoted by $\operatorname{supp}(\mathbf{c})$, is the set of coordinates at which $\mathbf{c}$ is nonzero.

represented as a tree in Figure 3.3.2, in which the parallel addition of subwords are emulated with the bitwise AND, the left shift and the word addition. The following operations, presented in Hex, are carried out at each step:

At 2 bits: $x = (x \& 0x5555) + ((x >> 1) \& 0x55555)$;
At 4 bits: $x = (x \& 0x3333) + ((x >> 2) \& 0x3333)$;
At 8 bits: $x = (x \& 0x0F0F) + ((x >> 4) \& 0x0F0F)$;
At 16 bits: $x = (x \& (x >> 8)) \& 0xFF$.

3.4 Linear Codes

A linear code is a type of block code used in error detection and correction in transmitting bits or data on a communications channel. A linear code of length n transmits blocks containing n bits (symbols).

3.4.1 Linear Code. A linear code C of length n and rank k over a finite field (or code alphabet) F_q is a subspace of the field F_q^n, where F_q has q elements. Such codes with parameter q are called q-ary codes. For example, for $q = 2$ the code is called a binary code, for $q = 3$ a ternary code, and for $q = 5$, a 5-ary code. The field F_2 is also represented by a *code alphabet* $A = \{0, 1\}$ and its elements are called *code elements*. A code over the code alphabet $A \in F_2$ (i.e., a binary alphabet) has elements $\{0, 1\}$, and over F_3 the elements $\{0, 1, 2\}$ and in general, a code alphabet $A = \{a_1, a_2, \ldots, a_q\}$ is the same as the finite field F_q. A q-ary word of length n over A is a sequence $\mathbf{a} = \{a_1, a_2, \ldots, a_q\}$, where each $a_k \in A$ for all k, $k = 1, \ldots, n$. In this sense, the sequence $\mathbf{a}$ can be treated like a vector. A q-ary block code of length n over A is a nonempty set C of q-ary words each of length n, and an element of C is called a *codeword* $\mathbf{c}$ in C. The number of codewords in C, denoted by $|C|$, is called the *size* of C. The *(information) rate* of a code C of length n is $\log_q |C|/n$, and a code of length n and size M is called an (n, M) code.

3.4.2 Dual Codes. If C is a linear code of length n over F_q, the *dual code* $C^\perp$ of C is the orthogonal complement of C, and is defined as

$$C^\perp = \{v \in F_q^n \mid \langle v, \mathbf{c} \rangle = 0 \text{ for all } \mathbf{c} \in C\},$$

where $\langle v, \mathbf{c} \rangle = \sum_{k=1}^{n} v_k c_k^p$, and p is the characteristic of F_q. In the language of linear algebra, the dual code is the annihilator of C with respect to the bilinear form $\langle \,, \, \rangle$. Note that $(C^\perp)^\perp = C$, i.e., the dual of a dual code is a dual code.

3.4.3 Dimension of a Code. The dimension of C, denoted by $\dim(C)$, is determined by $\dim(C) = \log_q |C|$. Moreover, $C^\perp$ is a linear code such that $\dim(C) + \dim(C^\perp) = n$. For example, let $C = \{0000, 1010, 0101, 1111\}$ for $q = 2$. Then $\dim(C) = \log_2 4 = 2$, and $C^\perp = \{0000, 1010, 0101, 1111\} = C$, so $\dim(C^\perp) = 2$. Note that a linear code of length n and dimension k over F_q

is sometimes called a q-ary (n, k, q) *code*, or if q is evident from the context, then an (n, k) *code*.

3.4.4 Self-Dual Codes and Types. If C is a linear code, then C is self-orthogonal, i.e., $C \subseteq C^\perp$, and C is self-dual, i.e., $C = C^\perp$. A self-dual code is one that is its own dual. It means that n is even and $\dim(C) = n/2$. Self-dual codes are classified into the following four types:

(i) TYPE I codes are binary self-dual codes that are not doubly-even; they are always even, i.e., every codeword has even Hamming weight.

(ii) TYPE II codes are binary self-dual codes that are doubly-even, i.e., the Hamming weight of every codeword is divisible by 4.

(iii) TYPE III codes are ternary self-dual codes; every codeword in this type has Hamming weight divisible by 3.

(iv) TYPE IV codes are self-dual codes over F_4; they are even codes, i.e., all its codewords are even.

An even code that is not doubly-even is strictly even. Examples of doubly-even codes are the extended binary Hamming code of block length 8 and the extended binary Golay code of block length 24; both these codes are self-dual codes.

3.4.5 Advantages of Linear Codes. Some advantages of using linear codes are:

(i) Being a vector space, a linear code can be completely described using a basis.

(ii) The distance of a linear code is equal to the smallest weights of its nonzero codewords.

(iii) The encoding and decoding procedures for a linear code are simpler and faster than those for nonlinear codes.

The Hamming distance is mostly used in coding theory, information theory, and cryptography. The popcount of a bit string is often required in these areas for which an efficient implementation is very desirable. While some processors have a single command to compute the popcount (see §3.4.6),[3] others use parallel operations on bit vectors. If these features are not available in a processor, then the best solution is found by adding counts in a tree pattern, as shown in the following example, or by using the six C++ codes given toward the end of this section.

---

[3] For example, to count the number of bits that are set, the bit-array data structure *bitset* in C++ STL has a count() method; the bit-array data structure BitSet in Java has a BitSet.cardinality() method; to count bits in 32-bit and 64-bit integers, the functions Integer.bitCount(int) and Long.bitCount(long) are available in Java, and also the BitInteger arbitrary-precision integer class has a BitInteger.bitCount() method. The C compiler GCC (starting with version 3.4, April 2004) has a built-in function popcount.

Example 3.4.1. To count the number of 1 bits in the 16-bit binary number $x = 0100111010111010$, we proceed as follows.

Binary	Decimal	Comment
$a = x \,\&\, 0101010101010101$		
$\quad = 0100010000010000$	$1, 0, 1, 0, 0, 1, 0, 0$	Every other bit from x
$b = x \gg 1 \,\&\, 000101010101010101$		
$\quad = 0000010101010001$	$0, 0, 1, 1, 1, 1, 1, 1$	Remaining bits from x
$c = a + b = 0100100101100101$	$1, 0, 2, 1, 1, 2, 1, 1$	# of 1s in each 2-bit parts of x
$d = c \,\&\, 00010000100100001$		
$\quad = 0000000100100001$	$0, 1, 2, 1$	Every other count from c
$e = c \gg 2 \,\&\, 0011001100110011$		
$\quad = 0001001000010001$	$1, 2, 1, 1$	Remaining counts from c
$f = d + e = 0001001100110010$	$1, 3, 3, 2$	# of 1s in each 4-bit parts of x
$g = f \,\&\, 0000111100001111$		
$\quad = 0000001100000010$	$3, 2$	Every other count in f
$h = f \gg 4 \,\&\, 00001111100001111$		
$\quad = 0000000100000011$	$1, 3$	Remaining counts from f
$i = g + h = 0000010000000101$	$4, 5$	# of 1s in each 8-bit parts of x
$j = i \,\&\, 0000000011111111$	5	Every other count from I
$k = i \gg 8 \,\&\, 0000000011111111$	4	Remaining counts from I
$wt\{x\} = j + k$	9	Number of 1s in x is 9

Thus, the Hamming weight of the string x is 9. The operation $x \gg n$ means to shift x to right by n bits, $x \& y$ means the bitwise AND of x and y, and $+$ is ordinary binary addition. ∎

An algorithm for problems of finding the Hamming weight of given strings in the binary case is known as the *population count* or *popcount*, which is discussed in the next section. However, a faster code is available if the given number has most of its bits 0, or most of its bits 1. Such codes are based on the fact that if x is a string, then $x \& (x - 1)$ is the same string x with the rightmost 1 bit set to 0 (Wagner [1960]).

Example 3.4.2. Given $x = 01000100001000$, we have

$$x - 1 = 01000100000111$$

$$x \& (x - 1) = 01000100000000.$$

Thus, if x originally had n bits of 1, then after n iterations the string x will be reduced to zero. Hence, the number of bits of 1 will be n. In this example it will be 3. Note bitwise operation for the result given above.[4] ∎

3.4.6 Popcount. This is a built-in function that is implemented using some additional hardware in many computers. Its purpose is to count the Hamming weight in a given string or word. This built-in function is mostly superior to many software implementations such as serial shifts, as explained in Example 3.4.1. However, there are a few software algorithms for this function, which achieve the same result without any need for extra hardware.

The simplest scheme for popcount is via serial shift that is generally implemented as follows:

```
set counter to 0
while the number = 0 do
    if the lowest list of number is one then
        increase counter by 1
        shift the number to the right one bit
    endif
endwhile
```

As defined in §3.3.4, the Hamming weight in the binary case is the total number of 1s in a string. Among many uses of the Hamming weight, one other application in coding theory is that of determining path lengths from node to query in Chord-distributed hash tables in Internet applications (see Stoica et al. [2003]). An algorithm for the tabular derivation of the Hamming weight in Example 3.4.1 is as follows: The following types and constants are used in the functions:

```
typedef unsigned __int64 uint64;     //for gcc, use uint64_t for __int64
const  uint64  m1  = 0x5555555555555555; //binary 0101...
const  uint64  m2  = 0x3333333333333333; //binary 00110011...
const  uint64  m4  = 0x0f0f0f0f0f0f0f0f; //binary 4 zeros, 4 ones, ...
const  uint64  m8  = 0x00ff00ff00ff00ff; //binary 8 zeros, 8 ones, ...
const  uint64  m16 = 0x0000ffff0000ffff; //binary 16 zeros,16 ones, ..
const  uint64  m32 = 0x00000000ffffffff; //binary 32 zeros, 32 ones
const  uint64  hff = 0xffffffffffffffff; //binary all ones
const  uint64  h01 = 0x0101010101010101; //the sum of 256
                          to the power of 0, 1, 2, 3, ...
```

In addition, the following five C code fragments can be used to determine the popcount in different kinds of computers, with slow or faster multiplication. For more details, see Knuth [2003].

CODE 1. This is the simplest code to implement; it helps understand the popcount function. It uses 24 arithmetic operations (right shift $\gg$, $+$, $\&$).

---

[4] The operation of subtracting 1 from a binary string **x** is carried out as follows: Change the rightmost string of 0s in **x** to 1s, and change the rightmost 1 in **x** to 0.

```
int popcount_1 (uint64 x)   {
x = (x & m1)+((x ≫ 1)& m1);//put count of each 2 bits into those 2 bits
x = (x & m2)+((x ≫2)& m2);//put count of each 4 bits into those 4 bits
x = (x & m4)+((x ≫4)& m4);//put count of each 8 bits into those 8 bits
x = (x & m8)+((x ≫8)& m8);//put count of each 16 bits into those 16 bits
x = (x & m16)+((x ≫16)& m16);//put count of each 32 bits into those
                                          32 bits
x = (x & m32)+((x ≫32)& m32);//put count of each 64 bits into those
                                          64 bits
}
```

CODE 2. This code works better when more bits in **x** are 0s. It uses 3 arithmetic operations and one comparison/branch per '1' bit in **x**. This algorithm is used in Example 3.4.2.

```
int popcount_2 (uint64 x)   {
    uint64 count;
    for (count=0; x!=0; count++)
      x &= x-1;
    return count;
}
```

CODE 3. This code uses fewer arithmetic operations than any other known code to implement popcount on machines with slow multiplication. It uses 17 arithmetic operations.

```
int popcount_3 (uint64 x)   {
x -=(x≫1)& m1;//put count of each 2 bits into those 2 bits
x = (x& m2) + ((x≫2)& m2);//put count of each 4 bits into those 4 bits
x = (x+(x≫4)& m4;//put count of each 8 bits into those 8 bits
x += x≫8;//put count of each 16 bits into their lowest 8 bits
x += x≫16;//put count of each 32 bits into their lowest 8 bits
x += x≫32;//put count of each 64 bits into their lowest 8 bits
return x & 0x7f;
}
```

CODE 4. This code uses fewer arithmetic operations to implement popcount on machines with fast multiplication. It uses 12 arithmetic operations.

```
int popcount_4 (uint64 x)   {
x -= (x≫1) & m1;//put count of each 2 bits into those 2 bits
x = (x & m2) + ((x≫2) & m2);//put count of each 4 bits into
                                          those 4 bits
x = (x+(x≫ 4))& m4;//put count of each 8 bits into those 8 bits
return (x*h01)≫56;//return left 8 bits of x+(x<<8)+(x<<16)+(x<<24)+...
}
```

Modern programming languages include certain intrinsic popcount functions, such as the __builtin_popcount function in the GCC C++ compiler or the bitset::count() method in the C++ STL that counts the number of bits that are set. Similarly Java has the BitSet.cardinality() method and the bitCount() methods on the Integer, Long and BigInteger classes. In Common Lisp, the function logcount returns the number of 1 bits for a given non-negative integer.

Efficient popcount implementations can also be provided in hardware. The popcount algorithm can be compiled using the Verilog hardware description language (HDL) used to model electronic systems. Early Cray supercomputers featured a popcount machine instruction. Intel Core processors introduced a POPCNT instruction with the SSE4.2 instruction set extension, first available in a Nehalem-based Core i7 processor released in November 2008. AMD also implemented POPCNT beginning with the Barcelona microarchitecture as part of their Advanced Bit Manipulation (ABM) functionality.

3.4.7 Parity Bits. A parity bit (or parity-check bit, or simply, check bit) is a binary digit 0 or 1 that is appended to ensure that the number of bits with values of 1 in a given set of bits is always even or odd. Parity bits are used as the simplest way to detect errors in transmission. The parity bits are all bit positions that are powers of 2, i.e., 2^j for $j = 0, 1, 2, \ldots$. Thus, they are the bit positions $1, 2, 4, 8, 16, 32, 64, \ldots$. There are two notations used for parity bits in Hamming and Golay codes: one, the *location notation* of the form 2^j, which denotes them as $p_1, p_2, p_4, p_8, p_{16}, \ldots, p_{2^j}, \ldots, j = 0, 1, 2, \ldots$, so that they conform to their bit locations; and the other, the *exponent notation* which uses the exponent j, $j = 0, 1, 2, \ldots$, as indices and denotes them as $p_0, p_1, p_2, p_3, p_4, \ldots$. Thus, in the location form, the parity bits p_1, p_2, p_4, p_8, $p_{16} \ldots$, or equivalently in the exponent form, the parity bits $p_0, p_1, p_2, p_3, p_4, \ldots$ are each located at the bit locations $1, 2, 4, 8, 16, \ldots$, respectively.

A number n is said to be of *even parity* if $2|n = 0$ and of *odd parity* if $2|n = 1$.[5] There are two variants of parity bits: *even parity bit* and *odd parity bit*. An even parity bit is set to 1 if the number of 1s in a given string is odd, thus making the total number of 1s, including the parity bit, even. An odd parity bit is set to 1 if the number of 1s in a given string is even, thus making the total number of 1s, including the parity bit, odd.

If an odd number of bits, including the parity bit, are changed in a transmission of a string of bits, then the parity bit will be incorrect and this will signal that an error has occurred. Thus, the parity bit provides an error detecting code, and not an error correcting code, as it will not tell which bit is corrupted. As a result, the data must be discarded entirely and re-transmitted again. On a noisy transmission line this repeated transmission may take a long time, may never reach completion, or may never occur. However, there is an

---

[5] The notation $a|b = c$ means that c is the remainder when a divides b.

underlying advantage in using the parity, because it is almost the best solution available that uses only a single bit of memory and only requires the XOR gates if using even parity, or NOT and XOR gates if using odd parity, to detect single errors.

Example 3.4.3. Assume that the data is a 4-bit word 1101, followed by the parity bit on the right.

TRANSMISSION USING EVEN PARITY:

Data to transmit: 1101

Compute parity bit value (using XOR): $1 \oplus 1 \oplus 0 \oplus 1 = 1$

Parity bit appended to the data: 11011

Data received: 11011

Compute overall parity: $1 \oplus 1 \oplus 0 \oplus 1 \oplus 1 = 0$

Correct transmission reported.

TRANSMISSION USING ODD PARITY:

Data to transmit: 1101

Compute parity bit value (using XOR): $\neg (1 \oplus 1 \oplus 0 \oplus 1) = 0$

Parity bit appended to the data: 11010

Data received: 11010

Compute overall parity: $1 \oplus 1 \oplus 0 \oplus 1 \oplus 0 = 1$

Correct transmission reported.

To detect a single-bit error,

TRANSMISSION USING EVEN PARITY:

Data to transmit: 1101

Compute parity bit value (using XOR): $1 \oplus 1 \oplus 0 \oplus 1 = 1$

Parity bit appended to the data: 11011

Data received: **1**0011

Compute overall parity: $1 \oplus 0 \oplus 0 \oplus 1 \oplus 1 = 1$

Incorrect transmission reported.

Further, suppose the following case occurs:

TRANSMISSION USING EVEN PARITY:

Data received: 11010

Compute overall parity: $1 \oplus 1 \oplus 0 \oplus 1 \oplus 0 = 1$

Correct transmission reported.

But which bit is in error cannot be determined.

Similarly, using odd parity, the following cases occur:

TRANSMISSION USING ODD PARITY:

Data to transmit: 1101

Compute parity bit value (using XOR): $1 \oplus 1 \oplus 0 \oplus 1 = 1$

Parity bit appended to the data: 11011

Data received: **10010**

Compute overall parity: $1 \oplus 0 \oplus 0 \oplus 1 \oplus 0 = 0$

Correct transmission reported, although it is actually incorrect. ∎

3.4.8 Uses of Parity Bits. Because of its simplicity, the parity bit is used in various hardware applications where an operation can be discarded and repeated in cases of difficulty, or where detecting an error is required. These situations occur in several situations, for example,

(i) SCSI and PCI buses, where parity bits are used to detect transmission errors.

(ii) Many microprocessor instruction caches, which include parity protection, and create cache data as a copy of main memory, to be discarded and re-fetched if found to be corrupted.

(iii) Serial data transmission, which generally has 7 data bits, followed by one even parity bit, and one or two stop bits. When transmitting 7-bit ASCII characters, the 8th bit can be used as a parity bit.

(iv) RAID levels, which use parity to provide fault tolerance by reconstructing missing data on a failed drive in the array.

Parity memory was a main requirement in the original IBM PC as this computer was designed by engineers for business, such as banks, airlines, stock brokers, and others. These engineers were familiar with the large mainframe computers. As the semiconductors used that time were not very reliable, it was required that every memory access contained accurate data and no errors were introduced either by hardware or by random electronic 'glitches'. On the other hand, Apple realized that the average home user would not be affected by an occasional random error that might be introduced, so they decided to design their computer without using parity memory modules. This reduced the cost of their machines because at that time memory was very expensive. Soon IBM PC clone manufacturers recognized the importance of non-parity memory and produced cheaper 386 and 486 models. Today, almost all PC machines and Apples sold contain non-parity memory, and only servers and very few other machines contain parity (or ECC) memory, as the soft error rate for modern-grade chips is negligibly small (about once in 10 years).

Parity checking is a simple method of detecting memory errors, and has no correction capabilities. A better error checking method is ECC (error correction codes), which include not only a single bit error correction, but even two, three and four bit detection. ECC can be implemented either on the module (ECC-on-SIMM, or EOS) or in the chipset, although ECC modules are very rare. Different 'hashing' algorithms (or codes) implement ECC that works on 8 bytes (64 bits) at a time and produces the result into an 8-bit ECC 'word'. Then the 8 bytes are read and again 'hashed' and the results are compared to the stored ECC word in the same manner as the parity checking is done. The only difference is that in parity checking each parity bit is associated with a single byte, whereas the ECC word is associated with the entire 8 bytes. Thus, the bit values in the parity checking and ECC will be different, and therefore ECC modules cannot be used in parity mode. But both parity and ECC modules can be used on any motherboard that does support the parity/ECC checking. Many early Pentium class chips did not have the ability to perform parity/ECC checking, and as such the feature is always set to 'disable' in the BIOS. While SIMMs (Single In-line Memory Modules) can be implemented on non-parity, parity, or ECC, DIMMs (Dual In-line Memory Modules) are either ECC or non-ECC.

3.4.9 Stop Bit. A stop bit is a character in asynchronous communication (or 'asynch' for short), in which data is transmitted without any external clock, and therefore it becomes necessary that the receiver must know when the byte being transmitted has ended. Without a stop bit, the receiver is likely to produce an error message and create problems at the transmitting computer. The receiver must know when the byte has ended in order to get ready to receive the next byte. A stop bit is always paired with a start bit that signals when to start receiving a new data. For example, network transmission is broken down into blocks of data, known as bytes, where a byte normally includes 8 bits of data plus a start bit and a stop bit, thus making a 10-bit character frame. Such bytes are then assembled and represent the original data to be transmitted.

Although the start and stop bits are prerequisites of modern communication technology, they are derived from older technologies. For example, the teletype machines in the 19th and early 20th centuries were required to be resynchronized after the start of each character, which were known as 'codes' and often included more than one word. The procedure was to transmit an additional stop bit after each 'code' was transmitted (or sometimes more than one stop bit) so that the receiver would know that it was time to resynchronize. In modern asynchronous transmission, only one stop bit is needed.

3.4.10 Parity-Check Matrix. As mentioned earlier, a basis for a linear code is required for an explicit description of a codeword C. Let S be a nonempty subset of a finite field F_q. Then a basis for $C = \langle S \rangle$ is a linear code generated by S. To find this basis, the algorithm is as follows: Form the matrix A whose columns are the words in S, put A in the row echelon form (REF) using a sequence of row operations, and locate the leading columns in the REF. Then the original columns of A corresponding to these leading columns form a basis for C.

Example 3.4.4. A basis for $C = \langle S \rangle \in F_2$, where

$$S = \{11101, 10110, 01011, 11010\},$$

is found by putting A in the REF:

$$A = \begin{bmatrix} 1 & 1 & 0 & 1 \\ 1 & 0 & 1 & 1 \\ 1 & 1 & 0 & 0 \\ 0 & 1 & 1 & 1 \\ 1 & 0 & 1 & 0 \end{bmatrix} \rightarrow \begin{bmatrix} 1 & 1 & 0 & 1 \\ 0 & 1 & 1 & 0 \\ 0 & 0 & 0 & 1 \\ 0 & 1 & 1 & 1 \\ 0 & 1 & 1 & 1 \end{bmatrix} \rightarrow \begin{bmatrix} 1 & 1 & 0 & 1 \\ 0 & 1 & 1 & 0 \\ 0 & 0 & 0 & 1 \\ 0 & 0 & 0 & 0 \\ 0 & 0 & 0 & 0 \end{bmatrix}.$$

Since the columns 1, 2, 4 of the REF are the leading columns, they form a basis for C, i.e., $\{11101, 10110, 11010\}$ is a basis for C. Alternatively, the matrix A can be put in a *reduced row echelon form* (RREF) by a sequence of elementary row operations. A matrix is row equivalent to a matrix in REF or in RREF; however, the RREF of a matrix is unique, while a matrix may have different REF. For the above example, the RREF form of the matrix A is given by

$$A = \begin{bmatrix} 1 & 1 & 1 & 0 & 1 \\ 1 & 0 & 1 & 1 & 0 \\ 0 & 1 & 0 & 1 & 1 \\ 1 & 1 & 0 & 1 & 0 \end{bmatrix} \rightarrow \begin{bmatrix} 1 & 1 & 1 & 0 & 1 \\ 0 & 1 & 0 & 1 & 1 \\ 0 & 0 & 1 & 1 & 1 \\ 0 & 0 & 0 & 0 & 0 \end{bmatrix} \rightarrow \begin{bmatrix} 1 & 0 & 0 & 0 & 1 \\ 0 & 1 & 0 & 1 & 1 \\ 0 & 0 & 1 & 1 & 1 \\ 0 & 0 & 0 & 0 & 0 \end{bmatrix}. \blacksquare$$

A basis for a linear code is used to describe the codewords explicitly. In this sense, a basis that is in a matrix form is called a *generator matrix*, whereas a matrix that represents a basis for the dual code is called a *parity-check matrix*. A generator matrix for a linear code C is denoted by $\mathbf{G}$; its rows form a basis for C. A parity-check matrix for a linear code C, denoted by $\mathbf{H}$ and defined by $\mathbf{H} : F_q^n \longrightarrow F_q^{n-k}$, is a generator matrix for the dual code $C^\perp$. If C is an (n, k, d)-linear code,[6] then its generator matrix $\mathbf{G}$ must be a $(k \times n)$ matrix and its parity-check matrix $\mathbf{H}$ an $(n - k) \times n$ matrix. The rows of both generator and parity-check matrices are linearly independent. A $(k \times n)$ matrix $\mathbf{G}$ is a generator matrix for an (n, k) linear code C iff the rows of $\mathbf{G}$ are linearly independent and $\mathbf{G}\mathbf{H}^T = \mathbf{0}$. Similarly, an $(n - k) \times n$ matrix $\mathbf{H}$ is

---

[6] A linear code of length n, rank k (i.e., having k codewords in its basis and k rows in its generator matrix), and of minimum Hamming distance d is usually denoted as an (n, k, d) code. Note that a different notation, (n, M, d), is used to denote a *nonlinear code* of length n, size M (i.e., having M codewords), and minimum Hamming distance d.

a parity-check matrix for an (n, k) linear code C iff the rows of $\mathbf{H}$ are linearly independent and $\mathbf{HG}^T = \mathbf{0}$.

Example 3.4.5. Using Example 3.4.4, where

$$C = \langle S \rangle = \{11101, 10110, 01011, 11010\},$$

the generator and the parity-bit matrices are

$$\mathbf{G} = \begin{bmatrix} 1 & 0 & 0 & | & 0 & 1 \\ 0 & 1 & 0 & | & 1 & 1 \\ 0 & 0 & 1 & | & 1 & 1 \end{bmatrix}, \quad \mathbf{H} = \begin{bmatrix} 0 & 1 & 1 & 1 & 0 \\ 1 & 1 & 1 & 0 & 1 \end{bmatrix}.$$

It is easy to verify that $\mathbf{HG}^T = \mathbf{0} = \mathbf{GH}^T$. ∎

When the matrix $\mathbf{G}$ is represented in the block matrix form $\mathbf{G} = [I_k \mid X]$, where I_k is the $k \times k$ identity matrix and X is some $k \times (n - k)$ matrix, then $\mathbf{G}$ is said to be in *standard form*.

The subspace definition of a linear code C also guarantees that the minimum Hamming distance d between any two codewords $\mathbf{c}$ and $\mathbf{c}_0$, $\mathbf{c} \neq \mathbf{c}_0$, is constant. Since the difference $\mathbf{c} - \mathbf{c}_0$ of these codewords in C is also a codeword, i.e., it is an element of the subspace C, and since $d(\mathbf{c}, \mathbf{c}_0) = d(\mathbf{c} - \mathbf{c}_0, \mathbf{0})$, we find that

$$\min_{\mathbf{c} \in C, \mathbf{c} \neq \mathbf{c}_0} d(\mathbf{c}, \mathbf{c}_0) = \min_{\mathbf{c} \in C, \mathbf{c} \neq \mathbf{c}_0} d(\mathbf{c} - \mathbf{c}_0, \mathbf{0}) = \min_{\mathbf{c} \in C, \mathbf{c} \neq \mathbf{0}} d(\mathbf{c}, \mathbf{0}) = d.$$

The distance d of a linear code C is connected to its parity-bit matrix $\mathbf{H}$ by the following property: If d is small, then (i) C has a non-negative distance d, i.e., C does not contain any nonzero word of weight $\leq (d-1)$ iff any $(d-1)$ columns of $\mathbf{H}$ are linearly independent, and (ii) C has a non-positive distance d, i.e., C contains a nonzero word of weight $\leq d$ iff $\mathbf{H}$ has d columns that are linearly independent. In other words, if C has distance d, then any $(d-1)$ columns of $\mathbf{H}$ have d columns that are linearly independent.

Example 3.4.6. Consider a binary linear code C with the parity-check matrix

$$\mathbf{H} = \begin{bmatrix} 1 & 0 & 1 & 0 & 0 \\ 1 & 1 & 0 & 1 & 0 \\ 0 & 1 & 0 & 0 & 1 \end{bmatrix}.$$

Note that there are no zero columns and no two columns of $\mathbf{H}$ sum up to $[0]^T$; thus, any two columns of $\mathbf{H}$ are linearly independent. But columns 1, 3, and 4 sum up to $[0]^T$, and so they are linearly dependent. Hence, $d(C) = 3$. ∎

Every linear code may not have a unique generator matrix, as the following example shows.

Example 3.4.7. Consider the binary linear code $C = \{000, 001, 1000, 101\}$. Since $\dim(C) = 2$, the number of bases for C is $\frac{1}{2!}(2^2 - 1)(2^2 - 2) = 3$, i,e,, all three bases for C are: $\{001, 100\}$, $\{001, 101\}$, $\{100, 101\}$, and its all six generator matrices are:

$$\begin{bmatrix} 0 & 0 & 1 \\ 1 & 0 & 0 \end{bmatrix}, \begin{bmatrix} 1 & 0 & 0 \\ 0 & 0 & 1 \end{bmatrix}, \begin{bmatrix} 0 & 0 & 1 \\ 1 & 0 & 1 \end{bmatrix}, \begin{bmatrix} 1 & 0 & 1 \\ 0 & 0 & 1 \end{bmatrix}, \begin{bmatrix} 1 & 0 & 0 \\ 1 & 0 & 1 \end{bmatrix}, \begin{bmatrix} 1 & 0 & 1 \\ 1 & 0 & 0 \end{bmatrix},$$

none of which are in standard form.[7] ∎

A simple method of creating a parity-check matrix for a given code is to derive it from its generator matrix; or conversely, the generator matrix can be derived from its parity-check matrix. Let the generator matrix for an $[n, k]$-linear code in the standard form be $\mathbf{G} = [I_k \,|\, X]$. Then the parity-check matrix can be calculated as $\mathbf{H} = [\neg X^T \,|\, I_{n-k}]$, where the negation ($\neg$) is performed mod q in the finite field F_q^n.

Example 3.4.8. Let a linear binary code have the generator matrix

$$\mathbf{G} = \begin{bmatrix} 1 & 0 & | & 1 & 0 & 1 \\ 0 & 1 & | & 1 & 1 & 0 \end{bmatrix}.$$

Then the parity-check matrix is

$$\mathbf{H} = \begin{bmatrix} 1 & 1 & | & 1 & 0 & 0 \\ 0 & 1 & | & 0 & 1 & 0 \\ 1 & 0 & | & 0 & 0 & 1 \end{bmatrix}.$$

Note that $\mathbf{Hc} = \mathbf{0}$ for any valid codeword $\mathbf{c}$. ∎

We note in passing that a linear $(n, k, d)_q$ code C is defined by

(i) A $k \times n$ generator matrix $\mathbf{G}$, so that the code C is generated by multiplying all vectors $\mathbf{x}$ of length n with $\mathbf{G}$; this yields the class of codes $C = \{\mathbf{x} \cdot \mathbf{G} : \mathbf{x} \in F_q^n\}$; and

(ii) An $(n-k) \times n$ parity-check matrix $\mathbf{H}$ such that $C = \{\mathbf{x} \in F_q^n : \mathbf{H} \cdot \mathbf{x}^T = \mathbf{0}\}$.

For an $(n, k, d)_q$ code C, the minimum Hamming distance d is defined by

$$d = \min_{\mathbf{c} \in C, \, \mathbf{c} \neq \mathbf{0}} \mathrm{wt}(\mathbf{c}),$$

where the codeword $\mathbf{c} \in C$; also, for an $(n, k, d)_q$ code C with parity-check matrix $\mathbf{H}$, d is the minimum number of linearly independent columns in $\mathbf{H}$.

3.4.11 Decoding. In coding theory and in communication theory, *decoding* describes the process of translating a received message into codewords of a

---

[7] If a matrix M contains k leading columns, then after permuting, the columns of M form a matrix $M' = (I_k \,|\, X)$, where I_k denotes the $k \times k$ identity matrix. Then a generator matrix of the form $(I_k \,|\, X)$ is said to be in standard form.

given code. In linear block codes the information bit stream is chopped into blocks of k bits. Each block is encoded to a larger block of n bits, and the coded bits are modulated and sent over the channel. The reverse procedure is done at the receiver, using a decoding method

$$\boxed{\text{Data Block}} \longrightarrow \boxed{\text{Encoder}} \longrightarrow \boxed{\text{Codeword}} \longrightarrow \boxed{\text{Decoder}} \longrightarrow \boxed{\text{Codeword}}$$
$$\quad\; {}_{k \text{ bits}} \qquad\qquad\qquad\qquad\qquad {}_{n \text{ bits}} \qquad\qquad\qquad\qquad {}_{n \text{ bits}}$$

There are $(n - k)$ random bits, and the code rate is $R_c = k/n$.

3.4.12 Decoding Methods. A list of different decoding methods is as follows: (a) Ideal observer decoding (§3.4.12.1); (b) Maximum likelihood decoding (§3.4.12.2); (c) Minimum distance decoding (§3.4.12.3); (d) Nearest neighbor decoding (§3.4.12.4); (e) Standard array decoding (§3.4.12.5); and (f) Syndrome decoding (§3.4.12.6). Details are given below.

3.4.12.1 Ideal Observer Decoding. Let $C \in F_2^n$ denote a binary code of length n, and let $d(\mathbf{c}_1, \mathbf{c}_2)$ denote the Hamming distance between two codewords $\mathbf{c}_1, \mathbf{c}_2 \in C$. Let $\mathbf{c}$ and $\mathbf{w}$ be the message (codeword) sent and received, respectively. Then the *ideal observer decoding* maximizes the probability $\mathfrak{P}(\mathbf{c} \text{ sent} \mid \mathbf{w} \text{ received})$. In other words, it chooses the codeword $\mathbf{c}$ that is most likely to be received as the codeword $\mathbf{w}$ after the transmission. Since the probability for each codeword may not be unique, there may be more than one codeword with an equal likelihood of selection as the codeword $\mathbf{w}$. In this situation both sender and receiver must agree on a decoding convention prior to sending the message. Two of such conventions are: (i) request that the codeword $\mathbf{c}$ be resent by implementing an automatic repeat request; and (ii) choose any random codeword from the set of most likely codewords. Thus, this decoding method remains non-unique.

3.4.12.2 Maximum Likelihood Decoding. Using the notation from the ideal observer decoding, the *maximum likelihood decoding* maximizes the probability

$$\mathfrak{P}(\mathbf{w} \text{ received} \mid \mathbf{c} \text{ sent}),$$

that is, it chooses the codeword $\mathbf{c}$ that is most likely to have been sent for the codeword $\mathbf{w}$ to be received. If all codewords are equally likely to be sent during transmission, then this method is equivalent to the ideal observer decoding (see §3.4.12.1), since

$$\mathfrak{P}(w \text{ received} \mid c \text{ sent}) = \frac{\mathfrak{P}(w \text{ received} \mid c \text{ sent})}{\mathfrak{p}(c \text{ sent})}$$

$$= \mathfrak{P}(c \text{ sent} \mid w \text{ received}) \frac{\mathfrak{p}(w \text{ received})}{\mathfrak{p}(c \text{ sent})} = \mathfrak{P}(c \text{ sent} \mid w \text{ received}).$$

Hence, this decoding method is similar to the ideal observer decoding and therefore non-unique, and requires the same kind of preconvention.

3.4.12.3 Minimum Distance Decoding. Let $\mathbf{c}$ and $\mathbf{w}$ be the codewords sent and received, respectively, in a code C. Then the *minimum distance decoding* picks a codeword $\mathbf{c} \in C$ to minimize the Hamming distance $d(\mathbf{c}, \mathbf{w}) = \#\{i : c_i \neq w_i\}$. If the probability $\mathfrak{p}$ of error on a discrete memoryless channel is less than 0.5, then the minimum distance decoding is equivalent to the maximum likelihood decoding (see §3.4.12.2), i.e., if $d(\mathbf{c}, \mathbf{w}) = d$, then

$$\mathfrak{P} \left(\mathbf{w} \text{ received} \,|\, \mathbf{c} \text{ sent} \right) = (1 - \mathfrak{p})^{n-d} \mathfrak{p}^d = (1 - \mathfrak{p})^n \left(\frac{\mathfrak{p}}{1 - \mathfrak{p}} \right)^d,$$

which is maximized by minimizing d since $\mathfrak{p} < 0.5$. As such, the minimum distance decoding is similar to the nearest neighbor decoding (see §3.4.12.4), and it can be automated using a standard array. This method works reasonably well if the probability $\mathfrak{p}$ that an error occurs is independent of the position of the code bit, and if the errors are independent events, i.e., an error in a bit in a message does not affect other bit positions. As such, this method may be suited for transmission over a binary symmetric channel, but not for media like a DVD where a single scratch on the disc causes errors in a large number of adjacent codewords. In this respect it provides a non-unique method.

3.4.12.4 Nearest Neighbor Decoding. The parameter d is closely related to the error correcting feature of the Hamming code. The following algorithm, known as the *nearest neighbor algorithm*, illustrates this relationship:

Input: A received vector $\mathbf{w} \in F_q^n$

Output: A codeword $\mathbf{c} \in C$ closest to $\mathbf{w}$.

- Enumerate the elements of the ball of (Hamming) radius t about $\mathbf{w}$, denoted by $B_t(\mathbf{w})$, where $\mathbf{w}$ is the received word. Set $\mathbf{c} = $ "Fail".

- For each $\mathbf{v} \in B_t(\mathbf{w})$, check if $\mathbf{v}$ in C. If so, put $\mathbf{c} = \mathbf{v}$ and break to the next step; otherwise discard $\mathbf{w}$ and move to the next

- Return C.

"Fail" is not returned unless $t > (d - 1)/2$. A linear code is said to be t *error-correcting* if there is at most one codeword in $B_t(\mathbf{w})$ for each $\mathbf{w} \in F_q^n$.

3.4.12.5 Standard Array Decoding. A *standard array* (or *Slepian array*) is a $q^{n-k} \times q^k$ array that lists all elements of a particular F_q^n field. Such arrays are used in decoding of linear codes, as they find the codeword for a received message. A standard array for a linear binary (n, k) code is created and organized as follows:

1. The first row lists all codewords beginning with $\mathbf{0}$ (zero vector) codeword

on the extreme left;

2. Each row is a coset with the coset leader in the first column; and

3. The entry in the i-th row and j-th column is the sum (mod 2) of the ith coset leader and the j-th codeword. Recall that the addition of two binary numbers mod 2 is equivalent to the bitwise XOR operation between them.

Example 3.4.9. Consider a $(5, 2)$ code $C = \{00000, 01010, 01100, 10101\} \in F_2^5$ (here $n = 5, q = 2, k = 2$). Then the standard array for C is

00000	01010	01100	10101
10000	11010	01101	10001
01000	01110	01110	11101
00100	01000	01000	10100
00010	01001	00100	00101
00001	10010	01001	01101
11000	11011	00110	00001
10100	11110	11101	11111

The first row is the codeword C. Starting with the second row, all columns contain the vectors of minimum weight, the first five being of weight 1 and the remaining two of weight 2. Each entry in these leftmost columns is the coset leader for the corresponding row. The remaining entries of each row are determined by the bitwise XOR between its coset leader and the corresponding codeword (in the first row), i.e., the entry in the i-th row and j-th column is the bitwise XOR operation between the i-th coset leader and the j-th codeword, such that each possible vector appears only once. This procedure stops when the q^{n-k}-th row is reached. ∎

Sometimes it is not possible to construct the coset leader of an array by the above method simply because that coset leader happens to be a vector in a previous array. This situation is resolved by choosing a different coset leader, as the following example shows.

Example 3.4.10. Consider the code $C = \{0000, 1011, 0101, 1110\} \in F_2^4$. Then listing this code in the first row and following the above procedure, we obtain the arrays as given in the table below.

0000	1011	0101	1110
1000	0011	1101	0110
0100	1111	0001	1010
0010	1001	0111	1100

Notice that in this example we cannot take 0001 as a coset leader, because it already occurs as a vector in the third row and third column, although this entry, being of weight 1, is perfectly qualified to be a coset leader. ∎

The standard array can be used as a decoding method since it is a form of the nearest neighbor decoding, but it is not very practical because decoding by the standard array method requires large storage; for example, a code with 32 codewords requires a standard array with 2^{32} entries.

Example 3.4.11. Consider the code C of the above example. Suppose the received code is the vector 0110 as a message. Then using row 2 in the above example, we find that $0110 \oplus 1000 = 1110$, i.e., we have received the codeword 1110.

Notice that the standard array decoding does not guarantee that all vectors are decoded correctly. Suppose that the received codeword is 1010. Then using its coset leader 0100 (third row), the message would be decoded as $1010 \oplus 0100 = 1110$, which also has the distance 1 from the codeword 1011. This situation would require that the message be sent again. This ambiguity, along with a very large storage requirement, suggests that the standard array decoding should be totally discouraged. In fact, the best method is the syndrome decoding (§3.4.12.6) which also uses the XOR operation, known as the XOR gate. ∎

3.4.12.6 Syndrome Decoding. For efficient decoding of a linear code of length n over a noisy channel, the notion of cosets becomes important if the number n is large. Let C be a linear code of length n over a field F_q, and let $\mathbf{u} \in F_q^n$ be any vector of length n. The *coset* of C determined by $\mathbf{u}$ is defined to be the set $C + \mathbf{u} = \{\mathbf{c} + \mathbf{u} : \mathbf{c} \in C\} = \mathbf{u} + C$. For example, let $q = 2$ and $C = \{000, 101, 010, 111\}$. Then

$$C + 000 = \{000, 101, 010, 111\}, \quad C + 100 = \{100, 001, 110, 011\},$$
$$C + 001 = \{001, 100, 011, 110\}, \quad C + 101 = \{101, 000, 111, 010\},$$
$$C + 010 = \{010, 111, 000, 101\}, \quad C + 110 = \{110, 011, 100, 001\},$$
$$C + 011 = \{011, 110, 001, 100\}, \quad C + 111 = \{111, 010, 101, 000\}.$$

Note that $C + 000 = C + 010 = C + 101 = C + 111 = C$; $C + 001 = C + 011 = C + 100 = C + 110 = F_2^3 \backslash C$.

Let C be a linear (n, k) code over F_q, and let $\mathbf{H}$ be its parity-check matrix. Then for any $\mathbf{w} \in F_q^n$, the *syndrome* of $\mathbf{w}$ is the word $S(\mathbf{w}) = \mathbf{w}\,\mathbf{H}^T \in F_q^n$. For $\mathbf{u}, \mathbf{v} \in F_q^n$, the following results hold: (i) $S(\mathbf{u} + \mathbf{v}) = S(\mathbf{u}) + S(\mathbf{v})$; (ii) $S(\mathbf{u}) = \mathbf{0}$ iff $\mathbf{u}$ is a codeword in C; and (iii) $S(\mathbf{u}) = S(\mathbf{v})$ iff $\mathbf{u}$ and $\mathbf{v}$ are in the same coset of C. A word of the least Hamming weight in a coset is called a *coset leader*.

The simplest kind of syndrome decoding is used in Hamming codes (see §3.4.12.6). Let $\mathbf{c} \in F_2^n$ be a linear codeword of length n and of minimum distance d with the parity-check matrix $\mathbf{H}$. Then $\mathbf{c}$ can be corrected up to $t = \lfloor (d-1)/2 \rfloor$ errors created by the channel. Let $\mathbf{c} \in F_2^n$ be a codeword sent over a noisy channel, and the error pattern $\mathbf{e} \in F_2^n$ occurs. Then $\mathbf{w} = \mathbf{c} + \mathbf{e}$ is the codeword received. If the minimum distance decoding (see §3.4.12.3) is used, it would look up the vector $\mathbf{w}$ in a table of size $|C|$ for the nearest match, i.e., an element $\mathbf{c} \in C$, not necessarily unique, such that $d(\mathbf{c}, \mathbf{w}) \le d(\mathbf{x}, \mathbf{w})$ for all $\mathbf{x} \in C$. Syndrome decoding uses the parity matrix to see if $\mathbf{Hc} = \mathbf{0}$ for all $\mathbf{c} \in C$ by defining the syndrome of the received codeword $\mathbf{w} = \mathbf{c} + \mathbf{e}$ as $\mathbf{Hw} = \mathbf{H}(\mathbf{c} + \mathbf{e}) = \mathbf{Hc} + \mathbf{He} = \mathbf{0} + \mathbf{He} = \mathbf{He}$, under the assumption that no more than t errors occurred during the transmission. The receiver (decoder) looks up the value $\mathbf{He}$ in a table, called the *syndrome lookup table* (see below) of the size $\sum_{i=0}^{t} \binom{n}{i} < |C|$ (for a binary code) against the pre-computed value of $\mathbf{He}$ for all possible error patterns $\mathbf{e} \in F_2^n$. It is very easy to decode $\mathbf{c}$ as $\mathbf{c} = \mathbf{w} - \mathbf{e}$, which always give a unique, but not necessarily accurate, value since $\mathbf{Hc} = \mathbf{Hx}$ iff $\mathbf{c} = \mathbf{x}$, and since the parity-check matrix $\mathbf{H}$ is a generator matrix for the dual code $C^\perp$ and, hence, of full rank.

3.4.13 Syndrome Lookup Table. A *syndrome lookup table* (SLT), or a *standard decoding array* (SDA), is a table that matches each coset leader with its syndrome. The following steps are needed to construct a syndrome lookup table in a complete nearest neighbor decoding:[8]

(i) List all the cosets for the code, choosing from each coset a word of least Hamming weight as coset leader $\mathbf{u}$; and

(ii) Find a parity-check matrix $\mathbf{H}$ for the code and, for each coset leader $\mathbf{u}$, calculate its syndrome $S(\mathbf{u}) = \mathbf{u}\,\mathbf{H}^T$.

Coset Leader $\mathbf{u}$	Syndrome $S(\mathbf{u})$
0000	00
0001	01
0010	10
1000	11

---

[8] A complete nearest neighbor decoding is defined as an error pattern or error string $\mathbf{e}$ of the least weight such that $\mathbf{e} = \mathbf{w} - \mathbf{v}$, where $\mathbf{v}$ and $\mathbf{w}$ are the transmitted and received codewords, respectively, in the same coset.

Then, for the binary linear code $C = \{0000, 1011, 0101, 1110\}$, using the coset leader in the above table, and choosing the words $0000, 0001, 0010, 1000$, a parity-check matrix for C is $\mathbf{H} = \begin{bmatrix} 1 & 0 & 1 & 0 \\ 1 & 1 & 0 & 0 \end{bmatrix}$.

Example 3.4.12. Let C be the linear binary (4,2) code with parity-check matrix $\mathbf{H}$ and generator matrix $\mathbf{G}$:

$$\mathbf{H} = \begin{bmatrix} 1 & 1 & 1 & 0 \\ 0 & 1 & 0 & 1 \end{bmatrix}, \quad \mathbf{G} = \begin{bmatrix} 1 & 0 & 1 & 0 \\ 0 & 1 & 1 & 1 \end{bmatrix}.$$

The corresponding cosets are

$$\text{message row:} \quad 00 \quad 10 \quad 01 \quad 11$$

$$\text{codewords:} \quad 0000 \quad 1010 \quad 0111 \quad 1101 \qquad \underbrace{\begin{bmatrix} 0 & 0 \end{bmatrix}}_{\text{syndrome}}$$

other cosets:

$$\left\{ \begin{array}{cccc} 1000 & 0010 & 0111 & 0101 \\ 0100 & 1110 & 0011 & 1001 \\ 0001 & 1011 & 0111 & 1100 \end{array} \quad \begin{array}{c} \begin{bmatrix} 1 & 0 \end{bmatrix} \\ \begin{bmatrix} 1 & 1 \end{bmatrix} \\ \underbrace{\begin{bmatrix} 0 & 1 \end{bmatrix}}_{\text{syndrome}} \end{array} \right. ,$$

where the first column denotes the coset leaders. Now suppose that $\mathbf{w} = 1110$ is received. Then we look where in the array this $\mathbf{w}$ occurs. But for large arrays this procedure will take a lot of time. Hence, we first find $S(\mathbf{w}) = \mathbf{H}\mathbf{w}^T = \begin{Bmatrix} 1 \\ 1 \end{Bmatrix}$, and then decide that the error is equal to the coset leader 0100 (second entry in the first column of coset leaders) that has the syndrome $[1\ 1]$. Thus, the original codeword was most likely the word 1010 and the original message was 10 (second symbol in the message row). ∎

This example also gives the following result: In a linear binary (n, k) code with parity-check matrix $\mathbf{H}$, the syndrome is the sum of those columns of $\mathbf{H}$ that correspond to the locations where errors have occurred.

3.5 Vector Operations

We will discuss vector summation and multiplications of two vectors in a linear space. Consider two vectors $\mathbf{x} = (x_1, x_2, \ldots, x_n)$ and $\mathbf{y} = (y_1, y_2, \ldots, y_n)$. Then their *sum*, defined by the operation XOR and denoted by $\oplus$ (see §2.2.2), is given by

$$\mathbf{x} \oplus \mathbf{y} = (x_1 \oplus y_1, x_2 \oplus y_2, \ldots, x_n \oplus y_n).$$

For example, in F_2 we have $1 \oplus 1 = 0, 1 \oplus 0 = 1, 0 \oplus 1 = 1, 0 \oplus 0 = 0$. If $a \in F_2$ is a scalar, then $a \oplus \mathbf{x} = (a \oplus x_1, a \oplus x_2, \ldots, a \oplus x_n)$. The *complement* of $\mathbf{x}$ is the vector $1 \oplus \mathbf{x}$. For example, $1 \oplus (000111) = (111000)$.

The *multiplication* of $\mathbf{x}$ and $\mathbf{y}$, denoted by $\mathbf{x} * \mathbf{y}$, is defined as

$$\mathbf{x} * \mathbf{y} = (x_1 * y_1, x_2 * y_2, \cdots, x_n * y_n).$$

Thus, for example, $0 * 0 = 0, 0 * 1 = 0, 1 * 0 = 0, 1 * 1 = 1$. Thus, for example, $(10011110) * (11100001) = (10000000)$. If $a \in F_2$ is a scalar, then $a * \mathbf{x} = (a * x_1, a * x_2, \dots, a * x_n)$. For example, $0 * (111001) = (000000)$.

The scalar product (or dot product) of these two vectors is defined as

$$\mathbf{x} \cdot \mathbf{y} = x_1 * y_1 + x_2 * y_2 + \cdots + x_n * y_n,$$

where the multiplication sign $*$ is often neglected. The result is a scalar. Geometrically, it is defined as $\mathbf{x} \cdot \mathbf{y} = |\mathbf{x}||\mathbf{y}| \cos \theta$, where θ is the angle between the vectors $\mathbf{x}$ and $\mathbf{y}$. This product has the following properties: (i) $\mathbf{x} \cdot \mathbf{y} = \mathbf{y} \cdot \mathbf{x}$; (ii) $(a\mathbf{x}) \cdot (b\mathbf{y}) = (ab)(\mathbf{x} \cdot \mathbf{y})$ for scalar a and b; (iii) $\mathbf{x} \cdot (\mathbf{y} + \mathbf{z}) = \mathbf{x} \cdot \mathbf{y} + \mathbf{x} \cdot \mathbf{z}$, where $\mathbf{z}$ is a vector; (iv) $\mathbf{x} \cdot \mathbf{x} = |\mathbf{x}|^2$; and (v) $\mathbf{x} \cdot \mathbf{y} \Leftrightarrow \mathbf{x} \perp \mathbf{y}$. In the sense of matrix algebra, if $\mathbf{x}$ is a column vector in $\mathbb{R}^n$, then the scalar product $\mathbf{x} \cdot \mathbf{y} = x_1 y_1 + x_2 y_2 + \cdots + x_n y_n$, where T denotes the transpose of a matrix, i.e., $\mathbf{x}^T$ is the row vector. The norm (length) $|\mathbf{x}| = \sqrt{\mathbf{x}^T \mathbf{x}} = \sqrt{x_1^2 + \cdots + x_n^2}$, and the distance by $|\mathbf{x} - \mathbf{y}| = \sqrt{(x_1 - y_1)^2 + \cdots + (x_n - y_n)^2}$. Other useful vector products are as follows.

3.5.1 Wedge Product. (also known as exterior product) Let $\mathbf{v}$ and $\mathbf{w}$ be any vectors in $\mathbb{R}^n$, such that (i) $\mathbf{v} \wedge \mathbf{v} = 0$, and (ii) $\mathbf{v} \wedge \mathbf{w} = -\mathbf{w} \wedge \mathbf{v}$, where the algebraic rules which apply to ordinary multiplication, like the associative law, also apply to $\wedge$, with the exception of (i) and (ii).[9] The property (ii) is known as the anti-commutative law. We will first define the concept of area in $\mathbb{R}^2$ and $\mathbb{R}^3$. For example in the Cartesian plane $\mathbb{R}^2$, which is a vector space, let the unit vectors along the coordinate axes be $\mathbf{e}_1 = (1, 0) = \mathbf{i}, \mathbf{e}_2 = (0, 1) = \mathbf{j}$. Suppose that $\mathbf{v} = v_1 \mathbf{e}_1 + v_2 \mathbf{e}_2$, $\mathbf{w} = w_1 \mathbf{e}_1 + w_2 \mathbf{e}_2$ are two vectors in $\mathbb{R}^2$. There is a unique parallelogram with $\mathbf{v}$ and $\mathbf{w}$ as two sides, and the area A of this parallelogram is given by

$$A(\mathbf{v}, \mathbf{w}) = \det\{\mathbf{v}, \mathbf{w}\} = |v_1 w_2 - v_2 w_1|.$$

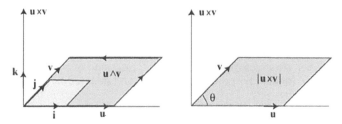

Figure 3.5.1. Wedge product and cross product.

---

[9] The notation for wedge product seems to be similar to the logical operation, but the context will keep any confusion away from this notation.

Now the wedge (exterior) product of $\mathbf{v}$ and $\mathbf{w}$ is

$$\mathbf{v} \wedge \mathbf{w} = (v_1\mathbf{e}_1 + v_2\mathbf{e}_2) \wedge (w_1\mathbf{e}_1 + w_2\mathbf{e}_2)$$
$$= v_1w_1\mathbf{e}_1 \wedge \mathbf{e}_1 + v_1w_2\mathbf{e}_1 \wedge \mathbf{e}_2 + v_2w_1\mathbf{e}_2 \wedge \mathbf{e}_1 + v_2w_2\mathbf{e}_2 \wedge \mathbf{e}_2$$
$$= (v_1w_2 - v_2w_1)\mathbf{e}_1 \wedge \mathbf{e}_2 \quad \text{(by property (ii))},$$

where the last expression is the determinant of the matrix $[\mathbf{v}\mathbf{w}]$, which may be positive or negative depending on the clockwise or counterclockwise orientation of the two vectors. Such an area A is called the *signed area* of the parallelogram, and the absolute value of the signed area is the value of the ordinary area. The signed area $A(\mathbf{v}, \mathbf{w})$ has the following properties: (i) $A(a\mathbf{v}, b\mathbf{w}) = abA(\mathbf{v}, \mathbf{w})$ for any scalars a and b; (ii) $A(\mathbf{v}, \mathbf{v}) = 0$ since the area degenerates to a single line; (iii) $A(\mathbf{v}, \mathbf{w}) = -A(\mathbf{w}, \mathbf{v})$ since the orientation of the parallelogram is reversed; (iv) $A\mathbf{v} + a\mathbf{w}, \mathbf{w}) = A(\mathbf{v}, \mathbf{w})$, since adding a multiple of $\mathbf{w}$ to $\mathbf{v}$ does not affect the base or the height of the parallelogram; and (v) $A(\mathbf{e}_1, \mathbf{e})_2) = 1$, since the area of the unit square is 1.

For vectors in $\mathbb{R}^3$, the wedge product is related to the cross product and triple product. Let the basis be $\mathbf{e}_1 = (1, 0, 0) = \mathbf{i}, \mathbf{e}_2 = (0, 1, 0) = \mathbf{j}, \mathbf{e}_3 = (0, 0, 1) = \mathbf{k}$. Then the wedge product of two vectors $\mathbf{u} = u_1\mathbf{e}_1 + u_2\mathbf{e}_2 + u_3\mathbf{e}_3$ and $\mathbf{v} = v_1\mathbf{e}_1 + v_2\mathbf{e}_2 + v_3\mathbf{e}_3$ is

$$\mathbf{u} \wedge \mathbf{v} = (u_1v_2 - u_2v_1)(\mathbf{e}_1 \wedge \mathbf{e}_2) + (u_3v_1 - u_1v_3)(\mathbf{e}_3 - \wedge\mathbf{e}_1) + (u_2v_3 - u_3v_2)(\mathbf{e}_2 \wedge \mathbf{e}_3).$$

With a third vector $\mathbf{w} = w_1\mathbf{e}_1 + w_2\mathbf{e}_2 + w_3\mathbf{e}_3$, the wedge product of three vectors is

$$\mathbf{u} \wedge \mathbf{v} \wedge \mathbf{w} = (u_1v_2w_3 + u_2v_3w_1 + u_3v_3w_2 - u_1v_3w_2 - u_2v_1w_3 - u_3v_2w_1)(\mathbf{e}_1 \wedge \mathbf{e}_2 \wedge \mathbf{e}_3),$$

or

$$\mathbf{u} \wedge \mathbf{v} \wedge \mathbf{w} = \begin{vmatrix} u_1 & u_2 & u_3 \\ v_1 & v_2 & v_3 \\ w_1 & w_2 & w_3 \end{vmatrix} (\mathbf{e}_1 \wedge \mathbf{e}_2 \wedge \mathbf{e}_3).$$

Note that the cross product $\mathbf{u} \times \mathbf{v}$ can be interpreted as a vector that is perpendicular to both $\mathbf{u}$ and $\mathbf{v}$ and whose magnitude is equal to the area of the parallelogram defined by the vectors. The triple product of $\mathbf{u}$, $\mathbf{v}$, and $\mathbf{w}$ is geometrically a signed volume which is equal to the determinant of the matrix with columns $\mathbf{u}$, $\mathbf{v}$, and $\mathbf{w}$.

Note that since there is no analogue of the binary cross product, the wedge product provides similar properties except that the wedge product of two vectors is a 2-vector instead of an ordinary vector.

3.5.2 Bar Product. The bar product of two linear codes C_1 and C_2 is defined as

$$C_1|C_2 = \{(\mathbf{c}_1|\mathbf{c}_1 + \mathbf{c}_2) : \mathbf{c}_1 \in C_1, \mathbf{c}_2 \in C_2\},$$

where $(a|b)$ denotes the concatenation of a and b. If the codewords in C_1 are of length n then the codewords in $C_1|C_2$ are of length $2n$. The bar product is useful in representing the Reed-Muller code $\mathrm{RM}(d, r)$ in terms of the Reed-Muller codes $\mathrm{RM}(d-1, r)$ and $\mathrm{RD}(d-1, r-1)$. The following two results relate to the properties of the bar product.

Theorem 3.5.1. The Hamming weight $\mathrm{wt}(C_1|C_2)$ of the bar product of linear codes C_1 and C_2 is equal to $\min\{2\,\mathrm{wt}(C_1), \mathrm{wt}(C_2)\}$.

PROOF. For all $\mathbf{c}_1 \in C_1$ we have $(\mathbf{c}_1|\mathbf{c}_1 + 0) \in C_1|C_2$, which has weight $2\,\mathrm{wt}(\mathbf{c}_1)$. Similarly, $(0|C_2) \in C_1|C_2$ for all $\mathbf{c}_2 \in C_2$ and has weight $\mathrm{wt}(C_2)$. Then minimizing over $\mathbf{c}_1 \in C_1, \mathbf{c}_2 \in C_2$ we get

$$\mathrm{wt}(C_1|C_2) \leq \min\{2\,\mathrm{wt}(C_1), \mathrm{wt}(C_2)\}.$$

Now, let $\mathbf{c}_1 \in C_1$ and $\mathbf{c}_2 \in C_2$ such that not both are zero. If $\mathbf{c}_2 \neq 0$, then $\mathrm{wt}(\mathbf{c}_1|\mathbf{c}_1 + \mathbf{c}_2) = \mathrm{wt}(\mathbf{c}_1) + \mathrm{wt}(\mathbf{c}_1 + \mathbf{c}_2) \geq \mathrm{wt}(\mathbf{c}_1 + \mathbf{c}_1 + \mathbf{c}_2) = \mathrm{wt}(\mathbf{c}_2) \geq \mathrm{wt}(C_2)$. But if $\mathbf{c}_2 = 0$, then $\mathrm{wt}(\mathbf{c}_1|\mathbf{c}_1 + \mathbf{c}_2) = 2\,\mathrm{wt}(\mathbf{c}_1) \geq 2\,\mathrm{wt}(C_1)$, and hence, $\mathrm{wt}(C_1|C_2) \geq \min\{2\,\mathrm{wt}(C_1), \mathrm{wt}(C_2)\}$ ∎

3.6 Sphere Packing

The basic notion of sphere packing of a vector space is based on the geometrical configuration of closest packing of spheres, which has been used in the energetic-synergetic geometry developed by R. Buckminster Fuller (1895–1983).[10] The three-dimensional geometrical model of energy configurations is presented by a symmetrical cluster of spheres. In such a structure, each sphere is conceived as an idealized model of an energy field, all of whose forces are in equilibrium, and whose vectors consequently are identical in length and in angular relationships. The term *closest-packing* was first used by the physicist and crystallographer Sir William Bragg. The geometrical presentation is based on the fact that if a unit sphere is completely surrounded by other spheres of equal radius and packed as closely together as possible, exactly 12 spheres make up the first surrounding layer. If a second layer, or shell, be clustered around the first by similar spheres, then 42 spheres are required to complete the second shell. The third layer will require 92 similar spheres, and so on for adding additional layers, ad infinitum. In this structure each layer is a complete and symmetrical enclosure of tangential spheres. The total number of spheres in any layer can be found by multiplying the second power (square) of the number of the layer by 10, and adding 2. Thus, the

---

[10] American inventor and architect who developed the so-called 'energetic-synergetic geometry' that led him to design the geodesic dome, a complete structure of a large dome.

number of spheres required in the nth layer is equal to $(10 \times n^2) + 2$, where $n \geq 1$. The first layer containing 12 spheres around the nucleus is presented in Figures 3.6.1(a), and the second layer of 42 spheres around the innermost shell in Figure 3.6.1(b).

In the two-dimensional case we require a total of $6n$ circular disks of the same radius to fill in the n-th layer, $n \geq 1$, where each layer is in the form of a hexagon. In fact, the hexagon in the n-th layer will have three alternate sides each filled by $n + 1$ disks while the remaining three sides each by $n - 1$ disk, making a total of $3(n + 1) + 3(n - 1) = 6n$ disks. Thus, the first, second, and subsequent layers surrounding the nucleus disk will require 6, 12, 18, 24, 30, 36, 42, ... disks to fill the two-dimensional space (Figure 3.6.2, which shows three consecutive layers of circular disks around the nucleus disk). The sphere-packing notion is used in determining bounds on linear codes, which are discussed in Chapter 6.

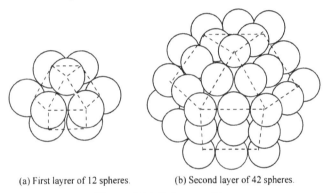

(a) First layrer of 12 spheres.   (b) Second layer of 42 spheres.

Figure 3.6.1 Sphere packing.

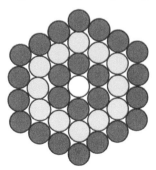

Figure 3.6.2. Disk packing.

A visual aspect of disk packing is obtained by packing pennies, say, in a rectangular boundary, is the number of neighbors a single penny (or a code-word) may have. In our example each penny packed in a rectangle will have

4 adjacent neighbors and 4 at the corners farther away. In the case of a hexagon, each penny will have 6 adjacent neighbors. However, as the dimension increases, the number of adjacent neighbors increases very rapidly. This observation translates into the number of ways the transmission noise can affect the receiver to choose a neighbor that may increase the error in a received code, and if the number of neighbors increases, the probability of total errors may also incease in block codes.

Sphere packing and block codes are related to each other in the sense that noise on a transmission line may affect a receiver to choose a neighboring symbol and thus cause an error. Sphere packing may be hard to visualize since block codes rely on three or more dimensions,[11] but its two-dimensional analogue may convey an idea of the full representation. We can lay a bunch of circular disks, say pennies, flat on a table and push them close together. The result is a hexagonal beehive pattern. In terms of higher dimensions, the question arises as to how many pennies can be packed into a circle on a tabletop or, in three dimensions, how many marbles (spheres) can be packed into a globe. Sometimes, instead of circle packing, hexagon packing is chosen for a code. For example, hexagon packing into a rectangular box will leave empty space around the corners. Thus, as the dimensions get larger, the percentage of empty space grows smaller; but if at certain dimensions the packing uses all the space for a code, that code is called a perfect code. As we will see in §7.1, there are very few perfect codes.

---

[11] For example, the Golay code $\mathcal{G}_{24}$ uses 24 dimensions. If used as a binary code, which it actually is, then the dimensions refer to the length of the codeword. But the coding theory often uses the N-dimensional sphere model.

4

Hamming Codes

4.1 Error Correcting Codes

Error detection and error correction are integral parts of many high-reliability and high-performance computer and transmission/storage devices. In data storage systems, memory caches are used to improve system reliability. The cache is generally placed inside the controller between the host interface and the disk array. Any reliable cache memory design must include error correction code (ECC) functions to safeguard the loss of data. Similarly, ECC is an important design aspect of many communication applications, such as satellite receivers. The significance of ECC lies in performance and cost efficiency by correcting any error and avoiding repeated retransmission of data.

When a message or data is transmitted through a channel, the data received depends on the properties of the resulting errors, which may be caused by the characteristics of the channel and the system. There are three major categories of errors that are encountered:

1. RANDOM ERRORS. These are bit errors that are independent of one another; they are generally caused by the noise in the channel. They are simply isolated erroneous bits in a message or data, caused by thermal (voltage) noise in communication channels.

2. BURST ERRORS. These are bit errors that occur sequentially in time and groups. Sometimes defects in the digital storage media cause these kinds of errors. They are difficult to correct by some codes, although block codes can handle this kind of errors effectively.

3. IMPULSE ERRORS. These are large blocks of data that are full of errors; they are typically caused by lightning strikes or major system failures. Impulse errors generally cause catastrophic failures in a communication system; they are so severe that they are not even recognized or detected by forward error correction.

In general, all simple error correction codes are not sufficiently efficient to detect and correct burst and impulse errors, and they fail to reconstruct the message in the case of catastrophic errors. In the current state of advancements in this field, the Reed-Solomon codes (see Chapter 13) were designed specifically to correct random and burst errors, and detect the presence of catastrophic errors by examining the message. If the number of errors per data is small, these errors can be totally corrected using the Reed-Solomon codes. These codes are therefore very useful in system design since they flag the unrecoverable message at the decoder. We will first discuss simple error detection and error correction codes, and slowly build up the analysis and description to finally reach the state-of-the-art aspects of modern coding theory.

4.1.1 Binary Linear Hamming Codes. These codes were discovered by R. W. Hamming and M. J. G. Golay. In particular, the Hamming code refers to the $(7,4)$ code introduced by Hamming in 1950[1] to provide a single error-correction and double error-detection (SEC-DED) code for errors introduced on a noisy communication channel. It was used to reduce the computer resonance and time that was wasted when the message was corrupted without the receiver realizing it and lending to the failure of communication. In general, all binary Hamming codes of a given length are equivalent. The dimension of a binary linear (n, k) code of length $n = 2^k - 1$ is $n - k$, where k is the number of data bits in the code, and its distance is $d = 3$, thus making it an exactly single-error correcting code.

4.2 Hamming (7,4) Code

This code has 4 data bits and 3 parity bits, hence the name. The parity bits are denoted by 2^r, $r = 0, 1, 2$, i.e., the bit numbers $1, 2, 4$. Thus, using the exponent form, the three parity bits, denoted by p_1, p_2, p_4, are added to every four data bits of message, denoted by d_1, d_2, d_3, d_4, forming a codeword $c = \{p_1, p_2, d_1, p_4, d_2, d_3, d_4\}$ that is used to detect all single-bit and two-bit errors and correct only a single-bit error. The algorithms for encoding and decoding are explained below. Although limited in its application, this code has been very effective in situations where excessive errors do not occur randomly in a transmission medium, that is, the Hamming distance between the transmitted and received words is at most 1, which can be corrected by this code.

---

[1] This code originated while Hamming was working as a theorist at Bell Telephone Laboratories in the 1940s on the Bell Model V computer, which was an electromechanical relay-based machine with cycle times in seconds. The input was made on punch cards, which were a constant source of read errors. The errors were manually corrected every weekday using special codes that flashed lights to warn the operators, but during weekends there were no operators to correct any errors, and the machine moved on to the next job, thus producing useless results that had to be started from scratch repeatedly. Frustrated by this situation, Hamming worked on the ECC problem for the next few years and published a series of algorithms in 1950, what is now known as the Hamming code.

4.2.1 Encoding and Decoding. The encoding part of the algorithm is described in Table 4.2.1 (where y=yes and n=no).

Table 4.2.1 Encoding of Hamming (7,4) Code

Bit Location	1	2	3	4	5	6	7
Codeword $\mathbf{c}$	c_1	c_2	c_3	c_4	c_5	c_6	c_7
Encoded Bit	p_1	p_2	d_1	p_4	d_2	d_3	d_4
p_1	y	n	y	n	y	n	y
p_2	n	y	y	n	n	y	y
p_4	n	n	n	y	y	y	y

The Venn diagram for Table 4.2.1, shown in Figure 4.2.1, is a geometrical representation of Table 4.2.1. It shows that the parity bit p_1 covers the data bits $1, 3, 5, 7$; the parity bit p_2 covers the data bits $2, 3, 6, 7$; and the parity bit p_4 covers the data bits $4, 5, 6, 7$. All these bits correspond to the entry 'y' in the above table. The Venn diagram is a visual means of establishing a relation between the parity bits and codeword. For smaller values of parity bits, it works fine, but as their number increases, the diagram becomes complicated and eventually becomes unintelligible even for $m = n - k > 4$.

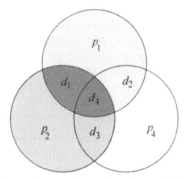

Figure 4.2.1. Venn diagram for 4 data bits and 3 parity bits.

Representing 'y' by 1 and 'n' by 0 in Table 4.2.1, the parity-check matrix $\mathbf{H}$ and the code-generator matrix $\mathbf{G}$ are defined, respectively, as

$$\mathbf{H} = \begin{bmatrix} 1 & 0 & 1 & 0 & 1 & 0 & 1 \\ 0 & 1 & 1 & 0 & 0 & 1 & 1 \\ 0 & 0 & 0 & 1 & 1 & 1 & 1 \end{bmatrix}, \quad \mathbf{G} = \begin{bmatrix} 1 & 1 & 1 & 0 & 0 & 0 & 0 \\ 1 & 0 & 0 & 1 & 1 & 0 & 0 \\ 0 & 1 & 0 & 1 & 0 & 1 & 0 \\ 1 & 1 & 0 & 1 & 0 & 0 & 1 \end{bmatrix}, \quad (4.2.1)$$

where $\mathbf{H}$ is a 3×7 matrix and $\mathbf{G}$ a 4×7 matrix. The entries in $\mathbf{H}$ represent the above table, or Figure 4.2.1, while the entries in $\mathbf{G}$ correspond to the following scheme: The first, second, and fourth rows in $\mathbf{G}$ represent the 'y' from p_1, p_2 and p_4, each under d_1, d_2, d_3, d_4, respectively, while the third, fifth, sixth, and seventh rows represent the identity matrix, since they are linearly independent, and inserted into $\mathbf{G}$ in that manner as part of the algorithm. Note that $\mathbf{HG}^T = \mathbf{0} \bmod 2$ (see also Moon [2005]).

4.2.2 Computation of Parity Bits. The information from Table 4.2.1, or equivalently from the above Venn diagram, is used to compute the parity bits p_1, p_2, p_4, given the data bits d_1, d_2, d_3, d_4, by the formula

$$
\begin{aligned}
p_1 &= \neg(d_1 \oplus d_2 \oplus d_4), \\
p_2 &= \neg(d_1 \oplus d_3 \oplus d_4), \\
p_4 &= \neg(d_2 \oplus d_3 \oplus d_4).
\end{aligned}
\tag{4.2.2}
$$

Then the codeword is $\mathbf{c} = \{p_1, p_2, d_1, p_4, d_2, d_3, d_4\}$. This is also one of the methods for encoding and decoding of Hamming (7,4) code, as shown in Example 4.2.2.

Example 4.2.1. Given the data bits $[1\,0\,1\,0]$, the three parity bits by formula (4.2.2) are $p_1 = \neg(1 \oplus 0 \oplus 0) = 0$, $p_2 = \neg(1 \oplus 1 \oplus 0) = 1$ $p_3 = \neg(0 \oplus 1 \oplus 0) = 0$. The codeword is $\mathbf{c} = [0\,1\,1\,0\,0\,1\,0]$. ∎

Example 4.2.2. Using the data from Example 4.2.1, the transmitted codeword is $[0\,1\,1\,0\,0\,1\,0]$. Suppose that the received word through a noisy channel is $\mathbf{w} = [0\ 1\ \mathbf{0}\ 0\ 0\ 1\ 0]$, where an error has occurred in the third bit (shown in boldface). The receiver uses formula (4.2.2) to check again the above parity bits $p_4\ p_2\ p_1 = 010$ as follows: $p_1 = \neg(0 \oplus 0 \oplus 0) = 1$, $p_2 = \neg(0 \oplus 1 \oplus 0) = 0$, $p_4 = \neg(0 \oplus 1 \oplus 0) = 0$. Note that two of these computed parity bits, $p_1\ p_2$, do not match with the original parity bits. Next, the bit in error is computed by the syndrome[2]: $0\,1\,0 \oplus 0\,0\,1 = (0\,1\,1)_2 = (3)_{10}$. Hence, the third bit is in error, which is corrected by flipping it or negating its value, and thus, the single-bit error is corrected. ∎

4.2.3 Syndrome Method. Syndrome was defined in §3.4.12.6. Let $\mathbf{d} = \{d_1, d_2, \ldots, d_k\}$ denote a dataword with data bits d_j, $j = 1, \ldots, k$. The codeword $\mathbf{c} = \{c_1, c_2, \ldots, c_n\}$ for a Hamming (n, k) code is obtained by solving $\mathbf{Hc}^T = \mathbf{0}$, where $\mathbf{H}$ is the parity-check matrix. Moreover, let $d_1 \cdots d_k$ be a given message over F_2. Then the coding scheme f is defined by $f : \{d_1 \cdots d_k\} \mapsto \{c_1 \cdots c_{k+1}\}$, where $c_i = d_i$ for $i = 1, \ldots, k$, and

$$
c_{k+1} = \begin{cases} 0 & \text{if } \sum_{i=1}^{k} d_i = 0, \\ 1 & \text{if } \sum_{i=1}^{k} d_i = 1. \end{cases}
\tag{4.2.3}
$$

This means that the sum of digits of any codeword $c_1 \cdots c_{k+1}$ is 0.

---

[2] Recall that the XOR of the expected error-checking bits with those actually received is called the syndrome of a received codeword.

Thus, for the Hamming (7,4) code, using $\mathbf{H}$ defined by (4.2.1), the values of $c_1, c_2, c_3.c_4 \in F_2^4$ are obtained by solving

$$c_1 + c_3 + c_5 + c_7 = 0,$$
$$c_2 + c_3 + c_6 + c_7 = 0,$$
$$c_4 + c_5 + c_6 + c_7 = 0,$$

which gives $c_5 = c_2 + c_3 + c_4, c_6 = c_1 + c_3 + c_4, c_7 = c_1 + c_2 + c_4$. The coding scheme in this case is the linear map from F_2^4 into F_2^7 given by

$$(d_1, d_2, d_3, d_4) \mapsto (d_1, d_2, d_3, d_4, d_1 + d_2 + d_3 + d_4, d_1 + d_3 + d_4, d_1 + d_2 + d_4).$$

If the sum of digits of the received word is 1, then the receiver knows that a transmission error has occurred.

If $\mathbf{c}$ is a codeword and $\mathbf{w}$ is the received word after transmission through a noisy channel, then $\mathbf{e} = \mathbf{w} - \mathbf{c} = \{e_1, \ldots e_n\}$ is called the *error word* or the *error vector*. All possible error vectors $\mathbf{e}$ of a received vector $\mathbf{w}$ are the vectors in the coset of $\mathbf{w}$. The most likely error vector $\mathbf{e}$ has the minimum weight in the coset of $\mathbf{w}$. Hence, we set $\mathbf{u} = \mathbf{w} - \mathbf{e}$ and decode $\mathbf{w}$.

Let $C \subseteq F_q^n$ be a linear (n, k) code. An element of minimum weight in a coset $\mathbf{u} + C$ is a *coset leader* of $\mathbf{u} + C$. If several vectors in $\mathbf{u} + C$ have minimum weight, we choose one of them as coset leader. Let $\mathbf{u}^{(1)}, \ldots, \mathbf{u}^{(n)}$ be the coset leaders of the cosets $\neq C$ and let $\mathbf{c}^{(1)}(= \mathbf{0}), \mathbf{c}^{(2)}, \ldots, \mathbf{c}^{(q^k)}$ be all codewords in C. Consider the following array:

$$
\begin{array}{cccc}
\mathbf{c}^{(1)} & \mathbf{c}^{(2)} & \cdots & \mathbf{c}^{(q^k)} \\
\mathbf{u}^{(1)} + \mathbf{c}^{(1)} & \mathbf{u}^{(1)} + \mathbf{c}^{(2)} & \cdots & \mathbf{u}^{(1)} + \mathbf{c}^{(q^k)} \\
\vdots & \vdots & & \vdots \\
\mathbf{u}^{(s)} + \mathbf{c}^{(1)} & \mathbf{u}^{(s)} + \mathbf{c}^{(2)} & \cdots & \mathbf{u}^{(s)} + \mathbf{c}^{(q^k)}
\end{array}
\left. \begin{array}{l} \\ \\ \\ \\ \\ \end{array} \right\}
$$

row of codewords

remaining cosets

$\underbrace{\phantom{\mathbf{u}^{(s)} + \mathbf{c}^{(1)}}}$
column of coset leaders

After receiving a word $\mathbf{w} = \mathbf{u}^{(j)} + \mathbf{e}^{(j)}$ the decoder decides that the error $\mathbf{e}$ is the corresponding coset leader $\mathbf{u}^{(j)}$ and decodes $\mathbf{w}$ as the codeword $\mathbf{v}^{(j)} = \mathbf{w}^{(j)} - \mathbf{e}^{(j)} = \mathbf{c}^{(j)}$. Thus, $\mathbf{w}$ is decoded as the codeword in the column of $\mathbf{w}$. The coset of $\mathbf{w}$ can be determined by calculating the *syndrome* of $\mathbf{w}$, which is defined as follows: Let $\mathbf{H}$ be the parity-check matrix of a linear (n, k) code C. Then the vector $S(\mathbf{w}) = \mathbf{H}\mathbf{w}^T$ of length $m = n - k$ is the *syndrome* of $\mathbf{w}$. Obviously, if $\mathbf{w}, \mathbf{z} \in F_q^n$, then (i) $S(\mathbf{w}) = \mathbf{0}$ iff $\mathbf{w} \in C$, and (ii) $S(\mathbf{w}) = S(\mathbf{z})$ iff $\mathbf{w} + C = \mathbf{z} + C$.

The decoding procedure is generally based on the coset-leader algorithm for error correction of linear codes, defined as follows: Let $C \subseteq F_q^n$ be a linear (n, k) code and let $\mathbf{w}$ be the received vector. To correct errors in $\mathbf{w}$, follow these steps: (i) Calculate $S(\mathbf{w})$; (ii) find the coset leader, say, $\mathbf{e}$, with syndrome equal to $S(\mathbf{w})$; and (iii) decode $\mathbf{w}$ as $\mathbf{v} = \mathbf{w} - \mathbf{e}$, where $\mathbf{v}$ is the codeword with minimum distance to $\mathbf{w}$.

Example 4.2.3. Consider the (7,4) Hamming code with parity-check matrix $\mathbf{H}$ defined by (4.2.1). If the syndrome of a received word $\mathbf{w}$ is, say, $S(\mathbf{w}) = [1\ 0\ 1]^T$, then an error has occurred in the fifth location since $(5)_{10} = (101)_2$. ∎

A syndrome lookup table for the Hamming (7,4) code is given in Table 4.2.2.

Table 4.2.2 Lookup Table for Hamming (7,4) Code

Coset Leader $\mathbf{u}$	Syndrome $S(\mathbf{u})$
0000000	000
1000000	001
0100000	010
0010000	011
0001000	100
0000100	101
0000010	110
0000001	111

4.2.4 Parity-Check Method. The algorithm of this code involves the creation of a set of parity bits which overlap so that a single-bit error in the data bits, where the bit in error is logically flipped in value, is detected and corrected. Using the same algorithm repeatedly, this code can also detect and correct multiple single-bit errors.

Example 4.2.4. Let the data bits to be transmitted be $\mathbf{d} = \{d_1, d_2, d_3, d_4\} = \{1, 0, 1, 0\}$, and assume that no error has occurred during the transmission. Then we compute

$$
\mathbf{c} = \mathbf{G}^T \mathbf{d} = \begin{bmatrix} 1 & 1 & 0 & 1 \\ 1 & 0 & 1 & 1 \\ 1 & 0 & 0 & 0 \\ 0 & 1 & 1 & 1 \\ 0 & 1 & 0 & 0 \\ 0 & 0 & 1 & 0 \\ 0 & 0 & 0 & 1 \end{bmatrix} \begin{Bmatrix} 1 \\ 0 \\ 1 \\ 0 \end{Bmatrix} = \begin{Bmatrix} 1 \\ 0 \\ 1 \\ 1 \\ 0 \\ 1 \\ 0 \end{Bmatrix} \pmod 2.
$$

Thus, the last column matrix is transmitted instead of the data bits $\mathbf{d}$. Now, the parity check is performed by computing $\mathbf{w} = \mathbf{c} + \mathbf{e}_i \pmod 2$, where $\mathbf{e}_i$ is the i-th unit vector, i.e., it is a zero vector with a 1 in the i-th place. For example, $\mathbf{e}_5 = [0\,0\,0\,0\,1\,0\,0]^T$, where T denotes the transpose of this column vector. After computing $\mathbf{H}\mathbf{w} = \mathbf{H}(\mathbf{c} + \mathbf{e}_i) = \mathbf{H}\mathbf{c} + \mathbf{H}\mathbf{e}_i$, we find that

$$\mathbf{H}\mathbf{w} = \mathbf{H}(\mathbf{c} + \mathbf{e}_i) = 0 + \mathbf{H}\mathbf{e}_i = \mathbf{H}\mathbf{e}_i,$$

since the data is error-free.

Now, suppose that the error occurs in bit 5 (d_2), i.e., 0 becomes 1. To detect this error and correct it, we compute

$$\mathbf{w} = \mathbf{c} + \mathbf{e}_5 = [1\,0\,1\,1\,0\,1\,0] + [0\,0\,0\,0\,1\,0\,0] = [1\,0\,1\,1\,1\,1\,0].$$

Then

$$\mathbf{H}\mathbf{w}^T = \begin{bmatrix} 1 & 0 & 1 & 0 & 1 & 0 & 1 \\ 0 & 1 & 1 & 0 & 0 & 1 & 1 \\ 0 & 0 & 0 & 1 & 1 & 1 & 1 \end{bmatrix} [1\,0\,1\,1\,1\,1\,0]^T = [1\,0\,1]^T = [p_1\,p_2\,p_4]^T,$$

where, to save space, a column vector is represented in the form of the transpose of a row vector. Note that $(p_4, p_2, p_1) = (101)_2 = (5)_{10}$, so the 5-th bit is corrupted. Also, the last vector is the 5-th column in the matrix $\mathbf{H}$. Hence, the corrected value of $\mathbf{w}$ is

$$\mathbf{w}_{\text{corrected}} = [1\,0\,1\,1\,\mathbf{0}\,1\,0], \text{ or } \mathbf{c} = [1\,0\,1\,1\,0\,1\,0].$$

After the corrupted bit is detected, the syndrome method for decoding is carried out as follows: Define a matrix $\mathbf{S}$ by

$$\mathbf{S} = \begin{bmatrix} 0 & 0 & 1 & 0 & 0 & 0 & 1 \\ 0 & 0 & 0 & 0 & 1 & 0 & 0 \\ 0 & 0 & 0 & 0 & 0 & 1 & 0 \\ 0 & 0 & 0 & 0 & 0 & 0 & 1 \end{bmatrix}.$$

Note that a 3×3 identity matrix is introduced in $\mathbf{S}$ in the last three rows on the right. Then the received value yields the correct data bits, which are given by $\mathbf{S}\mathbf{w} = [1\,0\,1\,0]^T$. ∎

4.2.5 Mapping Method. Consider 16 different messages each consisting of 4-bit strings, which are :

| 0000 | 0001 | 0010 | 0011 | 0100 | 0101 | 0110 | 0111 |

| 1000 | 1001 | 1010 | 1011 | 1100 | 1101 | 1110 | 1111 |.

Using the method of §4.2.4, these messages are mapped onto the following 16 valid codewords:

0	0000000	4	0101010	8	1001011	C	1100001
1	0000111	5	0101101	9	1001100	D	1100110
2	0011001	6	0110011	A	1010010	E	1111000
3	0011110	7	0110100	B	1010101	F	1111111.

These codewords, created using formula (4.2.2), are arranged such that the minimum distance between any two is 3. Since $k = 4, m = n - k = 3, n = 7$, the left side is $(k+m+1) \cdot (2^k) = (8)(16) = 128$, and the right side is $2^n = 128$, so the two sides match. Notice that 0000001 is not a valid dataword and, therefore, it is never transmitted! We will see in the following example as to what happens if this invalid dataword is received.

Example 4.2.5. Let us suppose that the dataword 0000000 is transmitted, but an error occurred in the last bit and it was received as $\mathbf{w} = 0000001$. The receiver then calculates the Hamming distance between $\mathbf{w}$ and all the above 16 datawords:

$$\text{dist}(\mathbf{w}, \mathbf{0}) = 1, \ \text{dist}(\mathbf{w}, \mathbf{1}) = 2, \ \text{dist}(\mathbf{w}, \mathbf{2}) = 2, \ \text{dist}(\mathbf{w}, \mathbf{3}) = 5,$$

$$\text{dist}(\mathbf{w}, \mathbf{4}) = 4, \ \text{dist}(\mathbf{w}, \mathbf{5}) = 3, \ \text{dist}(\mathbf{w}, \mathbf{6}) = 3, \ \text{dist}(\mathbf{w}, \mathbf{7}) = 4,$$

$$\text{dist}(\mathbf{w}, \mathbf{8}) = 3, \ \text{dist}(\mathbf{w}, \mathbf{9}) = 4, \ \text{dist}(\mathbf{w}, \mathbf{A}) = 4, \ \text{dist}(\mathbf{w}, \mathbf{B}) = 3,$$

$$\text{dist}(\mathbf{w}, \mathbf{C}) = 2, \ \text{dist}(\mathbf{w}, \mathbf{D}) = 5, \ \text{dist}(\mathbf{w}, \mathbf{E}) = 4, \ \text{dist}(\mathbf{w}, \mathbf{F}) = 6.$$

Looking at the minimum Hamming distance, which is 1 in $\text{dist}(\mathbf{w}, \mathbf{0})$, the received word $\mathbf{w}$ is corrected to $\mathbf{c} = \mathbf{0} = 0000000$. ∎

4.2.6 Systematic Hamming Codes. A binary Hamming (n, k) code is called a *systematic* code if there are k locations $I_1, \ldots, i_k$ with the property that, by restricting the codewords in these locations, we get all 2^k possible binary codewords of length k. The parity-check matrix and the generator matrix of the binary Hamming (7,4) code, denoted with the suffix s, are

$$\mathbf{H}_s = \begin{bmatrix} 1 & 1 & 0 & 1 & 1 & 0 & 0 \\ 1 & 0 & 1 & 1 & 0 & 1 & 0 \\ 0 & 1 & 1 & 1 & 0 & 0 & 1 \end{bmatrix}, \quad \mathbf{G}_s = \begin{bmatrix} 1 & 1 & 0 & 1 & 0 & 0 & 0 \\ 1 & 0 & 1 & 0 & 1 & 0 & 0 \\ 0 & 1 & 1 & 0 & 0 & 1 & 0 \\ 1 & 1 & 1 & 0 & 0 & 0 & 1 \end{bmatrix}. \quad (4.2.4)$$

For systematic codes, $\mathbf{Hc}^T = \mathbf{0}$ mod 2, but $\mathbf{Hw}^T \neq \mathbf{0}$ unless $\mathbf{w} = \mathbf{c}$, in which case the transmission is most probably error-free. The quantity $\mathbf{Hc}^T$ (or $\mathbf{Hw}^T$) is known as the syndrome for a codeword $\mathbf{c}$ (or a received word $\mathbf{w}$), which must be checked by the syndrome method given in §4.2.3.

Since every code is equivalent to its systematic code, the following method can be used to obtain the parity-check matrix from the generator matrix in the systematic form. If the code is nonsystematic, it must be multiplied by a permutation matrix to produce the equivalent systematic code. Then the unknown matrix, parity-check or generator, can be obtained from the generator or parity-check matrix, respectively, by this method, taking care that the matrix so obtained is permuted back to the order of the original matrix.

Some examples of of systematic q-ary Hamming codes are given in §4.7 for $q = 2, 3, 4$ and 5.

4.2.7 Codeword Sequences. So far we have represented codewords in the form $\mathbf{c} = \{p_1\, p_2\, d_1\, p_4\, d_2, d_3\, d_4\}$. There is another way to represent this sequence. In a Hamming (7,4) code, let the codeword $\mathbf{c} = \{c_1, c_2, \ldots, c_7\}$ be arranged such that the four data bits $\mathbf{d} = \{d_1, d_2, d_3, d_4\}$ are followed by the three parity bits $\mathbf{p} = \{p_1, p_2, p_3\}$, which are computed by formula (4.2.2). All 16 codewords for Hamming (7,4) code are presented in Table 4.2.3.

For any valid encoded word $\mathbf{c}$, the following checksums must be satisfied:

$$c_2 \oplus c_3 \oplus c_4 \oplus c_5 = 0, \quad c_1 \oplus c_3 \oplus c_4 \oplus c_6 = 0, \quad c_1 \oplus c_2 \oplus c_4 \oplus c_7 = 0.$$

If the value of one or more of these checksums is 1, then an error has definitely occurred. After the message is transmitted, let the received word be denoted by $\mathbf{w} = \{w_1, w_2, \ldots, w_7\}$. To determine and correct a single-bit error, we apply the above three checksums to $\mathbf{w}$. This yields seven possible sequences: 001, 010, 011, 100, 101, 110, 111 (syndrome), each one of which, when read as a decimal number from 1 through 7, will give us the location of the bit in error.

Table 4.2.3. All Codewords for Hamming (7,4) Code

Dataword $\mathbf{d}$	Codeword $\mathbf{c}$	Dataword $\mathbf{d}$	Codeword $\mathbf{c}$
0000	0000 000	1000	1000 011
0001	0001 111	1001	1001 100
0010	0010 110	1010	1010 101
0011	0011 001	1011	1011 010
0100	0100 101	1100	1100 110
0101	0101 010	1101	1101 001
0110	0110 011	1110	1110 000
0111	0111 100	1111	1111 111

Example 4.2.6. Suppose that we want to transmit the dataword 1100. This will be transmitted as the codeword $\mathbf{c} = 1100110$. Now suppose that the third bit of this codeword is changed from 0 to 1 during transmission, and the received word is $\mathbf{w} = 1110110$. We use the above checksums, which yield

$$1 \oplus 1 \oplus 0 \oplus 1 = 0, \quad 1 \oplus 1 \oplus 0 \oplus 1 = 1, \quad 1 \oplus 1 \oplus 1 \oplus 0 = 1.$$

Since the checksums are 011, this tells us that the third bit is in error, and we flip it from 1 to 0, and receive the corrected message. ∎

To summarize, parity checking is the basic idea of an ECC; it involves counting the number of 1s in a sequence of bits. Depending on the type of parity, checking the parity bit is created to make the total number of 1s either odd or even. Hamming codes are capable of not only detecting errors but also correcting them. As we have seen in Hamming (7,4) codes, the four data bits are arranged in a specific way so that when the parity checking is carried out, it can not only detect a single bit in error but also locates and corrects it. Table 4.2.4 represents this process.

To obtain the first parity check bit, a parity check is performed on each bit location from Table 4.2.4 that has a '1' in the least significant bit (lsb) location (farthest right bit) of its binary form (second column); to obtain the second parity check bit, a parity check is performed on every bit location that has a '1' in the middle location of its binary form; and finally, to obtain the third (the last) parity check bit, a parity check is performed on all bit locations with a '1" in the most significant bit (msb) location (farthest right bit) of its binary form. In other words, the first parity bit p_1 is generated from the bits 3, 5, and 7; the second p_2 from bits 3, 6, and 7; and the third and the last p_4 from bits 5, 6, and 7. This creates the parity bits of the block, and when the block is checked for errors, all bits, including the parity bits, are checked for parity and a syndrome is generated. In the case of the Hamming (7,4) code the syndrome is the binary number, three bits long, that results from each parity check on each grouping of bits. This syndrome is zero if there are no errors; but if there is an error, the syndrome will contain the information on the bit where the error has occurred, which is flipped and the error corrected.

Table 4.2.4 Process of Parity Checking

Bit Location	Binary Form	Type
1	001	Parity bit p_1
2	010	Parity bit p_2
3	011	Data bit d_1
4	100	Parity bit p_4
5	101	Data bit d_2
6	110	Data bit d_3
7	111	Data bit d_4

For any binary linear code C of dimension k and length n, the following two results are noteworthy:

(i) If $\mathbf{x}, \mathbf{y}$ are any two messages, then the codes $C(\mathbf{x})$ and $C(\mathbf{y})$ satisfy the relation: $C(\mathbf{x}) + C(\mathbf{y}) = C(\mathbf{x} + \mathbf{y})$.

(ii) For any binary linear code, the minimum distance is equal to the minimum Hamming weight of any nonzero codeword.

4.2.8 Multiple-Bit Errors. The Hamming code can be used to detect single- and double-bit errors, by checking the product $\mathbf{H}\,\mathbf{r}$, which becomes nonzero if errors have occurred, and they can be corrected, one at a time, by the above method. But the Hamming $(7, 4)$ code cannot distinguish between single-bit and double-bit errors.

4.3 Hamming (11,7) Code

The dataword in this code has 7 bits, with 4 parity bits. Using even parity, the parity bits are computed for each data bit and the dataword is encoded. The decoding is done by again checking the parity bits and using the syndrome method, as shown the following example.

Example 4.3.1. Let the seven data bits be 0110101. Using Table 4.3.1, the data bits are first placed in their respective locations and parity bits are created using even parity.

Table 4.3.1 Locations of Data and Parity Bits

	p_1	p_2	d_1	p_4	d_2	d_3	d_4	p_8	d_5	d_6	d_7
Data bits			0		1	1	0		1	0	1
p_1	1		0		1		0		1		1
p_2		0	0			1	0			0	1
p_4				0	1	1	0				
p_8								0	1	0	1

Thus, the codeword is $[1\,0\,0\,0\,1\,1\,0\,0\,1\,0\,1]$. Let us assume that the 10th bit gets corrupted during transmission, i.e., it changes from 0 to 1, so that the received word is $[10001100111]$. In decoding, we use the even parity and the result are as follows:

	p_1	p_2	d_1	p_4	d_2	d_3	d_4	p_8	d_5	d_6	d_7	
Codeword	1	0	0	0	1	1	0	0	1	1	1	Parity check
p_1	1		0		1		0		1	1	0	Pass
p_2		0	0			1	0		1	0		Pass
p_4				0	1	1	0			1		Fail
p_8								0	1	1	1	Fail

Hence, the parity bits $p_8\,p_4\,p_2\,p_1 = 1\,1\,0\,0 = (10)_{10}$, so the 10th bit is in error, which is corrected by flipping from 1 to 0. ■

4.4 General Algorithm

The general algorithm for a Hamming SEC code for any number of bits is as follows:

1. Number of bits starting from 1: bit # 1, 2, 3, 4, ...;

2. Write the bit numbers in binary: 1, 10, 11, 100, 101, ...;

3. All bit locations that are powers of 2 are parity bits;

4. All other bit locations are data bits;

5. Each data bit is included in a unique set of 2 or more parity bits, as determined by the binary form of its bit location. Thus,

(i) Parity bit # 1 covers all bit locations that have the lsb set: Bit # 1 (the parity bit itself), 3, 5, 7, 9, ...;

(ii) Parity bit # 2 covers all bit locations that have the second lsb bit set: Bit # 2 (the parity bit itself), 3, 6, 7, 10, 11, ...;

(iii) Parity bit # 4 covers all bit locations that have the third lsb bit set: Bit # 4 – 4, 12 – 15, 20 – 23, ...;

(iv) Parity bit # 8 covers all bit locations that have the fourth lsb bit set: Bit # 8 – 15, 24 – 31, 40 –47, ...;

(v) In general, each parity bit covers all bits where the binary AND of the parity location and the bit location is nonzero.

6. Parity can be even or odd; its type is irrelevant, although even parity is mathematically the purest of the two, but either parity can be used.

The above algorithm is presented in the following tabular form for 15 encoded bits (4 parity bits and 11 data bits).

Bit Location	1	2	3	4	5	6	7	8	9	10	11	12	13	14	15	$\cdots$
Encoded bits	p_1	p_2	d_1	p_4	d_2	d_3	d_4	p_8	d_5	d_6	d_7	d_8	d_9	d_{10}	d_{11}	$\cdots$
p_1	x		x		x		x		x		x		x		x	$\cdots$
p_2		x	x			x	x			x	x			x	x	$\cdots$
p_4				x	x	x	x					x	x	x	x	$\cdots$
p_8								x	x	x	x	x	x	x	x	$\cdots$

The errors are checked by the pattern of errors, called the syndrome, which identifies the bit in error. If all parity bits are correct, then there is no error. Otherwise, the sum of the bit locations in error identifies the bit in error, which is corrected by flipping it.

Example 4.4.1. Using the above algorithm, we will determine the parity check bits for the Hamming (11,7) code. Let the 7-bit dataword be 0110101. The Hamming scheme is explained in Table 4.4.1. Thus, the dataword with parity to be transmitted is: 10001100101. Assume that during transmission the last bit is corrupted and changed from 1 to 0. The codeword received then is: 10001100100. The Hamming code analyzes this codeword and corrects the single-bit error when the even parity check fails. This is presented in Table 4.4.2.

Table 4.4.1 Hamming Scheme

Bit Location	1	2	3	4	5	6	7	8	9	10	11
Encoded	p_1	p_2	d_1	p_4	d_2	d_3	d_4	p_8	d_5	d_6	d_7
Dataword[†]			0		1	1	0		1	0	1
p_1	1		0		1		0		1		1
p_2		0			1	0				0	1
p_4								0	1	0	1

[†] without parity.

Table 4.4.2 Hamming Code Analysis

Encoded	p_1	p_2	d_1	p_4	d_2	d_3	d_4	p_8	d_5	d_6	d_7	Parity check	Parity bit
Codeword	1	0	0	0	1	1	0	0	1	0	0		
p_1	1		0		1		0		1		0	Fail	1
p_2		0	0			1	0			0	0	Fail	1
p_4				0	1	1	0					Pass	0
p_8								0	1	0	0	Fail	1

This shows that the bit # in error is $p_8 + p_4 + p_2 + p_1 = 1011 = (8 + 2 + 1)_{10} = 11$, i.e., the 11th (last) bit in the codeword is in error and is flipped from 0 to 1, giving the original dataword that was transmitted. ∎

Example 4.4.2. (Hamming (15,11) code)[3] The above algorithm is used again to determine and test the codeword corresponding to the message (con-

---

[3] Note that for a SEC Hamming (n, k) code we must have $(n + 1)2^k = 2^n$ for $n > k > 1$. The two values of n and k of our interest are $n = 7, 15$ and $k = 4, 11$, respectively. In order to have a 1-bit correction code for 2^{11} words, note that since $(14 + 1)2^{11} = 30720 > 2^{14}$ but $(15 + 1)2^{11} = 32768 = 2^{15}$, the code must have 15 bits, of which 11 are data bits. This defines the Hamming (15,11) code, which is one of the so-called perfect codes (see §4.7, 6.5, and §7.1). It also satisfies the Hamming rule: $2^{n-k} \geq 1 + n$, where equality holds for $n = 15, k = 11$, i.e., $(15 + 1)2^{11} = 2^{15}$.

sisting of 11 data bits) 10101000101, which occupy the bit locations 3, 5, 6, 7, 9, 10, 11, 12, 13, 14, and 15. The entire scheme is presented in the form of a table as follows:

Bit Location	1	2	3	4	5	6	7	8	9	10	11	12	13	14	15
Encoded bits	p_1	p_2	d_1	p_4	d_2	d_3	d_4	p_8	d_5	d_6	d_7	d_8	d_9	d_{10}	d_{11}
Data bits	p_1	p_2	1	p_4	0	1	0	p_8	1	0	0	0	1	0	1

The parity check bits p_1, p_2, p_4, p_8 are computed as follows. Take the binary representation of each of the data bits 3, 6, 9, 13, 15 and XOR them together. Thus, $0011 \oplus 0110 \oplus 1001 \oplus 1101 \oplus 1111 = 1110$, which gives the parity check bits: $p_8 = 1, p_4 = 1, p_2 = 1, p_1 = 0$. and the 15-bit codeword is 011101011000101, which is transmitted on a noisy channel. Now, suppose that the data bit at location # 12 is flipped during transmission, i.e., it changes from 0 to 1, and the decoder receives the word 011101011001101. The decoder computes $(3)_2 \oplus (6)_2 \oplus (9)_2 \oplus (15)_2 = 0011 \oplus 0110 \oplus 1001 \oplus 1111 = 1100 = (12)_2$, which shows that the bit # 12 was flipped. Next, suppose that the parity bit at location # 4 is flipped, i.e., it changed from 1 to 0. The decoder received the word 011001011000101. It computes $(3)_2 \oplus (6)_2 \oplus (9)_2 \oplus (13)_2 \oplus (15)_2$, which is equal to 1110. The parity bits were received as 8241, i.e., 1010. Then $1110 \oplus 1010 = 0100 = (4)_2$, and this corrects the parity bit at location # 4. ∎

Notice that the rates R for the Hamming codes are as follows: Hamming (3,1): $R = 1/3 \approx 0.33$; Hamming (7,4)L $R = 4/7 \approx 0.57$; Hamming (15,11): $R = 11/15 \approx 0.73$; and Hamming (31,26): $R = 26/31 \approx 0.84$.

4.4.1. All Sixteen Possible Hamming (7,4) Codes. Since the data has 4 bits, there are exactly 16 possible transmitted codewords. As before, the $(7, 4)$ code has a 7-bit transmitted codeword $p_1 \, p_2 \, d_1 \, p_4 \, d_2 \, d_3 \, d_4$, whereas the $(8, 4)$ code has an 8-bit transmitted codeword $p_1 \, p_2 \, d_1 \, p_4 \, d_2 \, d_3 \, d_4 \, p_8$, which contains the data bits $d_1 \, d_2 \, d_3 \, d_4$. All these sixteen cases are presented in Table 4.3.2, in which the first part is a repetition of Table 4.2.3. Note that Hamming $(8,4)$ code is an extended $(7,4)$ code and is discussed in Chapter 5 (§5.2).

4.4.2 Hamming Code for Longer Data. The Hamming code for data longer than 4-bits can be used in an analogous manner for the detection of a single bit error. The method is as follows:

STEP 1. Mark all parity bit locations; they are $1, 2, 4, 8, 16, 32, \ldots$;

STEP 2. Mark all other bit locations for the data to be encoded; they are $3, 5, 7, 9, 10, 11, 12, 13, 4, 15, 17, \ldots$;

STEP 3. Since each parity bit calculates the parity for some of the bits in the codeword, the location of the parity bit determines the sequence of bits that it checks and skips alternately, as follows:

PARITY BIT LOCATION p_1: Check 1 bit, skip 1 bit, and repeat, i.e., check bit #s $1, 3, 5, 7, 9, 11, 13, 15, \ldots$;

PARITY BIT LOCATION p_2: Check 2 bits, skip 2 bits, and repeat, i.e., check bit #s $2, 3, 6, 10, 11, 14, 15, \ldots$;

PARITY BIT LOCATION p_4: Check 4 bits, skip 4 bits, and repeat, i.e., check bit #s $4, \ldots, 7, 12, \ldots, 15, 20, \ldots, 23, \ldots$;

PARITY BIT LOCATION p_8: Check 8 bits, skip 8 bits, and repeat, i.e., check bit #s $8, \ldots, 15, 24, \ldots, 31, 40, \ldots, 47, \ldots$;

PARITY BIT LOCATION p_{16}: Check 16 bits, skip 16 bits, and repeat, i.e., check bit #s $16, \ldots, 31, 48, \ldots, 63, 80, \ldots, 95, \ldots$;

PARITY BIT LOCATION p_{32}: Check 32 bits, skip 32 bits, and repeat, i.e., check bit #s $32, \ldots, 63, 96, \ldots, 127, 160, \ldots, 191, \ldots$; and so on.

In general, for the parity bit p_k, check 2^k bits, skip 2^k bits, and repeat, i.e., check bit #s $k, \ldots, (2k - 1), 3k, \ldots, (4k - 1), 5k, \ldots, (6k - 1), \ldots$ for $k = 1, 2, 4, 8, 16, \ldots$.

Table 4.3.2 All Sixteen Hamming (7,4) and (8,4) Codewords

Data Bits	Parity Bits (7,4)	(7,4) Codeword	(8,4) Codeword
0 0 0 0	0 0 0	0 0 0 0 0 0 0	0 0 0 0 0 0 0 0
0 0 0 1	1 1 1	1 1 0 1 0 0 1	1 1 0 1 0 0 1 0
0 0 1 0	0 1 1	0 1 0 1 0 1 0	0 1 0 1 0 1 0 1
0 0 1 1	1 0 0	1 0 0 0 0 1 1	1 0 0 0 0 1 1 1
0 1 0 0	1 0 1	1 0 0 1 1 0 0	1 0 0 1 1 0 0 1
0 1 0 1	0 1 0	0 1 0 0 1 0 1	0 1 0 0 1 0 1 1
0 1 1 0	1 1 0	1 1 0 0 1 1 0	1 1 0 0 1 1 0 0
0 1 1 1	0 0 1	0 0 0 1 1 1 1	0 0 0 1 1 1 1 0
1 0 0 0	1 1 0	1 1 1 0 0 0 0	1 1 1 0 0 0 0 1
1 0 0 1	0 0 1	0 0 1 1 0 0 1	0 0 1 1 0 0 1 1
1 0 1 0	1 0 1	1 0 1 1 0 1 0	1 0 1 1 0 1 0 0
1 0 1 1	0 1 0	0 1 1 0 0 1 1	0 1 1 0 0 1 1 0
1 1 0 0	0 1 1	0 1 1 1 1 0 0	0 1 1 1 1 0 0 0
1 1 0 1	1 0 0	1 0 1 0 1 0 1	1 0 1 0 1 0 1 0
1 1 1 0	0 0 0	0 0 1 0 1 1 0	0 0 1 0 1 1 0 1
1 1 1 1	1 1 1	1 1 1 1 1 1 1	1 1 1 1 1 1 1 1

Example 4.4.3. Let the 8-bit data bits be $\{d_1, \ldots, d_8\} = [0\,0\,1\,1\,0\,0\,1\,1]$. To compute the parity bits $\{p_1, p_2, p_4, p_8\}$, we follow the above steps:

For p_1, check bit # $1, 3, 5, 7, 9, 11$. This gives p_1 0 0 1 0 1, which has even parity (counting the number of 1s); hence, $p_1 = 0$;

For p_2, check bit # $2, 3, 6, 7, 10, 11$. This gives p_2 0 1 1 0 1, which has odd parity (counting the number of 1s); hence, $p_2 = 1$;

For p_4, check bit # $4, 5, 6, 7, 12$. This gives p_4 0 1 1 1, which has odd parity (counting the number of 1s); hence, $p_4 = 1$;

For p_8, check bit # $8, 9, 10, 1, 121$. This gives p_8 0 0 1 1, which has even parity (counting the number of 1s); hence, $p_8 = 0$.

Thus, the codeword is 0 1 0 1 0 1 1 0 1 0 1 1. Now, assume that the 9-th bit is corrupted in transmission. For decoding, we again follow the above steps:

For p_1, check bit # $1, 3, 5, 7, 9, 11$. This gives p_1 0 0 1 1 1, which has odd parity; hence, $p_1 = 1$;

For p_2, check bit # $2, 3, 6, 7, 10, 11$. This gives p_2 0 1 1 0 1, which has odd parity; hence, $p_2 = 1$;

For p_4, check bit # $4, 5, 6, 7, 12$. This gives p_4 0 1 1 1 1, which has odd parity; hence, $p_4 = 1$;

For p_8, check bit # $8, 9, 10, 1, 121$. This gives p_8 1 0 1 1, which has odd parity; hence, $p_8 = 1$.

Since the parity bits of p_1 and p_8 do not match with their encoded values, the error is in the bit $1 + 8 = 9$.

In all these calculations, F_2 arithmetic is used. For example, bit # 23 $= 1+2+4+16$ (as sum of powers of 2), so bit # 23 is checked by the check bits p_1, p_2, p_4, p_{16}. ∎

Example 4.4.4. Notice that in general a Hamming code is denoted as Hamming (n, k) code, where $n = 2^k - 1$ is the code length, and $m = n - k$ denotes the number of parity bits. The simplest (and most trivial) of all Hamming codes is Hamming $(3, 1)$ code, for which the parity-check and the generator matrix are, respectively,

$$\mathbf{H} = \begin{bmatrix} 1 & 0 & 1 \\ 0 & 1 & 1 \end{bmatrix}, \quad \mathbf{G} = [1\ 1\ 1]. \ \blacksquare$$

4.5 Hamming's Original Algorithm

In the interest of sheer academic curiosity it seems very desirable to go back in time and look at the original algorithm that was used by Hamming, although so much improvement and research have piled up on that outstanding and original concept during the past 60 years. The Hamming $(7,4,3)$ code checks a 7-bit message (dataword), which may have seven possible single-bit errors, and it has three parity bits, also known as the error control bits, which determine not only that an error has occurred but also the bit that has caused the error. The following algorithm, due to Hamming, appeared in Shannon's paper [Shannon, 1948] for encoding and decoding. This algorithm does not use a parity-check matrix or coset leaders to decode a received codeword. Curiously enough, the details are as follows: Let $\mathbf{c} = \{c_1, c_2, c_3, c_4, c_5, c_6, c_7\} \in \mathcal{K}^{7,4}$ where c_3, c_5, c_6, c_7 are the message (data) bits, and c_1, c_2, c_4 are the parity bits. In our notation this corresponds to $\mathbf{c} = \{p_1, p_2, d_1, p_4, d_2, d_3, d_4\}$. The parity bits are computed as follows:

1. c_1 is chosen to make $\alpha = c_1 + c_3 + c_5 + c_7 \equiv 0 \pmod 2$;

2. c_2 is chosen to make $\beta = c_2 + c_3 + c_6 + c_7 \equiv 0 \pmod 2$;

3. c_4 is chosen to make $\gamma = c_4 + c_5 + c_6 + c_7 \equiv 0 \pmod 2$.

Notice that α and β share c_3 and c_7; α and γ share c_5 and c_7; and β and γ share c_6 and c_7. After the codeword is received, the quantities α, β, and γ are computed: A value of 0 indicates that the no error has occurred, while a value of 1 flags the bit c_i that is in error.

The systematic generating matrix of this code is

$$\mathbf{G}_s = \begin{bmatrix} 1 & 1 & 1 & 0 & 0 & 0 & 0 \\ 1 & 0 & 0 & 1 & 1 & 0 & 0 \\ 0 & 1 & 0 & 1 & 0 & 1 & 0 \\ 1 & 1 & 0 & 1 & 0 & 0 & 1 \end{bmatrix}.$$

It generates the Hamming code C with the following 16 codewords:

$$C = \left\{ \begin{matrix} 0000000\ 1110000\ 1001100\ 0101010\ 1101001\ 0111100\ 1100110\ 1000011 \\ 0011001\ 1011010\ 0100101\ 0010110\ 1010101\ 0110011\ 0001111\ 1111111 \end{matrix} \right\}.$$

Assuming that only a single-bit error has occurred, the above error correcting algorithm works as follows:

Bit #		(α, β, γ)		Error		Bit #
1	$\longleftrightarrow$	$(1, 0, 0)$		1000000	$\longleftrightarrow$	1
2	$\longleftrightarrow$	$(0, 1, 0)$		0100000	$\longleftrightarrow$	2
3	$\longleftrightarrow$	$(1, 1, 0)$		0010000	$\longleftrightarrow$	3
4	$\longleftrightarrow$	$(0, 0, 1)$		0001000	$\longleftrightarrow$	4
5	$\longleftrightarrow$	$(1, 0, 1)$		0000100	$\longleftrightarrow$	5
6	$\longleftrightarrow$	$(0, 1, 1)$		0000010	$\longleftrightarrow$	6
7	$\longleftrightarrow$	$(1, 1, 1)$		0000001	$\longleftrightarrow$	7

Example 4.5.1. The following five codewords are sent in a message: $c_1 = 0001111$, $c_2 = 0011001$, $c_3 = 1010101$, $c_4 = 0111100$, $c_5 = 1001100$. The respective received words are denoted by w_i, and the decoded words by v_i, $i = 1, 2, 3, 4, 5$. The Hamming's original decoding algorithm works as follows:

---

[4] $\mathcal{K}^7$ in Shannon's paper refers to the field F_q^n in our notation, where $q = 2$ and $n = 7$.

c_i	w_i	(α, β, γ)	v_i	Errors Bit #	Detected	Corrected
0001111	0001111	$(0, 0, 0) \longleftrightarrow 0$	0001111	0	y	y
0011001	0011011	$(0, 1, 1) \longleftrightarrow 6$	0011001	1	y	y
1010101	0011101	$(1, 0, 1) \longleftrightarrow 5$	0011001	2	y	n
0111100	1001100	$(0, 0, 0) \longleftrightarrow 0$	1001100	3	n	n
1001100	1000011	$(0, 0, 0) \longleftrightarrow 0$	1000011	4	n	n

Note that the word w_3 with two error bits was decoded as c_3, but the fact that w_3 is not in the above list of C tells us that v_3 is not the right codeword c_3, yet there is no way to confirm it. ∎

4.6 Equivalent Codes

A binary linear code C_1 with generator matrix $\mathbf{G}_1$ is *equivalent* to the binary linear code C_2 with generator matrix $\mathbf{G}_2$ plus an associated column permutation for the second matrix, and is not equivalent otherwise.

Example 4.6.1. Let $\mathbf{G}_1$ be the Hamming generator matrix, C_1 the code list,

$$\mathbf{G}_1 = \begin{bmatrix} 1 & 1 & 0 & 1 & 0 & 0 & 0 \\ 1 & 0 & 1 & 0 & 1 & 0 & 0 \\ 0 & 1 & 1 & 0 & 0 & 1 & 0 \\ 1 & 1 & 1 & 0 & 0 & 0 & 1 \end{bmatrix}.$$

$C_1 = [[0000000], [1101000], [1010100], [1000011], [0110010],$
$\quad [0100101], [0011001], [0001110], [1110001], [1100110],$
$\quad [1011010], [1001101], [0111100], [0101011], [0010111],$
$\quad [1111111]].$

$$\mathbf{G}_2 = \begin{bmatrix} 1 & 0 & 1 & 1 & 0 & 0 & 0 \\ 0 & 1 & 0 & 1 & 1 & 0 & 0 \\ 0 & 0 & 1 & 0 & 1 & 1 & 0 \\ 0 & 0 & 0 & 1 & 0 & 1 & 1 \end{bmatrix}.$$

$C_2 = [[0000000], [1100010], [1011000], [1000101], [0110001],$
$\quad [0101100], [0010110], [0001011], [1110100], [1101001],$
$\quad [1010011], [1001110], [0111010], [0100111], [0011101],$
$\quad [1111111]].$ ∎

4.7 q-ary Hamming Codes

For $q \geq 2$, any nonzero vector $\mathbf{u} \in F_q^m$ generates a subspace $\mathbf{U}$ of dimension 1. Also, for $\mathbf{u}, \mathbf{v} \in F_q^m \backslash \{\mathbf{0}\}$ we will have $\mathbf{U} = \mathbf{V}$ iff there is a nonzero scalar $\alpha \in F_q \backslash \{0\}$ such that $\mathbf{u} = \alpha \mathbf{v}$. Thus, there are exactly $(q^m - 1)/(q - 1)$ distinct subspaces of dimension 1 in F_q^m. Let $m \geq 2$. Then a q-ary linear code, with parity-check matrix $\mathbf{H}$, such that the columns of $\mathbf{H}$ contain precisely one nonzero vector from each vector subspace of dimension 1 in F_q^m, is called a q-ary *Hamming code*. This code is a $[(q^m - 1)/(q - 1), (q^m - 1)/(q - 1) - m, 3]$ code, and it is a perfect SEC code. For $q = 2$, we get the usual binary Hamming code that has been discussed thus far.

Example 4.7.1. Parity check matrix (in standard and non-stantard form) for the 5-ary Hamming code of length $n = (5^3 - 1)(5 - 1) = 31$ is

$$\mathbf{H}_s = \begin{bmatrix} 1001234123400001234123412341234 \\ 0101111000012341111222233334444 \\ 0010000111111111111111111111111 \end{bmatrix},$$

$$\mathbf{H} = \begin{bmatrix} 0010000111111111111111111111111 \\ 0101111000012341111222233334444 \\ 1001234123400001234123412341234 \end{bmatrix}. \blacksquare$$

Example 4.7.2. Parity check and generator matrix of Hamming q-ary code with $q = 3, m = 3$ are, respectively,

$$\mathbf{H} = \begin{bmatrix} 1 & 0 & 0 & 1 & 2 & 1 & 2 & 0 & 0 & 1 & 2 & 1 & 2 \\ 0 & 1 & 0 & 1 & 1 & 0 & 0 & 1 & 2 & 1 & 1 & 2 & 2 \\ 0 & 0 & 1 & 0 & 0 & 1 & 1 & 1 & 1 & 1 & 1 & 1 & 1 \end{bmatrix},$$

$$\mathbf{G} = \begin{bmatrix} 2 & 2 & 0 & 1 & 0 & 0 & 0 & 0 & 0 & 0 & 0 & 0 & 0 \\ 1 & 2 & 0 & 0 & 1 & 0 & 0 & 0 & 0 & 0 & 0 & 0 & 0 \\ 2 & 0 & 2 & 0 & 0 & 1 & 0 & 0 & 0 & 0 & 0 & 0 & 0 \\ 1 & 0 & 2 & 0 & 0 & 0 & 1 & 0 & 0 & 0 & 0 & 0 & 0 \\ 0 & 2 & 2 & 0 & 0 & 0 & 0 & 1 & 0 & 0 & 0 & 0 & 0 \\ 0 & 1 & 2 & 0 & 0 & 0 & 0 & 0 & 1 & 0 & 0 & 0 & 0 \\ 2 & 2 & 2 & 0 & 0 & 0 & 0 & 0 & 0 & 1 & 0 & 0 & 0 \\ 1 & 2 & 2 & 0 & 0 & 0 & 0 & 0 & 0 & 0 & 1 & 0 & 0 \\ 2 & 1 & 2 & 0 & 0 & 0 & 0 & 0 & 0 & 0 & 0 & 1 & 0 \\ 1 & 1 & 2 & 0 & 0 & 0 & 0 & 0 & 0 & 0 & 0 & 0 & 1 \end{bmatrix}. \blacksquare$$

The q-ary systematic Hamming (n, k) code is defined the same way as in §4.2.6, except that we get all q^k possible q-ary codewords of length k.

Example 4.7.3. The generator matrix for the extended systematic Hamming (8,4) code is

$$\mathbf{G}_s = \begin{bmatrix} 0 & 1 & 1 & 1 & 1 & 0 & 0 & 0 \\ 1 & 0 & 1 & 1 & 0 & 1 & 0 & 0 \\ 1 & 1 & 0 & 1 & 0 & 0 & 1 & 0 \\ 1 & 1 & 1 & 0 & 0 & 0 & 0 & 1 \end{bmatrix} = \begin{bmatrix} 1 & 0 & 0 & 0 & 0 & 1 & 1 & 1 \\ 0 & 1 & 0 & 0 & 1 & 0 & 1 & 1 \\ 0 & 0 & 1 & 0 & 1 & 1 & 0 & 1 \\ 0 & 0 & 0 & 1 & 1 & 1 & 1 & 0 \end{bmatrix},$$

where the second form is obtained by rearranging the columns of the first. This code is discussed in the next chapter.

Example 4.7.4. By rearranging the columns the parity-check matrix of a binary Hamming (7,4) code can be reduced to different forms. Thus, the matrix $\mathbf{H}_s$ in (4.2.4) reduces to

$$\mathbf{H}_s = \begin{bmatrix} 1 & 0 & 0 & 1 & 0 & 1 & 1 \\ 0 & 1 & 0 & 1 & 1 & 1 & 0 \\ 0 & 0 & 1 & 0 & 1 & 1 & 1 \end{bmatrix}.$$

Again, it is easy to see that by rearranging the columns differently, the above matrix $\mathbf{H}_s$ reduces to the matrix $\mathbf{H}$ of (4.2.1). ∎

Example 4.7.5. As a preview of the Galois field algebra that is discussed in Chapter 8 and subsequently in Chapters 10 through 13, consider the generator polynomial $g(x) = x^3 + x + 1$, which is an algebraic representation of the dataword $[1\,0\,1\,1]$. Following any of the methods of this chapter it can be shown that the associated binary Hamming code is defined by the parity-check matrix $\mathbf{H}$ of the Example 4.7.4 and the generator matrix

$$\mathbf{G} = \begin{bmatrix} 1 & 1 & 0 & 1 & 0 & 0 & 0 \\ 0 & 1 & 1 & 0 & 1 & 0 & 0 \\ 0 & 0 & 1 & 1 & 0 & 1 & 0 \\ 0 & 0 & 0 & 1 & 1 & 0 & 1 \end{bmatrix}. \, \blacksquare$$

Note that for a binary (n, k) code, the parity-check matrix $\mathbf{H}$ is an $m \times n$ matrix, $m = n - k$, and the generator matrix $\mathbf{G}$ is a $k \times n$ matrix.

5

Extended Hamming Codes

5.1 SEC-DED Codes

The Hamming codes can be extended to a SEC-DED code by adding one additional parity-check bit. These codes, known as *extended Hamming codes*, use an extra parity bit to increase the Hamming distance between any two valid encodings. Recall that the Hamming distance between two strings of equal length is the number of bits at which the corresponding symbols differ (see §3.3.1). For example, the 8-bit strings 01010101 and 10101010 have a Hamming distance of 8 because all eight bits differ, but 10001000 and 10001001 have a Hamming distance of 1 because only the lsb differs. Encoding values with a Hamming distance greater than 1 require more bits to represent fewer unique values than the minimum 2^n values with n bits that the standard binary encoding allows. For example, encoding the two values $\{0, 1\}$ with a Hamming distance of 2 requires 2 bits, resulting in the valid encoding $\{00, 11\}$, and two invalid (erroneous) encodings $\{01, 10\}$. This example is identical to treating the first bit as the data bit and the second bit as an even parity bit. Note that a Hamming distance of 2 provides only the detection of a single error by noting an invalid encoding like 01 or 10, but it cannot correct it.

SEC-DED Hamming codes require a Hamming distance of 4 between all valid encodings so that they will be capable of forward error correction by examining logical values and determining whether an encoding is valid or not, and if it is invalid, it is capable of correcting a single erroneous bit, or detecting two erroneous bits and signaling an uncorrectable error. If three or more errors occur, SEC-DED code does not help and errors go undetected. However, the probability of three or more errors occurring simultaneously is very small.

The extended Hamming codes are, in fact, SEC-DED codes because of the following reasons: (i) for no error, the overall parity of the n-bit received word $\mathbf{w}$ will be even and the syndrome of the $(n-1)$-bit SEC part of $\mathbf{w}$ will be zero; (ii) for one error in any bit, the overall parity of $\mathbf{w}$ will be odd and the syndrome will be zero, but if a single error occurred in any other bit the syndrome will be nonzero; and (iii) for two errors the overall parity of $\mathbf{w}$ will be even whether the error is in the overall parity or in the SEC part of the received word (or both) and the syndrome in both cases will be nonzero, and in this case a double error will be detected. In all cases there is a parity-check bit that checks one of the two bit locations but not the other one. The parity of this check bit will be odd, thus giving zero syndrome. The other cases are similarly justified.

5.1.1 Extension of Hamming Codes. A Hamming code is extended by adding an additional parity-check bit either at the end or the beginning of each codeword. Hence, every extended Hamming code has an even number of 1s, and the minimum distance becomes 4. This feature enables an extended Hamming code to detect and correct a single-bit error and only detect up to three errors. An extension of a code always increases the length of the codewords, but the dimension of the code remains the same. Thus, by adding a parity bit at the end of the Hamming (7,4,3) code we obtain an (8,4,4) code, and the generator matrix $\mathbf{G}_7$ changes to $\mathbf{G}_8$:

$$\mathbf{G}_7 = \begin{bmatrix} 1 & 1 & 1 & 0 & 0 & 0 & 0 \\ 1 & 0 & 0 & 1 & 1 & 0 & 0 \\ 0 & 1 & 0 & 1 & 0 & 1 & 0 \\ 1 & 1 & 0 & 1 & 0 & 0 & 1 \end{bmatrix} \qquad \mathbf{G}_8 = \begin{bmatrix} 1 & 1 & 1 & 0 & 0 & 0 & 0 & 1 \\ 1 & 0 & 0 & 1 & 1 & 0 & 0 & 1 \\ 0 & 1 & 0 & 1 & 0 & 1 & 0 & 1 \\ 1 & 1 & 0 & 1 & 0 & 0 & 1 & 0 \end{bmatrix}.$$

The Hamming code generated by $\mathbf{G}_8$ still has 16 codewords (see Table 4.3.2).

Example 5.1.1. Consider Example 4.5.1 with 5 codewords, which is now analyzed by the extended Hamming (8,4,4) code, where α, β, γ are defined in §4.5. Notice that the words are now 8-bit long.

c_i	w_i	(α, β, γ)	v_i	Errors Bits	Detected	Corrected
00011110	00011110	$(0, 0, 0) \longleftrightarrow 0$	00011110	0	y	y
00110011	00110110	$(0, 1, 1) \longleftrightarrow 6$	00110011	1	y	y
10101010	00111010	$(1, 0, 1) \longleftrightarrow 5$	00110011	2	y	n
01111000	10011000	$(0, 0, 0) \longleftrightarrow 0$	10011001	3	y	n
10011001	10000111	$(0, 0, 0) \longleftrightarrow 0$	10000111	4	n	n

The codeword w_3 with two error bits was corrected by flipping the fifth bit as the value of (α, β, γ) is computed as 5, although the codeword c_3 and the decoded word v_3 are different. The error is detected in v_4 because of the wrong parity bit, but $c_4 \neq v_4$. Finally, in the case of the received codeword w_5, which has an even number of 1s, no error was detected because $(\alpha, \beta, \gamma) = (0,0,0)$; but here $v_5 = w_5 \neq c_5$. ■

In general, for every integer $m > 2$ there exists a Hamming $(2^m - 1,$ $2^m - m - 1, 3)$ code, since all Hamming codes have a minimum distance of 3, which means that all Hamming codes can detect and correct a single-bit error and detect double-bit errors. The Hamming (7,4,3) code is the simplest case with $m = 3$. If an extra parity bit is added, then the minimum Hamming distance increases to 4. This enables the Hamming (8,4) code to detect up to 3-bit errors, but correct only a single-bit error.

5.2 Hamming (8,4) Code

This is Hamming $(7, 4)$ code with an extra parity bit p_8. It is also known as Hamming code with an additional parity. This code is an extension of the $(7, 4)$ code, to correct a single-bit error in 4-bit data, for which the matrices $\mathbf{H}$ and $\mathbf{G}$ are modified accordingly, as explained in the next section. The Venn diagram in Figure 5.2.1 shows the locations of the parity bits and the data bits.

The matrices $\mathbf{H}$ (4×8) and $\mathbf{G}$ (4×8) for this code become

$$\mathbf{H} = \begin{bmatrix} 1 & 0 & 1 & 0 & 1 & 0 & 1 & 0 \\ 0 & 1 & 1 & 0 & 0 & 1 & 1 & 0 \\ 0 & 0 & 0 & 1 & 1 & 1 & 1 & 0 \\ 1 & 1 & 1 & 1 & 1 & 1 & 1 & 1 \end{bmatrix}, \quad \mathbf{G} = \begin{bmatrix} 1 & 1 & 1 & 0 & 0 & 0 & 0 & 1 \\ 1 & 0 & 0 & 1 & 1 & 0 & 0 & 1 \\ 0 & 1 & 0 & 1 & 0 & 1 & 0 & 1 \\ 1 & 1 & 0 & 1 & 0 & 0 & 1 & 0 \end{bmatrix}.$$

Note that the matrices $\mathbf{H}$ and $\mathbf{G}$ as given above are not in the systematic form. But using row operations we obtain their systematic form as (see Example 4.7.3):

$$\mathbf{H}_s = \begin{bmatrix} 0 & 1 & 1 & 1 & 1 & 0 & 0 & 0 \\ 1 & 0 & 1 & 1 & 0 & 1 & 0 & 0 \\ 1 & 1 & 0 & 1 & 0 & 0 & 1 & 0 \\ 1 & 1 & 1 & 0 & 0 & 0 & 0 & 1 \end{bmatrix}, \quad \mathbf{G}_s = \begin{bmatrix} 1 & 0 & 0 & 0 & 0 & 1 & 1 & 1 \\ 0 & 1 & 0 & 0 & 1 & 0 & 1 & 1 \\ 0 & 0 & 1 & 0 & 1 & 1 & 0 & 1 \\ 0 & 0 & 0 & 1 & 1 & 1 & 1 & 0 \end{bmatrix},$$

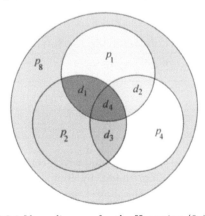

Figure 5.2.1 Venn diagram for the Hamming (8,4) code.

where the right 4×4 part in $\mathbf{H}$ and the left 4×4 part in $\mathbf{G}$ is the identity matrix I_4. Whereas the minimum distance of the Hamming (7,4) code is 3, in the case of Hamming (8,4) it increases to 4. Like the Hamming (7,4) code which is sometimes called as the (7,4,3) code, we can designate the Hamming (8,4) code as the (8,4,4) code. Since an extra parity bit p_8 is added, the (8,4) code has even parity. For example, the dataword 1011 is encoded as 01100110, where parity bits are $p_1 = 0$, $p_2 = 1$, $p_4 = 0$, $p_8 = 0$. Because of this additional parity, all Hamming codes are known as SEC-DED (Single-bit Error Correction Double-bit Error Detection) codes. These codes use additional check bits and an extra parity bit to increase the Hamming distance between two valid encodings. Since the 1970s they have been widely used to increase computer memory reliability, with the IBM 7030 and the IBM System/360, Model 85, after it was shown that a memory with ECC, compared to a memory without ECC or to two memories in a duplex configuration, greatly improved performance and reduced cost and size, as established by Allen [1966].

5.3 Hamming (13,8) Code

A (13,8) SEC-DED code for an 8-bit data $d_1, \ldots, d_8$ has the structure given in Table 5.3.1, with 5 parity bits $p_1, p_2, p_4, p_8, \widehat{p}$, where $\widehat{p}$ is an extra parity bit, and blank represents 'n' (no).

Table 5.3.1 Hamming (13,8) Code Scheme

Bit	0001	0010	0011	0100	0101	0110	0111	1000	1001	1010	1011	1100	1101
Index	1	2	3	4	5	6	7	8	9	10	11	12	13
Notation	p_1	p_2	d_1	p_4	d_2	d_3	d_4	p_8	d_5	d_6	d_7	d_8	$\widehat{p}$
p_1	y		y		y		y		y		y		
p_2		y	y			y	y			y	y		
p_4				y	y	y	y					y	
p_8								y	y	y	y	y	
$\widehat{p}$	y	y	y	y	y	y	y	y	y	y	y	y	

This scheme uses four check bits to uniquely identify the bit, if any, in a received word that has been corrupted, by comparing the stored check bits and parity bits against the ones that have been recomputed based on the data bits that are presented. The difference between the stored codeword and the recomputed word forms an index that points to the corrupted bit. The extra parity bit $\widehat{p}$ is used to detect noncorrectable double errors, which is explained below.

For k data bits, the minimum number of parity-check bits required for SEC and SEC-DED codes is given in Table 5.3.2, according to the *Hamming Rule*: $2^m \geq n + 1$.

Table 5.3.2 Parity-Check Bits m

# of Data Bits k	m for SEC	m for SEC-DED
1	2	3
2–4	3	4
5–11	4	5
12–26	5	6
27–57	6	7
58–120	7	8
121–247	8	9
248–502	9	10

5.4 Hamming (32,26) Code

Consider a Hammimg SEC-DED code for 32 data bits ($k = 32$). From Table 5.3.2 we find that for SEC a total of 6 parity bits and for SEC-DED 7 parity bits will be needed. We will also assume even parity of the parity-check bit and the bit being checked. First, for SEC: The Hamming $(38, 32)$ code requires six parity which are $p_0, \dots, p_5$ in the index notation, or $p_1, p_2, p_4, p_8, p_{16}, p_{32}$ in the location notation (see §3.4.7). These six parity bits combined with 32 data bits $d_1, \dots, d_{32}$ will give a Hamming code of length 38, and thus, e.g., in the location notation a codeword of 38 bits of the form:

$$\mathbf{c} = \{p_1, p_2, d_1, p_4, d_2, d_3, d_4, p_8, d_5, \dots, d_{11}, p_{16}, d_{12}, \dots, d_{26}, p_{32}, d_{27}, \dots, d_{32}\}$$

$$\equiv \{c_1, c_2, \dots, c_{38}\}.$$

This situation creates a problem for the Hamming scheme in its encoding and decoding. If the 32 data bits are moved into one word (register) and the 6 check bits into another word (register), it would create an irregular range of locations to be checked by each parity bit. However, this situation can be remedied as follows: Let the 6 parity bits check the locations according to the following scheme, in which each of the 32 data bit locations are checked by at least two check bits:

p_1 checks data bits: $d_1, d_2, d_4, d_6, d_8, d_{10}, \dots, d_{30}, d_{32}$

p_2 checks data bits: $d_1, d_3 - d_4, d_7 - d_8, d_{11} - d_{12}, \dots, d_{31} - d_{32}$

p_4 checks data bits: $d_1, d_5 - d_8, d_{13} - d_{16}, d_{21} - d_{24}, \dots, d_{29} - d_{32}$

p_8 checks data bits: $d_1, d_9 - d_{16}, d_{25} - d_{32}$

p_{16} checks data bits: $d_1, d_{17} - d_{32}$

p_{32} checks data bits: $d_2 - d_{32}$.

Notice that, for example, d_7 is checked by p_4 and p_{32}. Thus, if two data bits differ in one location, the codewords of length 38 will differ in three locations, so their maximum distance is 3. If two data bits differ in two bit locations, then, for example, at least one of the six parity bits $p_1 - p_{32}$ will check one of the locations but not the other, so the maximum distance is again 3. Thus, this SEC code has distance 3. The encoder generates a codeword of the form $\mathbf{c} = \{c_1, c_2, \ldots, c_{38}\}$, which is transmitted, of which 32 are data bits and 6 parity bits $\mathbf{p} = \{p_1, p_2, p_4, p_8, p_{16}, p_{32}\}$. Let $\mathbf{w} = \{w_1, \ldots, w_{38}\}$ denote the received word, and $\mathbf{s}$ the syndrome (determined by XOR-ing the parity bits $\mathbf{p}$ with the parity bits calculated from $\mathbf{w}$). The syndrome $\mathbf{s} = \{s_1 s_2 s_3 s_4 s_5 s_6\}$ for 32 received data bits and 6 parity bits is as follows:

Error in Bit #	Syndrome s	‖	Error in Bit #	Syndrome s
None	000000	‖	None	000000
d_1	111110	‖	p_1	100000
d_2	100001	‖	p_2	010000
d_3	010001	‖	p_4	001000
d_4	110001	‖	p_8	000100
d_5	001001	‖	p_{16}	000010
$\cdots$	$\cdots\cdots$	‖	p_{32}	000001
d_{31}	011111	‖		
d_{32}	111111	‖		

Suppose that the data bit d_2 is corrupted during transmission. Since d_2 is checked by parity bits p_1 and p_{32} (see above), the parity bit calculated by the encoder and decoder will differ in p_1 and p_{32}. Thus, the syndrome $p_1 \oplus p_{32} = 100000 \oplus 000001 = 100001$ yields the data bit d_2. If one of these two parity bits is corrupted, and no error occurs in data bits, then the parity bit received and the one calculated by the decoder will differ in the parity bit (one of p_1 and p_{32}) that was corrupted, and will be corrected as a single error by flipping its value. A similar method will detect and correct a single data bit error: The syndrome calculated for data bits will determine which data bit is in error. For example, if $\mathbf{s} = 000000$, no error has occurred; if $\mathbf{s} = 111110$, then d_1 is in error; in general, if $\mathbf{s} = $ xxxxx1, where xxxxx denotes nonzero bits, then the error is in w at location xxxxx. Let b denote the bit number. Then an algorithm for SEC is as follows:

```
if (s ⊕ (s − {111111})) = {000000} then ···    //No correction needed
else do
   if s = {0b0111} then b ← 0
   else b ← s ⊕ {0b0111}
   w ← w ⊕ (111111 ≪ 0b0000)    // complement bit b of w
end
```

For DED: One additional parity bit $\tilde{p}$ is added at the bit location 39. For a double-bit error, the syndrome must be nonzero. To compute parity of a received word, if the computer has popcount instruction (§3.3.6), which computes the number of 1s in a word, and hence equal the Hamming weight, the parity-check is computed as

$$p_0 = \text{pop } (\mathbf{w} \oplus 0\text{xAAAAAAAB}) \; \& \; 1;$$
$$p_1 = \text{pop } (\mathbf{w} \oplus 0\text{xCCCCCCCD}) \; \& \; 1;$$
$$\cdots \cdots \cdots .$$

5.5 Hamming (72,64) Code

The syndrome table for the Hamming (72,64) code is given in Table 5.5.1, where locations of each of the 64 data bits $(d_1, d_2, \ldots, d_{64})$ and each one of the eight (parity) check bits $(p_0, p_1, p_2, p_4, p_8, p_{16}, p_{32}, p_{64})$ are mapped in the reverse order. The location of each one of these bits is given by the row and column locations. For example, data bit d_{61} is at column 100 and row 1000, or location 1001000. The XOR operators are used for calculating the parity of each location which can be either even or odd, and this results in seven check bits. For example, the logical equation for first check bit p_0 is

$$p_0 = d_1 \oplus d_2 \oplus d_4 \oplus d_5 \oplus d_7 \oplus d_9 \oplus d_{11} \oplus d_{12} \oplus d_{14} \oplus d_{16} \oplus d_{18} \oplus d_{20} \oplus d_{22} \oplus$$
$$d_{24} \oplus d_{26} \oplus d_{27} \oplus d_{29} \oplus d_{31} d_{33} \oplus d_{35} \oplus d_{37} \oplus d_{39} \oplus d_{41} \oplus d_{43} \oplus d_{45} \oplus d_{47} \oplus$$
$$d_{49} \oplus d_{51} \oplus d_{53} \oplus d_{55} \oplus d_{57} \oplus d_{58} \oplus d_{60} \oplus d_{62} \oplus d_{64}.$$

In the above equation all data bits with a table cell location of 1 as the least significant bit (lsb) are chosen to determine p_0. For the logical equation for the check bit p_1 we choose the second lsb, for p_4 the third lsb, and so on.

Table 5.5.1 Syndrome for No-Bit Error Detection

111	110	101	100	011	010	001	000	
d_{64}	d_{63}	d_{62}	d_{61}	d_{60}	d_{59}	d_{58}	p_{64}	1000
d_{57}	d_{56}	d_{55}	d_{54}	d_{53}	d_{52}	d_{51}	d_{50}	0111
d_{49}	d_{48}	d_{47}	d_{46}	d_{45}	d_{44}	d_{43}	d_{42}	0110
d_{41}	d_{40}	d_{39}	d_{38}	d_{37}	d_{36}	d_{35}	d_{34}	0101
d_{33}	d_{32}	d_{31}	d_{30}	d_{29}	d_{28}	d_{27}	p_{32}	0100
d_{26}	d_{25}	d_{24}	d_{23}	d_{22}	d_{21}	d_{20}	d_{19}	0011
d_{18}	d_{17}	d_{16}	d_{15}	d_{14}	d_{13}	d_{12}	p_{16}	0010
d_{11}	d_{10}	d_9	d_8	d_7	d_6	d_5	p_8	0001
d_4	d_3	d_2	p_4	d_1	p_2	p_1	No Error	0000

CASE 1. If there is no bit error, the check bits match with the computed check bits of the data, and so all syndrome bits are 0s, which points to the

'No Error' location (see Table 5.5.2).

Table 5.5.2 Syndrome for No-Bit Error Detection

111	110	101	100	011	010	001	000	
d_{64}	d_{63}	d_{62}	d_{61}	d_{60}	d_{59}	d_{58}	p_{64}	1000
d_{57}	d_{56}	d_{55}	d_{54}	d_{53}	d_{52}	d_{51}	d_{50}	0111
d_{49}	d_{48}	d_{47}	d_{46}	d_{45}	d_{44}	d_{43}	d_{42}	0110
d_{41}	d_{40}	d_{39}	d_{38}	d_{37}	d_{36}	d_{35}	d_{34}	0101
d_{33}	d_{32}	d_{31}	d_{30}	d_{29}	d_{28}	d_{27}	p_{32}	0100
d_{26}	d_{25}	d_{24}	d_{23}	d_{22}	d_{21}	d_{20}	d_{19}	0011
d_{18}	d_{17}	d_{16}	d_{15}	d_{14}	d_{13}	d_{12}	p_{16}	0010
d_{11}	d_{10}	d_{9}	d_{8}	d_{7}	d_{6}	d_{5}	p_{8}	0001
d_{4}	d_{3}	d_{2}	p_{4}	d_{1}	p_{2}	p_{1}	No Error ←	**0000**

CASE 2. If there is a single bit error, several syndromes have odd parity. For example, if the data bit d_{29} is in error, then the parity bits p_0, p_2, p_{32} will have parity errors. This results in identifying d_{29} as the erroneous bit in the syndrome table (see Table 5.5.3, where the arrows indicate the columns and rows).

Table 5.5.3 Syndrome for One-Bit Error Detection

111	110	101	100	**011**	010	001	000	
d_{64}	d_{63}	d_{62}	d_{61}	d_{60}	d_{59}	d_{58}	p_{64}	1000
d_{57}	d_{56}	d_{55}	d_{54}	d_{53}	d_{52}	d_{51}	d_{50}	0111
d_{49}	d_{48}	d_{47}	d_{46}	d_{45}	d_{44}	d_{43}	d_{42}	0110
d_{41}	d_{40}	d_{39}	d_{38}	d_{37}	d_{36}	d_{35}	d_{34}	0101
d_{33}	d_{32}	d_{31}	d_{30}	$\boxed{d_{29}}$	d_{28}	d_{27}	p_{32}	**0100** ←
d_{26}	d_{25}	d_{24}	d_{23}	d_{22}	d_{21}	d_{20}	d_{19}	0011
d_{18}	d_{17}	d_{16}	d_{15}	d_{14}	d_{13}	d_{12}	p_{16}	0010
d_{11}	d_{10}	d_{9}	d_{8}	d_{7}	d_{6}	d_{5}	p_{8}	0001
d_{4}	d_{3}	d_{2}	p_{4}	d_{1}	p_{2}	p_{1}	No Error	0000

CASE 3. In the case when a double-bit error occurs, the error bit locations are either not marked at all or marked incorrectly. For example, if a double-bit error occurred at d_{29} and d_{23}, the resulting syndrome points to column 111 and row 0111, as shown in Table 5.5.4.

Table 5.5.4 Syndrome for Double-Bit Error Detection

↓

111	110	101	100	011	010	001	000		
d_{64}	d_{63}	d_{62}	d_{61}	d_{60}	d_{59}	d_{58}	p_{64}	1000	
d_{57}	d_{56}	d_{55}	d_{54}	d_{53}	d_{52}	d_{51}	d_{50}	**0111**	
d_{49}	d_{48}	d_{47}	d_{46}	d_{45}	d_{44}	d_{43}	d_{42}	0110	
d_{41}	d_{40}	d_{39}	d_{38}	d_{37}	d_{36}	d_{35}	d_{34}	0101	
d_{33}	d_{32}	d_{31}	d_{30}	$\boxed{d_{29}}$	d_{28}	d_{27}	p_{32}	0100	←
d_{26}	d_{25}	d_{24}	$\boxed{d_{23}}$	d_{22}	d_{21}	d_{20}	d_{19}	0011	←
d_{18}	d_{17}	d_{16}	d_{15}	d_{14}	d_{13}	d_{12}	p_{16}	0010	
d_{11}	d_{10}	d_9	d_8	d_7	d_6	d_5	p_8	0001	
d_4	d_3	d_2	p_4	d_1	p_2	p_1	No Error	0000	

5.5.1 Modified Hamming (72,64) Code. This is a distance 4 code that can correct all single-bit errors and detect all 2-bit errors, with the capability of detecting a high percentage of larger bit errors. Hardware encoding and decoding involves many additions mod 2, and their number is proportional to the number of 1s in the parity-check matrix, with the result that the parity-check matrix has two constraints. This feature makes the manufacturing hardware for encoding and decoding faster, less costly, and more reliable. The following guidelines are required: The total number of 1s in the parity-check matrix should be minimum, and the number of 1s in each row of the parity-check matrix should be divisible by the number of rows (Imai [1990]). If the parity-check matrix of an odd-weight column has an odd number of 1s in all its columns, a code with this feature would provide 'good discrimination of even number and odd number of errors by modulo 2 addition of the syndrome bits' [Imai, 1990], and it enables this code to detect a higher percentage of errors affecting more bits than detected by any other Hamming code.

A code is alternatively defined by its parity-check matrix. The 8×72 parity-check matrix **H** of rank 8 for the Hamming (72, 64) code is given by

$$
\begin{bmatrix}
100101001001010010000100001001000100001010101101101111011011001101100000 \\
010010010100100101000100100101000100101110110101011110110111100011000000 \\
011000010110000100100010000100100010010101011110110111101101100100110000 \\
000101100001011000010001011000010001111011100101110111101100001000110000 \\
010101011011110110111101101101100000001001010001000100100000000100100001000 \\
101110110101010101110111000110000000100100110010010010010011001010000100 \\
010101011110111011011110100100110000011000010001001000100100000100100010 \\
111011100101010111011110001000110000000101100110000100010010011000010001
\end{bmatrix}
$$

Here the codewords are the 2^{64} vectors that make up the null space[1] of **H**.

---

[1] The null space of a matrix is defined to be all the vectors X such that $\mathbf{H}\mathbf{X}^T = \mathbf{0}$, where $\mathbf{X}^T$ is the transpose of $\mathbf{X}$ and $\mathbf{0}$ is the $m \times 1$ zero matrix.

The encoding is carried by the 64×72 generator matrix $\mathbf{G}$, which is given below, where, to save space, we have partitioned it into four parts:

$$
\mathbf{G}_1 = \begin{bmatrix}
1001010010010100100010000100100 0 \\
0100100101001001010001001001010 0 \\
0110000101100001001000100001001 0 \\
0001011000010110000100010110000 1
\end{bmatrix};
$$

$$
\mathbf{G}_2 = \begin{bmatrix}
0101010110111011011111011011011 00 \\
1011101101010101011110111000110 0 \\
0101010111101110110110111101001 0011 \\
1110111001010101110111100010001 1
\end{bmatrix};
$$

$$
\mathbf{G}_3 = \begin{bmatrix}
0101010110111011011111011011011 00 \\
1011101101010101011110111000110 0 \\
0101010111101110110110111101001 0011 \\
1110111001010101110111100010001 1
\end{bmatrix};
$$

$$
\mathbf{G}_4 = \begin{bmatrix}
1001010010010100100010000100100 0 \\
0100100101001001010001001001010 0 \\
0110000101100001001000100001001 0 \\
0001011000010110000100010110000 1
\end{bmatrix},
$$

where each $\mathbf{G}_i$, $i = 1, 2, 3, 4$ is a 4×32 matrix. Then the generator matrix $\mathbf{G}$ can be written in the transpose form as

$$
\mathbf{G} = \begin{bmatrix} \mathbf{I}_{32} & \mathbf{G}_1 & \mathbf{I}_{32} & \mathbf{G}_3 \\ \mathbf{I}_{32} & \mathbf{G}_2 & \mathbf{I}_{32} & \mathbf{G}_4 \end{bmatrix}^T,
$$

where $\mathbf{I}_{32}$ is the 32×32 identity matrix. The rows of $\mathbf{G}$ are a set of basis vectors for the vector subspace of the code. The 2^{64} codewords are generated by all linear combinations of the rows, and a data vector is encoded in the code when a 1×64 data vector is multiplied by the generator matrix producing the 1×72 codeword. Each row $\mathbf{X}_i$ of the matrix $\mathbf{G}$ must satisfy the condition $\mathbf{X}_i \mathbf{H}^T = \mathbf{0}$, where $\mathbf{0}$ is a 64×8 zero matrix. This means that $\mathbf{H}\mathbf{G}^T = \mathbf{0}$, which leads to the methods for deriving generator matrices from parity-check matrices, or vice versa. This situation is trivial for systematic codes.

The scheme for decoding and correcting a single-bit error is as follows: Multiply the received vector $(\mathbf{w})$ by $\mathbf{H}^T$ and if the result (the syndrome) is zero, then the received word is a codeword and either no error has occurred or an undetectable error has occurred. As before, the syndrome method also locates the bit where the correctable error has occurred and corrects it. A single error correcting code has the property that the transpose of the syndrome matches one of the columns of the parity-check matrix, and the number of this matching column is the bit location in the codeword that is in error; the error is corrected by flipping this bit.

5.5.2 Encoding of Hamming (72,64) Code. Table 5.5.1 shows the generated check bits for the 64 data bits (DBs) numbered from 1 through 64.

Table 5.5.1 Parity Check Bits for Hamming (72,64) SEC-DED Code

DB	p_1	p_2	p_4	p_8	p_{16}	p_{32}	p_{64}	$\tilde{p}$	DB	p_1	p_2	p_4	p_8	p_{16}	p_{32}	p_{64}	$\tilde{p}$
1	y	y						y	33	y	y	y			y		y
2	y		y					y	34				y		y		y
3		y	y					y	35	y			y		y		y
4	y	y	y					y	36		y		y		y		y
5	y			y				y	37	y	y		y		y		y
6		y		y				y	38			y	y		y		y
7	y	y		y				y	39	y		y	y		y		y
8			y	y				y	40		y	y	y		y		y
9	y		y	y				y	41	y	y	y	y		y		y
10		y	y	y				y	42					y	y		y
11	y	y	y	y				y	43	y				y	y		y
12	y				y			y	44		y			y	y		y
13		y			y			y	45	y	y			y	y		y
14	y	y			y			y	46			y		y	y		y
15			y		y			y	47	y		y		y	y		y
16	y		y		y			y	48		y	y		y	y		y
17		y	y		y			y	49	y	y	y		y	y		y
18	y	y	y		y			y	50				y	y	y		y
19				y	y			y	51	y			y	y	y		y
20	y			y	y			y	52		y		y	y	y		y
21		y		y	y			y	53	y	y		y	y	y		y
22	y	y		y	y			y	54			y	y	y	y		y
23			y	y	y			y	55	y		y	y	y	y		y
24	y		y	y	y			y	56		y	y	y	y	y		y
25		y	y	y	y			y	57	y	y	y	y	y	y		y
26	y	y	y	y	y			y	58	y						y	y
27	y					y		y	59		y					y	y
28		y				y		y	60	y	y					y	y
29	y	y				y		y	61			y				y	y
30			y			y		y	62	y		y				y	y
31	y		y			y		y	63		y	y				y	y
32		y	y			y		y	64	y	y	y				y	y

5.6 Hsiao Code

This is an optimal minimum odd-weight-column single-error-correction, double-error-detection (SEC-DED) (n, k) code that provides an improvement upon the modified Hamming SEC-DED code. It was implemented by Hsaio [1970] and resulted in simplification of the hardware design.

The concept is based on the corollary to Peterson's theorem [Peterson 1961:33], which follows.

Theorem 5.6.1. *A code that is the null space of a matrix* **H** *has minimum weight (and minimum distance) at least* w *iff every combination of* $w - 1$ *or fewer columns of* **H** *is linearly independent.*

Based on this result, the minimum weight for a successful SEC-DED code is 4, which implies that three or fewer columns of the parity-check **H**-matrix are linearly independent. Thus, the conditions that the columns of the **H**-matrix must satisfy are as follows: (i) There are no all 0-columns; (ii) every column is distinct; and (iii) every column contains an odd number of 1s, which implies odd weights. The conditions (i) and (ii) give a distance-3 code, and condition (iii) guarantees that the code so generated has distance 4.

These conditions and the hardware construction require that the total number of 1s in each row of the **H**-matrix must correspond to the number of logic levels[2] that are necessary to generate the check bit or syndrome bit of that row. Let t_j denote the total number of 1s in the j-th row of the **H**-matrix and let c_j and s_j denote the check bit and the syndrome bit, respectively, specified by the j-th row. Then

$$l_{c_j} = \lceil \ln(t_j - 1) \rceil, \quad \text{and} \quad l_{s_j} = \lceil \ln t_j \rceil, \quad j = 1, 2, \ldots, m, \qquad (5.6.1)$$

where l_{c_j} (l_{s_j}) are the number of logic levels required to generate c_j (s_j) if only a v-input modulo-2 adder is used, and m is the total number of parity-check bits, and the v-input remains fixed for a given circuit. Hence, the minimum t_j is desirable for minimizing l_{s_j}. The SEC-DED code thus constructed always has fewer 1s in the **H**-matrix than the traditional Hamming SEC-DED codes described earlier.

The parity-check matrix **H** for a given (n, k) code must have the following constraints:

(i) Every column must have an odd number of 1s, so that all column vectors are of odd weight;

(ii) the total number of 1s in the **H**-matrix must be a minimum; and

---

[2] A logic level is determined by the time required for a signal to pass one transistor gate.

(iii) the number of 1s in each row of the **H**-matrix must be equal, or as close as possible, to the average number which is the total number of 1s in **H** divided by the number of rows, i.e., the total number of 1s in **H** is equal to the product of the average number and the number of parity bits. For such a code we have

$$2^{m-1} = m + k, \quad \sum_{j=0}^{m} \binom{m}{j} = 2^m,$$

which holds for an (n, k) code, $m = n - k$ and $n = 2^{m-1}$. Then

$$\sum_{\substack{j=1 \\ j \, odd}}^{\leq m} \binom{m}{j} = \frac{1}{2} \cdot 2^m = 2^{m-1} = m + k. \tag{5.6.2}$$

Thus, this code gives the same number of check bits as the standard Hamming SEC-DED code. The **H**-matrix is constructed as follows:

(i) All $\binom{m}{j}$ weight-1 columns are used for m check-bit location;

(ii) If $\binom{m}{3} \geq k$, select k weight-3 columns out of all possible $\binom{m}{j}$ combinations.

(iii) If $\binom{m}{3} < 3$, select all $\binom{m}{3}$ weight-3 columns. The left-over columns are then first selected from among all $\binom{m}{3}$ weight-5 columns, and so on, until all k columns have been specified.

(iv) If the codeword length $n = m+k$ is exactly equal to $\sum_{\substack{j=1 \\ j \, odd}}^{\leq m} \binom{m}{j}$ for some odd $j \leq m$, then each row of the **H**-matrix will have the following number of 1s:

$$\frac{1}{m} \sum_{\substack{j=1 \\ j \, odd}}^{\leq m} j \binom{m}{j} = 1 + \binom{m-1}{2} + \cdots + \binom{m-1}{j-1}, j \geq 1 \, odd. \tag{5.6.3}$$

(v) If $n \neq \sum_{\substack{j=0 \\ j \, odd}}^{m} \binom{m}{j}$, then the selection of $\binom{m}{j}$ cases of 1s in each row must be made arbitrarily close to the average number shown in Table 5.6.1, where the following notation is used: $j/\binom{m}{j}$ means that j out of all possible $\binom{m}{j}$ combinations are taken.

Table 5.6.1 Structure of **H** and **G**

n	k	m	Structure of **H**	Structure of **G**	Total # of 1s in **H** + **G**	Average # of 1s in Each Row of **H**
8	4	4	$4/\binom{4}{1} + 4/\binom{4}{3}$	$\binom{4}{1}$	20	5
13	8	5	$8/\binom{8}{3}$	$\binom{5}{1}$	29	5.8
14	9	5	$9/\binom{5}{3}$	$\binom{5}{1}$	32	6.4
15	10	5	$10/\binom{5}{3}$	$\binom{5}{1}$	35	7
16	11	5	$3/\binom{5}{3} + 4/\binom{5}{5}$	$\binom{5}{1}$	40	8
22	16	6	$16/\binom{6}{3}$	$\binom{6}{1}$	54	9
26	20	6	$20/\binom{6}{3}$	$\binom{6}{1}$	66	11
30	24	6	$20/\binom{6}{3} + 4/\binom{6}{5}$	$\binom{6}{1}$	86	14.3
47	40	7	$35/\binom{7}{3} + 9/\binom{7}{5}$	$\binom{7}{1}$	157	22.4
55	48	7	$35/\binom{7}{3} + 13/\binom{7}{5}$	$\binom{7}{1}$	177	25.3
72	64	8	$56/\binom{8}{3} + 8/\binom{8}{5}$	$\binom{8}{1}$	216	27
80	72	8	$56/\binom{8}{3} + 16/\binom{8}{5}$	$\binom{8}{1}$	256	32
88	80	8	$56/\binom{8}{3} + 24/\binom{8}{5}$	$\binom{8}{1}$	296	37
96	88	8	$56/\binom{8}{3} + 32/\binom{8}{5}$	$\binom{8}{1}$	336	42
104	96	8	$56/\binom{8}{3} + 40/\binom{8}{5}$	$\binom{8}{1}$	376	47
112	104	8	$56/\binom{8}{3} + 48/\binom{8}{5}$	$\binom{8}{1}$	416	52
120	112	8	$56/\binom{8}{3} + 56/\binom{8}{5}$	$\binom{8}{1}$	456	57
128	120	8	$56/\binom{8}{3} + 56/\binom{8}{5} + 8/\binom{8}{7}$	$\binom{8}{1}$	512	64

This table shows that the **H**-matrix of the above (8,4) code has 4 columns of single 1s and 4 columns of three 1s each column; and **G** (under bits $c_1, \ldots, c_4$) is a 4×4 identity matrix. Similarly, the (22,16) code has 6 columns of single 1s (check bits) and 16 columns of three 1s each in the parity-check matrix, and **G** is a 6×6 identity matrix.

Example 5.6.1. The matrix $\mathbf{H} + \mathbf{G}^T$ with parity-check matrix for 4 data bits and 4 check bits for SEC-DED (8,4) code is given by

$$
\begin{array}{c}
\quad 0\ \ 1\ \ 2\ \ 3\ \ 4\ \ 5\ \ 6\ \ 7\ \ c_1\ c_2\ c_3\ c_4 \\
\begin{array}{c} 1 \\ 2 \\ 3 \\ 4 \end{array}
\left[\begin{array}{cccccccccccc}
1 & 0 & 1 & 0 & 1 & 0 & 1 & 0 & 1 & 0 & 0 & 0 \\
0 & 1 & 1 & 0 & 0 & 1 & 1 & 0 & 0 & 1 & 0 & 0 \\
0 & 0 & 0 & 1 & 1 & 1 & 1 & 0 & 0 & 0 & 1 & 0 \\
0 & 0 & 1 & 0 & 1 & 1 & 0 & 1 & 0 & 0 & 0 & 1
\end{array}\right]
\begin{array}{l} 5\ 1s \\ 5\ 1s \\ 5\ 1s \\ 5\ 1s \end{array}
\end{array}
$$

A comparison with the matrix **H** for Hamming (8,4) code given in §5.2 shows a different approach from the above code which has 5 average number of 1s in each row, and total number of 1s is 20 (= 5×4). If a 3-way XOR gate is used ($v = 3$), the check bits and syndrome bits can be generated in two levels, which is a better and more cost-effective situation as compared to the traditional Hamming (8,4) code where three levels would be required to use the same kind of XOR gate.

Example 5.6.2. For SEC-DED (13,8) code the parity-check matrix for 8 data bits and 5 check bits, that is, $\mathbf{H} + \mathbf{G}^T$, is as follows:

$$
\begin{array}{c}
\begin{array}{cccccccccccccc}
 & 0 & 1 & 2 & 3 & 4 & 5 & 6 & 7 & c_1 & c_2 & c_3 & c_4 & c_5 \\
\end{array} \\
\begin{array}{c}
1 \\ 2 \\ 3 \\ 4 \\ 5
\end{array}
\begin{bmatrix}
1 & 0 & 0 & 1 & 1 & 0 & 1 & 1 & 1 & 0 & 0 & 0 & 0 \\
1 & 1 & 0 & 0 & 0 & 1 & 0 & 1 & 0 & 1 & 0 & 0 & 0 \\
1 & 1 & 1 & 1 & 0 & 0 & 1 & 0 & 0 & 0 & 1 & 0 & 0 \\
0 & 1 & 1 & 1 & 1 & 1 & 0 & 0 & 0 & 0 & 0 & 1 & 0 \\
0 & 0 & 1 & 0 & 1 & 1 & 1 & 1 & 0 & 0 & 0 & 0 & 1
\end{bmatrix}
\begin{array}{c}
6\,1s \\ 5\,1s \\ 6\,1s \\ 6\,1s \\ 6\,1s
\end{array}
\end{array}
$$

All these Hsiao SEC-DED codes can be successfully applied to the cache level of a memory hierarchy, in which a syndrome decoder and the check-bit generator circuit inside the memory chip are used for correction of a single-bit error. To obtain a memory chip with high degree of reliability, low cost, and greater memory capacity, additional cells are needed for testing bits. An SEC in the cache level is in fact a correction of single error at parallel transfer and data memorizing. The Hsiao code is very effective because of its properties mentioned above. These properties with column vectors of odd weight of 1s make the speed of the encoding/decoding circuit optimal. To illustrate this point we use the (13,8) code described above for the Tag part of the cache memory. This code has $m = 5$ control (parity) bits, denoted by $p_1, p_2, p_3, p_4, p_5 \equiv \hat{p}$, and $d = 8$ data bits $d_1\, d_2\, d_3\, d_4\, d_5\, d_6\, d_7\, d_8$; thus, the total number of bits is $n = 13$, and k and d satisfy the relation : $2^m > d + m + 1$. Although Hsiao (12,8) code works nicely for an SEC only, yet 5 control bits in the (13,8) code work better since the last bit is used in DED.

A typical codeword $\mathbf{c}$ used in this code has the following form:

$$\mathbf{c} = \{p_1\, p_2\, d_1\, p_3\, d_2\, d_3\, d_4\, p_4\, d_5\, d_6\, d_7\, d_8\, p_5\} = \{c_1, c_2, \dots, c_{13}\}.$$

The control bits are calculated with the following parity equations:

$$
\begin{aligned}
p_1 &= d_1 \oplus d_4 \oplus d_5 \oplus d_7 \oplus d_8, \\
p_2 &= d_1 \oplus d_2 \oplus d_6 \oplus d_8, \\
p_3 &= d_1 \oplus d_2 \oplus d_3 \oplus d_4 \oplus d_7, \\
p_4 &= d_2 \oplus d_3 \oplus d_4 \oplus d_5 \oplus d_6, \\
p_5 &= d_3 \oplus d_5 \oplus d_6 \oplus d_7 \oplus d_8.
\end{aligned}
\tag{5.6.4}
$$

Based on these, the following syndrome equations are used for decoding a received word:

$$
\begin{aligned}
s_1 &= p_1 \oplus d_1 \oplus d_4 \oplus d_5 \oplus d_7 \oplus d_8, \\
s_2 &= p_2 \oplus d_1 \oplus d_2 \oplus d_6 \oplus d_8,
\end{aligned}
$$

$$s_3 = p_3 \oplus d_1 \oplus d_2 \oplus d_3 \oplus d_4 \oplus d_7, \qquad\qquad (5.6.5)$$

$$s_4 = p_4 \oplus d_2 \oplus d_3 \oplus d_4 \oplus d_5 \oplus d_6,$$

$$s_5 = p_5 \oplus d_3 \oplus d_5 \oplus d_6 \oplus d_7 \oplus d_8.$$

This SEC-DED code is applied to cache Tag memory having a capacity of 16MB, as follows: When the cache Tag is read, it reads the data bits $\{d_1, \ldots, d_8\}$ as well as the control bits $\{p_1, \ldots, p_5\}$. The XOR gates implement Eq (5.6.4) to generate new control bits $\{p_1^* p_2^* p_3^* p_4^* p_5^*\}$ from data bits that were already read from the cache Tag. For example, the control bit p_1^* is determined from the equation $p_1^* = d_1 \oplus d_4 \oplus d_5 \oplus d_7 \oplus d_8$, which is implemented by using 4 XOR gates with 2 inputs, situated on 3 levels (see Novac et al. [2007]). The remaining control bits $p_2^*, \ldots, p_5^*$ are determined similarly. Finally, these so generated control bits $\{p_1^*, \ldots, p_5^*\}$ are compared with the control bits $\{p_1, \ldots, p_5\}$ read from the cache Tag, with 2 input XOR gates. As a result the following syndrome equations are obtained: $s_j = p_j \oplus p_j^*$, $j = 1, 2, 3, 4, 5$. Then 5 NOT gates are connected on each syndrome line, and 8 AND gates with 5 inputs that construct the syndrome decoder with the following equations:

$$d_1^* = s_1 \cdot s_2 \cdot s_3 \cdot \overline{s_4} \cdot \overline{s_5}, \qquad d_2^* = \overline{s_1} \cdot s_2 \cdot s_3 \cdot s_4 \cdot \overline{s_5},$$

$$d_3^* = \overline{s_1} \cdot \overline{s_2} \cdot s_3 \cdot s_4 \cdot s_5, \qquad d_4^* = s_1 \cdot \overline{s_2} \cdot s_3 \cdot s_4 \cdot \overline{s_5},$$

$$d_5^* = s_1 \cdot \overline{s_2} \cdot \overline{s_3} \cdot s_4 \cdot s_5, \qquad d_6^* = \overline{s_1} \cdot s_2 \cdot \overline{s_3} \cdot s_4 \cdot s_5,$$

$$d_7^* = s_1 \cdot \overline{s_2} \cdot s_3 \cdot \overline{s_4} \cdot s_5, \qquad d_8^* = s_1 \cdot s_2 \cdot \overline{s_3} \cdot \overline{s_4} \cdot s_5.$$

The output of the AND gates is then connected to an input of an XOR gate. To correct the data bit in error, 8 XOR gates with 2 inputs are used, which is achieved as follows:

$$d_{1\,\text{corrected}} = d_1 \oplus d_1^*, \quad d_{2\,\text{corrected}} = d_2 \oplus d_2^*,$$

$$d_{3\,\text{corrected}} = d_3 \oplus d_3^*, \quad d_{4\,\text{corrected}} = d_4 \oplus d_4^*,$$

$$d_{5\,\text{corrected}} = d_5 \oplus d_5^*, \quad d_{6\,\text{corrected}} = d_6 \oplus d_6^*,$$

$$d_{7\,\text{corrected}} = d_7 \oplus d_7^*, \quad d_{8\,\text{corrected}} = d_8 \oplus p_d^*. \qquad (5.6.6)$$

Thus, this code is implemented with 3 shift registers and one XOR gate.

Example 5.6.3. For SEC-DED (22,16) code the parity-check matrix for 16 data bits and 6 check bits, that is, $\mathbf{H} + \mathbf{G}$, is as follows, where the zero bit is represented by a blank:

$$
\begin{array}{c}
\quad\; 1\;\,2\;\,3\;\,4\;\,5\;\,6\;\,7\;\,8\;\,9\;10\;11\;12\;13\;14\;15\;16\;c_1\;c_2\;c_3\;c_4\;c_5\;c_6
\end{array}
$$

	1	2	3	4	5	6	7	8	9	10	11	12	13	14	15	16	c_1	c_2	c_3	c_4	c_5	c_6	
1	1	1	1	1	1	1			1			1		1									9 1s
2	1	1	1				1	1	1	1			1					1					9 1s
3	1			1		1	1	1				1	1	1		1							9 1s
4				1	1		1	1	1		1	1	1							1			9 1s
5	1		1			1	1	1	1	1											1		9 1s
6		1	1	1	1			1			1	1			1							1	9 1s

This matrix is constructed for the row $n = 22$ of Table 5.6.2. It has 16 columns corresponding to 16 possible combinations of 3-out-of-6 for the **H**-matrix, and 6 columns corresponding to the six possible combinations of 1-out-of-6 for the **G**-matrix. Thus, the total number of 1s in the above matrix is equal to $6 + (3 \times 16) = 54$, and the average number of 1s in each row is equal to $54/6 = 9$. This means that if a 3-way XOR gate is used ($v = 3$), the check bits and syndrome bits can be generated in two levels, which is a better and more cost-effective situation as compared with the traditional Hamming (22,16) code where three levels would be required to use the same kind of XOR gate. The hardware layout of the two-level encoder and decoder is available in Hsiao [1970:399].

In this code a 16-bit dataword and 6-bit parity bits are stored in the memory as follows:

Table 5.6.2 A 16-Bit Dataword

Bit Location	1	2	3	4	5	6	7	8	9	10	11	12	13	14	15	16	17	18	19	20	21	22
Bit Number	0	1	2	3	4	5	6	7	8	9	10	11	12	13	14	15	16	17	18	19	20	21
Data/Parity Bit	p_1	p_2	d_1	p_4	d_2	d_3	d_4	p_8	d_5	d_6	d_7	d_8	d_9	d_{10}	d_{11}	p_{16}	d_{12}	d_{13}	d_{14}	d_{15}	d_{16}	$\widehat{p}$

where $\widehat{p}$ is an extra parity bit to detect noncorrectable double errors.

The five parity bits $p_1, \dots, p_{16}$ are constructed as follows:

$$p_1 = d_1 \oplus d_2 \oplus d_4 \oplus d_5 \oplus d_7 \oplus d_9 \oplus d_{11} \oplus d_{12} \oplus d_{14} \oplus d_{16},$$

$$p_2 = d_1 \oplus d_3 \oplus d_4 \oplus d_6 \oplus d_7 \oplus d_{10} \oplus d_{11} \oplus d_{13} \oplus d_{14},$$

$$p_4 = d_2 \oplus d_3 \oplus d_4 \oplus d_8 \oplus d_9 \oplus d_{10} \oplus d_{11} \oplus d_{15} \oplus d_{16},$$

$$p_8 = d_5 \oplus d_6 \oplus d_7 \oplus d_8 \oplus d_9 \oplus d_{10} \oplus d_{11},$$

$$p_{16} = d_{12} \oplus d_{13} \oplus d_{14} \oplus d_{15} \oplus d_{16}.$$

These parity bits are created for single error detection and correction. One extra parity bit $\widehat{p}$, which is an overall parity bit at location 22 and is comprised of XOR-ing all the data bits $(d_1, \dots, d_{16})$ and the five parity bits $(p_1, \dots, p_{16})$, detects double errors. The syndrome is created by XOR-ing the parity bits read out of the memory with the newly created set of parity bits from the data stored in memory, as in the case of Hsiao (13,8) code discussed in Example 5.6.2; it provides a single-bit error correction where the value of the syndrome indicates the incorrect single bit location. Thus, if the syndrome is zero and $\hat{p} = 0$, then there is no error; if the syndrome is nonzero and $\hat{p} = 1$, then there is a single correctable error; if syndrome is nonzero and $\hat{p} = 0$, then there a noncorrectable double error; and finally if the syndrome is zero and $\hat{p} = 1$, then there is a parity error, i.e., the extra parity $\hat{p}$ is in error and can be corrected.

Example 5.6.4. The parity-check matrix $\mathbf{H}$ for the (39,32) SEC-DED code is (see §5.4):

$$
\mathbf{H}_1 =
\begin{matrix}
1 \\ 2 \\ 3 \\ 4 \\ 5 \\ 6 \\ 7
\end{matrix}
\begin{bmatrix}
1111111100000010000100101000001110000000 \\
0000100111111111001001001000010000100000 \\
0001000000010000111111110011011000100000 \\
0010001000100101100000000111111100001000 \\
0110010101001001000011110110100000000100 \\
1000011010001110111110000000010000000010 \\
1101100011110000010000010101000100000001
\end{bmatrix}
\begin{matrix}
15 \\ 15 \\ 15 \\ 15 \\ 15 \\ 14 \\ 14
\end{matrix}
$$

Notice that the numbers of 1s in rows 6 and 7 is 14.

Example 5.6.5. Parity-check matrix for the modified Hamming (72, 64) code is given in §5.5.1. For the SEC-DED (72, 64) code the parity-check matrix is:

$$\mathbf{H}_2 =$$

$$
\begin{matrix}
1 \\ 2 \\ 3 \\ 4 \\ 5 \\ 6 \\ 7 \\ 8
\end{matrix}
\begin{bmatrix}
111111110000111100001111000011000110100010001000100010001000000000010000000 \\
111100001111111100000000111100110110010001000100010001000100010001000000000100000 \\
001100001111000011111111100001110000000100010001000100010001001100100000100000 \\
110011110000000011100001111111110000000010001000100010001000101100000010000 \\
011010001000100110001000100000001111111100001110000000011110011000001000 \\
011001000100010001000100010000001111000011111111000011110000011111000000100 \\
000000010001000100010001000100110110011110000000011111110000111100000010 \\
000000010001000100010001000101100011000011110000111100001111111110000001
\end{bmatrix}
\begin{matrix}
27 \\ 27 \\ 27 \\ 27 \\ 27 \\ 27 \\ 27 \\ 27
\end{matrix}
$$

which has 27 1s in each row. Since each code is based on the assumption that the occurrence of multiple errors is statistically independent, these two versions are compared for a single-error correction and double-error detection. There are cases where triple errors yield syndrome patterns outside the columns of the code's parity-check matrix $\mathbf{H}$, which is the left part of the entire matrix. The SEC-DED (72,64) code is capable of correctly detecting a triple error because it will give a syndrome pattern that coincides with another column, which will force the decoder to perform a miscorrection. Since the probability of this miscorrection, denoted by $\mathfrak{p}_3$, is defined by

$\mathfrak{p}_3 = 4w(4)/\binom{72}{3}$, where $w(4)$ is the number of codewords of weight 4, it is obvious that among all possible $\binom{72}{3}$ triple error patterns, the number of miscorrection cases is $4w(4)$. Moreover, among all possible error patterns there are $4w(4)$ cases that will yield zero syndrome and so they cannot be detected. If we define the probability of undetected quadruple errors by $\mathfrak{p}_4 = w(4)/\binom{72}{4}$, and note that the probability of having one bit in error is very low, usually of the order of 10^{-5}, we can assume that the probability of detecting a large number of multiple errors is exceedingly small. The two versions of (72,64) code are compared in the following table for $\mathfrak{p}_3$ and $\mathfrak{p}_4$ for preassigned values of $w(4)$. Although both versions provide good results for detecting quadruple errors, the overall performance of the modified Hamming (72,64) code cannot be ignored (see Hsiao [1970:400]).

Code Used	$w(4)$	$\mathfrak{p}_3$ (%)	$\mathfrak{p}_4$ (%)
Modified Hamming (72,64)	11,313	75.88	98.9
SEC-DED (72,64): Version 1	8,392	56.28	99.19
SEC-DED (72,64): Version 2	8,408	56.39	99.18

5.6.1 Parity Encoder and Decoder. This encoder consists of XOR gates and a bit-error generator implemented in the lookup tables. A block diagram is shown in Figure 5.6.1, where an optional pipelined register is included in the circuit to provide better performance. The check bits together with the associated data bits are written in the memory as shown in Table 5.5.1 and both of them are read simultaneously during a memory read. This circuit contains XOR switches in FPGA lookup tables (LUTs) and detects any errors (single, double, triple, or none) introduced during a read/write session between the FPGA[3] and memory.

The decoder, shown in Figure 5.6.2, consists of three blocks: (i) syndrome generation; (ii) syndrome lookup table (LUT) and mask generation; and (iii) data generation. For the first block, the incoming 64 data bits (DBIN) together with the 8 parity (check) bits (PBIN) pass through the XOR gates to generate the 8-bit syndrome $s_1, s_2, \ldots, s_8$. Thus, for example,

$$s_1 = \text{DBIN1} \oplus \text{DBIN2} \oplus \cdots \oplus \text{DBIN64} \oplus \text{PBIN1},$$

and others are generated similarly. The next stage is to use these syndromes, one at a time, to look for the type of error and its location. An optional pipeline stage is again added in the circuit for better performance.

The syndrome lookup table (SLUT, or simply LUT) and mask generation are used as follows: For correcting a single-bit error, a 64-bit correction mask is created for each one of the 64 data bits, where each bit of this mask is

---

[3] Field-programmable gate array (FPGA) is an integrated (semi-conductor) device designed to be configured by the customer or designer after manufacturing.

generated based on the outcome of the syndrome from the previous stage. If all bits of the mask turn out to be zero, then no error is detected. In the case of the detection of a single-bit error, the corresponding mask conceals all bits except the one that is erroneous.

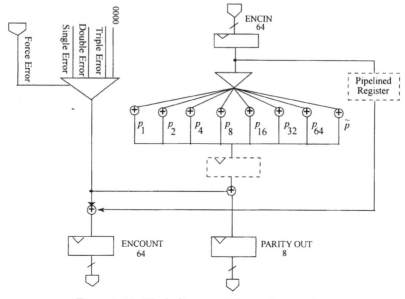

Figure 5.6.1. Block diagram of the parity encoder.

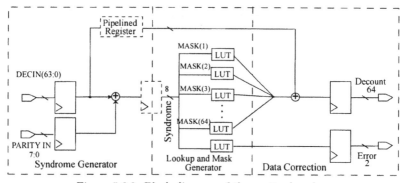

Figure 5.6.2. Block diagram of the parity decoder.

Subsequently the mask is XOR-ed with the original data, which reverses (i.e., corrects) the error bit and yields the correct state. If a double-bit error is detected, all mask bits become zero. The error type and the corresponding correction mask are created during the same state.

In the data correction block, the mask is XOR-ed together with the original incoming data (DBINs) to flip the error bit to the correct bit, if necessary. When there are no single-bit errors or double-bit errors, all mask bits become zero, and consequently all original incoming data bits (DNINs) pass through the error correcting process without any change in the original data. A design for each of the encoder and the decoder is given in Figure 5.6.1 and Figure 5.6.2, respectively. More details on the design and error diagnostics of this code can be found in Tam [2006].

5.7 Product Notes

The IBM Stretch Model 7030, manufactured in 1961, was the first computer that used a (72, 64) SEC-DED Hamming code, which is not a perfect code (Lin and Costello [1983]). The next-generation IBM machine was the Harvest, Model 7950, manufactured in 1962; it had 22-track tape drives with a (22,16) SEC-DED code. On modern machines, Hamming codes are not generally used, as these machines are equipped with some logical or electrical controls to minimize the depth of the parity-check trees and make them all the same length. Such computers do not use the Hamming technique of SEC, but instead depend on a hardware lookup.

There is no error checking system on modern personal computers (PCs), some of which may be equipped with a simple parity check, but server-class computers generally have ECC at the SEC-DED level. The modern construction of ECC memories is more complicated as the server memories generally have 16 or 32 information bytes (128 or 256 bits) checked as a single ECC codeword. For example, a DRAM chip may store 2, 3, or 4 bits in physically adjacent locations, and ECC is carried out on 4, 8, or 16 characters. Modern servers use ECC at different levels of cache memory, main memory, and non-memory areas, for example, buses.

5.8 Uses of Hamming Codes

Hamming codes are used in wireless communications, for example, wireless broadband with high error rate, which needs error correction rather than error detection. As seen above, these codes check each parity-check bit and record even or odd parity of every bit it covers. Any one-bit error in the data leads to an error in the check bit.

RAM subsystems also use Hamming codes with small codewords, so that they can easily decode and correct errors at high speeds. Hamming codes often ensure detection in fixed-length strings of multiple-bit errors, although detection of more than one bit error requires larger number of bits. Some examples of single-bit error correction are: (i) 4-bit data; requires 3 bits for ECC (8 entry table); (ii) 8-bit data: requires 5 bits for ECC or 1 bit for parity; (iii) 11-bit data: requires 4 bits for ECC (16 entry table); (iv) 16-bit data:

requires 6 bits for ECC or 2 bits for parity; (v) 32-bit data: requires 7 bits for ECC or 4 bits for parity; (vi) 64-bit (8 bytes) data: requires 8 bits for ECC and parity; and (vii) 128-bit (16 bytes) data: requires 9 bits for ECC or 16 bits for parity. Notice that an increase in the size of data bits requires fewer numbers of bits for ECC.

Hamming (72, 64) code has been used in the ECC module in Virtex-II, Virtex-II Pro, Virtex-4, or Virtex-5™ devices. The design detects and corrects all single-bit errors and detects double-bit errors in the data (see Tam [2006]). The modified Hamming (72, 64) code has an application in error correction on DRAM memory in personal computers to operate at high speeds; details for its hardware implementation are available in Croswell [2000]. An optimized Hamming (72, 64, 4) code has U.S. Patent # 3,623,155. The SEC-DED-S4ED code was devised by Shigeo Kaneda at Nippon Telegraph and Telephone; it is a Hamming (72,64) code that, besides correcting a single-bit error and detecting a double-bit error, also detects an error involving three or four bits as long as they belong to the same 4-bit nibble of the 72-bit word fetched from memory. "Anyone who has looked at a 72-pin memory stick, and found there were fewer than 36 chips on it, or a 168-pin DIMM, and found there were fewer than 72 chips on it, will appreciate the relevance of this property" (see http://www.quadibloc.com/crypto/mi0602.htm).

6

Bounds on Codes

There are some bounds that are placed on the size of error correcting codes. Some of the well-known bounds are discussed in the following sections.

6.1 Definitions

Let n be fixed, and let M denote the size of a code C (i.e., the number of codewords in C), and d the minimum distance between the codewords. For example, (i) for the code $C = \{00, 01, 10, 11\}$: $n = 2, M = 4; d = 1$; (ii) for the code $C = \{000, 011, 101, 110\}$: $n = 3, M = 4, d = 2$; and (iii) for the code $C = \{00000, 00111, 11111\}$: $n = 5, M = 3, d = 2$. Then for a q-ary (n, M, d) code, the *transmission rate* (also known as the information rate) is defined by $R(C) = \dfrac{\log_q M}{n}$, and the *relative minimum distance* by $\delta(C) = (d - 1)/n$. Note that this distance is also defined by d/n, but neater formulas are obtained by defining it as $(d - 1)/n$. For the binary repetition code $C = \{\underbrace{00\ldots0}_{n}\,\underbrace{11\ldots1}_{n}\}$, which is an $(n, 2, n)$ code, we find that

$$R(C) = \frac{\log_2 2}{n} = \frac{1}{n} \to 0 \text{ and } \delta(C) = \frac{n - 1}{n} \to 0 \text{ as } n \to \infty.$$

Recall that for the family of Hamming $(n, M, d) = (2^r - 1, 2^{n-r}, 3)$ codes C for all integers $r \geq 2$, we have $R(C) = \dfrac{\log_2 2^{n-r}}{n} = \dfrac{2^r - 1 - r}{2^r - 1} \to 1 \text{ as } r \to \infty$,

and $\delta(C) = \dfrac{2}{n} \to 0$ as $n \to \infty$.

DEFINITION 6.1.1. Let A be an alphabet of size q $(q > 1)$, and let n and d be given. Then $A_q(n, d)$ denotes the largest possible size of M such that

$$A_q(n, d) = \max\{M : \text{ there exists an } (n, M, d) \text{ code over } A\}.$$

Note that the quantity $A_q(n, d)$ depends only on n and d, and is independent of A.

DEFINITION 6.1.2. For a given $q = p$, where p is prime, and given values of n and d, let the quantity $B_q(n, d)$ denote the largest possible size q^k defined by

$$B_q(n, d) = \max\{q^k : \text{there exists an } (n, k, d) \text{ code over } F_q\}.$$

The following properties hold between $A_q(n, d)$ and $B_q(n, d)$:

(i) $B_q(n, d) \leq A_q(n, d) \leq q^n$ for all $1 \leq d \leq n$;

(ii) $B_q(n, d) = A_q(n, d) = q^n$; and

(iii) $B_q(n, n) = A_q(n, n) = q$.

Proof for these properties can be found in Ling and Xing [2004].

DEFINITION 6.1.3. The extended code for any code C over F_q, denoted by $\tilde{C}$, is defined as

$$\tilde{C} = \left\{ \left(c_1, \ldots, c_n, -\sum_{i=1}^{n} c_i \right) : (c_1, \ldots, c_n) \in C \right\}.$$

For $q = 2$, the extra coordinate $-\sum_{i=1}^{n} c_i = -\sum_{i=1}^{n} c_i$ that is added to the codeword is called the extra parity check bit. The following result holds:

Theorem 6.1.1. For an (n, M, d) code C over F_q, the extended code $\tilde{C}$ is an $(n + 1, M, d')$ code over F_q, where $d \leq d' \leq d + 1$. If C is linear, so is $\tilde{C}$, and in that case

$$\tilde{H} = \begin{bmatrix} & & 0 \\ & H & \vdots \\ & & 0 \\ \cdots & \cdots & \cdots \\ 1 \cdots 1 & \cdots & 1 \end{bmatrix}$$

is the parity-check matrix of $\tilde{C}$, where H is the parity-check matrix of C.

Two examples are: (a) $C = \{000, 110, 011, 101\}$ is a binary (3,2,2) code, but $\tilde{C} = \{0000, 1100, 0110, 1010\}$ is a (4,2,2) code; and (b) $C = \{000, 111, 011, 100\}$ is a binary (3,2,1) code, while $\tilde{C} = \{0000, 1111, 0110, 1001\}$ is a binary (4,2,2) code.

Theorem 6.1.2. If d is odd, then a binary (n, M, d) code exists iff a binary $(n + 1, M, d + 1)$ code exists, and in that case $A(n + 1, d + 1) = A(n, d)$. Moreover, if d is even, then $A(n, d) = A(n - 1, d - 1)$.

Determining the lower bounds is an important part of coding theory and extensive work has been done in this area. A couple of notable websites are http://www.eng.tau.ac.il/~litsyn/tableand/index.html, and http://www.win.tue.nl/~aeb/voorlincod.html. We will now present some well-known bounds and their properties, which are useful in the construction of linear codes.

6.2 Sphere-Packing Bound

The basic notion of sphere packing a vector space is discussed in §3.6. It will now be elaborated and used to determine bounds on linear codes.

DEFINITION 6.2.1. Let A be an alphabet of size $q > 1$. For any vector $\mathbf{a} \in A^n$ and any integer $r \geq 0$, a sphere of radius r and center $\mathbf{a}$ is denoted by $B_A(\mathbf{a}, r)$ and defined as the set $\{\mathbf{b} \in A^n : d(\mathbf{a}, \mathbf{b}) \leq r\}$.

Theorem 6.2.1. *For an integer $q \geq 1$, an integer $n > 0$ and an integer $r \geq 0$ define*

$$\mathbf{v}_q^n(r) = \begin{cases} \sum_{i=0}^{r} \binom{n}{i} (q-1)^i & \text{if } 0 \leq r \leq n, \\ q^n & \text{if } r \geq n. \end{cases}$$

Then a sphere of radius r in A^n contains exactly $\mathbf{v}_q^n(r)$ vectors, where A is an alphabet of size q.

PROOF. For a fixed vector $\mathbf{a} \in A^n$, determine the number of vectors $\mathbf{b} \in A^n$ such that $d(\mathbf{a}, \mathbf{b}) = m$, i.e., find the number of vectors in A^n at a distance m from $\mathbf{a}$. Since the number of ways to choose m coordinates out of n is $\binom{n}{m}$, there are exactly $q-1$ choices for each coordinate to be in $\mathbf{b}$. Thus, the total number of vectors at a distance m from $\mathbf{a}$ is given by $\binom{n}{m}(q-1)^m$, and then the theorem holds for $0 \leq r \leq n$. ∎

The sphere-packing bound is given by the following result.

Theorem 6.2.2. *For an integer $q > 1$ and integers n and d such that $1 \leq d \leq n$,*

$$\frac{q^n}{\sum_{i=0}^{d-1} \binom{n}{i} (q-1)^i} \leq A_q(n, d). \tag{6.2.1}$$

PROOF. Let $C = \{\mathbf{c}_1.\mathbf{c}_2, \ldots, \mathbf{c}_M\}$ be an optimal (n, M, d) code over A, where $|A| = q$; then $M = A_q(n, d)$. Because C has maximal size, then there will be no word in A^n that is at a distance at least d from every codeword of C, because if there were such a word, it would be an element of C and contained in an (n, M, d) code. Hence, for every vector $\mathbf{x} \in A^n$ there is at least one codeword $\mathbf{c}_i \in C$ such that $d(\mathbf{x}, \mathbf{c}_i) \leq d - 1$, i.e., $x \in B_A(\mathbf{c}_i, d-1)$; that is,

$$A^n \subseteq \bigcup_{i=1}^{M} B_A(\mathbf{c}_i, d-1). \tag{6.2.2}$$

Since $|A^n| = q^n$ and $\left| B_A(\mathbf{c}_i, d-1) \right| = \mathbf{v}_q^n(d-1)$ for any $i = 1, \ldots, M$, we have $q^n \leq M \mathbf{v}_q^n(d-1)$, which implies that $\dfrac{q^n}{\mathbf{v}_q^n(d-1)} \leq M = A_q(n, d)$. Note that since the spheres $B_A(\mathbf{c}_i, d-1)$, $i = 1, \ldots, M$, cover A^n, this bound is called sphere-covering or sphere-packing bound. ∎

The lower bounds for $A_q(n, d)$ given by the sphere-packing bounds for $d = 3, 5, 7$ are given in Tables 6.2.1, 6.2.2, and 6.2.3.

Table 6.2.1 Bounds for $A_2(n, 3)$

n	Sphere-Packing	Hamming	Singleton	Plotkin
3	2	2	2	2
4	2	3	4	2
5	2	5	8	4
6	3	9	16	8
7	5	16	32	16
8	7	28	64	
9	12	51	128	
10	19	93	256	
11	31	170	512	
12	52	315	1024	

Table 6.2.2 Bounds for $A_2(n, 5)$

n	Sphere-Packing	Hamming	Singleton	Plotkin
5	2	2	2	2
6	2	2	4	2
7	2	4	8	2
8	2	6	16	4
9	2	11	32	6
10	3	18	64	12
11	4	30	128	24
12	6	51	256	

Example 6.2.1. Using sphere-packing bounds, show that $A_2(5, 3) = 2$. By (6.2.1) we have $A_2(5, 3) \geq 2$. But since $A_2(5, 3) = A_2(4, 2)$ by Theorem 6.1.2, we must now show that $A_2(4, 2) \leq 2$. Let C be a binary $(4, M, 2)$ code and let $(\mathbf{c}_1, \mathbf{c}_2, \mathbf{c}_3, \mathbf{c}_4)$ be a codeword in C. Since $d(C) = 2$, the other codewords in C must be of the form

$$(\mathbf{c}_1^*, \mathbf{c}_2^*, \mathbf{c}_3, \mathbf{c}_4), \ (\mathbf{c}_1, \mathbf{c}_2^*, \mathbf{c}_3^*, \mathbf{c}_4), \ (\mathbf{c}_1, \mathbf{c}_2, \mathbf{c}_3^*, \mathbf{c}_4^*), \ (\mathbf{c}_1^*, \mathbf{c}_2, \mathbf{c}_3, \mathbf{c}_4^*),$$

where $\mathbf{c}_i^* = \begin{cases} 1 & \text{if } \mathbf{c}_i = 0 \\ 0 & \text{if } \mathbf{c}_i = 1 \end{cases}$. But since there are no pairs of these codewords that have distance 2 (or more), only one of them can be included in C. Hence,

$M \leq 2$, or $A_a(4,2) \leq 2$. This proves that $A_2(5,3) = A_2(4,2) = 2$. ∎

Table 6.2.3 Bounds for $A_2(n,7)$

n	Sphere-Packing	Hamming	Singleton	Plotkin
7	2	2	2	2
8	2	2	4	2
9	2	3	8	2
10	2	5	16	2
11	2	8	32	4
12	2	13	64	4

6.3 Johnson Bound

This bound, presented by S. Johnson [1962], considers a code C of length n such that $C \subset F_2^n$. Let d be the minimum distance of C, i.e., $d = \min\limits_{\mathbf{x,y} \in C, \mathbf{x} \neq \mathbf{y}} d(\mathbf{x,y})$, where $\mathbf{x, y} \in C$ are two codewords. Let $C_q(n,d)$ be the set of all q-ary codes of length n and minimum distance d, and let $C_q(n,d,\mu)$ denote the set of codes in $C_q(n,d)$ such that every element has exactly μ nonzero entries. Then the largest size of a code of length n and minimum distance d, denoted by $A_q(n,d)$, is defined by $A_q(n,d) = \max\limits_{C \in C_q(n,d)} |C|$, where $|C|$ is the number of elements in C. Similarly, if $A_q(n,d,\mu)$ defines the largest size of a code in $C_q(n,d,\mu)$, then $A_q(n,d,\mu) = \max\limits_{C \in C_q(n,d,\mu)} |C|$.

The Johnson bound for $A_q(n,d)$ is given by

CASE 1. If $d = 2t+1$, then

$$A_q(n,d) \leq \cfrac{q^n}{\sum\limits_{i=0}^{t} \binom{n}{i}(q-1)^i + \cfrac{\binom{n}{t+1}(q-1)^{t+1} - \binom{d}{t}A_q(n,d,d)}{A_q(n,d,t+1)}}. \quad (6.3.1)$$

CASE 2. If $d = 2t$, then

$$A_q(n,d) \leq \cfrac{q^n}{\sum\limits_{i=0}^{t} \binom{n}{i}(q-1)^i + \cfrac{\binom{n}{t+1}(q-1)^{t+1}}{A_q(n,d,t+1)}}. \quad (6.3.2)$$

Further, the Johnson bound for $A_q(n,d,\mu)$ is given by

CASE 1. If $d > 2\mu$, then $A_q(n,d,\mu) = 1$.

CASE 2. If $d < 2\mu$, then define ν as follows: If d is even, then $\nu = d/2$, and if d is odd, then $\nu = (d+1)/2$. Then

$$A_q(n,d,\mu) \leq \left\lfloor \frac{nq^*}{\mu} \left\lfloor \frac{(n-1)q^*}{\mu-1} \left\lfloor \cdots \left\lfloor \frac{(n-\mu+\nu)q^*}{\mu-\nu} \right\rfloor \cdots \right\rfloor \right\rfloor \right\rfloor, \quad (6.3.3)$$

where $q^* = q - 1$. A code for which equality holds in (6.3.1) is known as a nearly perfect code.

6.4 Gilbert–Varshamov Bound

This bound, also known as the *Gilbert–Shannon–Varshamov bound* (or GSV bound), determines the bound on the parameters of a linear code. Let $A_q(n,d)$ denote the maximum possible size of a q-ary code[1] C of length n and Hamming weight d. Then

$$A_q(n,d) \geq \frac{q^n}{\displaystyle\sum_{i=0}^{d-1} \binom{n}{i} (q-1)^i}. \qquad (6.4.1)$$

If q is a prime power, then the above bound becomes $A_q(n,d) \geq q^k$, where k is the greatest integer such that

$$q^k \geq \frac{q^n}{\displaystyle\sum_{i=0}^{d-2} \binom{n-1}{i} (q-1)^i}. \qquad (6.4.2)$$

PROOF. If the bound (6.4.1) holds, then there exists an $(n-k) \times n$ matrix H over F_q, such that every $d-1$ columns of H are linearly independent. To see this is true, the matrix H is constructed as follows: Let $\mathbf{c}_j$, $j = 2, \ldots, n$, denote the j-th column of H. Let $\mathbf{c}_1$ be any nonzero vector in F_q^{n-k}, let $\mathbf{c}_2$ be any vector not in the span of $\mathbf{c}_1$, and let, for any j, $2 \leq j \leq n$, $\mathbf{c}_j$ be any vector not in the linear span of $d-2$ (or fewer) of the vectors $\mathbf{c}_1, \ldots, \mathbf{c}_{j-1}$. Then the number of vectors in the linear span of $d-2$ or fewer of the vectors $\mathbf{c}_1, \ldots, \mathbf{c}_{j-1}$, $2 \leq j \leq n$, is

$$\sum_{i=0}^{d-2} \binom{j-1}{i} (q-1)^i \leq \sum_{i=0}^{d-2} \binom{n-1}{i} (q-1)^i < q^{n-k}. \qquad (6.4.3)$$

These inequalities imply that the vector $\mathbf{c}_j$ can always be found and the $(n-k) \times n$ matrix H properly constructed by the above procedure. Any $d-1$ of the columns of this matrix H are, therefore, linearly independent. The bound (6.4.2) follows from inequality (6.4.3). ∎

Note that the null space of the matrix H is a linear code over F_q of length n, distance at least d, and dimension at least k. Any desired linear code can be obtained from it by turning to a k-dimensional subspace. The Gilbert–Varshamov bound has been very useful for an infinite family of linear codes, in particular for certain codes with sufficiently large q. More on this bound can be found in MacWilliams and Sloane [1977].

---

[1] By a q-ary code we mean a code $C \subset F_q^n$. Note that $|F_q^n| = q^n$.

A corollary to the bound (6.4.2) is: For a prime $q > 1$ and $2 \leq d \leq n$,

$$B_q(n,d) \geq q^{n-\lceil \log_2(\mathbf{v}_q^{n-1}(d-2)+1) \rceil} \geq \frac{q^{n-1}}{\mathbf{v}_q^{n-1}(d-2)}. \qquad (6.4.4)$$

This can be proved by setting $k = n - \lceil \log_2(\mathbf{v}_q^{n-1}(d-2)+1 \rceil$. Then the bound (6.4.2) is satisfied, which implies that there exists a linear q-ary (n,k,d) code with $d \leq d_1$. Now, change $d_1 - d$ fixed linear coordinates to 0, which yields a linear q-ary (n, k, d) code. Further, recall that $B_q(n,d) \geq q^k$ (Property 6.1.2.(i)).

6.5 Hamming Bound

This bound, also known as the sphere-packing bound or the volume bound, places a limit on the parameters of a block code. Practically, it means packing balls in the Hamming metric into the space of all possible codewords. A message that is transmitted is composed in the form of a sequence of codewords, each of which belongs to an alphabet of q letters. Whereas each codeword contains n letters, the original message is of length $m < n$. Before transmission, the message is converted into an n-letter codeword by the encoder (an encoding algorithm), then transmitted over a noisy channel, and finally decoded by the decoder (receiver) through a decoding algorithm that interprets a received codeword as the valid n-letter string. Mathematically, there exist exactly q^m possible messages of length m, where each message can be regarded as a vector of m elements. The encoder converts this m-element vector into an n-element vector $(n > m)$. Although exactly q^m valid codewords are possible, one or more of the q^n erroneous codewords can be received, as the noisy channel may have distorted one or more letters during transmission. Let the maximum possible size of a q-ary block code of length n and minimum Hamming distance d be denoted by $A_q(n,d)$. Then

$$A_q(n,d) \leq \frac{q^n}{\sum\limits_{i=0}^{t} \binom{n}{i} (q-1)^i}, \qquad (6.5.1)$$

where $t = \lfloor d - 1/2 \rfloor$ is the number of errors that can be detected and corrected by the code.

PROOF. Let $C = \{\mathbf{c}_1, \ldots, \mathbf{c}_M\}$ be an optimal (n, M, d) code over A, where $|A| = q$. Then $M = A_q(n,d)$, and the packing spheres $S_A(\mathbf{c}_i, t)$ are disjoint. Thus, $\cup_{i=1}^M S_A(\mathbf{c}_i, t) \leq A^n$, where the union is that of disjoint sets. Since $|A^n| = q^n$ and $|S_A(\mathbf{c}_i, t)| = \mathbf{v}_q^n(t)$ for any $i = 1, \ldots, M$, we have $M\mathbf{v}_q^n(t) \leq q^n$, which implies that $A_q(n,d) = M \leq q^n/\mathbf{v}_q^n(t)$. ∎

Because error correcting codes consist of codewords with their disjoint decoding regions, larger decoding regions imply that more errors can be tolerated. However, if the number of codewords is larger, it implies that a larger number of distinct messages can be reliably transmitted. The dichotomy between these two desirable properties can be resolved only if the total volume of the decoding regions does not exceed the total volume of the vector space. This restriction is provided by the *sphere-packing bound* or *volume bound*, which can be stated as

$$q^n \geq q^k \sum_{i=0}^{t} \binom{n}{i} (q-1)^i. \qquad (6.5.2)$$

Note that the equality in the above two bounds holds for a perfect code, that is, a q-ary code that attains the Hamming (or sphere-packing) bound is called a *perfect code*. In other words, a perfect code has $\dfrac{q^n}{\sum_{i=0}^{t} \binom{n}{i}(q-1)^i}$
codewords. A perfect code may be regarded as a code in which the balls of Hamming radius $t = \lfloor (d-1)/2 \rfloor$ centered on codewords exactly fill out the space. In practice, most error correcting codes are imperfect; they leave 'holes' between the disjoint decoding regions. A quasi-perfect code is one in which the balls of Hamming radius t centered on codewords are disjoint and the balls of radius $t+1$ cover space, though not disjointly, i.e., they have some possible overlaps. Examples of perfect codes include binary repetition codes of odd length, codes that have only one codeword, and codes that use the entire field F_q^n. Such codes are sometimes called *trivial perfect codes*. However, it was proved in 1973 (see Hill [1988]) that any nontrivial perfect code over a prime-power alphabet contains parameters of a Hamming code or a Golay code (see Chapter 7). Similarly, a code in which the balls of Hamming radius t centered on codewords are disjoint, while those of radius $t+1$ cover the space, with some possible overlaps.

Example 6.5.1. It can be shown that the notion of perfect Hamming codes as equality in the right side of the inequality (6.5.1) is a generalization of the Hamming rule. In fact, the Hamming rule applies to the case $d = 3$ and the code has 2^k codewords, where k is the number of data bits, and $m = n - k$ the number of parity bits. The right-side inequality in (6.5.1) with $A_2(n, d) = 2^k$ and $d = 3$ becomes

$$2^k \leq \frac{2^n}{\binom{n}{0} + \binom{n}{1}} = \frac{2^n}{1+n}.$$

If we replace n by $m + k$ on the right side, we get $2^k \leq \dfrac{2^{m+k}}{1 + m + k}$, which after canceling 2^k on both sides gives $2^m \geq 1 + m + k$, or $2^{n-k} \geq 1 + n$, which is the Hamming rule. ∎

Example 6.5.2. Find the value of $A_2(n,d)$ if $n = d$, and show that for odd n, these codes are perfect. It is easy to see that the size of an (n,n) code is 2 since the code must contain a string of arbitrary bits and its one's complement. These codes are perfect for odd n if we can show that they achieve the upper bound in the inequality (6.5.1). To do this, note that an n-bit binary integer may be regarded as representing a choice from n objects where a 1-bit means to choose the corresponding object and a 0-bit means not to choose it. This means that there are 2^n ways to choose from 0 to n objects from n objects, which is equal to $\sum_{i=0}^{n} \binom{n}{i} = 2^n$. For odd n, as the index i ranging from 0 to $(n-1)/2$ will cover half of the terms in this sum, and since $\binom{n}{i} = \binom{n}{n-i}$ (by symmetry), we have $\sum_{i=0}^{(n-1)/2} \binom{n}{i} = 2^{n-1}$. Thus the upper bound in (6.5.1) is 2, and the (n,n) code achieves this upper bound for odd n. ∎

6.6 Singleton Bound

This bound is defined as follows: A linear code C satisfies the bound $A_q(n,d) \leq q^{n-d+1}$ for any integers $q > 1$, $n > 0$ and d such that $1 \leq d \leq n$. In particular, every linear (n,k,d) code C satisfies $k + d \leq n + 1$.

PROOF. To prove the first part, consider a linear (n, M, d) code C on an alphabet A of size q ($|A| = q$), where $M = A_q(n,d)$, and delete the last $d - 1$ coordinates from all the codewords of C. Because $d(C) = d$, the remaining codewords (after deleting the $d - 1$ coordinates) of length $n - d + 1$ are all distinct. Since the maximum codewords of length $n - d + 1$ are q^{n-d+1}, we have $A_q(n,d) = M \leq q^{n-d+1}$. The latter part follows from the first because of the definition $A_q(n,d) \geq q^k$. ∎

A code C for which $k + d = n + 1$ is called a *minimum distance separable code* (MDS code). Such codes, when available, are in some sense the best possible ones.

The MDS codes have the following properties: Let C be a linear (n, k, d) code over F_q, and let **G** and **H** be a generator and a parity-check matrix, respectively, for C. Then (i) C is an MDS code; (ii) every set of $n - k$ elements of **H** is linearly independent; (iii) every set of k columns of **G** is linearly independent; and (iv) the dual code $C^\perp$ is an MDS code.

To prove these properties, note that (i) and (ii) follow directly from the fact that if $d(C) = d$, then any $d - 1$ columns of **H** are linearly independent and **H** has d linearly independent columns; so take $d = n - k + 1$, and these two properties follow. Similarly properties (ii) and (iv) are equivalent, because **G** is also a parity-check matrix for $C^\perp$. Also, **H** is a generator matrix for $C^\perp$, so the length of $C^\perp$ is n and its dimension $n - k$. Now, to prove that $C^\perp$ is an MDS code, we must show that the minimum distance d' is $k + 1$. For this purpose, first let $d' \leq k$. Then there exists a code word $\mathbf{c} \in C^{\perp}$ that has at most k nonzero entries, and hence $n - k$ zero coordinates, where any permutation of these coordinates does not change the weight of the codewords. Thus, we can assume that the last $n - k$ coordinates of $\mathbf{c}$ are 0s. If we write $\mathbf{H}$ as $\mathbf{H} = (X|H')$, where X is some $(n - k) \times k$ matrix, and H' is a $(n - k) \times (n - k)$ matrix with linearly independent columns; then H' is invertible, and the rows of H' are linearly independent. If we use the 0-linear combination of the (linearly independent) rows of H', we will obtain 0s in all the last $n - k$ coordinates. Hence, the codeword $\mathbf{c} = \mathbf{0}$ (all zero entries). Therefore, $d' \geq k + 1$, and in view of the Singleton bound it implies that $vd' = k + 1$. Since $(C^{\perp})^{\perp} = C$, the property (iv) implies (i). $\blacksquare$

An MDS code C over F_q is said to be *trivial* iff C satisfies one of the following conditions: (i) $C = F_q^n$; (ii) C is equivalent to the code generated by $\mathbf{1} = \{1, 1, \ldots, 1\}$; or (iii) C is equivalent to the dual code generated by $\mathbf{1}$. Otherwise, the code C is nontrivial. The MDS codes for $q = 2$ are trivial. However, an interesting family of nontrivial MDS codes are the generalized Reed-Muller codes discussed in Chapter 12.

If C_1 and C_2 are two codes of length n each and if there exists a permutation p in the symmetric group S_n for which $(\mathbf{c}_1, \ldots, \mathbf{c}_n)$ is in C_1 iff $(\mathbf{c}_{p(1)}, \ldots, \mathbf{c}_{p(n)})$ is in C_2, then C_1 and C_2 are said to be *permutation equivalent*. In general, if there is an $n \times n$ monomial matrix $M : F_q^n \longrightarrow F_q^n$ that maps C_1 isomorphically on to C_2, then C_1 and C_2 are said to be *equivalent*. A linear code is permutation equivalent to a code that is in standard form.

6.7 Plotkin Bound

This bound is applicable to codes that have large d compared to n, although it is useful only in cases of a smaller range of values of d.

PLOTKIN BOUND. Let $q > 1$ be an integer, and let the integers n and d be such that $\nu n < d$, where $\nu = 1 - 1/q$. Then

$$A_q(n, d) \leq \left\lfloor \frac{d}{d - \nu n} \right\rfloor. \tag{6.7.1}$$

PROOF. Consider an (n, M, d) code C over an alphabet A of size q (i.e., $|A| = q$). For two different codewords $\mathbf{c}$ and $\mathbf{c}'$ of C, where $\mathbf{c} = \{c_1, \ldots, c_n\}$ and $\mathbf{c}' = \{c_1', \ldots, c_n'\}$, let D denote the double sum

$$D = \sum_{\mathbf{c} \in C} \sum_{\mathbf{c}' \in C} d(\mathbf{c}, \mathbf{c}').$$

Because $d \leq d(\mathbf{c}, \mathbf{c}')$ for $\mathbf{c}, \mathbf{c}' \in C$, $\mathbf{c} \neq \mathbf{c}'$, then

$$M(M - 1)d \leq D. \tag{6.7.2}$$

Again, let $\mathfrak{A}$ be an $M \times n$ array with rows composed of the M codewords in C, and let $n_{i,a}$, $i = 1, \ldots, n$ and $a \in \mathfrak{A}$, denote the number of entries in the i-th column of $\mathfrak{A}$ (these entries are equal to a). Then $\sum_{a \in \mathfrak{A}} n_{i,a} = M$ for every i, $i = 1, \ldots, n$, and

$$D = \sum_{i+1}^{n} \left(\sum_{\mathbf{c} \in C} \sum_{\mathbf{c}' \in C} d(\mathbf{c}, \mathbf{c}') \right) = \sum_{i=1}^{n} \sum_{a \in \mathfrak{A}} n_{i,a}(M - n_{i,a}) = M^2 n - \sum_{i=1}^{n} \sum_{a \in \mathfrak{A}} n_{i,a}^2.$$

If we apply the Cauchy-Schwarz inequality[2] with $m = q$ and $a_1 = \cdots = a_m = 1$, we find that

$$D \leq \sum_{i=1}^{n} q^{-1} \left(\sum_{a \in \mathfrak{A}} n_{i,a}^2 \right)^2 = M^2 \nu n. \tag{6.7.3}$$

Then the Plotkin bound (6.7.1) follows from (6.7.2) and (6.7.3). ∎

For $q = 2$, let C be a binary code of length n, i.e., $C \subseteq F_2^n$, and let d be the minimum distance of C, i.e. $d = \min_{\mathbf{c}_1, \mathbf{c}_2 \in C, \mathbf{c}_1 \neq \mathbf{c}_2} d(\mathbf{c}_1, \mathbf{c}_2)$, where $d(\mathbf{c}_1, \mathbf{c}_2)$ is the Hamming distance between the codewords $\mathbf{c}_1$ and $\mathbf{c}_2$. Let $A_2(n, d)$ denote the maximum number of possible codewords of length n and minimum distance d. Then the following four (Plotkin) bounds hold:

(i) If d is even and $2d > n$, then $A_2(n, d) \leq 2 \left\lfloor \dfrac{d}{2d - n} \right\rfloor$;

(ii) If d is odd and $2d > n$, then $A_2(n, d) \leq 2 \left\lfloor \dfrac{d + 1}{2d - n + 1} \right\rfloor$;

(iii) If d is even, then $A_2(2d, d) \leq 4d$;

(iv) If d is odd, then $A_2(2d + 1, d) \leq 4(d + 1)$.

Examples of the Plotkin bounds are given in Tables 6.2.1, 6.2.2, and 6.2.3.

6.8 Griesmer Bound

The Griesmer bound (Griesmer [1960]) is used for both linear and nonlinear codes. It places a bound on the length of binary codes of dimension k and minimum distance d. Let C be a linear (n, k, d) code, and let $\mathbf{c}$ be a codeword in C with $\min \text{wt}(\mathbf{c}) = d$.

GRIESMER BOUND. Let C be a linear q-ary (n, k, d) code where $k \geq 1$. Then

---

[2] For two sets of real numbers $\{a_1, \ldots, a_m\}$ and $b_1, \ldots, b_m\}$, this inequality is given by $\left(\sum_{j=1}^{m} a_j b_j \right)^2 \leq \left(\sum_{j=1}^{m} a_j^2 \right) \left(\sum_{j=1}^{m} b_j^2 \right)$, where equality holds for $a_j = c b_j$, c constant (see Abramowitz and Stegun [1972:11], where it is called the Cauchy inequality.).

$$n \geq \sum_{i=0}^{k-1} \left\lceil \frac{d}{q^i} \right\rceil. \tag{6.8.1}$$

PROOF. Note that the result is true for $k = 1$. Now, when $k > 1$ and $\mathbf{c} \in C$ is a codeword of minimum weight d, then by Theorem 3.3.2 of §3.3.5 the residual code $\mathrm{Res}(C, \mathbf{c})$ is an $(n - 1, k - 1, d')$ code, where $d' \geq \lceil d/q \rceil$. By mathematical induction let us assume that the Griesmer bound is true for $\mathrm{Res}(C, \mathbf{c})$. However, this yields

$$n - d \geq \sum_{i=0}^{k-2} \left\lceil \frac{d'}{q^i} \right\rceil \geq \sum_{i=0}^{k-2} \left\lceil \frac{d}{q^{i+1}} \right\rceil,$$

which implies that the Griesmer bound holds for all $k > 1$. ∎

For a binary linear code, we have $n \geq \sum_{i=0}^{k-1} \lceil d/2^i \rceil$. In the general case for a code over F_q^n, $q > 2$, we have $n \geq \sum_{i=0}^{k-1} \lceil d/q^i \rceil$.

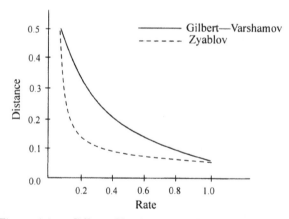

Figure 6.9.1. Gilbert–Varshamov and Zyablov bounds

6.9 Zyablov Bound

Zyablov [1971] proved the following result, which is useful in the construction of concatenated codes (see Chapter 13): Let $0 < R < 1$. Then it is possible to efficiently construct a code of rate R and distance

$$d_Z(R) = \max_{R \leq r \leq 1} \frac{1}{h} \left(1 - \frac{R}{r}\right)(1 - r). \tag{6.9.1}$$

The function $d_Z(R)$ is also called the *Zyablov trade-off curve*. For any $R \in (0,1)$, the value of $d_Z(R)$ is bounded away from 0. This leads to the result that asymptotically good codes of any desired rate $R \in (0,1)$ can be constructed in polynomial time. However, this bound is weaker than the Gilbert–Varshamov bound, as Figure 6.9.1 shows.

6.10 Bounds in F_2^n

Consider an n-dimensional sphere of radius r, and let codewords of Hamming distance equal to or less than r be at the center of the sphere. The question is as to how many codewords can be found in such a sphere of radius r from the word at the center of this sphere. The answer in the case of a (n,k,d) code is given by the binomial coefficient $\binom{n}{r} = \dfrac{n!}{r!\,(n-r)!}$, which is known as the *choice function*. The total number of codewords (points) in a sphere of radius r is equal to $\sum\limits_{i=0}^{r} \binom{n}{i}$, which is the sum of the points in the shells (surfaces) from the center to the outermost shell (surface) of the sphere.

Theorem 6.10.1. *The Hamming bound for a code $C \subset F_2^n$ of length n is given by*

$$A_2(n,d) = \frac{2^n}{\sum\limits_{i=0}^{t} \binom{n}{i}}, \tag{6.10.1}$$

where $|C|$ is the number of elements in C, and $t = \left\lfloor \dfrac{d-1}{2} \right\rfloor$.

PROOF. Since the code C is of length n and minimum distance d, assume that it contains M codewords. Surround each codeword by a sphere, such that all spheres have the same maximal radius and no two spheres have a common point (i.e., they do not touch each other). This radius is equal to $\dfrac{d-1}{2}$ if d is odd, and equal to $\dfrac{d-2}{2}$ if d is even (see Figure 6.10.1). Since each point (codeword) is in at most one sphere, the total number of points in M spheres must be at most equal to the total number of points in the space. Thus,

$$M \sum_{i=0}^{\lfloor (d-1)/2 \rfloor} \binom{n}{i} \le 2^n.$$

Since this inequality holds for any M, the bound (6.10.1) holds for for $M = A_2(n,d)$. ∎

Figure 6.10.1. Maximum radius for codewords in a sphere.

In terms of the space packing geometry (§3.6), this theorem states that the entire space of the field F_{2^m} can be packed by spheres where each sphere contains a valid codeword at its center as well as invalid codewords that are at a distance at most equal to the minimum distance of a code.

A lower bound for $A_2(n,d)$ is provided by (6.3.2) with $q = 2$, that is,

$$A_2(n,d) \geq \frac{2^n}{\sum_{i=0}^{d-1} \binom{n}{i}}. \tag{6.10.2}$$

To prove, note that since the code is of length n and minimum distance d, it has the maximum possible number of codewords, which is equal to $M = A_2(n,d)$. If we surround each codeword with spheres of radius $d-1$, then these spheres must cover all 2^n points (codewords) in the space, some of them possibly overlapping; but if they do not overlap, there would be a point that is at a distance $\geq d$ from all other points, and this is impossible because a point is in fact a codeword.

A strong form of the Gilbert–Varshamov bound that is applicable to linear codes[3] is as follows: $A_2(n,d) \geq 2^k$, where k is the largest integer such that

$$2^k < \frac{2^n}{\sum_{i=0}^{d-2} \binom{n-1}{i}}. \tag{6.10.3}$$

Since the Gilbert–Varshamov bound is a lower bound on linear codes, it is also a lower bound on the unrestricted codes considered in this section, and for large n it is the best known bound for both linear and unrestricted codes. Hence, combining (6.10.2) and (6.10.3) we have

$$\left\lfloor \frac{2^n}{\sum_{i=0}^{d-2} \binom{n-1}{i}} \right\rfloor \leq A_2(n,d) \leq \frac{2^n}{\sum_{i=0}^{\lfloor (d-1)/2 \rfloor} \binom{n}{i}}, \tag{6.10.4}$$

where the greatest integral power of 2 strictly less than its argument is taken. The Gilbert–Varshamov and Hamming bounds on $A_2(n,d)$ are given in Table 6.10.1. The best known bounds on $A_2(n,d)$ are given in Table 6.10.2.

Example 6.10.1. We will show that $A_2(n,d) = 2$ for $d > \dfrac{2n}{3}$. Since the two codewords can be all 0s and all 1s, we obviously have $A_2(n,d) \leq 2$. Let one of these codewords c_1 be all 0s. Then all other codewords must have more than $\dfrac{2n}{3}$ 1s. If the code has three or more additional codewords (other than c_1), then any two of them other than c_1 must have 1s in the same bit locations for more than $\dfrac{2n}{3} - \dfrac{n}{3} = \dfrac{n}{3}$ bit locations, of the form

---

[3] Recall that in a linear code, the sum (XOR) of any two codewords is also a codeword.

$$\underbrace{1111\ldots1111}_{>2n/3}\underbrace{0000\ldots0000}_{<n/3},$$

where the all 1s of the next codeword c_2 are pushed to the left. Now imagine that more than $\dfrac{2n}{3}$ 1s of c_3 are placed to minimize the overlapping of the 1s. Since c_2 and c_3 overlap in more than $\dfrac{n}{3}$ bit locations, they can differ in less than $n - \dfrac{n}{3} = \dfrac{2n}{3}$ bit locations, thus resulting in a minimum distance $< 2n/3$. ∎

Example 6.10.2. Let $m = n - k$. Find $\min m$ that satisfies the Hamming rule: $2^m \geq n + 1$. Since m and k are positive integers, we will treat them as real numbers. The iteration

$$m_0 = 0, \quad m_{i+1} = \log_2(m_i + k + 1), \quad i = 0, 1, \ldots,$$

converges from below very rapidly. In fact, $\lceil m_2 \rceil$ provides the required minimum.

Alternatively, define the *bitsize* as the size of k in bits so that, for example, bitsize(3) = 2, bitsize(4) = 3, and in general, bitsize$(i) = \lceil \log_2(i+1) \rceil = \lfloor \log_2(i) + 1 \rfloor$, where it is assumed that $\log_2(0) = -1$. Then it can be proved that bitsize$(k) \leq k \leq$ bitsize$(k) + 1$ for $k \geq 0$, by computing bitsize(k) for different values of k. ∎

Table 6.10.1 Gilbert–Varshmov and Hamming Bounds on $A_2(n,d)$ [4]

n	$d = 4$	$d = 6$	$d = 8$	$d = 10$	$d = 12$	$d = 14$	$d = 16$
6	4–5	2					
7	8–9	2					
10	32–51	4–11	2–3	2			
13	2^8–315	16–51	2–13	2–5			
16	2^{11}	2^6–270	8–56	2–16	2–6	2–3	2
19	2^{13} – 13797	2^8–1524	16–265	4–2^6	2–20	2–8	2–4
22	2^{16} – 95325	2^{10}–9039	2^6–1342	8–277	4–75	2–25	2–10
25	2^{19} – 671088	2^{12} – 55738	2^8–7216	32–1295	8–302	2–88	2–31
28	2^{22} – 793490	2^{15} – 354136	2^{10} – 40622	2^7 – 6436	16 – 1321	4 – 337	2 – 104

---

[4] A single number in an entry means that the lower and upper bounds given by (6.10.4) are the same. Note that the cases of $(n, d) = (7,3)$, (15,3), and (23,7) are perfect codes, and they achieve the upper bound in (6.10.4).

Example 6.10.3. To show that $A_2(2n, 2d) \geq A_2(n, d)$, note that in a code of length n and minimum distance d if we double up each 1 and each 0 in every codeword, the resulting code will be of length $2n$ and minimum distance $2d$. ∎

Table 6.10.2 Best Known Bounds on $A_2(n, d)$

n	$d = 4$	$d = 6$	$d = 8$	$d = 10$	$d = 12$	$d = 14$	$d = 16$
6	4	2					
7	8	2					
8	16	2	2				
9	20	4	2				
10	40	6	2	2			
11	72	12	2	2			
12	144	24	4	2	2		
13	256	32	4	2	2		
14	512	64	8	2	2	2	
15	2^{10}	128	16	4	2	2	
16	2^{11}	256	32	4	2	2	2
17	2720–3276	256–340	36–37	6	2	2	2
18	5312–6552	512–680	64–72	10	4	2	2
19	10496 –13104	2^{10} –1280	128–142	20	4	2	2
20	20480 –26208	2^{11} –2372	256–274	40	6	2	2
21	36864 –43688	2560–2^{12}	2^9	42–48	8	4	2
22	73728 –87376	2^{12} –6941	2^{10}	64–87	12	4	2
23	147456 –173015	2^{13}–13766	2^{11}	80–150	24	4	2
24	294912 –344308	2^{14} –24106	2^{12}	128–280	48	6	4
25	2^{19} –599184	2^{14} –48008	2^{12} –5477	192–503	52–56	8	4
26	2^{20} –1198368	2^{15} –84260	2^{12} –9672	384–859	64–98	14	4
27	2^{21} –2396736	2^{16} –157285	2^{13}–17768	2^9–1764	128–169	28	6
28	2^{22} –4793472	2^{17} –291269	2^{14} –32151	2^{10} –3200	178 –288	56	8

6.11 Reiger Bound

Burst errors are localized in short intervals rather than at random. They occur on certain communications channels, such as telephone lines and magnetic storage systems. A burst of length $l > 1$ is a binary vector whose elements are confined to l cyclically consecutive locations, where the first and the last locations are nonzero. A linear code C is called an *l-burst-error-correcting code* iff all burst errors of length $\leq l$ are located in distinct cosets of C. In fact, if all the burst errors of length l are located in distinct cosets, then each burst error is determined and corrected by its syndrome. Suppose two distinct burst errors $\mathbf{b}_1$ and $\mathbf{b}_2$ of length l or less lie in the same coset of C. Then the difference $\mathbf{c} = \mathbf{b}_1 - \mathbf{b}_2$ is a codeword. If burst error $\mathbf{b}_1$ is received, then $\mathbf{b}_1$ can be decoded by both $\mathbf{0}$ and $\mathbf{c}$. Let C be a linear (n, k) l-burst-error-correcting code. Then $n - k \geq 2l$ is called the *Reiger bound*, that is,

$$l \leq \left\lfloor \frac{n-k}{2} \right\rfloor. \tag{6.11.1}$$

For more on burst errors, see §12.8 and §13.5.

6.12 Krawtchouk Polynomials

The Krawtchouk polynomials $K_i(x; n)$, $0 \leq i \leq n$, (or simply denoted by $K_i(x)$) are a finite sequence of orthogonal polynomials, which for a given value of q are defined by

$$K_i(x; n) = \sum_{j=0}^{i} (-1)^j \binom{x}{j} \binom{n-x}{i-j} (q-1)^{i-j}, \tag{6.12.1}$$

so that $K_0(x; n) = 1$; $K_1(x; n) = -2x + n$; $K_2(x; n) = 2x^2 - 2nx + \binom{n}{2}$; $K_3(x; n) = \frac{4}{3}x^3 + 2nx^2 - \left(n^2 - n + \frac{2}{3}\right)x + \binom{n}{3}$, and so on. They are a special case of the Meixner polynomials of the first kind (see Nikiforov et al. [1991]). The properties of the polynomials $K_i(x; n)$ are

(i) $\sum_{i=0}^{\infty} K_i(x)z^i = [1 + (q-1)z]^{n-x} (1-z)^x$ for any variable z. The right side is the generating function of these polynomials.

(ii) $K_i(x) = \sum_{j=0}^{i} (-1)^j q^{i-j} \binom{n-i+j}{j} \binom{n-x}{i-j}$.

(iii) The polynomials $K_i(x)$ has the leading coefficient $\dfrac{(-q)^i}{i!}$ and the constant term $K_i(0) = \binom{n}{i}(q-1)^i$.

(iv) (Orthogonality:) $\sum_{j=0}^{n} \binom{n}{j}(q-1)^j K_i(j) = \delta_{il}\binom{n}{i}(q-1)^i q^n$, where δ_{il} is the Kronecker delta defined by $\delta_{il} = \begin{cases} 1 & \text{if } i = l, \\ 0 & \text{otherwise.} \end{cases}$

(v) $(q-1)^j \binom{n}{j} K_i(j) = (q-1)^i \binom{n}{i} K_j(i)$.

(vi) $\sum_{j=0}^{n} K_l(j) K_j(i) = \delta_{il} q^n$.

(vii) $\sum_{i=0}^{j} \binom{n-i}{n-j} K_i(x) = q^j \binom{n-x}{j}$.

(viii) For $q = 2$, we have $K_i(x) K_j(x) = \sum_{m=0}^{n} \binom{n-m}{(i+j+m)/2} \binom{m}{(i-j+m)/2} K_m(x)$.

(ix) A polynomial $f(x)$ of degree r can be represented as $f(x) = \sum_{i=0}^{r} f(i) K_i(x)$,

where $f(i) = q^{-n} \sum_{j=0}^{n} f(j) K_j(i)$. This is known as the Krawtchouk expansion of $f(x)$.

Note that the property (ii) follows easily from (i), which also gives $K_i(0) = \binom{n}{i}(q-1)^i$. The (integral) zeros of these polynomials are combinatorially important, especially in their association with the Hamming codes in which q is an integer. Let $x_{i,1}^n < x_{i,2}^n < \cdots < x_{i,i}^n$ denote the zeros of $K_i(x; n)$. The following results on the zeros of these polynomials is worth mentioning (see Chihara and Stanton [1987]):

(a) The zeros of $K_i(x; n)$ and $K_i(x; n+1)$ interlace, so do those of $K_{i-1}(x; n)$ and $K_i(x; n+1)$, and this interlacing is 'close', that is,

$$x_{i,j}^n < x_{i,j}^{n+1} < x_{i,j+1}^n < x_{i,j+1}^{n+1} \text{ for } j = 1, \dots, i-1,$$
$$x_{i,j}^{n+} < x_{i-1,j}^n < x_{i,j+1}^{n+1} \text{ for } j = 1, \dots, i-1,$$
$$x_{i,j}^n - x_{i,j}^{n+1} < 1 \text{ for } 1 \leq j \leq i \leq n. \tag{6.12.2}$$

(b) For $q = 2$, let $x + 1 < x_2$ be two consecutive zeros of $K_i(x; n)$. Then $x_1 + 2 < x_2$. Note that for $q = 2$, the zeros of $K_n(x : n)$ are $1, 3, \dots, 2n-1$.

(c) Let $x_{j,i}^n(q)$ be the j-th zero of $K_i(x; n)$. Then $x_{j,i}^n(q)$ is an increasing function of q.

(d) If $K_i(j, n) = 0$ for an integer j, then q divides $\binom{n}{i}$. Moreover, if $K_i(j; n) = 0$ for an integer m and $q = 2$, then $K_i(j \pmod{2^m}; n) \equiv 0 \pmod{2^{m+1}}$ for any integer $m \geq 1$.

The integral zeros for $q = 2$ are useful. A trivial set of such zeros is $n/2$ for even n. In general, the integral zeros of degrees 1, 2, and 3 are: (i) $\{1, i, 2i\}$ for $i \geq 1$; (ii) $\{3, i(3i \pm 1)/2, 3i^2 + 3i + 3/2 \pm (i + 1/2)\}$ for $i \geq 2$; and (iii) $\{2, i(i-1)/2, i^2\}$ for $i \geq 3$. The zeros in the set (i) are trivial, and those in (ii) are the pentagonal numbers. For zeros of degree 4, there is a list by Diaconis and Graham [1985] that lists them as follows:

$$(i, x, n) = (4, 1, 8)^*, (4, 3, 8)^*, (4, 5, 8)^*, (4, 7, 8)^*, (4, 7, 11), (4, 10, 17),$$

$(4, 30, 66)$, $(4, 36, 66)$, $(4, 715, 1521)$, $(4, 806, 1521)$, $(4, 7476, 15043)$, $(4, 7567, 15043)$,

where the starred values are trivial. No one is sure if this list is complete. However, for degree 4, Chihara and Stanton [*op. cit.*] have proved that if $K_i(x; n) = 0$ for $q = 2$ and for some integer x, then

(i) If $x \equiv 0 \pmod 4$, then $n \equiv 0, 1, 2, \text{or } 3 \pmod{32}$;

(ii) If $x \equiv 1 \pmod 4$, then $n \equiv 0, 1, 2, 3, \text{or } 8 \pmod{32}$;

(iii) If $x \equiv 2 \pmod 4$, then $n \equiv 0, 2, 3, \text{or } 17 \pmod{32}$;

(iv) If $x \equiv 3 \pmod 4$, then $n \equiv 2, 3, 8, \text{or } 17 \pmod{32}$;

and have found that all nontrivial integral zeros for $q = 2$ and $n \leq 700$ are

$(i, x, n) = (5,14,36), (5,22,67), (5,28,67), (5,133,289), (6,31,67), (6,155,345),$
$(14,47,98), (19,62,132), (23,31,67), (31,103,214), (34,254,514),$
$(61,86,177), (84,286, 576).$

6.13 Linear Programming Bound

The linear programming method, developed by Delsarte [1973], provides one of the best upper bounds on $A_q(n, d)$. It is based on the theory of Krawtchouk polynomials $K_i(x; n)$, defined in §6.12. Since the linear programming bound is an upper bound on $A_q(n, d)$, it is also applicable to nonlinear codes. As such, it depends on the distance between two codewords and not on the weight of each codeword. Let A be an alphabet of size q, and for an (n, M) code C over A let

$$A_j(C) = \frac{1}{M} \left| \{ \mathbf{u}, \mathbf{v} \in C^2 : d(\mathbf{u}, \mathbf{v}) = j \} \right| \text{ for all } 0 \leq j \leq n.$$

Then the sequence $\{ A_j(C) \}_{j=0}^n$ is called the *distance distribution* of C, which depends only on the size q of the code C and not on A. The linear programming bound is obtained if it is developed in the field F_q, so that the proofs for codes over A can be easily carried out for codes in F_q. It is known that $\sum_{j=0}^n A_j(C) K_i(j) \geq 0$ for a q-ary code C of length n, where $0 \leq i \leq n$ are integers (Ling and Xing [2004: 103]).

Theorem 6.13.1. (Linear programming bound) *For a given integer $q > 1$ and positive integers n and d, where $1 \leq d \leq n$, let $f(x) = 1 + \sum_{i=1}^n f(i) K_i(x)$ be a polynomial expansion, where $K_i(x)$ are the Krawtchouk polynomials, such that $f(i) \geq 0$ for $i = 1, \ldots, n$, and $f(j) \leq 0$ for $d \leq j \leq n$. Then $A_q(n, d) \leq f(0)$.*

PROOF. Consider the distance distribution of C. The condition $f(j) \leq 0$

for $d \leq j \leq n$ implies that $\sum_{j=d}^{n} A_j(C)f(i) \leq 0$. Then

$$f(0) = 1 + \sum_{i=1}^{n} f(i)K_i(0) \geq 1 - \sum_{i=1}^{n} f(i) \sum_{j=d}^{n} A_j(C)K_i(j)$$

$$= 1 - \sum_{j=d}^{n} A_j(C) \sum_{i=1}^{n} f(i)K_i(j) = 1 - \sum_{j=d}^{n} A_j(C)\left[f(j) - 1\right]$$

$$\geq 1 + \sum_{j=d}^{n} A_j(C) = M = A_q(n,d). \ \blacksquare$$

Example 6.13.1. To derive the Hamming, Singleton, and Plotkin bounds from the linear programming bound, we proceed as follows:

(i) (Hamming bound) Let $d = 2m + 1$, and $f(x) = \sum_{i=0}^{n} f(i)K_i(x)$, where

$$f(i) = \left\{ \frac{L_m(i)}{\sum_{j=0}^{m}(q-1)^j \binom{n}{j}} \right\}^2 \quad \text{for } i = 0, \dots, n,$$

and $L_m(x) = \sum_{i=0}^{m} K_i(x) = K_m(x - 1; n - 1)$ are the Lloyd polynomials. Obviously, $f(i) \geq 0$ for all $i = 0, 1, \dots, n$, and $f(0) = 1$. Then by property (iv) of the Krawtchouk polynomials (§6.12) we find that $f(0) = 0$ for $d \leq j \leq n$. Hence, by Theorem 6.13.2 we have

$$A_q(n,d) \leq f(0) = \frac{q^n}{\sum_{j=0}^{m}(q-1)^j \binom{n}{j}},$$

which is the Hamming bound (see §6.5).

(ii) (Singleton bound) Let $f(x) = q^{n-d+1} \prod_{j=d}^{n}\left(1 - \frac{x}{j}\right)$. Then using the Krawtchouk expansion of $f(x) = \sum_{i=0}^{n} f(i)K_i(x)$, we find that the coefficients $f(i)$ are given by

$$f(i) = \frac{1}{q^n} \sum_{j=0}^{n} f(j)K_j(i) = \frac{1}{\binom{n}{d-1}q^{d-1}} \sum_{j=0}^{d-1}\binom{n-j}{n-d+1}K_j(i)$$

$$= \frac{\binom{n-i}{d-1}}{\binom{n}{d-1}} \geq 0 \text{ by property (vii)of Krawtchouk polynomials, §6.12.}$$

Thus, $f(0) = 0$ for $d \leq j \leq n$, and by Theorem 6.12.2, $A_q(n,d) \leq f(0) = q^{n-d+1}$, which is the Singleton bound (see §6.6).

(iii) (Plotkin bound) We will derive this bound for $A_2(2l+1, l+1)$, i.e., we have $q = 2, n = 2l + 1$ and $d = l + 1$. Take $f(1) = \dfrac{l+1}{2l+1}$ and $f(2) = \dfrac{1}{2l+1}$. Then

$$f(x) = 1 + \frac{l+1}{2l+1} K_1(x) + \frac{1}{2l+1} K_2(x)$$
$$= 1 + \frac{l+1}{2l+1}(2l+1-2x) + \frac{1}{2l+1}\left[2x^2 - 2(2l+1)x + l(2l+1)\right].$$

It is obvious that $f(i) \geq 0$ for all $i = 1, \ldots, n$, and because $f(x)$ is a quadratic polynomial, $f(j) \leq 0$ for $d = l + 1 \leq j \leq n = 2l + 1$. Then, by Theorem 6.12.2, we have

$$A_2(2l+1, l+1) \leq f(0) = 1 + \frac{l+1}{2l+1}(2l+1) + \frac{1}{2l+1}l(2l+1) = 2(l+1),$$

which is the Plotkin bound (see §6.7). ∎

6.14 Stochastic Bounds for SEC-DED Codes

The iteration schemes of the Hamming and Golay codes which are two of the *symmetric* codes in the sense of Hamming [1950], also known as *symbol* codes in the sense of Golay [1949], involve the data symbols (bits) and check symbols (bits) whose combination is the codeword to be transmitted as a message over communication channels. The probability in the received message must be as low as the receiver decides, i.e., the probability per decoded symbol or for the entire symbol sequence is set by the receiver as low as possible, but the cost of receiving a more reliable message increases as the probability of errors decreases. Since the decoding procedures depend on iteration schemes used in SEC, their performance varies from one code to another. It is known that the Hamming and Golay SEC-DED extended codes perform the best for a binary noisy channel with a small and symmetric error probability.

Consider a binary channel that is used to transmit either a 0 or a 1 with a probability $\mathfrak{p}_0$ for the transmitted symbol to be received in error and a probability $1 - \mathfrak{p}_0 = \mathfrak{q}_0$ for it to be received with no error . It is assumed that the error probabilities for successive symbols are statistically independent. Let the receiver (decoder) divide the received symbol sequences into consecutive blocks of length n_j, $j \geq 1$. Under the independence assumption of successive transmission errors, the distribution of errors in the blocks will be binomial, such that the probability that no errors have occurred in a block is $\mathfrak{p}(0) = (1 - \mathfrak{p}_0)^{n_j}$, whereas a probability $\mathfrak{p}(i)$, defined by

$$\mathfrak{p}(i) = \binom{n_1}{i} \mathfrak{p}_0^i (1 - \mathfrak{p}_0)^{n_j - i}, \quad 1 \leq i \leq n_j, \, j \geq 1, \tag{6.14.1}$$

exists for the case that exactly i errors have occurred. If the expected value of errors $n_j \mathfrak{p}_0$ per received block is small, then a Hamming SEC will produce a smaller average number of errors $n_j \mathfrak{p}_i$ per block after correction. The Hamming SEC-DED code with m of the n locations in a block as parity (check) bits ($m = n - k$) and the remaining k as data bits satisfies the relation

$$m = \lfloor \log_2 n + 1 \rfloor. \tag{6.14.2}$$

The following results are due to Elias [1954], who introduced the concept of blocks of received words.

Theorem 6.14.1. Let $\mathfrak{p}_i$, $i \geq 1$, denote the average probability of errors per symbol after the j-th order check and correction in a SEC-DED code. Let $n_j = k_j + m_j$ denote the number of received words in the j-th block. Then at the j-th order check we have

(i) $\mathfrak{p}_1 < \mathfrak{p}_0$;

(ii) $\mathfrak{p}_1 \leq (n_1 - 1)\mathfrak{p}_0^2 < n_1 \mathfrak{p}_0^2$ for $n_1 \mathfrak{p}_0 \leq 3$;

(iii) $\mathfrak{p}_j \leq (n_j - 1)\mathfrak{p}_{j-1}^2 < n_j \mathfrak{p}_{j-1}^2$ for $n_j \mathfrak{p}_{j-1} \leq 3$, $j \geq 1$;

(iv) $\mathfrak{p}_i < \left(n_i^{2^0} \cdot n_{i-1}^{2^1} \cdots n_{i-j}^{2^j} \cdots n_1^{2^{j-1}} \right) \mathfrak{p}_0^{2^j}$ for $j \geq 1$;

(v) Let t_j denote the factor of symbols used for information (message) at the j-th order check and let E denote the equivocation[5] of the noisy binary channel. Then as $j \to \infty$, we have $t_\infty > 1 - 4E$ subject to the conditions $n_1 \mathfrak{p}_0 = \frac{1}{2}$ and $n_1 \geq 4$.

(vi) Let E_j denote the equivocation per sequence of n_j terms, and let $\mathfrak{q}_j$ be the probability that a check group of n_j digits is in error. Then

$$E_j < \mathfrak{q}_j n_j - \mathfrak{q}_j \log_2 \mathfrak{q}_j - (1 - \mathfrak{q}_j) \log_2 (1 - \mathfrak{q}_j).$$

PROOF. (i) follows from the definition of $\mathfrak{p}_0$ and $\mathfrak{p}_1$ under the independence assumption of successive transmission of errors.

(ii) Since SEC-DED (Hamming) decoder in the first-order correction leaves error-free blocks and corrects only single errors, it will not alter the number of errors when the check digit is even and may increase this number by at most 1 if it is even and greater than 1 when odd. Then the inequality for the expected number of errors per block after the first-order check is

$$n_1 \mathfrak{p}_1 \leq \sum_{\substack{2 \leq i \leq n_1 \\ i\,\text{even}}} i\, \mathfrak{p}(i) + \sum_{\substack{3 \leq i \leq n_1 \\ i\,\text{odd}}} (i+1)\mathfrak{p}(i) \leq \mathfrak{p}(2) + \sum_{i=3}^{n_1}(i+1)\mathfrak{p}(i)$$

$$\leq \sum_{i=0}^{n_1}(i+1)\mathfrak{p}(i) - \mathfrak{p}(0) - 2\mathfrak{p}(1) - \mathfrak{p}(2) \leq 1 + n_1 \mathfrak{p}_0 - \mathfrak{p}(0) - 2\mathfrak{p}(1) - \mathfrak{p}(2),$$

---

[5] In the sense of a formal or informal fallacy, where the slippage from one sense to the next alters the particular conclusion.

which after substitution into Eq (6.14.1) yields $\mathfrak{p}_1 \leq (n_1 - 1)\mathfrak{p}_0^2 < n_1\mathfrak{p}_0^2$, or $n_1\mathfrak{p}_1 \leq n_1(n_1 - 1)\mathfrak{p}_0^2$ for $n_1\mathfrak{p}_0 \leq 3$, which proves (ii). This result shows that the error probability per location can be reduced if n_1 is sufficiently small. The shortest code requires $n_1 = 4$, but (ii) implies that a reduction in errors is possible only if $\mathfrak{p}_0 < \frac{1}{3}$.

(iii) This is a generalization of (ii). At the second-order check ($j = 2$), the receiver (decoder) discards m_1 check digits, which leaves n_2 total received symbols with the reduced error probability $\mathfrak{p}_1$ per location. Note that the error probability after each stage of checking does not change even if some of the check bits are discarded. The receiver using the remaining check bits divides n_2 bits into blocks of k_2 data and m_2 parity bits in this block. This enables the correction of any single bit in the block. The choice of location of n_2 symbols in the block is important at this stage. The simplest method is to take several consecutive first-order blocks of m_1 adjacent check bits as a second-order block, although it may not always work because after the first-order check there may be two or more errors left, which the second-order check may not be able to correct. Thus, the inequality (iii) holds for the error probability per location after the j-th order check.

(iv) The proof depends on the construction of an algorithm for the assumed independence. Let us put each group of $n_1 \times n_2$ successive symbols in a rectangular array where each row of n_1 symbols are first checked by m_1 check bits, and then each column of already checked symbols are checked by m_2 check bits (Figure 6.14.1).

Figure 6.14.1 Rectangular array $n_1 \times n_2$.

The transmitter sends the n_1 information symbols in the first row, computes m_1 check bits and sends them (codewords) before it proceeds to the next row. The process repeats for the second row with n_2 received words, in which the transmitter computes the m_2 check bits for each column and writes them down in the last m_2 rows. The transmission is carried out by transmitting one row at a time, using the first n_1 of the locations in that row for the second-order check, and the last m_1 bits in the row for a first-order check of the second-order check bits. After the second-order check we have the inequality: $\mathfrak{p}_2 < n_2\mathfrak{p}_1^2 < n_2 n_1^2\mathfrak{p}_0^4$. This process continues for the third-order check with n_3 bits to be checked taken from the corresponding location in each of n_3

different $n_1 \times n_2$ rectangles, the n_4 bits in the fourth-order check from the corresponding locations in n_4 such collection of $n_1 \times n_2 \times n_3$ symbols each, and so on, such that the inequality (iii) holds at the j-th order check.

Since $n_1 \geq 4$, we have $n_j \geq 2^{j-1} n_1$, which gives the probability $\mathfrak{p}_j$ at the j-th stage as

$$
\begin{aligned}
\mathfrak{p}_j &< \left(n_1 \cdot 2^{j-1}\right)^{2^n} \cdots \left(n_1 \cdot 2^{k-i}\right)^{2^{i-1}} \cdots \left(n_1 \cdot 2^0\right)^{2^{j-1}} \mathfrak{p}_0^{2^j} \\
&< \frac{1}{n_1} (2 n_1 \mathfrak{p}_0)^{2^j} \cdot 2^{-(j+1)} \to 0 \quad \text{as } j \to \infty \text{ for any } n_1 \mathfrak{p}_0 \leq \tfrac{1}{2}.
\end{aligned}
\tag{6.14.3}
$$

Thus, the error probability vanishes in the limit. The inequality (iv) gives a much weaker approach to the limit value of zero for the threshold value $n_1 \mathfrak{p}_0 = \tfrac{1}{2}$ than for any smaller value of errors in the first-order check.

(v) Since the channel is not completely occupied at each stage, it is very likely that some information can also get through with check bits of different orders. Thus, the fraction of symbols used for information at the first stage is $t_1 = 1 - \dfrac{m_1}{n_1}$, and, in general, at the j-th stage is given by

$$
t_j = \prod_{i=1}^{j} \left(1 - \frac{m_i}{n_i}\right).
\tag{6.14.4}
$$

Thus, a lower bound on t_∞ can be obtained as follows using (6.14.2) and (6.14.4):

$$
t_\infty = \prod_{j=1}^{\infty} \left(1 - \frac{m_j}{n_j}\right) = \prod_{j=1}^{\infty} \left(1 - \frac{\log_2 n_j + 1}{n_j}\right) = \prod_{j=1}^{\infty} \left(1 - \frac{j+n}{2^{j+n-1}}\right).
\tag{6.14.5}
$$

Set $\sigma_j = \dfrac{m_j}{n_j}$ and $\sigma = \sum\limits_{j=1}^{\infty} \sigma_j$, where σ_j is monotone increasing in j, and $\sigma_j < 1$ for all Hamming codes for which $n_1 = 2^n \geq 4$. Then

$$
e^{-\sigma} > t_\infty > (1 - \sigma_1)^{\sigma/\sigma_1} > 1 - \sigma,
$$

which shows that $t_\infty > 0$ for $\sigma_1 < 1$ and $\sigma < \infty$. Also,

$$
\sigma = \sum_{j=1}^{\infty} \sigma_j = \sum_{j=1}^{\infty} \frac{j+n}{2^{j+n-1}} = \frac{n+2}{2^{n-1}} = \frac{2 \log_2(4 n_1)}{n_1},
$$

which at the threshold $(n_1 \mathfrak{p}_0 = \tfrac{1}{2})$ gives

$$
\sigma = 4 \mathfrak{p}_0 \log_2 \frac{2}{\mathfrak{p}_0} < 4 \left\{ \mathfrak{p}_0 \log_2 \frac{1}{\mathfrak{p}_0} + (1 - \mathfrak{p}_0) \log_2 \frac{1}{1 - \mathfrak{p}_0} \right\} = 4E.
\tag{6.14.6}
$$

Thus, for small $\mathfrak{p}_0$ we obtain (v).

(vi) Feinstein [1954] has shown that there exist ideal codes for which both the probability of error and the total equivocation vanishes in the limit as larger and larger symbol sequences are used in a code. This result holds for SEC-DED codes since every finite length message can be received without any infinite lag and zero equivocation with the error probability at each location. Let n_j denote the total number of bits (symbols) checked at the j-th stage. Of these bits k_j are data bits and the remaining m_j bits are check bits. Then since $n_1 \geq 2^2$, and $n_j = 2^{j-1}n_1 = 2^{j+1}$, as in the proof of (iv), we have

$$n_j = n_1^j 2^{j(j-1)/2}. \tag{6.14.7}$$

Multiplying the probability bound in (iv) we obtain a bound on the mean errors per n_j bits, which is also a bound on the fraction of sequences of n_j bits in error after the checking. Thus, in view of (6.14.7),

$$\mathfrak{q}_j < \mathfrak{p}_j n_j \leq \tfrac{1}{4} \left(\frac{n_1}{2}\right)^{j-1} (2n_1\mathfrak{p}_0)^{2j} \, 2^{j(j-1)/2},$$

which approaches zero as $j \to \infty$ under the condition that $n_1\mathfrak{p}_0 < \frac{1}{2}$. Then the equivocation E_j for each sequence of n_j terms attains the bound

$$E_j < \mathfrak{q}_j n_j + \mathfrak{q}_j \log_2 \frac{1}{\mathfrak{q}_j} + (1 - \mathfrak{q}_j) \log_2 \frac{1}{1 - \mathfrak{q}_j},$$

which is the inequality (vi). Note that $E_j \to 0$ as $j \to \infty$ for $n_1\mathfrak{p}_0 < \frac{1}{2}$, but not at the threshold. ∎

We have seen that a series of n_j of the type $n_j = 2^{j-1}n_1$ with $n_1 \geq 2^2$ increases rapidly with finite σ ($\sigma < +\infty$) and produces an error-free process for sufficiently small values of $n_1\mathfrak{p}_0$. If we choose an approximately geometric series such that

$$n_j \approx b^{j-1}n_1, \quad b > 1 \text{ and } b \text{ not an integer}, \tag{6.14.8}$$

then $\mathfrak{p}_j < \dfrac{1}{n_1} (bn_1\mathfrak{p}_0)^{2^j} b^{-(j+1)}$ (compare with Theorem 6.14.1(iii)) with a threshold at $n_1\mathfrak{p}_0 = 1/b$. Also, $\sigma = b^2 \mathfrak{p}_0 \log_2 \left(\dfrac{4}{b\,\mathfrak{p}_0}\right)$ (compare with (6.14.6)), and $\mathfrak{q}_j$ and E_j approach zero as $j \to \infty$ for $n_1\mathfrak{p}_0 < 1/b$, but $\max t_\infty$ is defined as in Theorem 6.1.1(v). For arbitrarily small $\mathfrak{p}_0$ the value of n_1 may be made very large, and in the limit the distribution of errors in the blocks becomes the Poisson distribution, and the probability $\mathfrak{p}(i)$ such that exactly i errors have occurred in a block is given by

$$\mathfrak{p}(i) = e^{-n_1\mathfrak{p}_0} \frac{(n_1\mathfrak{p}_0)^j}{i!}.$$

Using this result the average number of errors per block after SEC-DED decoding satisfies the inequality

$$n_j \mathsf{p}_j \leq 1 + n_j \mathsf{p}_{j-1} - e^{-n_j \mathsf{p}_{j-1}} \left\{ 1 + 2n_j \mathsf{p}_{j-1} + \frac{(n_j \mathsf{p}_{j-1})^2}{2!} + \frac{(n_j \mathsf{p}_{j-1})^4}{4!} + \cdots \right\}$$

$$\leq 1 - n_j \mathsf{p}_{j-1} \left(2e^{-n_j \mathsf{p}_{j-1}} - 1 \right) - \frac{1}{2} \left(1 - e^{-2n_j \mathsf{p}_{j-1}} \right). \qquad (6.14.9)$$

7

Golay Codes

7.1 Perfect Codes

Perfect codes have been introduced in §6.5 in connection with the Hamming bound (6.5.1) which for a q-ary code $C = (n, k, d)$ with $d \leq 2t + 1$ implies that

$$|C| \sum_{i=0}^{t} \binom{n}{i} (q-1)^i \leq q^n, \tag{7.1.1}$$

where $|C|$ $(= M)$ denotes the size of the code C (see §6.1) A code is said to be *perfect* if equality holds in (7.1.1). Since a t-error correcting code can correct up to t errors, a perfect t-error correcting code must be such that every codeword lies within a distance of t to exactly one codeword. In other words, if the code has $d_{\min} = 2t + 1$ that covers radius t, where the covering radius has the smallest number, then every codeword lies within a distance of the Hamming radius t to a codeword. In the case of a perfect code, the Hamming bound holds (see §6.1.4), which leads to a rigorous definition of a perfect code: If there is an (n, k) code with an alphabet of q elements, and if $d_{\min} = 2t + 1$, then the inequality (6.5.2) holds.

In fact, there are q^k codewords, and for each there are $\binom{n}{i}(q-1)^i$ codewords that differ from it in exactly i locations. Hence, the number on the right side of the inequality is the total number of codewords with distance at most t. Thus, the Hamming bound holds since the total number of allowable codewords cannot exceed the total number of possible codewords q^n. Obviously, a code is perfect exactly when it attains the equality in the Hamming bound. Only two binary Golay codes attain this equality: the (23,12) Golay code with $d_{\min} = 7$ and the (11,6) Golay code with $d_{\min} = 5$. As mentioned in §4.7, the other two binary ($q = 2$) codes that achieve equality in the Hamming bound are the $(2^r - 1, 2^r - 1 - r, 3)$ Hamming code, $r \geq 2$, with $d_{\min} = 3$, and the binary repetition code with two codewords, one all 0s and the other all 1s. In fact, perfect Hamming codes are the (3,1) code, (7,4) code, (15,11) code,

(31,26) code, and so on. For more information on other kinds of perfect codes, see Rhee [1989].

We will discuss only four kinds of linear perfect codes, three of which are binary and one ternary, namely, the repetition code, Hamming (7,4,3) code, Golay (23, 12, 3) code, and Golay (11,6,5) code.

7.1.1 Repetition Code. This is an $(n, 1, n)$ code, known as the *repetition code* of odd length. It is a perfect code which corrects $(n - 1)/2$ errors for odd n. Let $d = n = 2t + 1$. Then this repetition code can fix $(n - 1)/2$ errors. The Hamming bound for this code gives

$$2 \sum_{i=0}^{(n-1)/2} \binom{n}{i} = 2\left[\binom{n}{0} + \binom{n}{1} + \cdots + \binom{n}{(n - 1)/2}\right] = 2^n. \qquad (7.1.2)$$

Since equality is attained in (7.1.2), the repetition code is perfect.

7.1.2 Hamming (7,4,3) code. The Hamming (7,4,3) code is a linear, binary, single-error correcting code. As mentioned above, since all nontrivial single error correcting codes have the parameters $(2^r - 1, 2^r - r - 1, 3)$, we obtain this Hamming code for $r = 3$. The Hamming bound for the (7,4,3) code gives $2^4 \sum_{i=0}^{1} \binom{7}{i} = 2^4(1 + 7) = 2^7$. Thus, the Hamming (7,4,3) code is a perfect code.

Recall that a $k \times n$ matrix $\mathbf{G}$ of the (n, k, d) code C is the generator matrix of the code C if C is spanned by the rows of $\mathbf{G}$. If $\mathbf{G}$ is reduced in row to an echelon form, then C becomes equivalent to the code generated by $\mathbf{G} = [I_k \,|\, X]$, where I_k is the $k \times k$ identity matrix and X is a $k \times (n - k)$ matrix. The parity-check matrix of this code is $\mathbf{H} = \left[X^T \,|\, I_{n-k}\right]$, such that $\mathbf{c}\mathbf{H}^T = \mathbf{0}$ for all $\mathbf{c} \in C$.

The parity-check matrix $\mathbf{H}$ and the code generation matrix $\mathbf{G}$ for the Hamming (7, 4, 3) code are given in §4.2.

7.1.3 Binary Golay Codes. The linear, binary, 3-error correcting Golay code $\mathcal{G}_{23} = (23, 12, 7)$, with $n = 23, k = 12, d = 7$ is a perfect code. Golay [1949] noticed that $\binom{23}{0} + \binom{23}{1} + \binom{23}{2} + \binom{23}{3} = 2^{11} = 2^{23-12}$. This led him to conclude that a $(23, 12)$ perfect linear binary code existed, which could correct up to three errors. This discovery resulted in the development of the code $\mathcal{G}_{23}$, which is the only known code capable of correcting any combination of 3 or fewer random errors in a block of 23 elements (see Peterson [1961:70]).

The generator matrix of this code is

$$
\mathbf{G}_{23} = \begin{bmatrix}
0 & 1 & 1 & 1 & 1 & 1 & 1 & 1 & 1 & 1 & 1 \\
1 & 1 & 1 & 0 & 1 & 1 & 1 & 0 & 0 & 0 & 1 \\
1 & 1 & 0 & 1 & 1 & 1 & 0 & 0 & 0 & 1 & 0 \\
1 & 0 & 1 & 1 & 1 & 0 & 0 & 0 & 1 & 0 & 1 \\
1 & 1 & 1 & 1 & 0 & 0 & 0 & 1 & 0 & 1 & 1 \\
1 & 1 & 1 & 0 & 0 & 0 & 1 & 0 & 1 & 1 & 0 \\
1 & 1 & 0 & 0 & 0 & 1 & 0 & 1 & 1 & 0 & 1 \\
1 & 0 & 0 & 0 & 1 & 0 & 1 & 1 & 0 & 1 & 1 \\
1 & 0 & 0 & 1 & 0 & 1 & 1 & 0 & 1 & 1 & 1 \\
1 & 0 & 1 & 0 & 1 & 1 & 0 & 1 & 1 & 1 & 0 \\
1 & 1 & 0 & 1 & 1 & 0 & 1 & 1 & 1 & 0 & 0 \\
1 & 0 & 1 & 1 & 0 & 1 & 1 & 1 & 0 & 0 & 0
\end{bmatrix}. \tag{7.1.3}
$$

The parity-check matrix for this code is given by the matrix $H = [M I_{11}]$, where I_{11} is the 11×11 identity matrix and M is the 11×12 matrix

$$
M = \begin{bmatrix}
1 & 0 & 0 & 1 & 1 & 1 & 0 & 0 & 0 & 1 & 1 & 1 \\
1 & 0 & 1 & 0 & 1 & 1 & 0 & 1 & 1 & 0 & 0 & 1 \\
1 & 0 & 1 & 1 & 0 & 1 & 1 & 0 & 1 & 0 & 1 & 0 \\
1 & 0 & 1 & 1 & 1 & 0 & 1 & 1 & 0 & 1 & 0 & 0 \\
1 & 1 & 0 & 0 & 1 & 1 & 1 & 0 & 1 & 1 & 0 & 0 \\
1 & 1 & 0 & 1 & 0 & 1 & 1 & 1 & 0 & 0 & 0 & 1 \\
1 & 1 & 0 & 1 & 1 & 0 & 0 & 1 & 1 & 0 & 1 & 0 \\
1 & 1 & 1 & 0 & 0 & 1 & 0 & 1 & 0 & 1 & 1 & 0 \\
1 & 1 & 1 & 0 & 1 & 0 & 1 & 0 & 0 & 0 & 1 & 1 \\
1 & 1 & 1 & 1 & 0 & 0 & 0 & 0 & 1 & 1 & 0 & 1 \\
0 & 1 & 1 & 1 & 1 & 1 & 1 & 1 & 1 & 1 & 1 & 1
\end{bmatrix}. \tag{7.1.4}
$$

The $\mathcal{G}_{23}$ code can be generated either by the polynomial

$$
g_1(x) = 1 + x^2 + x^4 + x^5 + x^6 + x^{10} + x^{11},
$$

or by the polynomial

$$
g_2(x) = 1 + x^5 + x^6 + x^7 + x^9 + x^{11},
$$

where both polynomials $g_1(x)$ and $g_2(x)$ are factors of $x^{23} + 1$ in the Galois field GF(2)$[x]$, i.e., $x^{23} + 1 = (1+x)g_1(x) + g_2(x)$ (see Chapter 8 for polynomial generators).

There are different methods to decode the $(23, 12)$ binary Golay code so that its error-correcting capability at $t = 3$ can be maximized. The two well-known methods are the Kasami Decoder and the Systematic Search Decoder, which are explained in Lin and Costello [1983:102–106].

DEFINITION 7.1.1. For a q-ary (n, k) code C, let ν_i denote the number of codewords of weight i in C. Then $w_C(x) = \sum\limits_{i=0}^{n} \nu_i x^n$ determines the *weight enumerator* of the code C.

Example 7.1.1. To explain the weight enumerator of the Golay code C, note that since the Golay code is linear, we have $\mathbf{0} \in C$. Since for $d = 7$ and the code C has no codewords of weight 1, 2, 3, 4, 5, or 6, we find that $\nu_0 = 1$ and $\nu_1 = \nu_2 + \cdots = \nu_6 = 0$. Let us then consider ν_7. Since C is perfect, all codewords $\mathbf{c}_i \in C$ can be surrounded by balls of radius 3, which are pairwise disjoint, i.e., $\bigcup_{i=0}^{q} B(\mathbf{c}_i) = F_2^{23}$. This means that all 23-tuples of weight 4 in F_2^{23} are contained in a ball centered at a codeword of weight 8. Thus, $\binom{23}{4} = \nu_7 \binom{7}{3}$, which yields $\nu_7 = 253$. Next, let us consider ν_8 which can be calculated by looking at all 23-tuples of weight 5. All these tuples lie either in a ball entered at a codeword of weight 7 or in one centered at a codeword of weight 8. Thus, $\binom{23}{5} = \nu_7 \binom{7}{2} + \nu_8 \binom{8}{3}$, which yields $\nu_8 = 506$. However, the 23-tuples of weight 6 lie within balls of weight 7 in two ways: (i) by switching a single 1 to 0, and (ii) by switching two 1s and one 0. Thus,

$$\binom{23}{6} = \nu_7 \left[\binom{7}{1} + \binom{7}{2}\binom{16}{1} \right] + \nu_8 \binom{8}{2} + \nu_9 \binom{9}{3},$$

which yields $\nu_9 = \nu_{23} = 1$, $\nu_7 = \nu_{16} = 253$, $\nu_8 = \nu_{15} = 506$, $\nu_{11} = \nu_{12} = 1288$. All remaining coefficients in the weight enumerator are zero. Because the Golay code is perfect, all weight enumerators are uniquely determined. ∎

DEFINITION 7.1.2. For a q-ary code C, $\dim C = \log_2 q = \dfrac{\ln q}{\ln 2}$.

DEFINITION 7.1.3. Let a code C be a q-ary ECC of minimum distance $d = 2t$. The code C is said to be *quasi-perfect* if for every possible n-tuple $\mathbf{c}_0$ there exists a $\mathbf{c} \in C$ that differs from $\mathbf{c}_0$ in no more than t places (see §6.5). This means that, for some t, a quasi-perfect code has most codewords (vectors) of weight t or less, a few of weight $t + 1$, and none of weight greater than $t + 1$.

7.1.4 Extended Binary Golay Code. This code, denoted by $\mathcal{G}_{24}$, obtained from the perfect code $\mathcal{G}_{23}$ by adding an extra parity bit $\tilde{p}$ to each codeword in $\mathbf{G}_{23}$, is the nearly perfect (24,12,8) code, denoted by $\mathcal{G}_{24}$. The following result, proved in Sloane [1975], holds:

Theorem 7.1.1. *Let C be an (n, k) code whose minimum distance is odd. A new $(n + 1, k)$ code C' with the new minimum distance $d'_{\min} = d_{\min} + 1$ can be obtained from the code C by adding a 0 at the end of each codeword of even weight and a 1 at the end of each codeword of odd weight.*

Thus, the extended Golay code $\mathcal{G}_{24}$ can be generated by the 12×24 matrix

$\mathbf{G}_{24} = [I_{12}|X]$, where I_{12} is the 12×12 identity matrix and X is the matrix

$$
X = \begin{bmatrix}
1 & 1 & 0 & 1 & 1 & 1 & 0 & 0 & 0 & 1 & 0 & 1 \\
1 & 0 & 1 & 1 & 1 & 0 & 0 & 0 & 1 & 0 & 1 & 1 \\
0 & 1 & 1 & 1 & 0 & 0 & 0 & 1 & 0 & 1 & 1 & 1 \\
1 & 1 & 1 & 0 & 0 & 0 & 1 & 0 & 1 & 1 & 0 & 1 \\
1 & 1 & 0 & 0 & 0 & 1 & 0 & 1 & 1 & 0 & 1 & 1 \\
1 & 0 & 0 & 0 & 1 & 0 & 1 & 1 & 0 & 1 & 1 & 1 \\
0 & 0 & 0 & 1 & 0 & 1 & 1 & 0 & 1 & 1 & 1 & 1 \\
0 & 0 & 1 & 0 & 1 & 1 & 0 & 1 & 1 & 1 & 0 & 1 \\
0 & 1 & 0 & 1 & 1 & 0 & 1 & 1 & 1 & 0 & 0 & 1 \\
1 & 0 & 1 & 1 & 0 & 1 & 1 & 1 & 0 & 0 & 0 & 1 \\
0 & 1 & 1 & 0 & 1 & 1 & 1 & 0 & 0 & 0 & 1 & 1 \\
1 & 1 & 1 & 1 & 1 & 1 & 1 & 1 & 1 & 1 & 1 & 0
\end{bmatrix}. \tag{7.1.4}
$$

The generator matrix of this extended Golay code is also given by $\mathbf{G}'_{24} = [X|I_{12}]$. The extended Golay code has a minimum distance of 8; the weight of each codeword is a multiple of 4, and $\mathcal{G}_{24}$ is invariant under a permutation of coordinates that interchanges the two halves of each codeword. This code is not perfect, but simply quasi-perfect, because all spheres of radius 1 are disjoint, while every vector is at most a distance $t + 1$ from some code vector. There are $2^{12} = 4096$ binary codewords of length 24 in the extended Golay code $\mathcal{G}_{24}$, and it can also be used to correct up to three errors. Besides the zero codeword and the codeword consisting of twenty-four 1s, this extended Golay code consists of 759 words of (Hamming) weight 8 (i.e., they contain eight 1s in each); there are also 759 words of weight 16, and the remaining 2576 codewords (out of the total of 4096) are of weight 12. The codewords of weight 8 are called the *octads* and those of weight 12 are called the *dodecads*. Thus, adding all these codewords, we have a total of $1 + 759 + 2576 + 759 + 1 = 4096 = 2^{12}$ binary codewords. The extended code $\mathcal{G}_{24}$ is very useful in providing an easy protection against burst errors as the following result shows.

Theorem 7.1.2. *If a linear (n, k, d) code C can be used to correct burst errors of length t, then an extended code C' can be used to correct all burst errors of length up to $1 + (t - 1)m$, where m is a a positive integer.*

The codewords of a fixed weight 'hold' a 5-design, which means that we take the set $\{1, 2, \ldots, 24\}$ as a set of points and consider a codeword in $\mathcal{G}_{24}$ as the indicator function of a subset of points. These subsets form the blocks of the design. Thus, the octads form a 5-(24,8,1) design that is a Steiner $S(5, 8, 28)$ system. Moreover, the automorphism group of the Golay code, which the group of permutations of the coordinates that send codewords to codewords, is the Mathieu group M_{24}. The Golay code can also be used to construct the Leech lattice. For Mathieu groups and Steiner systems, see Appendix C.

7.1.5 Ternary Golay Code. The linear ternary Golay code $\mathcal{G}_{11} = (11, 6, 5)$, with $n = 11, k = 6, d = 5$ is a perfect code. Since a Hamming sphere with radius $t = 2$ over GF(3) contains $243 = 3^5$ vectors, and since $1 + 2\binom{11}{1} + 4\binom{11}{2} = 243$, there may be a perfect packing with $3^6 = 729$ spheres (codewords) of radius $t = 2$, the Golay code $\mathcal{G}_{11}$ over GF(3) has a minimum distance of 5, and can correct up to 2 errors. The generator matrix of this code is

$$\mathbf{G}_{11} = \begin{bmatrix} 1 & 0 & 0 & 0 & 0 & 0 & 1 & 1 & 1 & 1 & 1 \\ 0 & 1 & 0 & 0 & 0 & 0 & 0 & 1 & 2 & 2 & 1 \\ 0 & 0 & 1 & 0 & 0 & 0 & 1 & 0 & 1 & 2 & 2 \\ 0 & 0 & 0 & 1 & 0 & 0 & 2 & 1 & 0 & 1 & 2 \\ 0 & 0 & 0 & 0 & 1 & 0 & 2 & 2 & 1 & 0 & 1 \\ 0 & 0 & 0 & 0 & 0 & 1 & 1 & 2 & 2 & 1 & 0 \end{bmatrix}. \tag{7.1.5}$$

An extended ternary Golay code of this type is obtained by adding an extra zero-sum check digit to the code $\mathcal{G}_{11}$, which gives a code with parameters $(12, 6, 6)$. The generator matrix of this extended code is

$$\mathbf{G}_{12} = \begin{bmatrix} 1 & 0 & 0 & 0 & 0 & 0 & 0 & 1 & 1 & 1 & 1 & 1 \\ 0 & 1 & 0 & 0 & 0 & 0 & 1 & 0 & 1 & 2 & 2 & 1 \\ 0 & 0 & 1 & 0 & 0 & 0 & 1 & 1 & 0 & 1 & 2 & 2 \\ 0 & 0 & 0 & 1 & 0 & 0 & 1 & 2 & 1 & 0 & 1 & 2 \\ 0 & 0 & 0 & 0 & 1 & 0 & 1 & 2 & 2 & 1 & 0 & 1 \\ 0 & 0 & 0 & 0 & 0 & 1 & 1 & 1 & 2 & 2 & 1 & 0 \end{bmatrix}. \tag{7.1.6}$$

This code is a quasi-perfect code.

The following theorem provides the latest development in the area of perfect codes, due to van Lint and Tietavainen (see van Lint [1975]).

Theorem 7.1.2. *A perfect t-error correcting code C of length n over the finite field F_q satisfies one of the following conditions:*

(i) If $|C| = 1$, then $t = n$;

(ii) If $|C| = q^n$, then $t = 0$;

(iii) If $|C| = 2$, then $q = 2, n = 2t + 1$;

(iv) If $|C| = 3^6$, then $q = 3, t = 2, n = 11$;

(v) If $|C| = 2^{12}$, then $q = 2, t = 3, n = 23$;

(vi) If $|C| = q^{n-r}$, then $t = 1 \, n = \dfrac{r^r - 1}{q - 1}, r > 0$.

Of these, (i) and (ii) are trivial codes; (iii) represents all repetition codes; (iv) and (v) represent the Golay codes; and (vi) represents the Hamming codes.

Note that the ternary Golay code $\mathcal{G}_{11}$ satisfies the Hamming bound. The extended ternary Golay code $\mathcal{G}_{12}$ is a (12, 6, 6) code. Finally, a result of

Tiertäväinen [1974] and van Lint [1975, 1992] on perfect codes states that when $q \geq 2$ is prime, a nontrivial perfect code over f_q must have the same parameters as one of the Hamming or Golay codes.

7.2 Geometrical Representations

The extended binary Golay code $\mathcal{G}_{24}$ occupies a 12-dimensional subspace W of the vector space $V = F_2^{24}$ of 24-bit elements (words) such that any two distinct elements of W differ in at least 8 coordinates, i.e., any nonzero element of W has at least 8 non-zero coordinates. The possible non-zero coordinates elements of W are called codewords, and in $\mathcal{G}_{24}$ all codewords have the Hamming weights of 0, 8, 12, 16, or 24. The subspace W is unique up to relabeling the coordinates.

The perfect binary Golay code $\mathcal{G}_{23}$ has the geometrical significance that the spheres of radius 3 around codewords form a partition of the vector space W. Other geometrical properties of this code are: (i) the automorphism group of this perfect code is the Mathieu group M_{23}, where the automorphism group of the extended code $\mathcal{G}_{24}$ is the Mathieu group M_{24}; and (ii) the Golay codewords of Hamming weight 8 are elements of the Steiner system $S(5, 8, 24)$.

7.2.1 Constructions of Golay Codes. There are different geometrical methods of constructing the binary Golay codes, but we will present some important ones here.

A number of polyhedra provide geometrical construction of different Golay codes. An excellent account of this application is available in Conway and Sloane [1999], and readers interested in such constructions and related group theoretical discussions will find this book very useful. We will, however, provide some information on different polyhedra and their role in the Golay codes. The first four polyhedra described below are CW complexes, i.e., they have 2-dimensional faces, 1-dimensional edges, and 0-dimensional vertices.

1. CUBE. It has 6 faces, which provide a code of length 12 and dimension 6. Since the stabilizer of a face in the group of the cube is a cyclic 4-group, determination of the orbits of the faces under this group becomes easy. In fact, there are 3 orbits under a face stabilizer, which are (i) the face itself, (ii) the four adjacent faces, and (iii) the opposite face. This provides $2^3 = 8$ possible masks that can be easily generated and their minimum weight checked. One of them is the optimal (12,6,4) code.

2. ICOSAHEDRON. It provides 12 codes with minimum weight 8, and there are four doubly-even self-dual codes, all equivalent to one another, with the following weight distribution:

$$[1, 0, 0, 0, 0, 0, 0, 0, 285, 0, 0, 0, 21280, 0, 0, 0, 239970, 0, 0, 0, 525504, \ldots, 1].$$

There are four singly-even codes, all equivalent to one another, with the weight distribution:

[1, 0, 0, 0, 0, 0, 0, 0, 285, 0, 1024, 0, 11040, 0, 46080, 0, 117090, 0, 215040,
0, 267456, ... , 1].

In addition to these, there are four formally self-dual codes[1] with two weight
distributions:

[1, 0, 0, 0, 0, 0, 0, 0, 285, 0, 1024, 0, 11040, 0, 46040, 0, 117090, 0, 215040,
0, 267456, ... , 1],

and [1, 0, 0, 0, 0, 0, 0, 0, 285, 0, 960, 0, 11680, 0, 43200, 0, 124770, 0, 201600,
0, 283584, ... , 1],

where each pair with the same weight distribution are not only equivalent to
one another but to its dual as well.

3. DODECAHEDRON. The construction of the mask and encoding and
decoding of the Golay $\mathcal{G}_{12}$ is described in detail at the websites *http://www.
math.uic.edu/ ˜fields/DecodingGolayHTML/introduction.html, /encoding.html,
decoding.html*, and */example.html* at the same website (check *google.com*
also). Since this method involves paper constructions of the dodecahedron
and its mask with color-coded numbering, we will simply introduce the topic
and leave the interested reader to explore these websites.

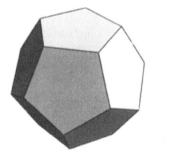 

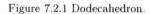
Figure 7.2.1 Dodecahedron.   Figure 7.2.2 Mask.

Fields [2009] has suggested a geometrical method of using the properties
of a dodecahedron for encoding and decoding of the extended Golay code $\mathcal{G}_{24}$.
As presented in Figure 7.2.1, a dodecahedron has 12 faces, which is one-half
of the length of $\mathcal{G}_{24}$. Thus, to obtain the required 24 bits, each face of the
dodecahedron will have to contain 2 bits. These 24 bits will be separated into
two parts: the first 12 bits are the data (information/message) bits, and the
other 12 bits are the parity check bits computed from the first 12 data bits,

---

[1] A code is said to be *formally self-dual* if it has the same weight distribution as its
dual.

as shown in the 'net' of the dodecahedron (Figure 7.2.3).

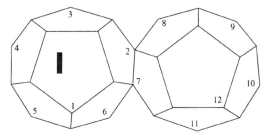

Figure 7.2.3 Dodecahedron net.

To compute the parity bits, the geometry of the dodecahedron is used as follows: On the dodecahedron, there are five faces adjacent to a given face with distance 1, five other faces with distance 2, and a single face (exactly opposite the original face) with distance 3 relative to the original face that is distance 0 from itself. Thus, we have 1, 5, 5, 1 number of faces at distances 0, 1, 2, 3, respectively, from a given face. The minimum weight of the resulting codewords must be at least 8. As an example, consider encoding a codeword where the data positions contain only a single 1. Then, to ensure that the codeword has the required eight 1s, we must have 7 1s among the parity bits. Of the 1, 5, 5, 1 faces of the dodecahedron, which add to 12, we will obtain 7 by eliminating one of the 5's. These 5 faces are known as the mask (Figure 7.2.2) that is used in encoding and decoding in this example. The encoding and decoding processes, and an example, are explained in Fields [loc. cit.]. The drawback to this process is that it relies heavily on human pattern recognition and thus, becomes costly and at times unreliable.

4. DODECADODECAHEDRON METHOD. A close relationship of a uniform dodecadodecahedron and the extended binary Golay code $\mathcal{G}_{24}$ can be established in terms of the Mathieu group M_{24}. The construction of this polyhedron, shown in Figure 7.2.4, again involves extensive and intricate paper models, including a design for its universal cover. The details are available at the website: http://homepages. wmich.edu/~drichter/golay.htm.

Figure 7.2.4 Dodecadodecahedron.

Other methods for constructing generator matrices for $\mathcal{G}_{23}$ and $\mathcal{G}_{24}$ are available in Conway and Sloane [1999]. Also check Wolfram Math World for articles on the Golay codes (http://mathworld.wolfram.com).

5. SOLID TETRAHEDRON. It has one 3-dimensional cell, four 2-dimensional cells, six 1-dimensional cells, and four 0-dimensional cells. All these cells can be used for parity and data locations and thus construct various codes. For example, consider a code in which the data bits are located on the vertices, whereas the parity check bits are on all other cells (faces, edges and the solid itself). Then we will need a mask for each type of parity check, which can be easily determined using the geometry. That is, the parity bits on each cell is the parity of the sum (XOR) of the data bits contained in that cell. For example, the (15×4) generator matrix of the $S(4,5,11)$ Steiner system is

$$
\begin{array}{cccccccccccccccc}
 & & & & & & & & & & & & & 1 \\
 & & & & & & & 1 & 1 & 1 & 2 & & 2 \\
 & & & & 1 & 1 & 1 & 2 & 2 & 3 & 2 & 2 & 3 & 3 & & 3 \\
1 & 2 & 3 & 4 & 2 & 3 & 4 & 3 & 4 & 4 & 3 & 4 & 4 & 4 & & 4 \\
\end{array}
$$

$$
\begin{bmatrix}
1 & 0 & 0 & 0 & 1 & 1 & 1 & 0 & 0 & 0 & 1 & 1 & 1 & 0 & 1 \\
0 & 1 & 0 & 0 & 1 & 0 & 0 & 1 & 1 & 0 & 1 & 1 & 0 & 1 & 1 \\
0 & 0 & 0 & 1 & 0 & 0 & 1 & 0 & 1 & 1 & 0 & 1 & 1 & 1 & 1 \\
\end{bmatrix},
$$

$$
\begin{array}{cccccccccccccccc}
1 & 1 & 0 & 1 & 2 & 1 & 2 & 1 & 2 & 1 & 2 & 3 & 2 & 2 & & 3
\end{array}
$$

where the columns corresponding to parity bits in various cells of the solid tetrahedron have been labeled over the matrix, and the row beneath the matrix gives the number of 1s in each column. The code associated with this system has weight distribution $[1,0,0,0,0,0,0,0,15,0,0,0,0,0,0,0]$ (see §B.4). The dual of this code is a $(15,11,3)$ code decribed in Example 4.4.2, and both of these codes are perfect and optimal.

7.3 Other Construction Methods

There are different methods for constructing binary Golay codes, described as follows.

1. CYCLIC CODE. As shown in §7.1.3, the binary Golay code $\mathcal{G}_{23}$ can be constructed by factorizing $x^{23} - 1$, i.e., this code is generated by $\{x^{11} + x^{10} + x^6 + x^5 + x^4 + x^2 + 1\}/\{x^{23} - 1\}$. This means that any integer from 0 to $2^{23} - 1$ is within distance of one of the codewords, such that up to three errors can be detected and corrected. The automorphism group is the Mathieu group M_{23}. The codewords of weight 7 are elements of an $S(4,7,23)$ Steiner system. In the case of the semi-perfect extended binary Golay code $\mathcal{G}_{24}$, any integer from 0 to $2^{24} - 1$ is within distance 4 of one of the codewords, and up to four errors can be detected but only up to three corrected. The automorphism group is the Mathieu group M_{24}. The codewords of weight 8 are elements of an $S(5,8,24)$

Steiner system. Thus, let $g(x) = x^{11} + x^{10} + x^6 + x^5 + x^4 + x^2 + 1$. Consider the powers of $g(x)$, namely $\{g(x)\}^k$ (mod $x^{23} - 1$). There are 253 polynomials with seven terms. By choosing 12 polynomial powers with a different initial term and appending 1 to each polynomial's coefficient list, we obtain a basis for the 4096 codewords.

2. QUADRATIC-RESIDUE CODE. To generate the code $\mathcal{G}_{23}$, let X denote the set of quadratic nonresidues (mod 23). Note that X is an 11-element subset of the cyclic group $Z/23$. Consider the translates $t + X$ of this subset. Increase each translate to a 12-element set S_t by adding an element ∞. Then we can label the basis elements of V by $0, 1, 2, \ldots, 22, \infty$, and define W as the span of the words S_t together with the word consisting of all basis vectors. The perfect code is obtained by excluding ∞ from W.

3. LEXICOGRAPHIC CODE. Let $V = F_2^{24}$ denote the space of 24-bit words. Let the vectors in V be ordered lexicographically, i.e., they are interpreted as unsigned 24-bit binary integers and take them in the usual ordering as follows: In the 12-dimensional subspace W of V, consisting of the words $w_1, \ldots, w_{12}$, start with $w_1 = 0$, then define $w_2, w_3, \ldots, w_{12}$ by the rule that w_n is the smallest integer which differs from all linear combinations of previous elements in at least 8 coordinates. Then W can be defined as the span of $w_1, \ldots, w_{12}$. For more on lexicode and lexicographic ordering, see §9.7 and §2.7, respectively.

4. WINNING POSITIONS IN THE GAME OF MOGHUL. A position in the Game of Moghul is a row of 24 coins, and each turn consists of flipping from one to seven coins such that the leftmost of the flipped coins goes from heads to tails. The losing positions are those with no legal move. If heads are taken as 1 and tails as 0, then moving to a codeword from the extended binary Golay code $\mathcal{G}_{24}$ guarantees it will be possible to force a win. In other words, start with the 24-bit codeword $(000 \ldots 000)$; add the first 24-bit word that has eight or more differences from all words in the list; repeat and generate all 4096 codewords. The codewords are winning positions in this game, played with 24 coins in a row. Each turn flips between one and seven coins such that the leftmost flipped coin goes from heads to tails. Last to move wins.

5. MIRACLE OCTAD GENERATOR. (Curtis [1977]) This method uses a 4×6 array of square cells to depict the 759 weight 8 codewords, or *octads*, of the extended binary Golay code $\mathcal{G}_{24}$. The remaining codewords are obtained by symmetric differences of subsets of 24 square cells, which is the same as binary addition. This yields a generator matrix of the form $[I_{12} \mid J_{12} - A]$, where I_{12} is a 12×12 identity matrix, J_{12} the 12×12 unit matrix (all elements 1s), and A is the adjacency matrix of the face graph of the dodecahedron, i.e., $A = [a_{ij}]$, where $a_{ij} = 1$ if the i-th and j-th faces share an edge of the dodecahedron (see §7.2.1(3) for details).

7.4 Finite-State Codes

Pollara et al. [1987] have shown that certain optimal finite-state codes can be constructed from known block codes. It is expected that such finite-state codes can be very useful in deep-space communication systems for ECC schemes so that power and antenna size requirements can be maintained within reasonable bounds.

A finite-state code is derived from the Golay $\mathcal{G}_{24}$ code using the following algorithm: Choose an (n, k_1) block code C_1 with minimum distance d_1, and then decompose C_1 into a disjoint union of cosets generated by an (n, k_2) subcode $C_2 \subset C_1$ with minimum distance d_2. Then an (n, k, m) finite-state code can be constructed by properly assigning these cosets to the edges of a 2^m-state completely connected graph. The Golay $\mathcal{G}_{24}$ code can be used in this construction provided it contains a subcode with minimum distance $d_2 > d_1 = 8$.

Theorem 7.4.1. (Pollara et al. [1987]) *The Golay $\mathcal{G}_{24}$ code has a $(24, 5)$ subcode with minimum distance 12.*

PROOF. The proof is based on the Turyn construction of the Golay $(24, 12)$ code $\mathcal{G}_{24}$ (see Turyn [1974] and MacWilliams and Sloane [1977:578]) and proceeds as follows: Let A be the $(7,3)$ code with codewords consisting of the zero word and the seven cyclic shifts of (1101000), i.e., it consists of the codewords: (0000000), (1101000), (1010001), (0100011), (1000110), (0001101), (0011010), and (0110100). Then the $(7,4)$ code $\mathcal{H} = A \cup \overline{A}$, where $\overline{A}$ is the $(7,3)$ code with complemented codeword of A, is the Hamming $(7,4)$ code. Next, let A^* denote the $(7,3)$ code obtained from A by reversing the order of its codewords, and let $\mathcal{H}^* = A^* \cup \overline{A}^*$. Let C and C^* be the extended $(8,4)$ codes obtained by adding an extra parity check bit to $\mathcal{H}$ and $\mathcal{H}^*$, respectively. Then both C and C^* have the minimum distance 4. Define a code G consisting of the vectors

$$|a+x|b+x|a+b+x|, \quad a,b \in C, \quad x \in C^*. \tag{7.4.1}$$

Then the code G is the Golay $(24,12)$ code $\mathcal{G}_{24}$ with minimum distance 8. Further, let a code $B \subset C$ consist of two codewords: (00000000) and (11111111). Then the construction defined in $(7.4.1)$ with $a, b \in B$ and $x \in A^*$ generates codewords of the form

$$|x|x|x|, \quad |x|\overline{x}|\overline{x}|, \quad |\overline{x}|x|\overline{x}|, \quad |\overline{x}|\overline{x}|\overline{x}|. \tag{7.4.2}$$

If we take codewords from the two distinct subcodes of the four subcodes $(7.4.2)$, they will be at minimum distance $8 + 8 = 16$ for fixed $x \in A^*$, and at distance $4 \times 3 = 12$ for $x \neq y \in A^*$. Moreover, codewords in the same subcode of the four subcodes $(7.4.2)$ are at minimum distance $4 \times 3 = 12$. Hence, using all these 8 possible choices for x a new $(24,5)$ subcode of the Golay G_{24} code is constructed with minimum distance 12. ∎

The above Golay code $\mathcal{G}_{24}$, presented as a trellis in Figure 7.4.1, where the edge x or $\bar{x}$ corresponds to 8 bits, consists of the union of 8 cosets D_i, $i = 0, 1, \ldots, 7$, defined by (7.4.2) with $x \in A^*$, such that each coset has 4 codewords and is represented by a trellis. This representation leads to the following result, which follows if we take $x = (00000000)$ in (7.4.2).

Corollary 7.4.1. *The Golay code $\mathcal{G}_{24}$ has a (24,2) subcode with minimum distance 16.*

Figure 7.4.2 shows the trellis representing the (24,2) subcode with 4 codewords.

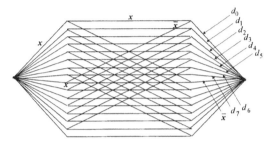

Figure 7.4.1 Trellis for Golay (24,5) code.

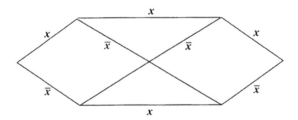

Figure 7.4.2 Trellis for the (24,2) subcode.

7.5 MacWilliams' Identity

We will first present some background material from the theory of integral transforms, and in particular from the discrete Fourier transform (see Appendix D). Let f be a function defined on F_p^n, where p is prime, such that value $f(\mathbf{u})$, where $\mathbf{u} \in F_p^n$, can be added to or subtracted from, and multiplied by a complex number. Define the *discrete Fourier transform* (DFT) $\tilde{f}$ of f by

$$\tilde{f}(\mathbf{u}) = \sum_{\mathbf{u} \in F_p^n} f(\mathbf{u}) \alpha^{\mathbf{u} \cdot \mathbf{v}}, \qquad (7.5.1)$$

where α denotes the p-th root of unity in the complex plane $\mathbb{C}$, i.e., $\alpha^p = 1$, but $\alpha^i \neq 1$ for all $0 < i < p$, and $\mathbf{u} \cdot \mathbf{v}$ is the scalar (dot) product in F_p^n. Let

$$C_i(\mathbf{v}) = \{\mathbf{u} \in C : \mathbf{u} \cdot \mathbf{v} = i\} \text{ for } \mathbf{v} \in F_p^n \text{ and } 0 \leq i \leq p - 1.$$

Then

(i) $C_i(\mathbf{v})$ is a coset of $C_0(\mathbf{v})$ in C for $1 \le i \le p-1$ iff $\mathbf{v} \notin C^{\perp}$. Also, if $\mathbf{v} \in C^{\perp}$,
then $c = C_0(\mathbf{v}) \cup C_1(\mathbf{v}) \cdots \cup C_{p-1}(\mathbf{v})$.

(ii) $\displaystyle\sum_{\mathbf{u} \in C} \alpha^{\mathbf{u} \cdot \mathbf{v}} = \begin{cases} |C| & \text{if } \mathbf{v} \in C^{\perp}, \\ 0 & \text{if } \mathbf{v} \notin C^{\perp}. \end{cases}$

(iii) For every $\mathbf{w} \in F_p^n$, $f(\mathbf{w}) = \dfrac{1}{p^n} \displaystyle\sum_{\mathbf{u} \in C} \tilde{f}(\mathbf{u}) \alpha^{-\mathbf{u} \cdot \mathbf{w}}$; and

(iv) $\displaystyle\sum_{\mathbf{u} \in C} f(\mathbf{v}) = \dfrac{1}{|C|} \sum_{\mathbf{u} \in C} \tilde{f}(\mathbf{u})$.

Let C be a linear code of length n over F_p^n, p prime. The *Hamming weight enumerator* of C is the homogeneous polynomial

$$w_C(x,y) = \sum_{\mathbf{c} \in C} x^{n-\mathrm{wt}(\mathbf{c})} y^{\mathrm{wt}(\mathbf{c})}. \tag{7.5.2}$$

If we set $f(\mathbf{c}) = x^{n-\mathrm{wt}(\mathbf{c})} y^{\mathrm{wt}(\mathbf{c})}$, then by the above property (iv) we have the *MacWilliams' identity*

$$w_{C^{\perp}}(x,y) = \frac{1}{|C|} w_C(x + (p-1)y, x - y), \tag{7.5.3}$$

which holds for the finite field F_q^n for all q.

The identity (7.5.3) holds for the code $C = C^{\perp} = \mathcal{G}_{24}$. Another interesting result on the extended Golay code $\mathcal{G}_{24}$ concerns the weight distribution and the number of codewords, as shown in Table 7.5.1.

Table 7.5.1 Weight Distribution

Weight	0	4	8	12	16	20	24
# of codewords	1	0	759	2576	759	0	1

Note that the 1-vector $\{\mathbf{1}\}$ is in the extended Golay code $\mathcal{G}_{24}$, and hence, the code $\mathcal{G}_{24}$ does not have any codewords of weight 20.

7.6 Golay's Original Algorithm

The contribution by Golay [1949], which is simply a half-page publication, has been very significant in generalizing the Hamming code that could correct only a single error. This generalization produced a code that has found many applications, including the provision of the error control for the Jupiter fly-by of Voyager 1 and 2 launched by NASA Deep Space Missions during the summer of 1977, and the Saturn mission in 1979, 1980, and 1981 fly-bys from Cape Canaveral, Florida, to conduct closeup studies of Jupiter and Saturn,

Saturn's Rings, and the larger moons of the two planets. These missions transmitted hundreds of color pictures of Jupiter and Saturn within a constrained telecommunication bandwidth. The Golay (24, 12, 8) code was used to transmit color images that required three times the amount of data. Although this code is only three-error correcting, it could transmit data at a much higher rate than the Hadamard code (see §9.4) that was used during the Mariner Jupiter-Saturn missions. The new NASA standards for automatic link establishment (ALE) in high-frequency (HF) radio systems specify the use of the extended Golay code $\mathcal{G}_{24}$ for forward error correcting (FEC). Since this code is a (24,12) block code encoding 12 data bits to produce 24-bit codewords, it is systematic, i.e., the 12 data bits remain unchanged in the codeword, and the minimum Hamming distance between two codewords is eight, i.e., the number of bits by which any pair of codewords differ is eight.

Golay extended Shannon's case of blocks of seven symbols (see Shannon [1948]) to those of $2^n - 1$ binary symbols, one or none of which can be in error when transmitted in message coding. This was further generalized by Golay [1949] to coding schemes of blocks of $\dfrac{p^n - 1}{p - 1}$, where p is a prime number, which are transmitted with no errors with each block consisting of n redundant symbols (parity check bits, in modern terms) that are designed to remove errors. Let the parity bits be denoted by X_n and the data bits by Y_k. While encoding, these bits are governed by the congruent relation E_m, defined by

$$E_m = X_m + \sum_{k=1}^{(p^n - 1)/(p-1)\, -n} a_{mk} Y_k \equiv 0 \quad \mod p. \qquad (7.6.1)$$

In the decoding process, the E_m are recalculated with the symbols received and their equivalence set determines a number (mod p), that determines the bit in error during transmission and corrects it.

Table 7.6.1 Golay (23, 12,7) Code

	Y_1	Y_2	Y_3	Y_4	Y_5	Y_6	Y_7	Y_8	Y_9	Y_{10}	Y_{11}	Y_{12}
X_1	1	0	0	1	1	1	0	0	0	1	1	1
X_2	1	0	1	0	1	1	0	1	1	0	0	1
X_3	1	0	1	1	0	1	1	0	1	0	1	0
X_4	1	0	1	1	1	0	1	1	0	1	0	0
X_5	1	1	0	0	1	1	1	0	1	1	0	0
X_6	1	1	0	1	0	1	1	1	0	0	0	1
X_7	1	1	0	1	1	0	0	1	1	0	1	0
X_8	1	1	1	0	0	1	0	1	0	1	1	0
X_9	1	1	1	0	1	0	1	0	0	0	1	1
X_{10}	1	1	1	1	0	0	0	0	1	1	0	1
X_{11}	0	1	1	1	1	1	1	1	1	1	1	1

The extended code is obtained as follows: The $n \times \dfrac{p^n - 1}{p - 1}$ matrix obtained above with the coefficients of X_n and Y_k is repeated p-times horizontally, while an $(n+1)$st row is added consisting of $\dfrac{p^n - 1}{p - 1}$ zeros, followed by $p - 1$ number of 1s; finally, a column of n zeros with a 1 for the lowest term is added.

The following result is then established: A lossless coding scheme exists in a binary system iff there exist three or more numbers in a line of Pascal's triangle, which add up to an exact power of 2. In a restricted search, only two cases are found: first, the first three numbers of line 90, which add up to $2^{12} = 4096$, and the second, the first four numbers of line 23, which add up to $2^{11} = 2048$. However, the first case does not provide a lossless coding for the following reason: if such a scheme were to exist, let us denote by r the number of E_m sets that correspond to one error and have an odd number of of 1s, and denote by $90 - r$ the remaining even number of sets. Then the odd sets corresponding to two transmission errors could be formed by re-entering term-by-term all the combinations of one even and one odd set associated with each error, and this would number $r(90 - r)$, which would yield $r + r(90 - r) = 2^{11} = 2048$, which is impossible for integral values of r.

The matrix a_{mk} for the second case has 12 binary symbols and is given in Table 7.6.1, while Table 7.6.2 has 6 ternary symbols for lossless coding.

Table 7.6.2 Golay (11, 6,5) Code

	Y_1	Y_2	Y_3	Y_4	Y_5	Y_6
X_1	1	1	1	2	2	0
X_2	1	1	2	1	0	2
X_3	1	2	1	0	1	2
X_4	1	2	0	1	2	1
X_5	1	0	2	2	1	1

According to Golay, the ternary coding scheme (Table 7.6.2) will always result in a rate loss, but the binary scheme (Table 7.6.1) has a comparatively lower rate loss with smaller probability of errors.

7.7 Structure of Linear Codes

The algebraic structure of linear binary codes starts with the assumption that the symbols of the message and the coded message are elements of the same finite field. Coding means to encode a block of k message symbols $a_1 a_2 \cdots a_k, a_i \in F_q$ into a codeword $c_1 c_2 \cdots c_n$ of n symbols $c_j \in F_q$, where $n > k$. Such a codeword is regarded as an n-dimensional row vector $\mathbf{c} \in F_q$. A coding scheme is a function $f : F_q^k \mapsto F_q^n$, whereas a decoding scheme is a function $g : F_q^n \mapsto F_q^k$, as presented in Figure 7.7.1.

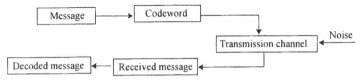

Figure 7.7.1 A communication system.

A simple coding scheme consists of encoding each block $a_1 a_2 \cdots a_k$ of message symbols into a codeword of the form $a_1 a_2 \cdots a_k c_{k+1} \cdots c_n$, where the first k symbols are the original message symbols and the remaining $n - k$ symbols in F_q are control symbols. For such a coding scheme, let $\mathbf{H}$ be a given $(n - k) \times n$ matrix of the form $\mathbf{H} = [A, I_{n-k}]$, with entries in F_q, where A is an $(n - k) \times k$ matrix and I_{n-k} is the identity matrix of order $n - k$. Then the control symbols are calculated by solving the system of equations $\mathbf{H}\mathbf{c}^T = \mathbf{0}$ for the codeword $\mathbf{c} \in F_q^n$. This equation is called the *parity-check equation*. The set C of all n-dimensional vectors $c \in F_q^n$ such that $\mathbf{H}\mathbf{c}^T = \mathbf{0}$ is called a *linear code* over F_q. For $q = 2$, C is called a *binary code*. If $\mathbf{H}$ is of the form $[A, I_{n-k}]$, then C is called a *systematic code*.

Example 7.7.1. (Parity-check code) Let $a_1 \cdots a_k$ be a given message over F_2. Then the coding scheme f is defined by $f : a_1 \cdots a_k \mapsto b_1 \cdots b_{k+1}$, where $b_i = a_i$ for $i - 1, \ldots, k$, and

$$
b_{k+1} = \begin{cases} 0 & \text{if } \sum_{i=1}^{k} a_i = 0, \\ 1 & \text{if } \sum_{i=1}^{k} a_i = 1. \end{cases}
$$

This means that the sum of digits of any codeword $b_1 \cdots b_{k+1}$ is 0. If the sum of digits of the received word is 1, then the receiver knows that a transmission error has occurred. Let $n = k + 1$; then this code is a linear binary $(n, n - 1)$ code with parity-check matrix $\mathbf{H} = [1\,1\,\cdots\,1]$. ∎

Example 7.7.2. (Repetition code) In a repetition code, each codeword consists of only one message symbol a_1 and $(n - 1)$ control symbols $c_2 = \cdots + c_n = a_1$. It means that a_1 is repeated $n - 1$ times. This is a linear $(n, 1)$ code with parity-check matrix $\mathbf{H} = [-1|I_{n-1}]$, in which case the parity-check equations $\mathbf{H}\mathbf{c}^T = \mathbf{0}$ imply that

$$
\mathbf{c}^T = \begin{bmatrix} I_k \\ -A \end{bmatrix} \mathbf{a}^T = \left[\mathbf{a}(I_k, -A^T) \right]^T,
$$

where $\mathbf{a} = a_1 \cdots a_k$ is the original message and $\mathbf{c} = c_1 \cdots c_n$ is the codeword. ∎

The matrix $\mathbf{G} = [I_k | - A^T]$ of a linear (n, k) code with parity-check matrix $\mathbf{H} = [A, I_{n-k}]$ is called the *canonical generator matrix*. Since $\mathbf{H}\mathbf{c}^T = \mathbf{0}$ and

$\mathbf{c} = \mathbf{aG}$, it follows that $\mathbf{H}$ and $\mathbf{G}$ are related by

$$\mathbf{HG}^T = \mathbf{0}, \tag{7.7.1}$$

as in §4.2.1. The code C is then equal to the row space of the matrix $\mathbf{G}$. In general, any $k \times n$ matrix $\mathbf{G}$ whose row space is equal to C is called a *generator matrix* of C.

7.7.1 Decoding of Linear Codes. If $\mathbf{c}$ is a codeword and $\mathbf{w}$ is the received word after transmission through a noisy channel, then $\mathbf{e} = \mathbf{w} - \mathbf{c} = \mathbf{e}_1 \cdots \mathbf{e}_n$ is called the *error word* or the *error vector*. All possible error vectors $\mathbf{e}$ of a received vector $\mathbf{w}$ are the vectors in the coset of $\mathbf{w}$. The most likely error vector $\mathbf{e}$ has the minimum weight in the coset of $\mathbf{w}$. Hence, we decode $\mathbf{w}$ as $\mathbf{c} = \mathbf{w} - \mathbf{e}$.

Let $C \subseteq F_q^n$ be a linear (n, k) code and let F_q^n/C be the factor space. An element of minimum weight in a coset $\mathbf{a} + C$ is called a *coset leader* of $\mathbf{a} + C$. If several vectors in $\mathbf{a} + C$ have minimum weight, we choose one of them as coset leader. Let $\mathbf{a}^{(1)}, \ldots, \mathbf{a}^{(n)}$ be the coset leaders of the cosets $\neq C$ and let $\mathbf{c}^{(1)} = \mathbf{0}, \mathbf{c}^{(2)}, \ldots, \mathbf{c}^{(q^k)}$ be all codewords in C. Consider the following array:

$\mathbf{c}^{(1)}$	$\mathbf{c}^{(2)}$	$\cdots$	$\mathbf{c}^{(q^k)}$	$\}$ row of codewords
$\mathbf{a}^{(1)} + \mathbf{c}^{(1)}$	$\mathbf{a}^{(1)} + \mathbf{c}^{(2)}$	$\cdots$	$\mathbf{a}^{(1)} + \mathbf{c}^{(q^k)}$	
$\vdots$	$\vdots$		$\vdots$	remaining cosets
$\mathbf{a}^{(s)} + \mathbf{c}^{(1)}$	$\mathbf{a}^{(s)} + \mathbf{c}^{(2)}$	$\cdots$	$\mathbf{a}^{(s)} + \mathbf{c}^{(q^k)}$	

$\underbrace{\phantom{\mathbf{a}^{(s)} + \mathbf{c}^{(1)}}}_{\text{column of coset leaders}}$

After receiving a word $\mathbf{w} = \mathbf{a}^{(j)} + \mathbf{c}^{(j)}$, the decoder decides that the error $\mathbf{e}$ is the corresponding coset leader $\mathbf{a}^{(j)}$ and decodes $\mathbf{w}$ as the codeword $\mathbf{c} = \mathbf{w} - \mathbf{e} = \mathbf{c}^{(j)}$. Thus, $\mathbf{w}$ is decoded as the codeword in the column of $\mathbf{w}$. The coset of $\mathbf{w}$ can be determined by calculating the *syndrome* of $\mathbf{w}$, which is defined as follows: Let $\mathbf{H}$ be the parity-check matrix of a linear (n, k) code C. Then the vector $S(\mathbf{w}) = \mathbf{Hw}^T$ of length $n - k$ is the syndrome of $\mathbf{w}$. Obviously, if $\mathbf{w}, \mathbf{z} \in F_q^n$, then (i) $S(\mathbf{w}) = \mathbf{0}$ iff $\mathbf{w} \in C$; and (ii) $S(\mathbf{w}) = S(\mathbf{z})$ iff $\mathbf{w} + C = \mathbf{z} + C$.

The decoding procedure is generally based on the COSET-LEADER ALGORITHM for error correction of linear codes, defined as follows: Let $C \subseteq F_q^n$ be a linear (n, k) code and let $\mathbf{w}$ be the received vector. To correct errors in $\mathbf{w}$, follow these steps:

(i) Calculate $S(\mathbf{w})$;

(ii) Find the coset leader, say $\mathbf{e}$, with syndrome equal to $S(\mathbf{w})$; and

(iii) Decode $\mathbf{w}$ as $\mathbf{c} = \mathbf{w} - \mathbf{e}$, where $\mathbf{c}$ is the codeword with minimum distance to $\mathbf{w}$.

8

Galois Fields

8.1 Finite Fields

We have applied the notion of a finite field as a finite vector space in previous chapters. As a structure, a field has two operations, denoted by $+$ and $*$, which are not necessarily ordinary addition and multiplication. Under the operation $+$ all elements of a field form a commutative group whose identity is denoted by 0 and inverse of a by $-a$. Under the operation $*$, all elements of the field form another commutative group with identity denoted by 1 and the inverse of a by a^{-1}. Note that the element 0 has no inverse under $*$. There is also a *distributive identity* that links $+$ and $*$, such that $a * (b + c) = (a * b) + (a * c)$ for all field elements a, b and c. This identity is the same as the *cancellation property*: If $c \neq 0$ and $a * c = b * c$, then $a = b$. The following groups represent fields:

1. The set $\mathbb{Z}$ of rational numbers, or the set $\mathbb{R}$ of real numbers, or the set $\mathbb{C}$ of complex numbers, are all infinite fields under ordinary addition and multiplication.

2. The set F_p of integers mod p, where p is a prime number, can be regarded as a group under $+$ (ordinary addition followed by remainder on division by p). All its elements except 0 form a group under $*$ (ordinary multiplication followed by remainder on division by p), with the identity 1, while the inverse of an element $a \neq 0$ is not obvious. However, it is always possible to find the unique inverse x using the extended Euclidean algorithm, according to which, since p is prime and $a \neq 0$, the $\gcd(p, a) = 1$. Thus, the inverse x is the inverse of a. However, x might be a negative integer; in that case, just add p to obtain it as an element of F_p.

There exists a unique finite field with p^n elements for any integer $n > 1$. This field is denoted by $\mathrm{GF}(p^n)$ and is known as a Galois field. In particular, for $p = 2$, it has 2^n elements for $n > 1$. The new American Encryption

Standard (AES) uses $2^8 = 256$ elements. Of these 256 elements, all elements are represented as all possible 8-bit strings; addition is the bitwise XOR operation (or bitwise addition mod 2), the zero element is $\mathbf{0} = 00000000$, and the identity element is $\mathbf{1} = 00000001$. However, multiplication is not easy: an element is regarded as a polynomial of degree 7 with coefficients in the field F_2, which are 0 or 1, and use a special technique for multiplication of these polynomials, which is explained below.

8.1.1 Extension Fields. The ASCII (American Standard Code for Information Interchange) uses the 8-bit binary values (see Table A.1) in which each number, letter, and other characters correspond to the keys on computer keyboards. Thus, the sentence "A B; " when typed with these keys has the value

$$\text{``01000001 \ 010000010 \ 00100000 \ 00111011''}.$$

The field of binary numbers, having strings of 0s and 1s, is denoted by F_{2^m}. Since the digital data is transmitted in multiples of eight bits (2^3), we can regard each message as having the values from a field of 2^m bits, where $m > 0$ is an integer. The field of our choice must be F_{2^m}, which has the number of elements 2^m. This can be generalized to a field F_{p^m}, where p is prime. However, the field F_{2^m} has a serious drawback, namely that it does not meet the criterion for inverses. For example, consider the field F_{2^4}. It is easy to see that this field has no multiplicative inverse, i.e., $2b \bmod 16 = 1$ has no solution b. Two elements of a field are multiplicative inverses of each other if their product yields 1. Thus, F_{2^4} is not a field with multiplication as defined in F_p. This situation forces us to look for a different arithmetic for multiplication, such as the polynomial arithmetic, since $F_{2^m} \simeq F_2[x]/ < g(x) >$, where $g(x)$ is a minimal polynomial with root α. The new arithmetic replaces the non-prime 2^m with a 'prime polynomial' which is irreducible $p(x)$, and gives us an extension field $\mathrm{GF}(2^m)$.

We construct the extension field $\mathrm{GF}(2^m)$ of $F_2[x]$ using the above isomorphism and find a minimal polynomial with a root in F_{2^4}. Since we need 16 elements in $\mathrm{GF}(2^m)$, the degree of this polynomial must be 4 because $2^4 = 16$. In addition, we require another condition to be fulfilled: This minimal polynomial must be primitive in $\mathrm{GF}(2^4)$. Thus, every element in F_{2^4} must be expressible as some power of $g(x)$. Under these two restrictions, there are only two choices that may be made from primitive polynomials among all possible nonzero elements of $\mathrm{GF}(2^4)$, namely $x^4 + x^3 + 1$ and $x^4 + x + 1$, of which either may be used. The process to generate the elements is as follows: Start with three initial elements of the extension field and perform all arithmetic operations mod $g(x) = x^4 + x^3 + 1$ (which we have chosen). The initial elements are $0, 1, \alpha$. Raising α to successive powers identifies α^2 and α^3 as members of the extension field; note that α^4 is not an element of F_{2^4}. Since

$g(\alpha) = 0 = \alpha^4 + \alpha^3 + 1$, we have the identity $\alpha^4 = \alpha^3 + 1$,[1] and we use this identity to reduce each power greater than 3.

Example 8.1.1. In order to multiply 1011 by 1101, we use the polynomial representation of these numbers and get

$$(x^3 + x + 1)(x^3 + x^2 + 1) = x^6 + x^5 + x^4 + 3x^3 + x^2 + x + 1.$$

Using the identity $\alpha^4 = \alpha^3 + 1$, the expression on the right side reduces to x, i.e., $(x^3 + x + 1)(x^3 + x^2 + 1) = x$, which gives the result as $(1011) \times (1101) = 0010$. ∎

The above identity can be used to reduce every power of α, which will build an entire field $GF(2^4)$. These fields of 16 elements are presented below in Tables 8.1.1 and 8.1.2.

Table 8.1.1 Field of 16 Elements

(a) Generated by $x^4 + x + 1$ (b) Generated by $x^4 + x^3 + 1$

Power	4-bit Form	Polynomial	4-bit Form	Polynomial
0	0000	0	0000	0
1	0001	1	0001	1
α	0010	α	0010	α
α^2	0100	α^2	0100	α^2
α^3	1000	α^3	1000	α^3
α^4	0011	$\alpha + 1$	1001	$\alpha^3 + 1$
a^5	0110	$\alpha^2 + \alpha$	1011	$\alpha^3 + \alpha + 1$
α^6	1100	$\alpha^3 + \alpha^2$	1111	$\alpha^3 + \alpha^2 + \alpha + 1$
α^7	1011	$\alpha^3 + \alpha + 1$	0111	$\alpha^2 + \alpha + 1$
α^8	0101	$\alpha^2 + 1$	1110	$\alpha^3 + \alpha^2 + \alpha$
α^9	1010	$\alpha^3 + \alpha$	0101	$\alpha^2 + 1$
α^{10}	0111	$\alpha^2 + \alpha + 1$	1010	$\alpha^3 + \alpha$
α^{11}	1110	$\alpha^3 + \alpha^2 + \alpha$	1101	$\alpha^3 + \alpha^2 + 1$
α^{12}	1111	$\alpha^3 + \alpha^2 + \alpha + 1$	0011	$\alpha + 1$
α^{13}	1101	$\alpha^3 + \alpha^2 + 1$	0110	$\alpha^2 + \alpha$
α^{14}	1001	$\alpha^3 + 1$	1100	$\alpha^3 + \alpha^2$

Note that if instead of a primitive polynomial like $x^4 + x + 1$ or $x^4 + x^3 + 1$, a nonprimitive polynomial is used to generate the field of 16 elements in $GF(16)$, the capability of encoding and decoding messages is destroyed or at least impaired. However, since some types of codes need nonprimitive polynomials, we provide an example of generating elements in $GF(16)$ by the

---

[1] Recall that in a binary field, addition is equivalent to multiplication.

nonprimitive polynomial $x^4 + x^3 + x^2 + x + 1$, which is presented in Table 8.1.2.

Table 8.1.2 Field of 16 Elements
Generated by $x^4 + x^3 + x^2 + x + 1$

Power	4-bit Form	Polynomial	$\parallel$	Power	4-bit Form	Polynomial
0	0000	0	$\parallel$	α^7	0100	α^2
1	0001	1	$\parallel$	α^8	1000	α^3
α	0010	α	$\parallel$	α^9	1111	$\alpha^3 + \alpha^2 + \alpha + 1$
α^2	0100	α^2	$\parallel$	α^{10}	0001	1
α^3	1000	α^3	$\parallel$	α^{11}	0010	α
α^4	1111	$\alpha^3 + \alpha^2 + \alpha + 1$		α^{12}	0100	α^2
a^5	0001	1	$\parallel$	α^{13}	1000	α^3
α^6	0010	α	$\parallel$	α^{14}	1111	$\alpha^3 + \alpha^2 + \alpha + 1$

Table 8.1.3 Field of 32 Elements
Generated by $x^5 + x^2 + 1$

Power	4-bit Form	Polynomial	$\parallel$	Power	4-bit Form	Polynomial
0	00000	0	$\parallel$	α^{15}	11111	$\alpha^4 + a^3 + \alpha^2 + \alpha + 1$
1	00001	1	$\parallel$	α^{16}	11011	$\alpha^4 + a^3 + \alpha + 1$
α	00010	α	$\parallel$	α^{17}	10011	$\alpha^4 + \alpha + 1$
α^2	00100	α^2	$\parallel$	α^{18}	00011	$\alpha + 1$
α^3	01000	α^3	$\parallel$	α^{19}	00110	$\alpha^2 + \alpha$
α^4	10000	α^4	$\parallel$	α^{20}	01100	$\alpha^3 + \alpha^2$
a^5	00101	$\alpha^2 + 1$	$\parallel$	α^{21}	11000	$\alpha^4 + \alpha^3$
α^6	01010	$\alpha^3 + \alpha$	$\parallel$	α^{22}	10101	$\alpha^4 + \alpha^2 + 1$
α^7	10100	$\alpha^4 + \alpha^2$	$\parallel$	α^{23}	01111	$\alpha^3 + \alpha^2 + \alpha + 1$
α^8	01101	$\alpha^3 + \alpha^2 + 1$	$\parallel$	α^{24}	11110	$\alpha^4 + \alpha^3 + \alpha^2 + \alpha$
α^9	11010	$\alpha^4 + \alpha^3 + \alpha$	$\parallel$	α^{25}	11001	$\alpha^4 + \alpha^3 + 1$
α^{10}	10001	$\alpha^4 + 1$	$\parallel$	α^{26}	10111	$\alpha^4 + \alpha^2 + \alpha + 1$
α^{11}	00111	$\alpha^2 + \alpha + 1$	$\parallel$	α^{27}	01011	$\alpha^3 + \alpha + 1$
α^{12}	01110	$\alpha^3 + \alpha^2 + \alpha$	$\parallel$	α^{28}	10110	$\alpha^4 + \alpha^2 + \alpha$
α^{13}	11100	$\alpha^4 + \alpha^3 + \alpha^2$	$\parallel$	α^{29}	01001	$\alpha^3 + 1$
α^{14}	11101	$\alpha^4 + a^3 + \alpha^2 + 1$	$\parallel$	α^{30}	10010	$\alpha^4 + \alpha$

Tables 8.1.1 and 8.1.2 can be easily generated for GF(16) if the following rules are followed: (i) multiplication by an element α will produce a left shift

of the previous binary value, and (ii) if a 1 is shifted out of the fourth position (since this amounts to a polynomial of degree 4), the result is reduced by subtracting the generator polynomial from the result. Recall that subtraction in binary arithmetic is the XOR function. The table for FG(32) is similarly generated and is given in Table 8.1.3. The other two tables generated by the polynomials $x^5 + x^4 + x^3 + x^2 + 1$ and $x^5 + x^4 + x^2 + x + 1$ can be similarly constructed in GF(32).

8.2 Construction of Galois Fields

The Galois Theorem answers all questions completely about the existence and uniqueness of finite fields. First, the number of elements q in a finite field must be a prime power, i.e. $q = p^r$, where p is prime and $r \in \mathbb{Z}^+$. Second, for each prime power $q = p^r$, there exists a field of order q, and it is unique up to isomorphism. The construction is as follows: Let F_0 be a field of integers mod p. Choose an irreducible polynomial $f(x)$ of degree r over F_0, of the form

$$f(x) = x^r + c_{r-1}x^{r-1} + \cdots + c_1 x + c_0, \qquad (8.2.1)$$

with the leading coefficient equal to 1. Let the elements of F be of the form $x_0 + x_1 a + x_2 a^2 + \cdots + x_{r-1}a^{r-1}$, where a is chosen such that $f(a) = 0$, and $x_0, \ldots, x_{r-1} \in F_0$. Note that this construction is very similar to that for the complex numbers of the form $x + iy$, where x, y are real numbers and $i^2 + 1 = 0$. The number of expressions of the above form is p^r, since there are p choices for each of the r coefficients $x_0, \ldots, x_{r-1}$. Addition of these expressions is element-wise, but to multiply them, notice that by (8.2.1) we have $a^r = -c_{r-1}a^{r-1} - \cdots - c_1 a - c_0$; thus, a^r, and similarly any higher power of a, can be reduced to the required form. Then the irreducibility of the polynomial f shows that this construction is a field, and although there are different choices for irreducible polynomials, the fields so constructed will be all isomorphic.

Example 8.2.1. To construct a field of order $9 = 3^2$, where $p = 3$ and $r = 2$, we take the polynomial $f(x) = x^2 + 1$, which is irreducible over the field of integers mod 3. The elements of this field are all expressions of the form $x + ya$, where $a^2 = 2$ and $x, y = 0, 1, 2$. Then

$$(2 + a) + (2 + 2a) = 4 + 3a = 1,$$
$$(2 + a)(2 + a) = 4 + 6a + 2a^2 = 4 + 0 + 4 = 8 = 2 \mod 3. \blacksquare$$

Some properties of the Galois field $F = GF(q)$, where $q = p^r$, p prime, are as follows:

(i) The elements $0, 1, 2, \ldots, p - 1 \in F$ form a subfield F_0 which is isomorphic to the integers mod p; it is known as the *prime subfield of F*.

(ii) The *additive group* of $GF(q)$ is an elementary Abelian p-group, since $x + \cdots + x = (1 + \cdots + 1)x = 0x = 0$. The sum is the direct sum of r cyclic groups of order p.

Thus, F is a vector space of dim r over F_1; that is, there is a *basis* $(a_1, \ldots, a_r)$ such that every element $x \in F$ can be written uniquely in the form $x = x_1 a_1 + \cdots + x_r a_r$ for some $a_1, \ldots, a^r \in F_0 = \{0, 1, \ldots, p-1\}$.

(iii) The *multiplicative group* of $GF(q)$ is a cyclic group; that is, there exists an element g, called a *primitive root*, such that every nonzero element of F can be written uniquely in the form g^j for some j, $0 \leq j \leq q-2$; also, $g^{q-1} = g^0 = 1$.

(iv) Let α be an element in a finite field F_q. Then α is a primitive element of F_q if $F_q = \{0, \alpha, \alpha^2, \ldots, \alpha^{q-1}\}$. For example, consider the field F_4, where α is a root of the irreducible polynomial $1 + x + x^2 \in F_4$. Then $\alpha^2 = -(1 + \alpha) = 1 + \alpha^2, \alpha^3 = \alpha(\alpha^2) = \alpha(1 + \alpha) = \alpha + \alpha^2 = \alpha + 1 + \alpha = 1$, and so $F_4 = \{0, \alpha, 1 + \alpha, 1\} = \{0, \alpha, \alpha^2, \alpha^3\}$, and α is a primitive element.

(v) Assume that q is odd. Then the cyclic group of order $q-1$ has the property that exactly one half of its elements are *squares*, i.e., they are even powers of a primitive element. These squares are also known as *quadratic residues*, and the non-squares as *quadratic non-residues*.

(vi) An automorphism of F is a one-to-one mapping $x \mapsto x^\pi$ from F onto F, such that $(x + y)^\pi = x^\pi + y^\pi$, $(xy)^\pi = x^\pi y^\pi$ for all $x, y \in F$. In general, the map $\sigma : x \mapsto x^p$ is an automorphism of F, called the *Frobenius automorphism*. The elements of F fixed by this automorphism are exactly those contained in the prime subfield F_0. Further, the group of automorphims of F is cyclic of order r, generated by σ; that is, every automorphism has the form $x \mapsto x^{p^j}$ for some value of j, $0 \leq j \leq r-1$.

(vii) Since F has bases of size r as a vector space F_0, these bases can be chosen with certain additional properties. The simplest type of basis is one of the form $\{1, a, a^2, \ldots, a^{r-1}\}$, where a is the zero of an irreducible polynomial of degree r. In general, a basis of the form $\{a, a^\sigma, a^{\sigma^2}, \ldots, a^{\sigma^{r-1}}\}$, where σ is the Frobenius automorphism, is called a *normal basis*. Such a basis always exists, and in particular the automorphism group of F has a simple form relative to a normal basis, as the basis elements are themselves permuted cyclically by the automorphism.

(vii) If the field $GF(p^r)$ has a subfield $GF(p_1^s)$, where p_1 is prime, then $p_1 = p$ and s divides r. Conversely, if s divides r, then $GF(p^r)$ has a unique subfield of order p^s.

(viiii) Addition in $GF(q)$ is easy if a basis has already been chosen; thus,

$$(x_1 a_1 + \cdots + x_r a_r) + (y_1 a_1 + \cdots + y_r a_r)$$

$$= (x_1 + y + 1)a_1 + \cdots + (x_r + y_r)a_r,$$

that is, addition is 'coordinate-wise'. Similarly, multiplication is easy if a primitive root g has already been chosen, so that

$$\left(g^j\right) \cdot \left(g^k\right) = g^{j+k},$$

where the exponent is reduced mod $q - 1$, if necessary. Note that in order to perform these two operations we need a table to tell us how to translate between the two representations.[2]

Example 8.2.2. For the field $GF(9)$, using an element a satisfying $a^2 = 2$ over the integers mod 3, we find that $g = 1 + a$ is a primitive element, and the 'table' is as follows:

g^0	g^1	g^2	g^3	g^4	g^5	g^6	g^7
1	$a+1$	$2a$	$2a+1$	2	$2a+2$	a	$a+2$

As an example, we get $(a + 2)(2a + 2) = g^7 \cdot g^5 = g^{12} = g^4 = 2.$ ∎

8.3 Galois Field of Order p

For prime p, let F_p be the set $\{0, 1, \dots, p-1\}$ of integers and let $\phi : Z/(p) \mapsto F_p$ be the mapping defined by $\phi([a]) = a$ for $a = 0, 1, \dots, p - 1$, where $Z/(p)$ is the ring of integers. Then F_p under the field structure defined by ϕ is a finite field, called the *Galois field of order p*. The mapping ϕ is an isomorphism so that $\phi([a] + [b]) = \phi([a]) + \phi([b])$ and $\phi([a][b]) = \phi([a])\phi([b])$. The finite field F_p has the zero element 0, identity 1, and its structure is exactly like that of $Z/(p)$. Hence, it follows the ordinary arithmetic of integers with reduction modulo p.

Example 8.3.1. (i) The field $Z/(5)$ is isomorphic to $F_5 = \{0, 1, 2, 3, 4\}$, where the isomorphism is given by: $[0] \to 0, [1] \to 1, [2] \to 2, [3] \to 3, [4] \to 4$. The operations $+$ and $*$ in F_5 are defined as in Table 8.3.1

Table 8.3.1 Addition and Multiplication in F_5

+	0	1	2	3	4		*	0	1	2	3	4
0	0	1	2	3	4		0	0	0	0	0	0
1	1	2	3	4	0		1	0	1	2	3	4
2	2	3	4	0	1		2	0	2	4	1	3
3	3	4	0	1	2		3	0	3	1	4	2
4	4	0	1	2	3		4	0	4	3	2	1

(ii) In the finite field F_2 with elements 0 and 1 (known as the binary elements), which we will encounter mostly, the two operations are defined as

---

[2] This table is like the table of logarithms, since if $g^j = x$, we can regard g as the exponential and j as the 'logarithm' of x.

+	0	1
0	0	1
1	1	0

*	0	1
0	0	0
1	0	1 ∎

If b is any nonzero element of the ring Z of integers, then the additive order of b is infinite, i.e., $nb = 0$ implies that $n = 0$. However, in the ring $Z/(p)$, where p is prime, the additive order of every nonzero element b is p, i.e., $pb = 0$, and p is the least positive integer for which this equality holds. Thus, if R is any ring and if there exists a positive integer n such that $nr = 0$ for every $r \in R$, then the least such positive integer n is called the *characteristic* of R and R is said to have (positive) characteristic n. If no such integer n exists, then R is said to have characteristic 0. This leads to an important result: A ring $R \neq \{0\}$ of positive characteristic having an identity and no zero divisors must have a prime characteristic. Thus, a finite field has a prime characteristic. Moreover, if R is a communicative ring of prime characteristic p, then

$$(a \pm b)^{p^n} = a^{p^n} \pm b^{p^n} \quad \text{for } a, b \in R \text{ and } n \in \mathbb{N}.$$

PROOF. Since $\binom{p}{i} = \dfrac{p(p-1)\cdots(p-i+1)}{1 \cdot 2 \cdot \,\cdots\, \cdot i} \equiv 0 \,(\mathrm{mod}\ p)$ for all $i \in \mathbb{Z}$, $0 < i < p$, by the binomial theorem we get

$$(a+b)^p = a^p + \binom{p}{1}a^{p-1}b + \cdots + \binom{p}{p-1}ab^{p-1} + b^p = a^p + b^p,$$

and by induction on n the proof follows for the $+$ part of the identity. Therefore, the second identity follows from $a^{p^n} = [(a-b)+b]^{p^n} = (a-b)^{p^n} + b^{p^n}$. ∎

Let R be a commutative ring with identity. An element $a \in R$ is called a *divisor* of $b \in R$ if there exists $c \in R$ such that $ac = b$. A *unit* of R is a divisor of the identity; two elements $a, b \in R$ are said to be *associates* if there is a unit $\varepsilon \in R$ such that $a = b\varepsilon$. An element $c \in R$ is called a *prime element* if it is not a unit and if it has only the units of R and the associates of c as divisors. An ideal $P \neq R$ of the ring R is called a *prime ideal* if for $a, b \in R$ we have $ab \in P$ if either $a \in P$ or $b \in P$. An ideal $M \neq R$ of R is called a *maximal ideal* of R if for any ideal J of R the property $M \subseteq J$ implies $J = R$ or $J = M$. Also, R is said to be a *principal ideal domain* if R is an integral domain and if every ideal J of R is principal, i.e., if there is a generating element a for J such that $J = (a) = \{ra : r \in R\}$. Thus, if R is a commutative ring with identity, then

(i) An ideal M of R is a maximal ideal iff R/M is a field;

(ii) An ideal P of R is a prime ideal iff R/P is an integral domain;

(iii) Every maximal ideal of R is a prime ideal; and

(iv) If R is a principal ideal domain, then $R/(c)$ is a field iff c is a prime element of R.

Proofs of these results can be found in most of books on modern algebra.

8.4 Prime Fields

Let p be prime. Then all integers mod p, which consist of the integers $0, 1, \ldots, p-1$ with addition and multiplication performed mod p, is a finite field of order p, denoted by F_p, where p is called the *modulus* of F_p. For any integer a, a mod p will denote the unique integer remainder r, $0 \le r \le p-1$, which is obtained upon dividing a by p. This operation is known as *reduction modulo p*.

Example 8.4.1. The elements of the prime field F_{29} are $\{0, 1, \ldots, 28\}$, and some examples of arithmetic operations on this field are:

(i) Addition: $15 + 20 = 6$, since 35 mod 29 = 6.

(ii)Subtraction: $15 - 20 = 24$, since -5 mod 29 = 24.

(iii) Multiplication: $15 * 20 = 10$, since 300 mod 29 = 10.

(iv) Inversion: $15^{-1} = 2$, since $15 * 2$ mod 29 = 1. ∎

8.5 Binary Fields

These finite fields of order 2^r are denoted by F_q, where $q = 2^r$. A useful method to construct these fields is to use a polynomial basis representation, where the elements of F_q are the binary polynomials, that is, they are polynomials with coefficients in the field $F_3 = \{0, 1\}$ of degree at most $r - 1$:

$$F_{2^r} = \left\{ a_{r-1}x^{r-1} + a_{r-2}x^{r-2} + \cdots + a_2 x^2 + a_1 x + a_0 \ : \ a_i \in \{0, 1\} \right\}.$$

We choose an irreducible polynomial $f(x)$ of degree r, which means that $f(x)$ cannot be factored as a product of binary polynomials of degree less than r. The addition of field elements is the usual addition of polynomials, with coefficients mod 2. Multiplication of field elements is performed modulo $f(x)$, which is known as the reduction polynomial. Thus, for any binary polynomial $p(x)$, where $p(x)$ is mod $f(x)$, the division of $p(x)$ by $f(x)$ yields a unique remainder polynomial $r(x)$ of degree less than r, and this operation is known as *reduction mod* $f(x)$. Also note that $-a = a$ for all $a \in F_{2^r}$.

Example 8.5.1. For $r = 3$, the elements of F_8 are the following 8 binary polynomials of degree at most 2:

$$0; \quad 1; \quad x; \quad x+1; \quad x^2; \quad x^2+1; \quad x^2+x; \quad x^2+x+1.$$

For $r = 4$, the elements of F_{16} are the following 16 binary polynomials of degree at most 3:

0	x^2	x^3	$x^3 + x^2$
1	$x^2 + 1$	$x^3 + 1$	$x^3 + x^2 + 1$
x	$x^2 + x$	$x^3 + x$	$x^3 + x^2 + x$
$x + 1$	$x^2 + x + 1$	$x^3 + x + 1$	$x^3 + x^2 + x + 1$ ∎

Example 8.5.2. The arithmetic operations in F_{2^4} with the reduction polynomial $x^4 + x + 1$ are as follows:

(i) Addition: $(x^3 + x^2 + 1) + (x^3 + x + 1) = x^3 + x$.

(ii) Subtraction: $(x^3 + x^2 + 1) - (x^2 + x + 1) = x^3 + x$, since $-1 = 1$ in F_{16}.

(iii) Multiplication: $(x^3 + x^2 + 1) * (x^3 + x + 1) = x^5 + x + 1 = x^2 + 1$ mod $(x^4 + x + 1)$, since the remainder upon dividing $x^5 + x + 1$ by $x^4 + x + 1$ is $x^2 + 1$.

(iv) Inversion: $(x^3 + x^2 + 1)^{-1} = x^2$, since $(x^3 + x^2 + 1) * x^2 = 1$ mod $(x^4 + x + 1)$. ∎

Example 8.5.3. (Isomorphic Fields) There are three irreducible binary polynomials of degree 4 in the field F_{2^4}, namely $f_1(x) = x^4 + x + 1$, $f_2(x) = x^4 + x^3 + 1$, and $f_3(x) = x^4 + x^3 + x^2 + x + 1$. Each of these three reduction polynomials can be used to construct the field F_{2^4} as follows: Let us denote the three resulting fields by G_1, G_2, and G_3. Then the field elements of G_1, G_2 and G_3 are the same 16 binary polynomials of degree at most 3, as listed in Example 8.5.1. Although these fields may appear to be different, for example, $x^3 * x = x + 1$ in G_1, $x^3 * x = x^3 + 1$ in G_2, and $x^3 * x = x^3 + x^2 + x + 1$ in G_3, all fields of a given order are isomorphic, i.e., the differences are only in labeling of the elements. For example, an isomorphism between G_1 and G_2 may be established by finding $c \in G_2$ such that $f_1(c) = 0 \,(\text{mod} f_2)$ and then extending $x \mapsto c$ to an isomorphism $\phi : G_1 \to G_2$; in this case, the choices for c are: $x^2 + x, x^2 + x + 1, x^3 + x^2$, and $x^3 + x^2 + 1$. ∎

8.5.1 Polynomial Basis Representation. The polynomial basis representation for binary fields, discussed above, can be generalized to an extension field. Let p be a prime and $r \geq 2$. Let $F_p[x]$ denote the set of all polynomials in x with coefficients from F_p. As before, let $f(x)$ be the reduction polynomial, which is an irreducible polynomial of degree r in $F_p[x]$. The elements of F_{p^r} are the polynomials in $F_p[x]$ of degree at most $r - 1$ of the form

$$F_{p^r} = \left\{ a_{r-1}x^{r-1} + a_{r-2}x^{r-2} + \cdots + a_2x^2 + a_1x + a_0 : a_i \in F_{p^r} \right\}, \quad (8.5.2)$$

where addition of field elements is the usual addition of polynomials, with coefficients arithmetic performed in F_p, while their multiplication is performed modulo the reduction polynomial $f(x)$.

Example 8.5.4. (Extension field) For $p = 251$ and $r = 5$ the polynomial $f(x) = x^5 + x^4 + 12x^3 + 9x^2 + 7$ is irreducible in $F_{251}[x]$. This $f(x)$ will be the reduction polynomial for the construction of F_{251^5}, which is a finite field of order 251^5. The elements of the field F_{251^5} are the polynomials of $F_{251}[x]$ of degree at most 4. Some examples of arithmetic operations in F_{125} are as follows: Let $a = 123x^4 + 76x^2 + 7x + 4$ and $b = 196x^4 + 12x^3 + 225x^2 + 76$. Then

(i) Addition: $a + b = 68x^4 + 12x^3 + 50x^2 + 7x + 80$.

(ii) Subtraction : $a - b = 178x^4 + 239x^3 + 102x^2 + 7x + 179$,

(iii) Multiplication: $a * b = 117x^4 + 151x^3 + 117x^2 + 182x + 217$.

(iv) Inversion: $a^{-1} = 109x^4 + 111x^3 + 250x^2 + 98x + 85$.

8.6 Arithmetic in Galois Field

For performing arithmetic operations of addition or subtraction and multiplication or division in binary and nonbinary cases, it is important to keep the following material ready to use.

Before we describe encoding and decoding methods for RS codes, a short review of arithmetic in $\text{GF}(q = 2^m)$ is in order. Assuming that we are working in blocks of 8 bits (one byte or symbol), if we happen to work in GF(251) we will not use all the advantages of 8-bit arithmetic. In addition to being wasteful, this situation will definitely create problems in the case when the data contains a block of 8 bits that corresponds to the number 252. Thus, working in a field $\text{GF}(2^m)$ with 2^m elements appears to be natural. Arithmetic in this field is done by finding irreducible polynomials $f(x)$ of degree m and doing all arithmetic in $F_2[f(x)]$ where all coefficients are modulo 2, arithmetic is done modulo $f(x)$, and $f(x)$ cannot be factored over GF(2).

Example 8.6.1. Consider $\text{GF}(2^8)$, which has an irreducible polynomial $f(x) = x^8 + x^6 + x^5 + x + 1$. A byte can be regarded as a polynomial in this field by letting the lsb represent x^0 and the i-th lsb represent x^i. Thus, the byte (10010010) represents the polynomial $x^7 + x^4 + x$. Addition in $\text{GF}(2^m)$ is carried out by XOR-ing two bytes. For example, (10010010) $\oplus$ (10101010) = 00111000, which in terms of polynomials states that $(x^7 + x^4 + x) + (x^7 + x^5 + x^3 + x) = x^5 + x^4 + x^3$. The operation of subtraction is the same thing as addition. Multiplication is modulo $f(x)$: For example, $(x^4 + x) * (x^4 + x^2) = x^8 + x^6 + x^5 + x^3$, which must be reduced modulo $f(x)$. Since we have $f(x) = 0$, i.e., $x^8 = x^6 + x^5 + x + 1$, the above product reduces to $(x^4 + x) * (x^4 + x^2) = x^8 + x^6 + x^5 + x^3 = (x^6 + x^5 + x + 1) + x^6 + x^5 + x^3 = x^3 + x + 1$, because $x^6 + x^6 = 0$ and $x^5 + x^5 = 0$. This product can be written in terms of the bytes as (00010010) * (00010100) = 00001011. Division is similar to multiplication. ■

It would be very inconvenient if we had to carry out addition and multiplication for every $f(x)$ in the field $F_2^m[f(x)]$ each time it is needed. However, addition and multiplication lookup tables for all possible 256 pairs are either available or must be prepared for the field of interest. These arithmetic operations in GF(2^m) for $2 \leq m \leq 8$ can be easily performed with great speed using memory and processing in software or hardware. Some of the hardware circuitry is presented for certain fields under consideration throughout this book.

8.6.1 Addition in Extension Field GF(2^m). Each of the 2^m elements of GF(2^m) can be represented as a distinct polynomial of degree $\leq (m-1)$. Let the polynomial $a_i(x)$ denote each of the nonzero elements of GF(2^m), of which at least one of the m coefficients of $a_i(x)$ is nonzero. For $i = 0, 1, 2, \ldots, 2^m - 2$, let

$$\alpha^i = a_i(x) = a_{i,0} + a_{i,1}(x) + a_{i,2}x^2 + \cdots + a_{i,m-1}x^{m-1}. \tag{8.6.1}$$

As an example, consider the case $m = 3$. This is the Galois field GF(2^3), and the mapping of the eight elements α^i, $0 \leq i \leq 7$, and the zero element in terms of the basis elements $\{x^0, x^1, x^2\}$, as defined by Eq (8.6.1), is given in Table 8.6.1. Notice that each row in this table contains a sequence of binary values representing the coefficients $a_{i,0}, a_{i,1}$ and $a_{i,2}$ of Eq (8.6.1). Some useful formulas are as follows:

$$+1 = -1; \quad \alpha^i = -\alpha^i \text{ for } i = 0, 1, \ldots, k;$$

$$\alpha^i \alpha^j = \alpha^{i+j} \text{ for any } i \text{ and } j \ (i, j = 0, 1, \ldots, k);$$

$$1\alpha^i = \alpha^0 \alpha^i = \alpha^i; \quad \alpha^i + \alpha^i = 0 \text{ (see formula (2.2.1))}; \tag{8.6.2}$$

$$2\alpha^i = \alpha^i + \alpha^i = 0; \quad 2n\alpha^i = n(2\alpha^i) = 0 \text{ for } n = 1, 2, \ldots.$$

Table 8.6.1 Basis Element and Field Elements for $m = 3$

	x^0	x^1	x^2
0	0	0	0
$\alpha^0 = 1$	1	0	0
α^1	0	1	0
α^2	0	0	1
α^3	1	1	0
α^4	0	1	1
α^5	1	1	1
α^6	1	0	1
$\alpha^7 = \alpha^0 = 1$	1	0	0

8.6.2 Primitive Polynomials. An irreducible polynomial $f(x)$ of degree m is said to be *primitive* if the smallest positive integer n for which $f(x)$ divides $x^n + 1$ is $n = 2^m - 1$. A table for primitive polynomials over F_2 is given in

Appendix C (Table C.2).

Example 8.6.2. Consider the irreducible polynomials (a) $1 + x + x^2$, and (b) $1 + x + x^2 + x^3 + x^4$. Both polynomials are of degree 4. In (a) the polynomial will be primitive if we determine whether it divides $x^n + 1 = x^{2^m - 1} + 1 = x^{15} + 1$, but does not divide $x^n + 1$ for values of $n \in [1, 15)$. In fact, it is easy to verify that $1 + x + x^4$ divides $x^{15} + 1$, but after repeated divisions it can be verified that $1 + x + x^4$ will not divide $x^{15} + 1$ for any $n \in [1, 15)$. Thus, $1 + x + x^4$ is a primitive polynomial. In (b) the polynomial $1 + x + x^2 + x^3 + x^4$ obviously divides $x^{15} + 1$, but it divides $x^{15} + 1$ for some $n < 15$. So this polynomial is not primitive, although it is irreducible. ∎

8.6.3 Simple Test for Primitiveness. An irreducible polynomial can be easily tested for primitiveness using the fact that at least one of its roots must be a primitive element. For example, to find the $m - 3$ roots of $f(x) = 1 + x + x^3$, we will list the roots in order. Clearly, $\alpha^0 = 1$ is not a root since $f(\alpha^0) = 1$. Using Table 8.6.3 we check if α^1 is a root: Since $f(\alpha) = 1 + \alpha + \alpha^3 = 1 + \alpha^0 = 0 \pmod 2$, so α is a root. Next we check α^2. Since $f(\alpha^2) = 1 + \alpha^2 + \alpha^6 = 1 + \alpha^0 = 0$, so α^2 is a root. Next, we check α^3: Since $f(\alpha^3) = 1 + \alpha^3 + \alpha^9 = 1 + \alpha^3 + \alpha^2 = \alpha + \alpha^2 = \alpha^4 \neq 0$, so α^3 is not a root. Finally we check α^4: Since $f(\alpha^4) = 1 + \alpha^4 + \alpha^{12} = 1 + \alpha^4 + \alpha^5 = 1 + \alpha^0 = 0$, so α^4 is a root. Hence the roots of $f(x) = 1 + x + x^3$ are $\{\alpha, \alpha^2, \alpha^4\}$, and since each of these roots will generate seven nonzero elements in the field, each of them is a primitive element. Because we only need one root to be a primitive element, the given polynomial is primitive. Some primitive polynomials are listed in Table 8.6.2.

Table 8.6.2 Some Primitive Polynomials

m	Primitive Polynomial		m	Primitive Polynomial
3	$1 + x + x^3$	‖	14	$1 + x + x^6 + x^{10} + x^{14}$
4	$1 + x + x^4$	‖	15	$1 + x + x^{15}$
5	$1 + x^2 + x^5$	‖	16	$1 + x + x^3 + x^{12} + x^{16}$
6	$1 + x + x^6$	‖	17	$1 + x^3 + x^{17}$
7	$1 + x^3 + x^7$	‖	18	$1 + x^7 + x^{18}$
8	$1 + x^2 + x^3 + x^4 + x^8$	‖	19	$1 + x + x^2 + x^5 + x^{19}$
9	$1 + x^4 + x^9$	‖	20	$1 + x^3 + x^{20}$
10	$1 + x^3 + x^{10}$	‖	21	$1 + x^2 + x^{21}$
11	$1 + x^2 + x^{11}$	‖	22	$1 + x + x^{22}$
12	$1 + x + x^4 + x^6 + x^{12}$	‖	23	$1 + x^5 + x^{23}$
13	$1 + x + x^3 + x^4 + x^{13}$	‖	24	$1 + x + x^2 + x^7 + x^{24}$

Example 8.6.3. Consider the primitive polynomial $f(x) = 1 + x + x^3$ ($m = 3$). There are $2^3 = 8$ elements in the field defined by $f(x)$. Solving $f(x) = 0$ will give us its roots. But the binary elements (0 and 1) do not satisfy this equation because $f(0) = 1 = f(1)$ (mod 2). Since, by the fundamental theorem of algebra, a polynomial of degree m must have exactly m zeros, the equation $f(x) = 0$ must have three roots, and these three roots lie in the field $GF(2^3)$. Let $\alpha \in GF(2^3)$ be a root of $f(x) = 0$. Then we can write $f(\alpha) = 0$, or $1 + \alpha + \alpha^3 = 0$, or $\alpha^3 = -1 - \alpha$. Since in the binary field $+1 = -1$, α^3 is represented as $\alpha^3 = 1 + \alpha$. This means that α^3 is expressed as a weighted sum of α-terms of lower orders. Using $\alpha^3 = 1 + \alpha$, we get $\alpha^4 = \alpha * \alpha^3 = \alpha * (1 + \alpha) = \alpha + \alpha^2$ and $\alpha^5 = \alpha * \alpha^4 = \alpha * (\alpha + \alpha^2) = \alpha^2 + \alpha^3 = \alpha^2 + 1 + \alpha = 1 + \alpha + \alpha^2$, (using $\alpha^3 = 1 + \alpha$). Further, $\alpha^6 = \alpha * \alpha^5 = \alpha * (1 + a + a^2) = \alpha + \alpha^2 + \alpha^3 = 1 + \alpha^2$; and $\alpha^7 = \alpha * \alpha^6 = \alpha * (1 + \alpha^2) = \alpha + \alpha^3 = 1 = \alpha^0$. Thus, the eight finite field elements of $GF(2^3)$ are $\{0, \alpha^0, \alpha^1, \alpha^2, \alpha^3, \alpha^4, \alpha^5, \alpha^6\}$, which are similar to a binary cyclic code. ∎

Addition and multiplication tables for the above eight field elements of $GF(2^3)$ are given in Tables 8.6.3 and 8.6.4. Note that in some tables 1 and α are written as α^0 and α^1, respectively. These tables are also reproduced in Appendix C, §C.3, for ready reference.

Table 8.6.3 (Addition)

+	1	α	α^2	α^3	α^4	α^5	α^6
1	0	α^3	α^6	α	α^5	α^4	α^2
α	α^3	0	α^4	1	α^2	α^6	α^5
α^2	α^6	α^4	0	α^5	α	α^3	1
α^3	α	1	α^5	0	α^6	α^2	α^4
α^4	α^5	α^2	α	α^6	0	1	α^3
α^5	α^4	α^6	α^3	α^2	1	0	α
α^6	α^2	α^5	1	α^4	α^3	α	0

Table 8.6.4 (Multiplication)

×	1	α	α^2	α^3	α^4	α^5	α^6
1	1	α	α^2	α^3	α^4	α^5	α^6
α	α	α^2	α^3	α^4	α^5	α^6	1
α^2	α^2	α^3	α^4	α^5	α^6	1	α
α^3	α^3	α^4	α^5	α^6	1	α	α^2
α^4	α^4	α^5	α^6	1	α	α^2	α^3
α^5	α^5	α^6	1	α	α^2	α^3	α^4
α^6	α^6	1	α	α^2	α^3	α^4	α^5

8.6.4 Gauss Elimination Method. This is an example of an application of binary arithmetic operations in a Galois field. The traditional Gauss elimination method, as found in numerical analysis, is performed on the infinite set of real numbers $\mathbb{R}$. However, in $GF(2^m)$, especially where multiplication/division is performed, the traditional rules do not apply. The operation of division is not defined over all elements in $GF(2^m)$; for example, $\frac{3}{2}$ is not defined in $GF(4)$, and thus, Gauss elimination becomes unsolvable in many cases. For positive integer m, the $GF(2^m)$ field contains 2^m elements that range from 0 to 2^{m-1}. Of the four arithmetic operations, addition and subtraction are XOR operations. For example, in $GF(16)$ we have $11 + 6 = 1011 \oplus 0110 = 1101 = 13$, and $11 - 6 = 1011 \oplus 0110 = 1101 = 13$. For multiplication and division, if $m \leq 16$, we use the Galois field logarithm and antilogarithm functions, denoted by glog and galog, respectively, and construct two tables, one for each of these functions of length $2^m - 1$, where these functions are integers such that

glog(i) is defined over $i = 1, \ldots, 2^{m-1}$, and it maps the index to its logarithm in $GF(2^m)$; and

galog(i) is defined over $i = 0, 1, \ldots, 2^m - 2$, and it maps the index to its antilogarithm (also called inverse logarithm) in $GF(2^m)$.

Thus, multiplication is changed to addition, where XOR applies, by these two formulas,

$$\text{glog}[\text{galog}(i)] = i, \quad \text{and} \quad \text{galog}[\text{glog}(i)] = i. \tag{8.6.3}$$

An algorithm for the construction of glog and galog tables for $m = 4, 8, 16$ is as follows:

```
if w = 4 p := 023 // octal
else
if w = 8 p := 0435 // octal
else
if w = 16 p := 0210013 // octal

x := 1 << w // Note: same as set x to 2 to the power w
b := 1

for log := 0 to log < x - 1
glog[b] := log
galog[log] := b
b := b << 1 // shift the bits of b left 1 position and store in
b again
if (b and x) // Note: if both bits are 1
b := b xor  p // Note: set b to the xor of b and p
end
```

For example, these tables for $m = 4$ are as follows.

Table 8.6.5 Logarithm Tables for $GF(2^4)$

i	0	1	2	3	4	5	6	7	8	9	10	11	12	13	14	15
glog (i)	–	0	1	4	2	8	5	10	3	14	9	7	6	13	11	12
galog(i)	1	2	4	8	3	6	12	11	5	10	7	14	15	13	9	–

Since $\mathrm{glog}(0) = -\infty$, it is not defined in $\mathrm{GF}(2^m)$, just as in the case of regular logarithm.

Example 8.6.4. Using Tables 8.6.5 the multiplication/division arithmetic in $\mathrm{GF}(2^4)$ is as follows:

$4 * 7 = \mathrm{galog}[\mathrm{glog}(4) + \mathrm{glog}(7)] = \mathrm{galog}[2 + 10] = \mathrm{galog}(12) = 15$,

$14 * 9 = \mathrm{galog}[\mathrm{glog}(14) + \mathrm{glog}(9)] = \mathrm{galog}[11 + 14] = \mathrm{galog}(25) = \mathrm{galog}(10) = 7$,

$\frac{4}{7} = \mathrm{galog}[\mathrm{glog}(4) - \mathrm{glog}(7)] = \mathrm{galog}[2 - 10] = \mathrm{galog}(-8) = \mathrm{galog}(7) = 11$,

$\frac{14}{9} = \mathrm{galog}[\mathrm{glog}(14) - \mathrm{glog}(9)] = \mathrm{galog}[11 - 14] = \mathrm{galog}(5) = 6$. ∎

8.7 Polynomials

Let R be an arbitrary ring. A polynomial over R is an expression of the form $f(x) = \sum_{i=0}^{n} a_i x^i$, where n is a non-negative integer, with coefficients a_i, $0 \leq i \leq n$, which are elements of R, and x is a symbol not belonging to R, called an *indeterminate* over R. The polynomial $f(x)$ may also be given by the equivalent form $f(x) = a_0 + a_1 x + \cdots + a_n x^n + 0 x^{n+1} + \cdots + 0 x^{n+h}$, where $h > 0$ is an integer. Two polynomials $f(x) = \sum_{i=0}^{n} a_i x^i$ and $g(x) = \sum_{i=0}^{n} b_i x^i$ are considered equal iff $a_i = b_i$ for $0 \leq i \leq n$. The sum of $f(x)$ and $g(x)$ over R is defined by $f(x) + g(x) = \sum_{i=0}^{n} (a_i + b_i) x^i$, whereas the product of $f(x) = \sum_{i=0}^{n} a_i x^i$ and $g(x) = \sum_{j=0}^{m} b_j x^j$ over R is defined by

$$f(x)g(x) = \sum_{k=0}^{n+m} c_k x^k, \quad \text{where } c_k = \sum_{\substack{i+j=k \\ 0 \leq i \leq n, 0 \leq j \leq m}} a_i b_j.$$

A ring formed by the polynomials over R with the above operations is called the *polynomial ring* over R and is denoted by $R[x]$. The zero element of $R[x]$ is the polynomial, called the *zero polynomial*, all of whose coefficients are zero, and it is denoted by 0. If $f(x)$ is not a zero polynomial over R, then $a_n \neq 0$, in which case a_n is called the *leading coefficient* of $f(x)$ and a_0 is called the *constant term*, while n is called the *degree* of the polynomial which is denoted by $\deg(f(x)) = \deg(f)$. Polynomials of degree ≤ 0 are called *constant polynomials*. If R has the identity 1 and the leading term is 1, then $f(x)$ is called a *monic polynomial*. Let $f(x) \in R[x]$. Then $\deg(f + g) \leq \max\{\deg(f), \deg(g)\}$, and $\deg(fg) \leq \deg(f) + \deg(g)$, where equality holds if R is an *integral domain*.

8.7.1 Division Algorithm. Let $g \neq 0$ be a polynomial in the polynomial ring $F[x]$. Then for any $f \in F[x]$ there exist polynomials $q, r \in F[x]$ such that $f = qg + r$, where $\deg(r) < \deg(g)$.

Let $f_1, \ldots, f_n$ be polynomials in $F[x]$ not all of which are 0. Then there exists a uniquely determined monic polynomial $d \in F[x]$ such that (i) d divides each f_j, $1 \leq j \leq n$; (ii) any polynomial $c \in F[x]$ dividing each f_j, $1 \leq j \leq n$, divides d; and (iii) d can be expressed as $d = \sum_{j=1}^{n} b_j f_j$, where $b_j \in F[x]$ for $i \le j \le n$. The monic polynomial d is called the *greatest common divisor* of $f_1, \ldots, f_n$, written $d = \gcd(f_1, \ldots, f_n)$. If $d = 1$, then the polynomials $f_1, \ldots, f_n$ are said to be *relatively prime*. They are called *pairwise relatively prime* if $\gcd(f_i, f_j) = 1$ for $1 \le i < j \le n$.

This algorithm involves the process of long division, which is explained in the example that determines the quotient and the remainder.

Example 8.7.1. For the two polynomials $f(x) = x^{15} + x^{11} + x^{10} + x^9 + x^8 + x^7 + x^5 + x^3 + x^2 + x + 1$ and $g(x) = x^{21} + x^{15}$, we will find the quotient $q(x)$ and the remainder $r(x)$ first by long division:

$$
\begin{array}{r}
x^6 + x^2 + x \\
\hline
\end{array}
$$

$$x^{15} + x^{11} + x^{10} + x^9 + x^8 + x^7 + x^5 + x^3 + x^2 + x + 1 \left| \, x^{21} + x^{15} \right.$$

$$\underline{x^{21} + x^{17} + x^{16} + x^{15} + x^{14} + x^{13} + x^{11} + x^9 + x^8 + x^7 + x^6}$$

$$x^{17} + x^{16} + x^{14} + x^{13} + x^{11} + x^9 + x^8 + x^7 + x^6$$
$$\underline{x^{17} + x^{13} + x^{12} + x^{11} + x^{10} + x^9 + x^7 + x^5 + x^4 + x^3 + x}$$

$$x^{16} + x^{14} + x^{12} + x^{10} + x^8 + x^6 + x^5 + x^4 + x^3 + x$$
$$\underline{x^{16} + x^{12} + x^{11} + x^{10} + x^9 + x^8 + x^6 + x^4 + x^3 + x^2 + x}$$

$$x^{14} + x^{11} + x^9 + x^5 + x$$

Thus, the quotient $q(x) = x^6 + x^2 + x$ and the remainder $r(x) = x^{14} + x^{11} + x^9 + x^5 + x$. Since in coding theory only the remainder polynomial is important, we will now give another simpler method for the above division which is carried out in polynomial algebra of the extension field without showing the quotient:

$$1000111110101111\,\big|\,000000000100001000000000000000$$
$$\underline{1000111110101111}$$
$$1101101011110000$$
$$\underline{1000111110101111}$$
$$1010101010111110$$
$$\underline{1000111110101111}$$
$$100101000100010$$

The last line is the remainder $r(x) = x^{14} + x^{11} + x^9 + x^5 + x$ as above. ∎

8.7.2 Euclidean Algorithm. This algorithm is used to compute the gcd of two polynomials $f, g \in F[x]$ as follows: Without loss of generality, suppose that $g \ne 0$ and that g does not divide f. Then using the division algorithm repeatedly we get

$$f = q_1 g + r_1, \qquad 0 \le \deg(r_1) < \deg(g)$$
$$g = q_2 r_1 + r_2, \qquad 0 \le \deg(r_2) < \deg(r_1)$$
$$r_1 = q_3 r_2 + r_3, \qquad 0 \le \deg(r_3) < \deg(r_2)$$

$$\vdots$$

$$r_{s-1} = q_s r_{s-1} + r_s, \quad 0 \le \deg(r_s) < \deg(r_{s-1})$$
$$r_{s-1} = q_{s+1} r_s,$$

where $q_1, \dots, q_{s+1}$ and $r_1, \dots, r_s$ are polynomials in $F[x]$. The above process stops after finitely many steps since $\gcd(g)$ is finite.

Example 8.7.2. Consider $f(x) = 2x^6 + x^3 + x^2 + 2 \in F_3[x]$ and $g(x) = x^4 + x^2 + 2x$, both in $F_3[x]$. By the Euclidean algorithm

$$2x^6 + x^3 + x^2 + 2 = (2x^2 + 1)(x^4 + x^2 + 2x) + x + 2,$$
$$x^4 + x^2 + 2x = (x^3 + x^2 + 2x + 1)(x + 2) + 1,$$
$$x + 2 = (x + 2)\,1.$$

Thus, $\gcd(f, g) = 1$, and f and g are relatively prime. ∎

The prime elements of a ring $F[x]$ are called irreducible polynomials. More precisely, a polynomial $p \in F[x]$ is said to be *irreducible over* F (or *irreducible in* $F[x]$ or *prime in* $F[x]$) if p has a positive degree and $p = bc$, where $b, c \in F[x]$, implies that either b or c is a constant polynomial. Thus, a polynomial of positive degree is irreducible over F if it allows only trivial factorizations. The irreducibility (or reducibility) of a given polynomial depends on the field under consideration. For example, the polynomial $x^2 - 2 \in \mathbb{Z}[x]$ is irreducible over the field $\mathbb{Z}$ of rational numbers, but $x^2 - 2 = (x + \sqrt{2})(x - \sqrt{2})$ is reducible over the field $\mathbb{R}$ of real numbers. Irreducible polynomials are important for the structure of the ring $F[x]$ since the polynomials in $F[x]$ can always be written uniquely as a product of irreducible polynomials. One can easily show that if an irreducible polynomial $p \in F[x]$ divides a product $f_1 \cdots f_n$ of polynomials in $F[x]$, then at least one of the factors f_j is divisible by p. The *unique factorization theorem in* $F[x]$ states that any polynomial $f \in F[x]$ of positive degree can be written in the form $f = a\,p_1^{e_1} \cdots p_k^{e_k}$, where $a \in F$, p_j are distinct monic irreducible polynomials in $F[x]$ for $1 \le j \le k$, and $e_j, 1 \le j \le k$ are positive integers. This factorization is unique irrespective of the order in which the factors occur. A proof of this theorem can be found in any book on modern algebra.

Example 8.7.3. Since a nonzero monic polynomial in F_2 is monic, there are $2^4 = 16$ polynomials in $F_2[x]$ of degree 4, and such a polynomial is reducible over F_2 iff it has a divisor of degree 1 or 2. Thus, we compute all products $(a_0 + a_1 x + a_2 x^2 + x^3)(b_0 + x)$ and $(a_0 + a_1 x + x^2)(b_0 + b_1 x + x^2)$, $a, b \in F_2$, and obtain all reducible polynomials over F_2 of degree 4. Comparing with all 16 polynomials of degree 4, we find the following three irreducible polynomials: $f_1(x) = x_x^4 + 1, f_2(x) = x^4 + x^3 + 1, f_3(x) = x^4 + x^3 + x^2 + x + 1 \in F_2[x]$. ∎

An element $b \in F$ is a zero (or root) of the polynomial $f \in F[x]$ iff $f(b) = 0$, i.e., iff $x - b$ divides $f(x)$. The following result is useful:

Theorem 8.7.1. (Lagrange Interpolation Formula) *For $n \geq 0$ let $a_0, \ldots, a_n$ be $n + 1$ distinct elements of F, and let $b_1, \ldots, b_n$ be $n + 1$ arbitrary elements of F. Then there exists one polynomial $f \in F[x]$ of degree $\leq n$ such that $f(a_i) = b_i$ for $i = 0, 1, \ldots, n$, and this polynomial is given by*

$$f(x) = \sum_{i=0}^{n} b_i \prod_{k+0, k \neq i}^{n} (a_i - a_k)^{-1}(x - a_k). \tag{8.7.1}$$

8.7.3 Subfields. For every prime p, the residue class ring $Z/(p)$ of integers forms a finite field with p elements, which may be identified with the Galois field F_p of order p. Let F be a finite field. Then F has p^n elements, where p is called the *characteristic* of F and n the *degree* of F over its prime subfield. An important result is: If F is a finite field with q elements, then every a in F satisfies $a^q = a$. Moreover, if K is a subfield of F, then the polynomial $x^q - x$ in $K[x]$ has factors in $F[x]$ as

$$x^q - x = \prod_{a \in F} (x - a),$$

and F is a *splitting field* of $x^q - x$ over K. If F_q is a finite field with $q = p^n$ elements, then every subfield of F_q has order p^m, where m is a positive divisor of n; conversely, if m is a positive divisor of n, then there is exactly one subfield of F_q with p^n elements.

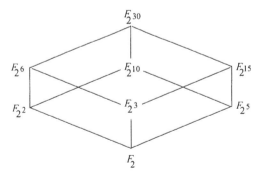

Figure 8.7.1 Subfields of $F_{2^{30}}$.

Example 8.7.4. To find subfields of $F_{2^{30}}$, first list all positive divisors of 30, which are 1, 2, 3, 5, 6,10, 15, and 30. These subfields are contained as shown in Figure 8.7.1, where it is found that the containment relations are the same as the divisibility relations among the positive divisors of 30. ∎

For the subfields of a finite field, see §8.2, Property (vii). For a finite field F_q, let F_q^* denote the *multiplicative group* of nonzero elements of F_q. Then F_q^* is a cyclic group.[3] The generator of the cyclic group F_q^* is called a *primitive element* of F_q. Finally, for every finite field F_q and every positive integer n, there exists an irreducible polynomial in $F_q[x]$ of degree n.

8.8 Polynomial Codes

Consider the polynomial ring $F_2[x]$ with coefficients in the field F_2. The laws of F_2 are followed for addition and multiplication. For example, $(1+x+x^2)+(1+x+x^3) = x^2+x^3$ and $(1+x+x^2)*(1+x+x^3) = 1+x^4+x^5$.

If a polynomial $f(x) \in F_2[x]$ is identified with the binary vector (sequence) of the coefficients of f, i.e., $1+x+x^3 \leftrightarrow (1101)$, we have the following definition: Given the generator polynomial $g(x) = a_0 + a_1 x + \cdots + a_k x^k \leftrightarrow (a_0 a_1 \ldots a_k)$, the polynomial code $K : F_2^m \mapsto F_2^n$ ($n = m + k$) maps $\mathbf{x} = p(x) \in F_2^m$ to $\mathbf{y} = q(x) \in F_2^n$ by $q(\mathbf{x}) = g(x)p(x)$. The last equality can be written as $(q_0 q_1 \ldots q_{n-1}) = (p_0 p_1 \ldots p_{m-1})\, G$, which is an $m \times n$ matrix:

$$G = \begin{bmatrix} a_0 & a_1 & \cdots & a_k & 0 & \cdots & 0 & 0 \\ 0 & a_0 & a_1 & \cdots & a_k & 0 & \cdots & 0 \\ \cdots & \cdots & \cdots & \cdots & \cdots & \cdots & \cdots & \cdots \\ 0 & \cdots & 0 & a_0 & a_1 & \cdots & \cdots & a_k \end{bmatrix}.$$

Thus, a polynomial code is a matrix code, albeit not necessarily normalized.

Example 8.8.1. Let $K : F_2^2 \mapsto F_5^2$, $g(x) = 1 + x^2 + x^3$. Then

$$\mathbf{x} = (x_0 x_1) \leftrightarrow x_0 + x_1 x \curvearrowright (x_0 + x_1 x)(1 + x^2 + x^3)$$
$$= x_0 + x_1 x + x_0 x^2 + (x_0 + x_1)x^3 + x_1 x^4$$
$$\leftrightarrow (x_0, x_1, x_0, x_0 + x_1, x_1) = \mathbf{y}.$$

Thus,

$$(00) \curvearrowright (00000), (01) \curvearrowright (01011), (10) \curvearrowright (10110), (11) \curvearrowright (11101),$$

which shows the datawords (00), (01), (10), and (11) generate the codewords (00000), (01011), (10110), and (11101), respectively, under the generator polynomial $g(x) = 1 + x^2 + x^3$. ∎

---

[3] A multiplicative group G is said to be cyclic if there is an element $a \in G$ such that for any $b \in G$ there is an integer j with $b = a^j$. The element a is called a generator of the cyclic group and we write $G = < a >$.

9

Matrix Codes

Since most ECCs are matrix codes, we present their common features about encoding and decoding and discuss certain special codes, which will be useful later in defining other important codes.

9.1 Matrix Group Codes

We use modulo 2 arithmetic. Let F_2^n denote the binary n-tuples of the form $\mathbf{a} = (a_1, a_2, \ldots, a_n) : a_i = 0$ or 1 for $i = 1, \ldots, n$. Then the properties of matrix codes are:

1. An (m, n) code K is a one-to-one function $K : X \subset F_2^m \mapsto F_2^n$, $n \geq m$.

2. The *Hamming distance* $d(\mathbf{a}, \mathbf{b})$ between $\mathbf{a}, \mathbf{b}$ in F_2^n is defined by

$$d(\mathbf{a}, \mathbf{b}) = \sum_{i=1}^{n} (a_i + b_i)$$

   $=$ number of coordinates for which a_i and b_i are different.

3. A code $K : F_2^m \mapsto F_2^n$ is a *matrix code* if $\mathbf{x} = \mathbf{x}' A$, where $\mathbf{x}'$ denotes the (unique) complement of $\mathbf{x}$ and A is an $m \times n$ matrix.

4. The range of the code K is the set of *codewords*.

5. A code $K : F_2^m \mapsto F_2^n$ is a *group code* if the codewords in F_2^n form an additive group.

6. The *weight* of a code is the minimum distance between the codewords; it is equal to the minimal number of rows in the control (decoding) matrix $\mathbf{R}$ whose sum is the zero vector. Thus, the *weight* of $\mathbf{a}$ is defined as $\mathrm{wt}(\mathbf{a}) = \sum_{i=1}^{n} a_i$ (ordinary sum).

7. A code *detects* k errors if the minimum distance between the codewords (the weight) is at least $k + 1$. A code *corrects* k errors if the minimum distance between the codewords (the weight) is at least $2k + 1$.

9.2 Encoding and Decoding Matrices

The following diagram illustrates the different aspects of encoding and decoding process.

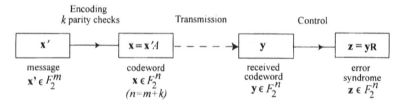

The encoding matrix is $A = [I_m, Q]$ (normalized form), where A is an $m \times n$ matrix, I_m is an $m \times m$ identity matrix, and Q an $m \times k$ matrix. The control matrix is $\mathbf{R} = \begin{bmatrix} Q \\ I_k \end{bmatrix}$, where Q is an $m \times k$ matrix and I a $k \times k$ matrix. Note that $\mathbf{R}^T$ is the same as the parity-check matrix $\mathbf{H}$.

9.3 Decoding Procedure

The following steps are followed to decode a message:

1. If $\mathbf{z} = \mathbf{0}$, then $\mathbf{y}$ is a codeword. The decoded message $\mathbf{x}'$ includes the first m coordinates of $\mathbf{y}$.

2. If $\mathbf{z} \neq \mathbf{0}$, find $\mathbf{y}_1$ of minimal weight (the coset leader) with the same error syndrome $\mathbf{z}$. Then $\mathbf{y} + \mathbf{y}_1$ is probably the transmitted codeword and the first m coordinates make the message $\mathbf{x}'$.

To find $\mathbf{y}_1$: Write $\mathbf{z}$ as a sum of a minimum number of rows of the matrix $\mathbf{R}$, say $\mathbf{z} = \mathbf{r}_{i_1} + \cdots + \mathbf{r}_{i_s}$. Then $\mathbf{y}_1 = \mathbf{e}_{i_1} + \mathbf{e}_{i_2} + \cdots + \mathbf{e}_{i_s}$, where $\mathbf{e}_k = (0, \ldots, 0, 1, 0, \ldots, 0)$ with the '1' on the k-th place. Alternatively, a decoding table may be used.

In a Hamming code, the rows of $\mathbf{R}$ are formed by all binary m-tuples $\neq \mathbf{0}$, i.e., $n = 2^m - 1$. Example of Hamming codes are:

$$\mathbf{R}_3^T = \mathbf{H}_3 = \begin{bmatrix} 1 & 1 & 0 \\ 1 & 0 & 1 \end{bmatrix}, \quad \mathbf{R}_7^T = \mathbf{H}_7 = \begin{bmatrix} 1 & 1 & 1 & 0 & 1 & 0 & 0 \\ 1 & 1 & 0 & 1 & 0 & 1 & 0 \\ 1 & 0 & 1 & 1 & 0 & 0 & 1 \end{bmatrix}.$$

Example 9.3.1. Given $\mathbf{H} = \begin{bmatrix} 1 & 0 & 0 & 1 & 1 & 0 \\ 0 & 1 & 0 & 1 & 0 & 1 \\ 0 & 0 & 1 & 1 & 1 & 1 \end{bmatrix}$, where $m = 3, k = 3, n = 6$, decode (a) $\mathbf{y} = (110011)$, (b) $\mathbf{y} = (011111)$, and (c) $\mathbf{y} = (111111)$, and find the message. The solution is as follows:

$$\text{Control matrix } \mathbf{R} = \begin{bmatrix} 1 & 1 & 1 & 1 & 0 & 0 \\ 1 & 0 & 1 & 0 & 1 & 0 \\ 0 & 1 & 1 & 0 & 0 & 1 \end{bmatrix};$$

Since $\mathbf{R}^T$ is a 6×3 matrix, so wt $\mathbf{R}^T = 3$ and row(1)+row(4)+row(5)=(000); thus, the code detects two errors and corrects one error (in view of properties 6 and 7 of §9.1).

ANS. (a) $\mathbf{z} = y\mathbf{R}^T = (000)$. Thus, $\mathbf{y}$ is a codeword. Message $\mathbf{x}' = (110)$.

(b) $\mathbf{z} = \mathbf{y}\mathbf{R}^T = (101) = \mathbf{r}_2$ (second row of $\mathbf{R}^T$); $\mathbf{y}_1 = \mathbf{e}_2 = (010000)$, which yields $\mathbf{y}_1\mathbf{R}^T = (101)$. Thus, $\mathbf{y} + \mathbf{y}_1 = (001111)$ and $\mathbf{x}' = (001)$.

(c) $\mathbf{z} = \mathbf{y}\mathbf{R}^T = (011) = \mathbf{r}_1 + \mathbf{r}_2 = \mathbf{r}_5 + \mathbf{r}_6$. No codeword of weight 1 has this error syndrome. The error has weight ≥ 2. No 'safe' correction is possible. ∎

9.4 Hadamard Codes

This code, named after the mathematician Jacques Hadamard [1893], is one of the family of $(2^n, n + 1, 2^{n-1})$ codes. It is used for single error detection and correction, although it is not very successful for large n. The code is based on Hadamard matrices, which were invented by Sylvester [1867] and initially called anallagmetic pavement.

9.4.1 Hadamard Matrices. A Hadamard matrix $\mathfrak{H}_n$ is an $n \times n$ matrix with elements $+1$ and -1 such that $\mathfrak{H}_n\mathfrak{H}_n^T = nI_n$, where I_n is the $n \times n$ identity matrix, since $\mathfrak{H}_n^{-1} = (1/n)\,\mathfrak{H}_n^T$. Also, since $\det(\mathfrak{H}_n\mathfrak{H}_n^T) = n^n$, we have $\det(\mathfrak{H}_n) = n^{n/2}$, although the following theorem of Hadamard that imparted the name to the matrix $\mathfrak{H}_n$ states that $\det(\mathfrak{H}_n) \leq n^{n/2}$.

Theorem 9.4.1. *Let $M = (m_{ij})$ be a real matrix with elements satisfying the condition that $|m_{ij}| \leq 1$ for all i, j. Then $\det(\mathfrak{H}_n) \leq n^{n/2}$, where equality holds iff $M = \mathfrak{H}_n$.*

PROOF. Let $m_1, \ldots, m_n$ be the rows of M. Then, geometrically $|\det(M)|$ is the volume of the parallelopiped with sides $m_1, \ldots, m_n$, and so $|\det M| \leq |m_1| \cdots |m_n|$, where m_i is the (Euclidean) length of x_i, $i = 1, \ldots, n$, and equality holds iff $m_1, \ldots, m_n$ are mutually orthogonal. Since, by hypothesis, $|m_i| \leq (m_{i1}^2 + \cdots + m_{in}^2)^{1/2} \leq n^{1/2}$, where equality holds iff $|m_{ij}| = 1$ for all $j = 1, \ldots, n$, the result follows since a Hadamard matrix is real with elements $|1|$ and rows mutually orthogonal. ∎

Since changing the signs of rows or columns leaves the Hadamard matrix unaltered, we assume that $\mathfrak{H}_n$ is normalized, i.e., all entries in the first row and first column are $+1$. Thus, the order n of a Hadamard matrix can only be 1, 2, or a multiple of 4. This is established by the following theorem, which determines the orders n for which Hadamard matrices exist; it provides a necessary condition.

Theorem 9.4.2. *If a Hadamard matrix of order n exists, then $n = 1$ or 2 or $n \equiv 0 \,(\text{mod } 4)$.*

PROOF. First note that changing the sign of every element in a column of a Hadamard matrix gives another Hadamard matrix. Thus, by changing the signs of all columns for which the element in the first row is -1, we may assume that all elements in the first row are $+1$. Since every other row is orthogonal to the first, we see that each further row has k elements $+1$ and k elements -1, where $n = 2k$. If $n > 2$, the first three rows will become as in Figure 9.4.1, with $n = 4a$. ∎

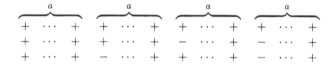

Figure 9.4.1. Three rows of a Hadamard matrix.

The entries $+1$ and -1 in Hadamard matrices represent the polarity of the bits 1 and 0, i.e., these bits when transmitted electronically are represented by voltages that are multiples of $+1$ and -1, respectively. This so-called *polarity rule* implies that the positive and negative voltages represent the bit 1 and 0, respectively. It is still a conjecture that there exists a Hadamard matrix of every order divisible by 4. Therefore, it suggests that the necessary condition in Theorem 9.4.2 is sufficient. As documented in Kharaghani and Tayfeh-Rezaie [2005], the smallest multiple of 4 for which no Hadamard matrix has been constructed was 428.[1] According to Seberry et al. [2005], the smallest known order is 668, and Doković [2008] claims to have constructed Hadamard matrices of order 764.

The simples construction of new Hadamard matrices from an old one is achieved through the Kronecker (tensor) product. Thus, if $A = (a_{ij})$ and $B = (b_{kl})$ are matrices of size $m \times n$ and $p \times q$, respectively, the Kronecker product $A \otimes B$ is an $mp \times nq$ matrix made up of $p \times q$ blocks, where the (i, j) block is $a_{ij}B$. Then

Theorem 9.4.3. *The Kronecker product of Hadamard matrices is a Hadamard matrix.*

The SYLVESTER MATRIX $S(k)$ of order 2^k is the iterated Kronecker product of k copies of the Hadamard matrix $\begin{bmatrix} +1 & +1 \\ +1 & -1 \end{bmatrix}$ of order 2; thus, for example,

$$S(2) = \begin{bmatrix} +1 & +1 & +1 & +1 \\ +1 & -1 & +1 & -1 \\ +1 & +1 & -1 & -1 \\ +1 & -1 & -1 & +1 \end{bmatrix}.$$

---

[1] This matrix can be downloaded from Neil Sloane's library of Hadamard matrices (http://www2.research.att.com/~njas/hadamard/).

By Theorem 9.4.3, the matrix $S(2)$ is a Hadamard matrix. Sylvester matrices can be constructed from a Hadamard matrix $\mathfrak{H}_n$ of order n. Then the partitioned matrix $\begin{bmatrix} \mathfrak{H}_n & \mathfrak{H}_n \\ \mathfrak{H}_n & -\mathfrak{H}_n \end{bmatrix}$ is a Hadamard matrix of order $2n$. Applying this procedure recursively we obtain the following sequence of matrices, also called *Walsh functions*:

$$\mathfrak{H}_1 = [1]; \quad \mathfrak{H}_2 = \begin{bmatrix} 1 & 1 \\ 1 & -1 \end{bmatrix}$$

and in general,

$$\mathfrak{H}_{2^k} = \begin{bmatrix} \mathfrak{H}_{2^{k-1}} & \mathfrak{H}_{2^{k-1}} \\ \mathfrak{H}_{2^{k-1}} & -\mathfrak{H}_{2^{k-1}} \end{bmatrix} = \mathfrak{H}_2 \otimes \mathfrak{H}_{2^{k-1}} \text{ for } k \geq 2, k \in \mathbb{N},$$

where $\otimes$ denotes the Kronecker (tensor) product. Sylvester [1867] constructed Hadamard matrices of order 2^k for every non-negative integer k by this procedure. Sylvester matrices are symmetric, with elements of the first row and first column all positive, while the elements in other rows and columns are evenly divided between positive and negative. Sylvester matrices can also be described as follows: If we index the rows and columns by all k-tuples over the binary field GF(2), we can take the entries in row $a = (a_1, \ldots, a_k)$ and column $b = (b_1, \ldots, b_k)$ to be $(-1)^{a \cdot b}$, where $a \cdot b = \sum a_i b_i$ is the dot (scalar) product of vectors. Thus, we can regard the index $(a_1, \ldots, a_k)$ as being the binary (base 2) representation of an integer $\sum a_i 2^{k-i}$ in the range $[0, 2^{k-1}]$. Moreover, the Sylvester matrix $S(k)$ is the character table of the abelian group of order 2^k.

Another construction of Sylvester's Hadamard matrices, which leads to the *Walsh code*, is as follows: If we map the elements of the Hadamard matrix using the group homomorphism $\{1, -1, \times\} \mapsto \{0, 1, \times\}$, we construct the matrix F_n, which is the $n \times 2^n$ matrix whose columns consist of all n-bit numbers arranged in ascending counting order. Thus, the matrix F_n is defined recursively by

$$F_1 = [0\ 1]; \quad F_n = \begin{bmatrix} 0_{1 \times 2^{n-1}} & 1_{1 \times 2^{n-1}} \\ F_{n-1} & F_{n-1} \end{bmatrix}.$$

It can be proved by induction that the image of a Hadamard matrix under this homomorphism is given by $\mathfrak{H}_{2^n} = F_n^T F_n$. This construction shows that the rows of the Hadamard matrix $\mathfrak{H}_{2^n}$ can be regarded as a length 2^n linear ECC of rank n and minimum distance 2^{n-1} with the generating function F_n. This ECC is also known as a Walsh code, while in contrast the Hadamard code is constructed directly from the Hadamard matrix $\mathfrak{H}_{2^n}$ by a method described in §9.4.3.

The PALEY MATRIX. Let q be a prime power congruent to 3 mod 4. Since in the Galois field GF(q), one-half of the nonzero elements are quadratic residues or squares, and the remaining one-half are quadratic non-residue or non-squares. In particular, since $+1$ is a square and -1 a non-square, we define the quadratic characteristic function of $\mathrm{GF}(q)$ as

$$\chi(x) = \begin{cases} 0 & \text{if } x = 0, \\ +1 & \text{if } x \text{ is a quadratic square,} \\ -1 & \text{if } x \text{ is a quadratic non-square.} \end{cases}$$

The Paley matrix is constructed as follows: Consider a matrix A whose rows and columns are indexed by the elements of $\mathrm{GF}(q)$, with the (x, y) element of the form $a_{xy} = \chi(y - x)$. Such a matrix A is skew-symmetric, with zero diagonal and ± 1 elsewhere. By replacing the diagonal zeros by -1 and bordering A with a row and column of $+1$, we obtain the Paley matrix, which is simply a Hadamard matrix of order $q + 1$.

There are two methods of constructing Hadamard matrices of order 36:

METHOD 1. USING LATIN SQUARES: Let $L = (l_{ij})$ be a Latin square of order 6, which is a 6×6 array with entries $1, 2, \ldots, 6$ such that each entry occurs exactly once in each row or column of the array. Let $\mathfrak{H}$ be a matrix with rows and columns indexed by the 36 cells of the array in the following manner: An element at the location corresponding to a pair (h, h') of distinct cells is defined to be $+1$ if h and h' lie in the same row, or in the same column, or have the same element; all other elements including the diagonal are -1. Then $\mathfrak{H}$ is a Hadamard matrix.

METHOD 2. USING STEINER TRIPLE SYSTEMS: Let S be a Steiner triple system of order 15, i.e., S is a set of 'triples' or 3-element subsets of $\{1, \ldots, 15\}$ such that any two distinct elements of this set are contained in a unique triple. There are 35 triples, and two distinct triples have at most one point in common. Let A be a matrix with rows and columns indexed by the triples, with entry at the location (t, t') being -1 if t and t' meet in a single point; all other entries including the diagonal ones are $+1$. If A is bordered by a row and column of $+1$'s, the resulting matrix $\mathfrak{H}$ is a Hadamard matrix.

The Sylvester and Paley matrices of order 4 and 8 are equivalent, while they are not equivalent for larger orders n for which both types exist with $n = p+1$, where p is a Mersenne prime. However, the equivalence of Hadamard matrices is maintained by the following operations: (a) permuting rows, and changing the sign of some rows; (b) permuting columns, and changing the sign of some columns; and (c) transposition. Thus, two Hadamard matrices $\mathfrak{H}_1$ and $\mathfrak{H}_2$ are said to be *equivalent* if one can be obtained from the other by operations of the types (a) and (b). If $\mathfrak{H}_2 = P^{-1}\mathfrak{H}_1 Q$, where P and Q are monomial matrices[2] with nonzero entries ± 1. Hence, the automorphism group of a Hadamard matrix $\mathfrak{H}$ is the group consisting of all pairs (P, Q) of monomial matrices with nonzero entries ± 1 such that $P^{-1}\mathfrak{H}Q = \mathfrak{H}$, where the group

---

[2] These matrices have just one nonzero element in each row or column.

operation is defined by $(P_1, Q_1) \circ (P_2, Q_2) = (P_1 P_2, Q_1 Q_2)$. Recall that there is an automorphism $(-I, -I)$ that lies at the center of the automorphism group. According to Hall [1962], there exists, up to equivalence, a unique Hadamard matrix $\mathfrak{H}$ of order 12. Also, if G is an automorphism of $\mathfrak{H}$, and Z is the central subgroup generated by $(-I, -I)$, then G/Z is isomorphic to the sporadic simple Mathieu group M_{12} and has its two 5-transitive representations on the rows and columns. Finally, the map $(P, Q) \mapsto (Q, P)$ provides an outer automorphism of M_{12} interchanging these two representations.

9.4.2 Nonlinear Codes. The existence of a Hadamard matrix $\mathfrak{H}_n$ implies the existence of nonlinear codes with the following parameters:

$(n, \lfloor d/(2d - n) \rfloor, d)$ for d even and $d \leq n < 2d$;

$(2d, 4d, d)$ for d even;

$(n, 2\lfloor (d + 1)/(2d + 1 - n) \rfloor, d)$ for d odd and $d \leq n < 2d + 1$; and

$(2d + 1, 4d, +4, d)$ for d odd.

All these four codes were constructed by Levenshtein [1961], and they are all optimal codes due to the Plotkin bound (§6.7).

Example 9.4.1. It can be shown that there does not exist any nonlinear code with parameters $(16, 8, 6)$. To prove it, assume on the contrary that there exists a binary linear code C with parameters $(16, 8, 6)$. Let C' be the residual code of C with respect to a codeword of weight 6 (for the definition of a residual code see §3.3.5). Then, by Theorem 3.1, C' is a binary linear code with parameters $(10, 7, d')$, where $3 \leq d' \leq 4$. However, by Theorem 3.2, we have $d' \geq \lceil 6/2 \rceil = 3$, and since only the binary MDS codes are trivial, we have $d' = 3$. The Hamming bounds (6.5.1) and (6.5.2) for C' show that k cannot be 7, which implies that such a code C' cannot exist. ■

Example 9.4.2. If $\mathfrak{H}_n$ is a Hadamard matrix of order n, then

$$\begin{bmatrix} \mathfrak{H}_n & \mathfrak{H}_n \\ \mathfrak{H}_n & -\mathfrak{H}_n \end{bmatrix}$$

is a Hadamard matrix of order 2^n. Hence, by Theorem 9.4.3, Hadamard matrices of orders $2^h(q + 1)$, $h = 1, 2, \ldots$, and prime powers $q \equiv 3 \pmod 4$ can be obtained as follows: Since $\mathfrak{H}_1 = [1]$, starting with the matrices $\mathfrak{H}_2$ and $\mathfrak{H}_4$ described in the next section we can obtain all Hadamard matrices of orders 2^h, $h \geq 0$. See Walsh functions defined above. ■

9.4.3 Hadamard Graph. The binary logical elements 1 and 0 are represented electronically by voltage $+1$ and -1, respectively, and they can be graphically represented by a black square for 1 and a white square for 0 (or -1). A Hadamard matrix of order $4n + 4$ corresponds to a Hadamard design $(4n + 3, 2n + 1, n)$ and a Hadamard matrix $\mathfrak{H}_n$ gives a graph on $4n$ vertices known as a Hadamard graph. If $\mathfrak{H}_n$ and $\mathfrak{H}_m$ are known then $\mathfrak{H}_{nm}$ can be obtained by replacing all 1s by $\mathfrak{H}_n$ and all -1's by $-\mathfrak{H}_m$. Hadamard matrices of the lowest orders are:

$\mathfrak{H}_1 = [1];$ ■ ;

$$\mathfrak{H}_2 = \begin{bmatrix} 1 & 1 \\ 1 & -1 \end{bmatrix} = \quad \blacksquare \quad ;$$

$$\mathfrak{H}_4 = \begin{bmatrix} \mathfrak{H}_2 & \mathfrak{H}_2 \\ \mathfrak{H}_2 & -\mathfrak{H}_2 \end{bmatrix} = \begin{bmatrix} \begin{bmatrix} 1 & 1 \\ 1 & -1 \end{bmatrix} & \begin{bmatrix} 1 & 1 \\ 1 & -1 \end{bmatrix} \\ \begin{bmatrix} 1 & 1 \\ 1 & -1 \end{bmatrix} & -\begin{bmatrix} 1 & 1 \\ 1 & -1 \end{bmatrix} \end{bmatrix} = \begin{bmatrix} 1 & 1 & 1 & 1 \\ 1 & -1 & 1 & -1 \\ 1 & 1 & -1 & -1 \\ 1 & -1 & -1 & 1 \end{bmatrix} =$$

■ ,

where these matrices are represented as tilings on the right. The matrix $\mathfrak{H}_8$ can be similarly constructed from $\mathfrak{H}_4$. However, for $n \leq 100$, Hadamard matrices with $n = 12, 20, 36, 44, 52, 60, 68, 76, 84, 92, 100$ cannot be constructed from lower-order Hadamard matrices. Note that a complete set of Walsh functions of order n gives a Hadamard matrix $\mathfrak{H}_{2^n}$.

9.4.4 Decoding. Hadamard matrices have been used to construct error correcting codes, e.g., the Reed–Muller codes (see Chapter 12). Let $\mathfrak{H}$ be an Hadamard matrix of order 2^n. Then the codewords are constructed by taking the rows of $\mathfrak{H}$ and $-\mathfrak{H}$ as codewords, where each -1 is replaced by 0. This procedure generates 2^{n+1} codewords, each having a length of 2^n. Since the rows of an Hadamard matrix are orthogonal, the minimum Hamming distance is 2^{n-1}, and therefore, it can correct up to $t = 2^{n-2} - 1$ errors. The decoding for this code is as follows: After a codeword $\mathbf{c}$ is received, transform it to a vector $\mathbf{c}'$ by changing all 0s to -1; then compute $\mathbf{c}'\mathfrak{H}^T$; then take the entry with the maximum absolute value as a codeword: if this value is positive, the codeword came from $\mathfrak{H}$; if negative, it came from $-\mathfrak{H}$.

The justification for this decoding is as follows: If there were no errors, the matrix $\mathbf{c}'\mathfrak{H}^T$ would consist of 0s and one entry of $+2^n$ or -2^n. If there are errors in $\mathbf{c}$, each error that occurs changes a 0 into 2. Thus, the 0s can become at most $2t = 2^{n-1} - 2$ and the maximum decrease at most $2^n - 2t = 2^{n-1} + 2$. The maximum for the row with an error will be larger in absolute value than the other values in that row.

The Hadamard code is found to be optimal for $n \leq 6$. The (32, 6, 16) Hadamard code was used during the NASA Mariner 9 mission in 1971 to correct picture transmission error with 6-bit datawords, which represented 64 grayscale values. The matrix $\mathfrak{H}$ of the (32, 6, 16) Hadamard code for the Reed–Muller code (1,5) is defined as follows:

$$\mathfrak{H} = \begin{bmatrix}
1111111111111111111111111111111111 \\
1010101010101010101010101010101010 \\
1100110011001100110011001100110011 \\
1111000011110000111100001111000011 \\
1011101110111011101110111011101110 \\
1111000011110000111100001111000011 \\
1010011010010110101001101001011001 \\
1100001111000011110000111100001111 \\
1001001010010010110010010101001001 \\
1111111100000000111111110000000000 \\
1010101001010101101010100101010101 \\
1100110000110011110011000011001100 \\
1001100101100110100110010110011001 \\
1111000000001111111100000000001111 \\
1010010101010101101001010101010101 \\
1100001100111100110000110011110001 \\
1111111111111111000000000000000000 \\
1010101010101010101010101010101010 \\
1100110011001100110011001100110011 \\
1001100110011001011001100110011001 \\
1111000011110000000011110000111100 \\
1010010110100101010110100101101001 \\
1100001111000011001111000011110011 \\
1001011010010110011010010110100101 \\
1111111110000000000000011111111111 \\
1010101001010101101010101111001100 \\
1001100101001100011001101001011000 \\
1111000000001111000011111111100000 \\
1010010101011010010110101010010101 \\
1100001100111100001111001100001111 \\
1001011001101001011010011001011010 \\
1001011001101001011010011001011010 \\
0000000000000000000000000000000000 \\
0101010101010101010101010101010101 \\
0011001100110011001100110011001100 \\
0110011001100110011001100110011001 \\
0000111100001111000011110000111100 \\
0101101001011010010110100101101001 \\
\end{bmatrix}$$

$\cdots$ continued

$$\begin{bmatrix}
0011110011110011110011110011110011110 \\
0110100101101001011010010110100101101001 \\
0000000011111111000000001111111 \\
0101010110101010010101011010101010 \\
0011001111001100001100111100110 \\
0110011010011001011001101010011001 \\
0000111111110000000011111111110000 \\
0101101010100101010110101010100101 \\
0011110011000011001111001100011 \\
0110100110010111011010011001011 \\
0000000000000000111111111111111 \\
0101010101010101101010101010101010 \\
0011001100110011110011001100110 \\
0110011010011001100110010110011 \\
0000111111110000011110000000001111 \\
0101101010100101101001010101011010 \\
0011110011000011110000110011110 \\
0110100110010110100101011001001
\end{bmatrix}$$

This matrix can be represented as a Hadamard graph in the tiling format at follows:

In addition to error correcting, the Hadamard code is used in the following areas: (i) Olivia MFSK, which is an amateur-radio digital protocol designed to work in difficult conditions (low signal-to-noise ratio plus multipath propagation) on shortwave bands; (ii) BRR (Balanced Repeated Replication), which is a technique used by statisticians to estimate the variance of a statistical estimator; and (iii) Coded aperture spectrometry, which is a technique for measuring the spectrum of light where the mask element is often a variant of a Hadamard matrix. For other applications of Hadamard matrices, see Seberry et al. [2005].

9.5 Hadamard Transform

Hadamard transform, also known as Walsh-Hadamard transform, Fourier-Walsh transform, or Walsh functions, is a Fourier-related transform. It performs an orthogonal, symmetrical, linear operation on 2^m real numbers. The computation involves no multiplication. This transform is its own inverse, so it is an involuntary transform. Its construction is based on a size-2 discrete Fourier transform (DFT, see Appendix D). It is a multi-dimensional DFT of size $\underbrace{2 \times 2 \times \cdots \times 2 \times 2}_{m-\text{times}}$. The Hadamard transform, denoted by $\mathcal{H}_m$, is a $2^m \times 2^m$ matrix, which is a Hadamard matrix scaled by a normalization factor. It transforms 2^m real numbers x_n into 2^m real numbers X_k. It can be defined in two ways, as follows:

(i) RECURSIVE DEFINITION. Let a 1×1 Hadamard transform $\mathcal{H}_0$ be defined by the identity $\mathcal{H}_0 = 1$. Then the transform $\mathcal{H}_m$ is defined for $m > 0$ by

$$\mathcal{H}_m = \frac{1}{\sqrt{2}} \begin{bmatrix} \mathcal{H}_{m-1} & \mathcal{H}_{m-1} \\ \mathcal{H}_{m-1} & -\mathcal{H}_{m-1} \end{bmatrix}, \tag{9.5.1}$$

where the normalization factor $1/\sqrt{2}$ is sometimes omitted. The matrix (9.5.1) is composed of entries $+1$ or -1.

(ii) BINARY REPRESENTATION. The (k, n)-th element in the matrix (9.5.1) can be defined by

$$k = 2^{m-1}k_{m-1} + 2^{m-2}k_{m-2} + \cdots + 2k_1 + k_0,$$
$$n = 2^{m-1}n_{m-1} + 2^{m-2}n_{m-2} + \cdots + 2n_1 + n_0, \tag{9.5.2}$$

where k_j and n_j, $j = 0, \ldots, m-1$, are the bits 0 or 1 of k and n, respectively, while the element at the top left entry is defined with $k_j = 0$. Thus,

$$(\mathcal{H}_m)_{k,n} = \frac{1}{2^{m/2}}(-1)^{\sum\limits_{j=0}^{m-1} k_j n_j}, \tag{9.5.3}$$

which is the multidimensional $\underbrace{2 \times 2 \times \cdots \times 2 \times 2}_{m-\text{times}}$ DFT, with inputs and outputs as multidimensional arrays indexed by k_j and n_j, respectively.

Example 9.5.1. $\mathcal{H}_0 = 1 = \mathfrak{H}_1$,

$$\mathcal{H}_1 = \frac{1}{\sqrt{2}} \begin{bmatrix} 1 & 1 \\ 1 & -1 \end{bmatrix} = \mathfrak{H}_2,$$

$$\mathcal{H}_2 = \frac{1}{2} \begin{bmatrix} 1 & 1 & 1 & 1 \\ 1 & -1 & 1 & -1 \\ 1 & 1 & -1 & -1 \\ 1 & -1 & -1 & 1 \end{bmatrix} = \mathfrak{H}_4,$$

which is a size-2 DFT, and the Sylvester matrix $S(2)$ (see §9.4.1).

$$\mathcal{H}_3 = \frac{1}{2\sqrt{2}} \begin{bmatrix} 1 & 1 & 1 & 1 & 1 & 1 & 1 & 1 \\ 1 & -1 & 1 & -1 & 1 & -1 & 1 & -1 \\ 1 & 1 & -1 & -1 & 1 & 1 & -1 & -1 \\ 1 & -1 & -1 & 1 & 1 & -1 & -1 & 1 \\ 1 & 1 & 1 & 1 & -1 & -1 & -1 & -1 \\ 1 & -1 & -1 & 1 & -1 & -1 & -1 & 1 \\ 1 & 1 & -1 & -1 & -1 & -1 & -1 & -1 \\ 1 & -1 & -1 & 1 & -1 & 1 & 1 & -1 \end{bmatrix} = \mathfrak{H}_8.$$

In general,

$$\mathcal{H}_n = [h_{i,j}] = \frac{1}{2^{n/2}}(-1)^{i \cdot j}, \tag{9.5.4}$$

where $i \cdot j$ denotes the bitwise dot product of the binary representation of the numbers i and j. Thus, for example, $h_{32} = (-1)^{3 \cdot 2} = (-1)^{(1,1) \cdot (1,0)} = (-1)^{1+0} = (-1)^1 = -1$. The first row and first column of $\mathcal{H}_n$ is denoted by h_{00}. The rows of the Hadamard matrices are called Walsh functions. ∎

9.6 Hexacode

The hexacode is a linear code of length 6 and dimension 3 over the Galois field $GF(4) = \{0, 1, \alpha, \alpha^2\}$ of 4 elements; it is defined by

$$H = \{a, b, c, f(1), f(\alpha), f(\alpha^2) : f(x) = ax^2 + bx + c; a, b, b \in GF(4)\}.$$

Thus, H consists of 45 codewords of weight 4, 18 codewords of weight 6, and the zero word. The full automorphism group of H is $3S_6$. This characterization gives a generator matrix for the hexacode:

$$G = \begin{bmatrix} 1 & 0 & 0 & 1 & \alpha^2 & \alpha \\ 0 & 1 & 0 & 1 & \alpha & \alpha^2 \\ 0 & 0 & 1 & 1 & 1 & 1 \end{bmatrix},$$

which is equivalent to the Miracle Octad Generator of R. T. Curtis (see Conway and Sloane [1999]).

Example 9.6.1. Let $F_4 = \{0, 1, \alpha, \alpha^2\}$. Then the following linear codes over F_4 are self-dual with respect to the Hermitian inner product:

(i) $C_1 = \{(0,0), (1,1), (\alpha, \alpha), (\alpha^2, \alpha^2)\}$;

(ii) Let C_2 be a linear code in C_4 with the generator matrix

$$G = \begin{bmatrix} 1 & 0 & 0 & 1 & \alpha & \alpha \\ 0 & 1 & 0 & \alpha & 1 & \alpha \\ 0 & 0 & 1 & \alpha & \alpha & 1 \end{bmatrix}$$

Then C_2 is a hexacode which is a (6,3,4) code over F_4. Also, C_2 is an MDS 4-ary code. Let C_2' be the code obtained from C_2 by deleting the last coordinate from every codeword. Then C_2' is a Hamming code over F_4.

(iii) Let C_3 be the code over F_4 with the generator matrix

$$G = \begin{bmatrix} 1 & 0 & 1 & 1 \\ 0 & 1 & \alpha & \alpha^2 \end{bmatrix}.$$

Then C_3 is an MDS code, and so is $C_3^\perp$. Thus, C_3 is a self-dual MDS code. ∎

9.7 Lexicodes

Consider an (n, k, d) code, where n denotes the number of bits, k the dimension (k bits, or 2^k symbols), and d the Hamming distance. The lexicodes are constructed by iterating through the integers and accepting any integer that is specified distance from all other accepted integers. An (n, k, d) lexicode can encode up to k bits in a total of n bits by including $n - k$ additional check bits. It can detect up to $d - 1$ error bits and correct up to $\frac{d-1}{2}$ error bits. The values for a lexicode of dimension d differ by at least d bits. The list of codewords (powers of 2) for $d \geq 3$ for small values of k is as follows:

$d = 3$ up to infinity;
$d = 5$ up to $(2149, 2122, 5)$;
$d = 7$ up to $(305, 275, 7)$, including the Golay code;
$d = 9$ up to $(108, 78, 9)$;
$d = 11$ up to $(69, 38, 11)$;
$d = 13$ up to $(60, 26, 13)$;
$d = 15$ up to $(62, 25, 15)$;
$d = 17$ up to $(52, 14, 17)$;
$d = 19$ up to $(54, 14, 19)$;
$d = 21$ up to $(47, 8, 21)$;
$d = 23$ up to $(41, 3, 23)$;
$d = 25$ up to $(52, 6, 25)$;
$d = 27$ up to $(48, 3, 27)$;
$d = 29$ up to $(56, 4, 29)$;
$d = 31$ up to $(31, 1, 31)$;
$d = 33$ up to $(69, 6, 33)$.

The codewords 2^k form a partial basis of a lexicode. To obtain a full basis, an identity matrix is concatenated to the left and the check bits are scattered among the data bits, where the lsb is kept on the right; the data bit for a row is always the rightmost location to the left of all other used bits; the first row using a given check bit places the column for that check bit immediately to the left of the data bit for the previous row; and the check bits themselves are correctly ordered in the tables.

Any two equal-length bit arrays have a Hamming distance d if they differ in d bits. In other words, the XOR of the two arrays contains d number of 1s. Conversely, two distinct positive integers have a Hamming distance d if their binary representations have Hamming distance d. A set of symbols have a minimum Hamming distance d if the minimum Hamming distance between any two distinct symbols is d. We will now find a set of 2^k n-bit numbers with a minimum Hamming distance d. We know that the Hamming codes construct sets of symbols with minimum Hamming distance $d = 3$ (see Chapter 4). Since $2^0 + 2^1 + 2^2 + 2^3 = 1 + 2 + 4 + 8 = 15$, the Hamming code for 15 is the XOR of the Hamming codes for $1, 2, 4, 8$:

	3	2	1	c	0	b	a
1:	0	0	0	0	1	1	1
2:	0	0	1	1	0	0	1
4:	0	1	0	1	0	1	0
$\oplus$ 8:	1	0	0	1	0	1	1
15:	1	1	1	1	1	1	1

Since the Hamming codes for powers of 2 are the basis for the entire set of Hamming codes, such codes can be derived for all other numbers by XOR-ing together the codes for the powers of 2. Thus, by stacking the codewords for the powers of 2 on top of one another we obtain the generating matrix for Hamming codes, and by multiplying it by the binary representation of some data bits yields the matching codeword. The algorithm for finding sets of numbers with a Hamming distance d is as follows:

Start with the set $\{0\}$; then count upward, adding any number to the set that has Hamming distance at least d from every element in the set.

This method produces a set that is closed under XOR, i.e., it is a linear set with basis as powers of 2. Such sets are known as *lexicodes* which are sets of binary linear codes that maintain the values in lexicographic order. All Hamming codes are lexicodes.

The above algorithm will automatically separate check bits from data bits, since every bit that is a high bit of the representation of power of 2 is a data bit and will be 0 in the representation of all other powers of 2. In addition, by rearranging the bit locations for all values in a set with Hamming distance d, another set of values with the same Hamming distance is obtained. If the check bits are placed on the left, the lexicographic order is still preserved.

The representations of powers of 2 may not always have only a d-bit set. Also, if the search is limited to check bits with $d - 1$ bits set, we may not always get the same set of codes or as compact. However, there is a way to obtain sets that have d bits and are as compact, but some powers of 2 would set data bits owned by previous powers of 2. Usually the lexicodes are not the smallest possible codes, just as there are no smallest possible smallest linear codes, except the Hamming codes.

Another feature of lexicodes allows us to construct an extended lexicode by adding an extra check bit provided the original lexicode has the Hamming distance d, where d is odd. Then the lexicode is extended to one with the Hamming distance $d + 1$.

The lexicodes can be produced by grouping the data bits into bytes (sets of 8 bits), then placing the check bits for all 256 ($= 2^8$) combinations of these 8 bits in a 256-term array (one array per byte). The check bits for all data bits can be XOR-ed together to generate the final set of check bits. As in the case of Hamming codes, bytes or words can be substituted for bits. Thus, if the original code had n bits, the new code will also take up n bits; and for all bit locations j, $1 \leq j \leq n$, the j-th bit location of those n words forms an instance of the original n-bit code. Such codes are useful in burst errors since the bits of each code are spread over many words, and they are easily computed since they use representations of powers of 2 that have only a d-bits set. A set with Hamming distance d can correct $(d - 1)/2$ bit errors. For sets with Hamming distance $d > 3$, there are values with Hamming distance greater than $(d - 1)/2$. Those values have more errors than the number of errors that can be corrected. This is true for high values of d, and the problem becomes equivalent to a close sphere packing in high dimensions covering only a small part of the whole space (see §3.6).

9.8 Octacode

There are several equivalent definitions of an octacode, denoted by $\mathcal{O}_8$, as described below. For more details, see Conway and Sloane [1999]. The three useful definitions for this code are as follows:

(i) The code $\mathcal{O}_8$ is the unique self-dual code of length 8 over F_4 with minimum Lee distance 6. It has symmetrized weight enumerator

$$x^8 + 16y^8 + 14x^4z^4 + z^8 + 112xy^2z(x^2 + z^2). \tag{9.8.1}$$

(ii) The code $\mathcal{O}_8$ is obtained by forming a cyclic code of length 7 over F_4 with the generator polynomial $g(x) = x^3 + 3x^2 + 2x + 3$, and adding an overall parity check bit to make the sum of the coordinates zero. Note that in this construction the polynomial $g(x)$ is a divisor of $x^7 - 1 \pmod 4$ and reduces to $x^3 + x^2 + 1 \pmod 2$. Hence, the code $\mathcal{O}_8$ may be regarded as an F_4-analog of a Hamming code, and $\mathcal{O}_8$ then becomes a code of length 8 with the generator matrix

$$
\begin{array}{cccccccc}
\infty & 0 & 1 & 2 & 3 & 4 & 5 & 6
\end{array}
$$

$$
\mathbf{G} = \begin{bmatrix}
3 & 3 & 2 & 3 & 1 & 0 & 0 & 0 \\
3 & 0 & 3 & 2 & 3 & 1 & 0 & 0 \\
3 & 0 & 0 & 3 & 2 & 3 & 1 & 0 \\
3 & 0 & 0 & 0 & 3 & 2 & 3 & 1
\end{bmatrix}, \tag{9.8.2}
$$

or equivalently,

$$
\mathbf{G} = \begin{bmatrix}
2 & 0 & 0 & 0 & 1 & 3 & 1 & 1 \\
2 & 0 & 1 & 1 & 2 & 2 & 3 & 1 \\
0 & 1 & 0 & 1 & 2 & 3 & 0 & 1 \\
1 & 1 & 1 & 1 & 1 & 1 & 1 & 1
\end{bmatrix}, \tag{9.8.3}
$$

where the order-2 codewords in (9.8.2) and (9.8.3) are generated by

$$
\begin{bmatrix}
2 & 2 & 2 & 0 & 2 & 0 & 0 & 0 \\
2 & 0 & 2 & 2 & 0 & 2 & 0 & 0 \\
2 & 0 & 0 & 2 & 2 & 0 & 2 & 0 \\
2 & 0 & 0 & 0 & 2 & 2 & 0 & 2
\end{bmatrix}
\quad \text{and} \quad
\begin{bmatrix}
0 & 0 & 0 & 0 & 2 & 2 & 2 & 2 \\
0 & 0 & 2 & 2 & 0 & 0 & 2 & 2 \\
0 & 2 & 0 & 2 & 0 & 2 & 0 & 2 \\
2 & 2 & 2 & 2 & 2 & 2 & 2 & 2
\end{bmatrix},
$$

respectively. These matrices are the generator matrices for the binary Hamming (8,4) code multiplied by 2 except for the elementary row and column operations (see §5.2).

(iii) The code $\mathcal{O}_8$ is the 'glue code' used in the 'holy construction' of the Leech lattice from 8 copies of the face-centered cubic lattice. The relationship between these lattices and codes over F_4 depends on the result that the quotient of A_3^*/A_3 of the dual lattice A_3^* by A_3 is isomorphic to F_4.

The $\mathcal{O}_8$ code is useful in establishing some results in nonlinear codes, in particular the Nordstrom-Robinson code.

9.9. Simplex Codes

Let $\{c_1, \ldots, c_m\}$ be a set of m binary codewords of length n, with components ± 1, and define the relationship between any two components $(c_{i1}, \ldots, c_{iN})$ and $(c_{j1}, \ldots, c_{jN})$ by $\rho(c_i, c_j) = \frac{1}{n}\sum_{k=1}^{N} c_{ik}c_{jk}$. Then for $i \neq j$

$$
\max_{i \neq j} \rho(c_i, c_j) \geq \begin{cases} -\dfrac{1}{m-1} & \text{if } m \text{ is even,} \\ -\dfrac{1}{m} & \text{if } m \text{ is odd.} \end{cases} \tag{9.9.1}
$$

A set of binary codewords that attains the bound in (9.9.1) is called a *simplex code*. In fact, the dual of a binary Hamming $(n, 2)$ code, denoted by $S(n, 2)$, is called a binary simplex code, and in general, the dual of a q-ary Hamming (n, q) code, denoted by $S(n, q)$, is called a q-ary simplex code. They are projective, linear $(s, n, {}^{n-1})$ codes over F_q^n, with $s = \dfrac{q^n - 1}{q - 1}$. For $n = 1$ the $(1, 1, 1)$ code $S(1, q)$ is a repetition code. The simplex codes exist for all n

(see Theorem 9.9.1), and satisfy the Griesmer bound; thus, they are linear codes with the lowest possible length s for given n and distance q^{n-1}. The code $S(2, q)$ is an extended Reed–Solomon code $RS(2, q)$ (see Chapter 13).

Theorem 9.9.1. *Simplex codes exist for all values of m.*

PROOF. If $\{c_1, \ldots, c_m\}$ is a simplex code with $m = 2n$, then $\{c_1, \ldots, c_{m-1}\}$ is a simplex code with $m = 2n - 1$. Hence, it will suffice to prove this theorem for even m. If we construct the incidence matrix of all subsets of a $2n$-set, such a matrix will contain 0s and 1s with $2n$ rows and $\binom{2n}{n}$ columns, in which each row will correspond to an element of the $2n$-set, and each column will have n 1s that are the elements of the corresponding n-subset. Thus, for example, for $n = 3$ (or $m = 6$), the incidence matrix in the binary lexicographic order is

$$
\begin{bmatrix}
1 & 1 & 1 & 1 & 1 & 1 & 1 & 1 & 1 & 1 & & & & & & & & & & \\
1 & 1 & 1 & 1 & & & & & & & 1 & 1 & 1 & 1 & 1 & 1 & & & & \\
1 & & & & 1 & 1 & 1 & & & & 1 & 1 & 1 & & & & 1 & 1 & 1 & \\
& 1 & & & 1 & & & 1 & 1 & & 1 & & & 1 & 1 & & 1 & 1 & & 1 \\
& & 1 & & & 1 & & 1 & & 1 & & 1 & & 1 & & 1 & 1 & & 1 & 1 \\
& & & 1 & & & 1 & & 1 & 1 & & & 1 & & 1 & 1 & & 1 & 1 & 1
\end{bmatrix}.
$$

Note that each column contains 3 1s, and the blanks are all 0s. Let a denote the maximum number of agreements and b the maximum number of disagreements in any two rows. Then, in this example $a = 8$ and $b = 12$, and $(a - b)/(a + b) = -\frac{1}{5}$. In general, for $m = 2n$, each element belongs to $\binom{2n-1}{n-1}$ of the n-sets. Thus, each codeword has weight $\binom{2n-1}{n-1}$. Each pair of elements will be a 2-subset of $\binom{2n-2}{n-2}$ of the n-sets. Any two codewords will both have 1s in $\binom{2n-2}{n-2}$ of the columns. Then, in the general case, $a = 2\binom{2n-1}{n-1} - 2\binom{2n-2}{n-2}$ and $b = \binom{2n}{n} - a$. Hence,

$$
\begin{aligned}
\frac{a - b}{a + b} &= \frac{\dbinom{2n}{n} - 4\dbinom{2n-1}{n-1} + 4\dbinom{2n-2}{n-2}}{\dbinom{2n}{n}} \\[2mm]
&= \frac{2n(2n-1) - 4n(2n-1) + 4n(n-1)}{2n(2n-1)} \\[2mm]
&= \frac{-1}{2n-1} = -\frac{1}{m-1}. \quad \blacksquare
\end{aligned}
$$

Note that the codewords have the same weight. Any two of them overlap (i.e., have two 1s) in the same number of locations; any three of them overlap in the same number of locations. For more details on these codes, see MacWilliams and Sloane [1977: §1.9], and Bienbauer [2004: §2.5 and 3.4].

The relation between a binary Hamming code and a simplex code is as follows: The dual of the binary Hamming (7,4) code is called a *binary simplex code*, sometimes denoted by $S(r, 2)$ for $n = 2^r - 1$. If $\mathbf{G}$ is the generator matrix of the simplex code $S(r, q)$, then for a given nonzero vector $\mathbf{v} \in F_q^r$, there are exactly $(q^r - 1)/(q - 1)$ columns $\mathbf{c}$ of $\mathbf{G}$ such that $\mathbf{v} \cdot \mathbf{c} = 0$. Every nonzero codeword of $S(r, q)$ has weight $q^r - 1$. Also, the dual of a q-ary Hamming code is called a *q-ary simplex code*.

9.9.1 n-Simplex. In an n-dimensional Euclidean space $\mathbb{R}^n$, an n-dimensional *simplex* (or, an n-simplex) is the convex body spanned by any $m + 1$ distinct points. An n-simplex is said to be *regular* if all $\binom{n+1}{2}$ edges, which are the lines that connect pairs of vertices, have the same length. There always exists an n-simplex for all $n \geq 1$, and for $n \geq 5$ there are exactly three regular hypersolids in $\mathbb{R}^n$: the regular simplexes, the hypercube, and the cross-polytope (Coxeter [1973]). Although the set of vertices for an n-dimensional hypercube consists of 2^n point vectors of the form $(\pm 1, \pm 1, \pm 1, \ldots, \pm 1)$, and the set of vertices for the n-dimensional cross-polytope is $(\pm 1, 0, 0, \ldots, 0), (0, \pm 1, 0, \ldots, 0), (0, 0, \pm 1, \ldots, 0), \ldots, (0, 0, 0, \ldots, \pm 1)$, there is, in general, no easy way to define the set of vectors for the n-simplex. The simplest n-simplex is the polygon in $\mathbb{R}^n$ with $n + 1$ vertices at $(0, 0, \ldots, 0), (0, 1, \ldots, 0), \ldots, (0, 0, \ldots, 1)$.

A (m, k, λ) *cyclic difference set* $D = \{s_1, s_2, \ldots, s_k\}$ is set of k residues (mod m) such that for each nonzero residue d (mod m), there exist λ solutions (x, y) of the equation $x - y \equiv d$ (mod m). A necessary and sufficient condition for the existence of this set D is that (van Lint and Wilson [1992])

$$k(k-1) = \lambda(m-1). \tag{9.9.3}$$

The simplest (and trivial) cyclic difference set D is obtained for $k = 1, \lambda = 0$ so that $k - \lambda = 1$.

The simplex codes are constructed as follows: Let $\alpha_0, \alpha_1, \ldots, \alpha_n$ be $n + 1$ unit vectors in $\mathbb{R}^n$, which represent the vertices of a regular simplex. Then, for every pair $\alpha_i = \{a_1, a_2, \ldots, a_n\}$ and $\beta_j = \{b_1, b_2, \ldots, b_n\}$, the scalar (dot) product $\alpha_i \cdot \beta_i = \sum_{k=1}^{n} a_k b_k$ must satisfy the condition

$$a_i b_j = \begin{cases} -1/n & \text{for } 0 \leq i < j \leq n, \\ 1 & \text{for } 0 \leq i = j \leq n. \end{cases} \tag{9.9.4}$$

Conversely, any set of $n + 1$ unit vectors in $\mathbb{R}^n$, which satisfy the condition (9.9.4) for the scalar product of every pair of vertices, represents the $n + 1$ vertices of a regular n-simplex (Weber [1987]).

Theorem 9.9.2. Let the $n + 1$ unit vectors be defined by

$$\alpha_0 = \frac{1}{\sqrt{m}} [1, 1, 1, \ldots, 1], \; \alpha_1 = \frac{1}{\sqrt{m}} [-a, b, b, \ldots, b],$$

$$\alpha_2 = \frac{1}{\sqrt{m}} [b, -a, b, \ldots, b], \; \cdots, \; \alpha_m = \frac{1}{\sqrt{m}} [b, b, b, \ldots, -a], \tag{9.9.5}$$

where α_i, $i = 2, \ldots, n$, represents a cyclic shift of α_1, and the real numbers a and b are defined by

$$\{a, b\} = \left\{ \frac{1 \pm (n-1)\sqrt{n+1}}{n}, \frac{-1 \pm \sqrt{n+1}}{n} \right\}. \qquad (9.9.6)$$

For every positive integer $n \geq 1$, we can use scheme (9.9.5) to locate the $n+1$ vertices of the n-simplex on the unit hypersphere. In particular, if $n = m^2 - 1$ for some integer m, then a and b are rational numbers defined by

$$\{a, b\} = \left\{ \frac{m^2 + m - 1}{m + 1}, \frac{1}{m + 1} \right\} \quad \text{or} \quad \left\{ \frac{-m^2 + m - 1}{m - 1}, \frac{-1}{m - 1} \right\}.$$

PROOF. From conditions (9.9.4) we obtain the following system of three equations:

$$\alpha_0 \cdot \beta_i = \frac{1}{n}\left[-a + (n-1)b\right] = -\frac{1}{n} \quad \text{for } 1 \leq i \leq n,$$

$$\alpha_i \cdot \beta_j = \frac{1}{n}\left[-2ab(n-2)b^2\right] = -\frac{1}{n} \quad \text{for } 1 \leq i < j \leq n,$$

$$\alpha_i \cdot \beta_i = \frac{1}{n}\left[a^2 + (n-1)b^2\right] = 1 \quad \text{for } 1 \leq i \leq n,$$

solving which we obtain the values of a and b given by (9.9.6). ∎

Example 9.9.1. For $n = 8$, using the notation $-$ for $-a$ and '+' for $+b$, where $\{a, b\} = \{11/4, 1/4\}$ or $\{-5/2, -1/2\}$, the scheme of Theorem 9.9.2 yields

$$\alpha_0 = \frac{1}{2\sqrt{2}}\,[1\ 1\ 1\ 1\ 1\ 1\ 1\ 1], \qquad \alpha_1 = \frac{1}{2\sqrt{2}}\,[-\ +\ +\ +\ +\ +\ +\ +],$$

$$\alpha_2 = \frac{1}{2\sqrt{2}}\,[+\ -\ +\ +\ +\ +\ +\ +], \quad \alpha_3 = \frac{1}{2\sqrt{2}}\,[+\ +\ -\ +\ +\ +\ +\ +],$$

$$\alpha_4 = \frac{1}{2\sqrt{2}}\,[+\ +\ +\ -\ +\ +\ +\ +], \quad \alpha_5 = \frac{1}{2\sqrt{2}}\,[+\ +\ +\ +\ -\ +\ +\ +],$$

$$\alpha_6 = \frac{1}{2\sqrt{2}}\,[+\ +\ +\ +\ +\ -\ +\ +], \quad \alpha_7 = \frac{1}{2\sqrt{2}}\,[+\ +\ +\ +\ +\ +\ -\ +],$$

$$\alpha_8 = \frac{1}{2\sqrt{2}}\,[+\ +\ +\ +\ +\ +\ +\ -].$$

These values of α_i for $0 \leq i \leq 8$ are the vertices of a regular simplex in an 8-dimensional space. ∎

A generalization of Theorem 9.9.2 is as follows:

Theorem 9.9.3. Assume that a (m, k, λ) cyclic difference set exists, and let

$$\alpha_0 = \frac{1}{\sqrt{m}} [1, 1, 1, \ldots, 1],$$

$$\alpha_1 = \frac{1}{\sqrt{m}} [\beta_1, \beta_2, \beta_3, \ldots, \beta_m],$$

$$\alpha_2 = \frac{1}{\sqrt{m}} [\beta_m, \beta_1, \beta_2, \ldots, \beta_{m-1}],$$

$$\vdots$$

$$\alpha_m = \frac{1}{\sqrt{m}} [\beta_2, \beta_3, \beta_4, \ldots, \beta_1], \quad \text{where } \beta_j = \begin{cases} -a & \text{if } j \in D, \\ b & \text{if } j \notin D, \end{cases} \tag{9.9.7}$$

and $\alpha_2, \ldots, \alpha_m$ are all the other cyclic shifts of α_1. Then the real numbers a and b are given by

$$\{a, b\} = \left\{ \frac{1}{m} \pm \frac{m-k}{m} \sqrt{\frac{m+1}{k-\lambda}}, -\frac{1}{m} \pm \frac{k}{m} \sqrt{\frac{m+1}{k-\lambda}} \right\}. \tag{9.9.8}$$

PROOF. Note that the $m + 1$ vectors $\alpha_0, \alpha_1, \ldots, \alpha_m$ are the position vectors of the $m + 1$ vertices of a m-dimensional regular simplex on the unit hypersphere. Eqs (9.9.4) yield the following system of equations:

$$\alpha_i \cdot \beta_j = \frac{1}{m} \left[ka^2 + (m-k)b^2 \right] = 1 \quad \text{for } 1 \leq i \leq n,$$

$$\alpha_0 \cdot \beta_j = \frac{1}{m} [-ka + (m-k)b] = -\frac{1}{m} \quad \text{for } 1 \leq i \leq n,$$

$$\alpha_i \cdot \beta_j = \frac{1}{m} \left[\lambda a^2 - 2(k-\lambda)ab + (m - 2k + \lambda)b^2 \right] = -\frac{1}{m} \quad \text{for } 1 \leq i < j \leq n, \tag{9.9.9}$$

of which the first two simultaneous equations have two pairs of real roots a and b that are given by (9.9.8). If we multiply the third equation by m^3, we obtain

$$\lambda(ma)^2 - 2(k-\lambda)(ma)(mb)(m - 2k + \lambda)(mb)^2 + m^2 = 0. \tag{9.9.10}$$

Since

$$(ma)^2 = 1 + (m-k)^2 \frac{m+1}{k-\lambda} \pm 2(m-k) \sqrt{\frac{m+1}{k-\lambda}},$$

$$(mb)^2 = 1 + k^2 \frac{m+1}{k-\lambda} \mp 2k \sqrt{\frac{m+1}{k-\lambda}},$$

$$(ma)(mb) = -1 + k(m-k) \frac{m+1}{k-\lambda} \pm k \sqrt{\frac{m+1}{k-\lambda}} \mp (m-k) \sqrt{\frac{m+1}{k-\lambda}},$$

we can write the left side of (9.9.8) as $R + S\sqrt{\dfrac{m+1}{k-\lambda}}$, where R and S denote the rational parts. Then

$$S \pm 2\lambda(m-k) \mp 2(k-\lambda) \pm 2(k-\lambda)(m-k) \mp 2k(m-k) \pm 2k(k-\lambda) = 0,$$

$$R = \lambda\Big[1 + (m-k)^2\frac{m+1}{k-\lambda}\Big] - 2(k-\lambda) - 2(k\lambda)\Big[-1 + k(m-k)\frac{m+1}{k-\lambda}\Big]$$

$$+ (m - 2k + \lambda)\Big[1 + k^2\frac{m+1}{k-\lambda}\Big] + m^2$$

$$= \frac{m+1}{k-\lambda}\Big[\lambda m(m-1) + mk - mk^2\Big] = 0 \quad \text{by (9.9.3).} \;\blacksquare$$

9.9.2 M-Sequences. The maximum-length shift register sequences, in short called the M-sequences, for the Hamming and simplex codes of order $m = 5$ are given as follows:

$$\alpha^0\alpha^1\alpha^2\alpha^3\alpha^4\alpha^5 \qquad \cdots \qquad \cdots \qquad \alpha^{30}$$

$$\begin{bmatrix}
0 & 0 & 0 & 0 & 1 & 0 & 0 & 1 & 0 & 1 & 1 & 0 & 0 & 1 & 1 & 1 & 1 & 0 & 0 & 0 & 1 & 1 & 0 & 1 & 1 & 1 & 0 & 1 & 0 & 1 \\
0 & 0 & 0 & 1 & 0 & 0 & 1 & 0 & 1 & 1 & 0 & 0 & 1 & 1 & 1 & 1 & 0 & 0 & 0 & 1 & 1 & 0 & 1 & 1 & 1 & 0 & 1 & 0 & 1 & 0 \\
0 & 0 & 1 & 0 & 0 & 1 & 0 & 1 & 1 & 0 & 0 & 1 & 1 & 1 & 1 & 0 & 0 & 0 & 1 & 1 & 0 & 1 & 1 & 1 & 0 & 1 & 0 & 1 & 0 & 0 \\
0 & 1 & 0 & 0 & 1 & 0 & 1 & 1 & 0 & 0 & 1 & 1 & 1 & 1 & 0 & 0 & 0 & 1 & 1 & 0 & 1 & 1 & 1 & 0 & 1 & 0 & 1 & 0 & 0 & 0 \\
1 & 0 & 0 & 1 & 0 & 1 & 1 & 0 & 0 & 1 & 1 & 1 & 1 & 0 & 0 & 0 & 1 & 1 & 0 & 1 & 1 & 1 & 0 & 1 & 0 & 1 & 0 & 0 & 0 & 0
\end{bmatrix}.$$

This matrix is $\mathbf{H}$ of the Hamming (31,26) code, or $\mathbf{G}$ of the simplex code. Each row is an M-sequence, in which each nonzero 5-tuple occurs exactly once cyclically. Row 1 of this matrix (and thus each row) is generated by the polynomial $f(x) = x^5 + x^2 + 1$, and the respective shift register sequences are presented in Figure 9.9.1.

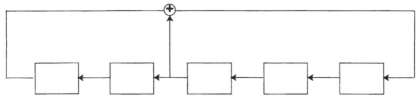

Figure 9.9.1. Shift register sequence.

9.10 Block Codes

A block code is a kind of channel coding, as it adds redundancy to a message that can be decoded with a minimum number of errors, theoretically zero errors, provided the transmission rate (in bits per second) does not exceed the channel capacity. A block code is of fixed length (unlike source coding like Huffman codes or channel coding methods like convolution codes). In general, a block code takes a k-digit information (data) word and transforms it into an n-digit codeword. Thus, a block code encodes strings formed from an alphabet set A into codewords by encoding each letter of A separately. Let $(k_1, \ldots, k_m)$ be a sequence of natural numbers each less than $|A|$ (length of the set A). If $A = (a_1, \ldots, a_n)$, and a given dataword $\mathbf{d}$ is written as $\mathbf{d} = a_{k_1} \cdots a_{k_n}$, then the codeword $\mathbf{c}$ corresponding to $\mathbf{d}$ can be written as $\mathbf{c}(\mathbf{d})$ and defined by $\mathbf{c}(\mathbf{d}) = \mathbf{c}(a_{k_1}) \cdots \mathbf{c}(a_{k_m})$. Recall that the Hamming distance determines a fixed correction capability of a fixed codeword length n. The information rate for a binary block code consisting of r codewords of length n is given by $\log_2(r)/n$; but if the first k bits of a codeword are independent data bits, then the information rate becomes $\log_2(2^k)/n = k/n$.

A q-ary block code of length n over an alphabet A is a nonempty set C of q-ary codewords having the same length n, where the number of codewords in C are denoted by $|C|$ and called the size of C, and the information rate of a codeword C of length n is defined to be $\log_q |C|/n$. A code of length n and size M is called an (n, M) code.

Block codes differ from convolution codes (see §16.4.2) in that the data is encoded in discrete blocks, not continuously. In block codes the message is broken up into suitable parts (blocks of data bits) and parity check bits are appended to each block; the purpose of parity check bits is to detect and correct errors. The concatenation of data bits and the parity check bits constitutes a codeword. The code is linear since each codeword is a linear combination of one or more codewords. The codewords so generated are also called *vectors*, which are regulated by the rules of linear algebra.

Some block codes are cyclic in nature. Since then a lot of research has been published on algebraic coding structures and cyclic codes (Chapter 10) for both error detection and error correction and burst-error correction. Some important classes of codes among cyclic codes are the Hamming codes and Golay codes that have been reviewed thus far.

We will be studying linear block codes. These codes differ from conventional codes (generally Hamming and Golay codes) in that the data is encoded in discrete blocks, and not continuously. The main feature is to break data bits into blocks (parts), appending parity bits to each block, so that a block of data bits together with parity bits is called a codeword. Recall that a code is linear when each codeword is a linear combination of a finite number of codewords. For this reason, from an algebraic point of view, codewords are symbols that are referred to as vectors, and form a vector space. Another feature of some block codes is their cyclic nature, i.e., any cyclic shift of a codeword is again a codeword. Thus, linear cyclic block codewords can be added together and shifted cyclically in any manner, and this will still yield a codeword. Hence, since sets of codewords, being members of a vector space, may also be generated by polynomial division, there are two methods of performing computations: either by linear algebraic methods or by polynomial arithmetic in a Galois field, as already considered in the previous chapter.

10

Cyclic Codes

10.1 Definition

Cyclic codes were first studied by Prange [1957]. They are a special class of linear codes defined as follows: A subset S of F_q^n is *cyclic* (or of cyclic order, or a cyclic shift of one position) if

$$\{a_0, a_1, \ldots, a_{n-1}\} \in S \to \{a_{n-1}, a_0, a_1, \ldots, a_{n-2}\} \in S.$$

This can be regarded as the conversion of a combinatorial structure S into an algebraic structure, also in S. The cyclic shift of r positions is defined as

$$\{a_0, a_1, \ldots, a_{n-1}\} \in S \to \{a_{n-r}, \ldots, a_{n-1}, a_0, a_1, \ldots, a_{n-r-1}\} \in S.$$

See §C.1 for cyclic permutations. A linear code C is called a *cyclic code* if C is a cyclic set. The elements of S can be the codewords $\mathbf{c}_n \in F_q^n$. Then the cyclic shift of r positions of codewords in F_q^n is

$$\{\mathbf{c}_0, \mathbf{c}_1, \ldots, \mathbf{c}_{n-1}\} \in F_q^n \to \{\mathbf{c}_{n-r}, \ldots, \mathbf{c}_{n-1}, \mathbf{c}_0, \mathbf{c}_1, \ldots, \mathbf{c}_{n-r-1}\} \in F_q^n.$$

Cyclic codes were the first linear codes, generated by using shift registers as described in §2.2.6–2.2.10.

Example 10.1.1. The following codes C are cyclic codes:

(i) $\mathbf{0}$, and $\lambda \cdot \mathbf{1}$, both in F_q^n, and F_q^n itself;

(ii) Binary linear (3, 2, 2) code $C = \{000, 011, 101, 110\}$;

(iii) Code $S(3, 2) = \{0000000, 0111001, 01001110, 0010111, 1001011, 1011100,$
$1100101, 1110010\}$;

(iv) Binary parity-check code; and

(v) Repetition codes (see §7.1.1). ∎

A cyclic code C is *invariant* under a right cyclic shift, that is,

$$\mathbf{c} = \{c_0, c_1, c_2, \ldots, c_{n-2}, c_{n-1}\} \in C \text{ iff } \hat{\mathbf{c}} = \{c_{n-1}, c_0, c_1, \ldots, c_{n-2}\} \in C.$$

Thus, since C is invariant under one right cyclic shift, it is invariant under any number of right cyclic shifts. Similarly, C is also invariant under any number of left cyclic shifts if it is invariant under one left cyclic shift. This means that the linear code C is cyclic when it is invariant under all cyclic shifts.

Thus, for linear (n, k) cyclic codes over the field F_q, we will require that $\gcd(n, q) = 1$. Let $(x^n - 1)$ be the ideal generated by $x^n - 1 \in F_q[x]$. Then all elements of $F_q[x]/(x^n - 1)$ can be represented by polynomials of degree less than n, and this residue class ring is isomorphic to F_q^n as a vector over F_q. Such an isomorphism is given by

$$\{a_0, a_1, \ldots, a_{n-1}\} \longleftrightarrow a_0 + a_1 x + \cdots + a_{n-1} x^{n-1}. \tag{10.1.1}$$

The polynomial $a(x) = a_0 + a_1 x + \cdots + a_{n-1} x^{n-1}$ is called the *code polynomial*. If $\mathbf{c} \in C$ is a codeword, then $c(x)$ is called the *associated code polynomial*. Then a shifted codeword $\hat{\mathbf{c}}$ has the associated code polynomial

$$\hat{c}(x) = c_{n-1} + c_0 x + c_1 x^2 + \cdots + c_i x^{i+1} + \cdots + c_{n-2} x^{n-1} = xc(x) - c_{n-1} \left(x^n - 1 \right).$$

Thus, $\hat{c}(x) = xc(x) \pmod{x^n - 1}$, and $c(x) \in C \pmod{x^n - 1}$ iff $xc(x) \in C \pmod{x^n - 1}$. Since the cyclic code C is invariant under additional shifts, we have $x^i c(x) \in C \pmod{x^n - 1}$ for all i, which for any $a_i \in F$ gives $a_i x^i c(x) \in C \pmod{x^n - 1}$, and in general, $\sum_{i=0}^{d} a_i x^i c(x) \in C \pmod{x^n - 1}$. Hence, for every polynomial $a(x) = \sum_{i=0}^{d} a_i x^i \in F[x]$, the product $a(x)c(x) \pmod{x^n - 1}$ belongs to C.

In view of the isomorphism (10.1.1), the elements of $F_q[x]/(x^n - 1)$ are denoted either as polynomials of degree $< n$ modulo $x^n - 1$ or as vectors (or words) over F_q. The operation of multiplication of polynomials modulo $x^n - 1$ is defined in the usual way: If $f \in F_q[x]/(x^n - 1)$, $g_1, g_2 \in F_q[x]$, then $g_1 g_2 = f$ means that $g_1 g_2 \equiv f \mod (x^n - 1)$.

The combinatorial structure of a cyclic code is converted into an algebraic structure by the following correspondence relation:

$$\pi : \begin{cases} F_q^n \to F_q[x]/(x^n - 1), \\ (a_0, a_1, \ldots, a_{n-1}) \mapsto a_0 + a_1 x + \cdots + a_{n-1} x^{n-1}. \end{cases}$$

Note that $q = 2^m$ for binary cyclic codes. The correspondence π is a linear mapping F_q of the vector space V over F_q. For brevity of notation, we will henceforth write $F_q[x]/(x^n - 1)$ for F_q^n and the polynomial $c(x) = \sum_{i=0}^{n-1} c_i x^i$ for a vector $\mathbf{c} = (c_0, c_1, \ldots, c_{n-1})$.[1]

Example 10.1.2. Consider the cyclic code (ii) of Example 10.1.1. For this code, $\pi = \{0, 1 + x, 1 + x^2, x + x^2\} \subset F_2[x]/(x^3 - 1)$. ∎

Example 10.1.3. Consider the linear transformation $\pi : F_q^n \to F_q[x]/(x^n - 1)$, which implies that

---

[1] Note that $F_q[x]/(x^n - 1)$ is a ring, but not a field unless $n = 1$.

$$\{a_0, a_1, \ldots, a_{n-1}\} \to a_0 + a_1 x + \cdots + a_{n-1} x^{n-1}.$$

Then in terms of codewords, $\mathbf{c} = \{\mathbf{c}_0, \mathbf{c}_1, \ldots, \mathbf{c}_{n-1}\} \to \mathbf{c}(x) = \sum_{i=0}^{n-1} \mathbf{c}_i x^i$.

If $\mathbf{c} = \{000, 011, 101, 110\}$, then $\pi(\mathbf{c}) = \{0, 1+x, 1+x^2, x+x^2\} \subset F_2[x]/(x^3 - 1)$. Note that such $\pi(\mathbf{c})$ is sometimes written, with decreasing order of terms, as $\pi(\mathbf{c}) = \{0, x+1, x^2+1, x^2+x\}$, and for good reason, as the position of 1s and 0s in $\mathbf{c}$ directly lend to the algebraic structure on the right. ∎

By the division algorithm (§2.1.2), if a polynomial $f(x)$ of degree n is divided by another polynomial $g(x)$ of degree $\leq n$, then we have $f(x) = q(x)g(x) + r(x)$, where $q(x)$ is the quotient and the degree of the remainder $r(x)$ is less than that of $g(x)$.

Theorem 10.1.1. *Let C be a nonempty cyclic code of length n over F, and let $g(x)$ be a monic polynomial of minimum degree in C. Then $g(x)$ is uniquely determined in C and*

$$C = \{q(x)g(x) : q(x) \in F[x]_{n-r}\}, \quad r = \deg\{g(x)\}.$$

In particular, C has dimension $n - 1$. Moreover, the polynomial $g(x)$ divides $x^n - 1$ on $F[x]$.

PROOF. Let $C_0 = \{q(x)g(x) : q(x) \in F[x]_{n-r}\}$. Then $C_0 \subseteq C$, and C_0 contains those multiples of the code polynomial $g(x)$ whose degree is less than n. Thus, C_0 is a vector space of dimension $n - r$, and $g(x)$ is a unique polynomial of degree r in C_0. Since by the division theorem, $r(x) = c(x) - q(x)g(x) \in C$, suppose that $r(x) \neq 0$. Then it would have a scalar monic multiple (of degree $< r$) that belongs to C. However, this is a contradiction of the original choice of $g(x)$, and therefore $r(x) = 0$ and $c(x) = q(x)g(x)$. Let $x^n - 1 = h(x)g(x) + s(x)$ for some $s(x)$ of degree $< \deg\{g(x)\}$. Then, $s(x) = [-h(x)]g(x) \pmod{x^n - 1}$ is in C. Again, let $s(x) \neq 0$. Then using the above argument we can show that $s(x) = 0$ and $x^n - 1 = h(x)g(x)$. ∎

The polynomial $g(x)$ is called the *generator polynomial* for the code C, and the polynomial $h(x \in F[x]$, defined by $h(x)g(x) = x^n - 1$, is called the *parity-check polynomial*.

Example 10.1.4. Let C be a cyclic code of length 7. The polynomial $x^7 - 1$ is factorized into irreducible polynomials as $x^7 - 1 = (x - 1)(x^3 + x + 1)(x^3 + x^2 + 1)$. Since all the codes are binary, the minus sign is replaced by the plus sign, thus giving $x^7 + 1 = (x + 1)(x^3 + x + 1)(x^3 + x^2 + 1)$. Since it contains 3 irreducible factors, there are $2^3 = 8$ cyclic codes including $\mathbf{0} \in F_2^7$. The 8 generator polynomials $g(x)$ are

(i) $1 = 1$; (ii) $x + 1 = x + 1$; (iii) $x^3 + x + 1 = x^3 + x + 1$; (iv) $x^3 + x^2 + 1 = x^3 + x^2 + 1$; (v) $(x+1)(x^3+x+1) = x^4 + x^3 + x^2 + 1$; (vi) $(x+1)(x^3+x^2+1) = x^4 + x^2 + x + 1$; (vii) $(x^3 + x + 1)(x^3 + x^2 + 1) = x^6 + x^5 + x^4 + x^3 + x + 1$; and (viii) $(x + 1)(x^3 + x + 1)(x^3 + x^2 + 1) = x^7 + 1$.

Note that the polynomial 1 in (i) generates all F_2^7. In (ii) we find the parity-check code. In (iii) and (iv) the polynomials are of degree 3, and so they generate (7,4) codes that are, in fact, the Hamming codes. In (v) and (vi) the (7,3) codes are the duals of the Hamming codes. In (vii) we find the repetition code. In (viii) the polynomial generates the **0** code. ∎

10.2 Construction of Cyclic Codes

A cyclic (n, k) code C can be constructed by multiplying each message of k coordinates (identified as a polynomial of degree $< k$) by a fixed polynomial $g(x)$ of degree $n - k$ with $g(x)$ a divisor of $x^n - 1$. Then the polynomials $g(x), xg(x), \dots, x^{k-1}g(x)$ will correspond to codewords of C. Given $g(x) = g_0 + g_1 c + \dots + g_{n-k}x^{n-k}$, a generator matrix of C is given by

$$\mathbf{G} = \begin{bmatrix} g_0 & g_1 & \cdots & g_{n-k} & 0 & 0 & 0 & \cdots & 0 \\ 0 & g_0 & g_1 & \cdots & g_{n-k-1} & g_{n-k} & 0 & \cdots & 0 \\ \vdots & \vdots & \vdots & & & & & & \vdots \\ 0 & 0 & 0 & \cdots & 0 & g_0 & g_1 & \cdots & g_{n-k} \end{bmatrix}. \quad (10.2.1)$$

The rows of $\mathbf{G}$ are linearly independent and $\text{rank}(\mathbf{G}) = k = \dim(C)$. If

$$h(x) = (x^n - 1)/g(x) = h_0 + h_1 x + \dots + h_k x^k,$$

then the parity-check matrix for C is

$$\mathbf{H} = \begin{bmatrix} 0 & 0 & \cdots & 0 & 0 & h_k & h_{k-1} & \cdots & h_1 & h_0 \\ 0 & 0 & \cdots & 0 & h_k & h_{k-1} & h_{k-2} & \cdots & h_0 & 0 \\ \vdots & \vdots & & & & & & & & \vdots \\ h_k & h_{k-1} & \cdots & h_1 & h_0 & 0 & 0 & \cdots & 0 & 0 \end{bmatrix}. \quad (10.2.2)$$

Note that if $\mathbf{H}$ is taken as the generator matrix, then the corresponding code is the dual code $C^\perp$ of C, which is again cyclic. While discussing cyclic codes, the terminology of vectors $\{a_0, a_1, \dots, a_{n-1}\}$ and the polynomials $a_0 + a_1 x + \dots + a_{n-1}x^{n-1}$ over F_q is used synonymously. Thus, we can regard C as a subset of the factor ring $F_q[x]/(x^n - 1)$.

Theorem 10.2.1. *The linear code C is cyclic if and only if C is an ideal of $F_q[x]/(x^n - 1)$.*

PROOF. If C is an ideal and $\{a_0, a_1, \dots, a_{n-1}\} \in C$, then so is

$$x\left(a_0 + a_1 x + \dots + a_{n-1}x^{n-1}\right) = \{a_{n-1}, a_0, \dots, a_{n-2}\} \in C.$$

Conversely, if $\{a_{n-1}, a_0, a_1, \dots, a_{n-2}\} \in C$, then for every $a(x) \in C$ we have $xa(x) \in C$, and thus, also $x^2 a(x) \in C, x^3 a(x) \in C$ and so on. Hence, $b(x)a(x) \in C$ also for any polynomial $b(x)$. This shows that C is an ideal. ∎

Every ideal of $F_q[x]/(x^n - 1)$ is principal in the sense that every nonzero ideal C is generated by the monic polynomial of lowest degree in the ideal, say, $g(x)$, where $g(x)$ divides $x^n - 1$. Hence, if $C = \{g(x)\}$ is a cyclic code, then $g(x)$ is called the generator polynomial of C and $h(x) = (x^n - 1)/g(x)$ the parity-check polynomial of C.

10.3 Methods for Describing Cyclic Codes

There are three methods of describing cyclic codes, which are as follows.

METHOD 1: Let $x^n - 1 = f_1(x)f_2(x) \cdots f_m(x)$ be the decomposition of $x^n - 1$ into monic irreducible factors over F_q. There are no multiple factors since we have assumed that $\gcd(n, q) = 1$. If $f_i(x)$ is irreducible over F_q, then $\{f_i(x)\}$ is a maximal ideal and the cyclic code generated by $f_i(x)$ is called a *maximal cyclic code*. The code generated by $(x^n - 1)/f_i(x)$ is called an *irreducible cyclic code*. This gives the method to determine all cyclic codes of length n over F_q: Keep the factor $x^n - 1$ as above and take any of the $2^m - 2$ nontrivial monic factors of $x^n - 1$ as a generator polynomial.

If $h(x)$ is the parity-check polynomial of a cyclic code $C \subseteq F_q[x]/(x^n - 1)$ and $v(x) \in F_q[x]/(x^n - 1)$, then the received polynomial $v(x) \in C$ iff $v(x)h(x) \equiv 0 \mod (x^n - 1)$. A message polynomial $a(x) = a_0 + a_1x + \cdots + a_{n-1}x^{n-1}$ is *encoded* by C into $w(x) = a(x)g(x)$, where $g(x)$ is the generator polynomial of C. If the received polynomial $v(x)$ is divided by $g(x)$, and if there is a nonzero remainder, we know that an error has occurred.

The canonical generator matrix of C is obtained as follows: Let $\deg\{g(x)\} = n - k$. Then there are unique polynomials $a_j(x)$ and $r_j(x)$ with $\deg\{r_j(x)\} < n - k$ such that $x^j = a_j(x)g(x) + r_j(x)$. Hence, $x^j - r_j(x)$ is a code polynomial, and so is $g_j(x) = x^k \left(x^j - r_j(x)\right)$ modulo $x^n - 1$. The polynomials $g_j(x), j = n - k, \ldots, n - 1$, are linearly independent and form the canonical generator matrix $(I_k, -R)$, where I_k is the $k \times k$ identity matrix and R is the $k \times (n - k)$ matrix whose i-th row is the vector of coefficients of $r_{n-k-1+i}(x)$.

Example 10.3.1. For $n = 7, q = 2$, the polynomial $x^7 - 1$ is decomposed into irreducible monic polynomials $x^7 - 1 = (x + 1)(x^3 + x + 1)(x^3 + x^2 + 1)$. Thus, $g(x) = x^3 + x^2 + 1$ generates a cyclic (7,4) code with parity-check polynomial $h(x) = (x + 1)(x^3 + x + 1) = x^4 + x^3 + x^2 + 1$. The corresponding canonical generator matrix and parity-check matrix are, respectively,

$$
\mathbf{G} = \begin{bmatrix} 1 & 0 & 0 & 0 & 1 & 0 & 1 \\ 0 & 1 & 0 & 0 & 1 & 1 & 1 \\ 0 & 0 & 1 & 0 & 1 & 1 & 0 \\ 0 & 0 & 0 & 1 & 0 & 1 & 1 \end{bmatrix}, \quad \mathbf{H} = \begin{bmatrix} 1 & 1 & 1 & 0 & 1 & 0 & 0 \\ 0 & 1 & 1 & 1 & 0 & 1 & 0 \\ 1 & 1 & 0 & 1 & 0 & 0 & 1 \end{bmatrix}. \blacksquare
$$

METHOD 2: Recall that if $f \in F_q[x]$ is a polynomial of the form $f(x) = f_0 + f_1x + \cdots + f_kx^k$, $f_0 = 0, f_k = 1$, then the solutions of the linear recurrence relation $\sum_{j=0}^{k} f_j a_{i+j} = 0$, $i = 0, 1, \ldots$, are periodic (of period n). The set of n-tuples of the first n terms of each possible solution, considered as polynomials $x^n - 1$ modulo $(x^n - 1)$, is the ideal generated by $g(x)$ in $F_q[x]/(x^n - 1)$, where $g(x)$ is the reciprocal polynomial of $(x^n - 1)/f(x)$ of degree $n - k$.[2] These linear recurrence relations provide a method to generate codewords of cyclic codes, and this generation process can be implemented using feedback shift registers (see §2.2.9).

Example 10.3.2. Let $f(x) = x^3 + x + 1$, which is an irreducible factor of $x^7 - 1$ over F_2 (see Example 10.3.1). The associated linear recurrence relation is $a_{i+3} + a_{i+1} + a_i = 0$ which produces a (7,3) cyclic code. Suppose that this relation encodes 111 as 1110010. The generator polynomial is the reciprocal polynomial of $(x^7 - 1)/f(x)$, which is $g(x) = x^4 + x^3 + x^2 + 1$. ∎

METHOD 3: We will describe cyclic codes by prescribing certain roots of all code polynomials in a suitable extension field of F_q[3] and use the property that all code polynomials are multiples of $g(x)$. Let $\alpha_1, \ldots, \alpha_s$ be elements of a finite extension field of F_q and $p_i(x)$ be the minimal polynomial of α_i over F_q for $i = 1, 2, \ldots, s$. Let $n \in \mathbb{N}$ be such that $\alpha_i^n = 1, i = 1, 2, \ldots, n$. Define $g(x) = \text{lcm}\{p_1(x), \ldots, p_s(x)\}$. Then $g(x)$ divides $x^n - 1$. If $C \subseteq F_q^n$ is the cyclic code with generator polynomial $g(x)$, then $v(x) \in C$ iff $v(\alpha_i) = 0, i = 1, 2, \ldots, s$. The following result establishes a relationship between cyclic codes and Hamming codes:

Theorem 10.3.1. *The binary cyclic code of length $n = 2^m - 1$ for which the generator polynomial is the minimal polynomial over F_q of a primitive element of F_{2^m} is equivalent to the binary (n, k) Hamming code, where $k = n - m$.*

PROOF. Let α denote a primitive element of F_{2^m} and let $p(x) = (x - \alpha)(x - \alpha^2) \cdots (x - \alpha^{2^{m-1}})$ be the minimal polynomial of α over F_2. For a cyclic code C generated by $p(x)$, we construct an $m \times (2^m - 1)$ matrix $\mathbf{H}$ with the j-th element as $[c_0, c_1, \ldots, c_{m-1}]^T$, where

$$\alpha^{j-1} = \sum_{i=0}^{m-1} c_i \alpha^i, \quad j = 1, 2, \ldots, 2^m - 1, \quad c_j \in F_2.$$

If $\mathbf{a} = \{a_0, a_1, \ldots, a_{n-1}\}$ and $a(x) = a_0 + a_1 x + \cdots + a_{n-1} x^{n-1} \in F_2[x]$, then the vector $\mathbf{H a}^T$ corresponds to the element $a(\alpha)$ expressed in the basis $(1, \alpha, \ldots, \alpha^{m-1})$. Hence, $\mathbf{H a}^T = \mathbf{0}$ only when $p(x)$ divides $a(x)$. This means

---

[2] Let $f(x) = a_n x^n + a_{n-1} x^{n-1} + \cdots + a_1 x + a_0 \in F_q[x]$, $a_n \neq 0$. Then the reciprocal polynomial f^* of f is defined by $f^*(x) = x^n f(1/x) = a_0 x^n + a_1 x^{n-1} + \cdots + a_{n-1} x + a_n$. Both f and f^* are of the same order.

[3] If L is an extension of a field K, then L is viewed as a vector space over K. The elements of L form an abelian group under addition, and each vector $\alpha \in L$ can be multiplied by a scalar $r \in K$ such that $r\alpha \in L$ and the laws of multiplication by scalars are satisfied. If L is finite-dimensional, then L is called a finite extension of K.

that $\mathbf{H}$ is a parity-check matrix of C. Since the columns of $\mathbf{H}$ are a permutation of the binary representation of the numbers $1, 2, \ldots, 2^m - 1$, the proof is complete. ∎

Example 10.3.3. The polynomial $x^4 + x + 1$ is primitive over F_2 and thus has a primitive element α of F_{2^4} as a root.[4] Using the vector notation for 15 elements $\alpha^j \in F_{2^4}^*, j = 0, 1, \ldots, 14$, expressed in the basis $\{1, \alpha, \alpha^2, \alpha^3\}$, if we form a 4×15 matrix with these vectors as columns, then we get the parity-check matrix of a code that is equivalent to the (15,11) Hamming code. Let a message $(a_0, a_1, \ldots, a_{10})$ be encoded into a code polynomial $w(x) = a(x)(x^4 + x + 1)$, where $a(x) = a_0 + a_1 x + \cdots + a_{10} x^{10}$. Suppose that the received polynomial contains one error, so that $w(x) + x^{e-1}$ is received instead of $w(x)$. Then the syndrome is $w(\alpha) + \alpha^{e-1} = \alpha^{e-1}$, and so the decoder concludes that there is an error in the e-th location. ∎

Theorem 10.3.2. *Let $C \subseteq F_q[x]/(x^n - 1)$ be a cyclic code with generator polynomial $g(x)$, and let $\alpha_1, \ldots, \alpha_{n-k}$ be the roots of g. Then $f \in F_q[x]/(x^n - 1)$ is a code polynomial iff the coefficient vector $(f_0, \ldots, f_{n-1})$ of f is in the null space of the matrix*

$$\mathbf{H} = \begin{bmatrix} 1 & \alpha_1 & \alpha_1^2 & \cdots & \alpha_1^{n-1} \\ \vdots & \vdots & \vdots & & \vdots \\ 1 & \alpha_{n-k} & \alpha_{n-k}^2 & \cdots & \alpha_{n-k}^{n-1} \end{bmatrix}. \tag{10.3.1}$$

PROOF. Let $f(x) = f_0 + f_1 x + \cdots + f_{n-1} x^{n-1}$. Then $f(\alpha_i) = f_0 + f_1 \alpha_i + \cdots + f_{n-1} \alpha_i^{n-1} = 0$ for $1 \leq i \leq n - k$. This implies that

$$\begin{bmatrix} 1, \alpha_i, \ldots, \alpha_i^{n-1} \end{bmatrix} \begin{bmatrix} f_0, f_1, \ldots, f_{n-1} \end{bmatrix}^T = 0 \quad \text{for } 1 \leq i \leq n - k,$$

iff $\mathbf{H} \begin{bmatrix} f_0, f_1, \ldots, f_{n-1} \end{bmatrix}^T = \mathbf{0}$. ∎

A SIMPLER METHOD. Let α be a primitive n-th root of unity in F_{q^m} and let the generator polynomial g be the minimal polynomial of α over F_q. Since g divides $f \in F_q[x]/(x^n - 1)$ iff $f(a) = 0$, we can replace the matrix $\mathbf{H}$ in (10.3.1) by

$$\mathbf{H} = \begin{bmatrix} 1 & \alpha & \alpha^2 \cdots \alpha^{n-1} \end{bmatrix}.$$

Recall that in the case of cyclic codes the syndrome is a column vector of length $n - k$, but now we can replace it by $S(\mathbf{w}) = \mathbf{H}\mathbf{w}^T$, where $\mathbf{w}$ is the received word, and then $S(\mathbf{w}) = w(\alpha)$ since $\mathbf{w} = [w_0, w_1, \cdots, w_{n-1}]$ can be regarded as a polynomial $w(x)$ with coefficients w_i.

Now if we use the notation $\mathbf{c}$ for the transmitted codeword and $\mathbf{w}$ for the received word, and if write $c(x)$ and $w(x)$, respectively, for the corresponding

---

[4] This field is also written as F_{16}, but we use the notation F_{2^4} to remind us that arithmetic operations of addition and multiplication are performed over F_2.

polynomials, and if we suppose that $e^{(j)}(x) = x^{j-1}, 1 \leq j \leq n$, is an error polynomial with a single error, so that we have $\mathbf{w} = \mathbf{c} + \mathbf{e}^{(j)}$ as the received word, then $w(a) = c(a) + e^{(j)}(\alpha) = e^{(j)}(\alpha) = a^{j-1}$, where $e^{(j)}(\alpha)$ is called the *error-location number*. Thus, $S(\mathbf{w}) = \alpha^{j-1}$ determines the error uniquely, since $e^{(i)}(\alpha) \neq e^{(j)}(\alpha)$ for $1 \leq i \leq n, i \neq j$.

Example 10.3.4. (Correction of 2 errors.) Let $\alpha \in F_{2^4}$ be a root of $x^4 + x + 1 \in F_2[x]$. Then α and α^3 have the minimal polynomials $m^{(1)}(x) = x^4 + x + 1$ and $m^{(3)}(x) = x^4 + x^3 + x^2 + x + 1$ over F_2, respectively. Both $m^{(1)}(x)$ and $m^{(3)}(x)$ are divisors of $x^{15} - 1$. Thus, we can define a binary cyclic code C with generator polynomial $g = m^{(1)}m^{(3)}$. Since g divides $f \in F_2[x]/(x^{15} - 1)$ iff $f(\alpha) = f(\alpha^3) = 0$, we can replace the matrix $\mathbf{H}$ in (10.3.1) simply by

$$\mathbf{H} = \begin{bmatrix} 1 & \alpha & \alpha^2 & \cdots & \alpha^{14} \\ 1 & \alpha^3 & \alpha^6 & \cdots & \alpha^{42} \end{bmatrix}.$$

Since the minimum distance of C is ≥ 5, we can correct up to 2 errors. Again, because C is a cyclic (15,7) code, let $S_1 = \sum_{i=0}^{14} v_i \alpha^i$ and $S_3 = \sum_{i=0}^{14} v_i \alpha^{3i}$ denote the components of $S(\mathbf{w}) = \mathbf{H}\mathbf{w}^T$, then $\mathbf{v} \in C$ iff $S_1 = S_3 = 0$. Using the binary representation for the element of F_{2^4}, we can determine $\mathbf{H}$ by the following method: The first four entries of the first column are the coefficients in $1 = 1 \cdot \alpha^0 + 0 \cdot \alpha^1 + 0 \cdot \alpha^2 + 0 \cdot \alpha^3$, the first four entries in the second column are the coefficients in $\alpha = 0 \cdot \alpha^0 + 1 \cdot \alpha^1 + 0 \cdot \alpha^2 + 0 \cdot \alpha^3$, and so on; the last four entries in the first column are the coefficients in $1 = 1 \cdot \alpha^0 + 0 \cdot \alpha^1 + 0 \cdot \alpha^2 + 0 \cdot \alpha^3$, the last four entries in the second column are the coefficients in $\alpha^3 = 0 \cdot \alpha^0 + 0 \cdot \alpha^1 + 0 \cdot \alpha^2 + 1 \cdot \alpha^3$, and so on. We have used $\alpha^4 + \alpha + 1 = 0$ in all the calculations. Thus,

$$\mathbf{H} = \begin{bmatrix} 1 & 0 & 0 & 0 & 1 & 0 & 0 & 1 & 1 & 0 & 1 & 0 & 1 & 1 & 1 \\ 0 & 1 & 0 & 0 & 1 & 1 & 0 & 1 & 0 & 1 & 1 & 1 & 1 & 0 & 0 \\ 0 & 0 & 1 & 0 & 0 & 1 & 1 & 0 & 1 & 0 & 1 & 1 & 1 & 1 & 0 \\ 0 & 0 & 0 & 1 & 0 & 0 & 1 & 1 & 0 & 1 & 0 & 1 & 1 & 1 & 1 \\ 1 & 0 & 0 & 0 & 1 & 1 & 0 & 0 & 0 & 1 & 1 & 0 & 0 & 0 & 1 \\ 0 & 0 & 0 & 1 & 1 & 0 & 0 & 0 & 1 & 1 & 0 & 0 & 0 & 1 & 1 \\ 0 & 0 & 1 & 0 & 1 & 0 & 0 & 1 & 0 & 1 & 0 & 0 & 1 & 0 & 1 \\ 0 & 1 & 1 & 1 & 1 & 0 & 1 & 1 & 1 & 0 & 1 & 1 & 1 & 1 & 1 \end{bmatrix}.$$

Now suppose that the received vector $\mathbf{w} = \{w_0, \ldots, w_{14}\}$ has at most two errors, say, $e(x) = x^{a_1} + x^{a_2}$, $0 \leq a_1, a_2 \leq 14$, $a_1 \neq a_2$. Then we have $S_1 = \alpha^{a_1} + \alpha^{a_2}$ and $S_3 = \alpha^{3a_1} + \alpha^{3a_2}$. Let $\nu_1 = \alpha^{a_1}, \nu_2 = \alpha^{a_2}$ be the error location numbers. Then $S_1 = \nu_1 + \nu_2, S_3 = \nu_1^3 + \nu_2^3$, and thus, $S_3 = S_1^3 + S_1^2 \nu_1 + S_1 \nu_1^2$, which gives $1 + S_1 \nu_1^{-1} + (S_1^2 + S_3 S_1^{-1})\nu_1^{-2} = 0$. If two errors have occurred during transmission, then ν_1^{-1} and ν_2^{-1} are the roots of the polynomial

$$s(x) = 1 + S_1 x + \left(S_1^2 + S_3 S_1^{-1}\right) x^2. \tag{10.3.2}$$

However, if only one error has occurred, then $S_1 = \nu_1$ and $S_3 = \nu_1^3$. Hence $S_1^3 + S_3 = 0$, or

$$s(x) = 1 + S_1 x. \qquad (10.3.3)$$

If no error occurred, then $S_1 = S_3 = 0$ and the correct codeword $\mathbf{w}$ has been recovered.

This analysis leads to the method that is described by the following steps: (i) Evaluate the syndrome $S(\mathbf{w}) = \mathbf{H}\mathbf{w}^T$ of the received vector $\mathbf{w}$; (ii) determine $s(x)$; and (iii) find the errors from the roots of $s(x)$. The polynomial in (10.3.3) has a root in F_{2^4} whenever $S_1 = 0$. If $s(x)$ in (10.3.2) has no roots in F_{2^4}, then we know that the error $e(x)$ has more than two error locations and cannot be corrected by the given $(15,7)$ code.

Example 10.3.5. As a specific example of this method, suppose $\mathbf{w} = 100111000000000$ is the received word. Then $S(\mathbf{w}) = \begin{bmatrix} S_1 \\ S_3 \end{bmatrix}$ is given by

$$S_1 = 1 + \alpha^3 + a^4 + \alpha^5 = \alpha^2 + \alpha^3, \; S_3 = 1 + \alpha^9 + \alpha^{12} + \alpha^{15} = 1 + \alpha^2.$$

For the polynomial $s(x)$ in (10.3.2) we get

$$s(x) = 1 + (\alpha^2 + \alpha^3)x + \left[1 + \alpha + \alpha^2 + \alpha^3 + (1 + \alpha^2)(\alpha^2 + \alpha^3)^{-1}\right]x^2$$
$$= 1 + (\alpha^2 + \alpha^3)x + (1 + \alpha + \alpha^3)x^2.$$

The roots of $s(x)$ are determined by trial and error as α and α^7. Hence, $\nu_1^{-1} = \alpha, \nu_2^{-1} = \alpha^7$, giving $\nu_1 = \alpha^{14}, \nu_2 = \alpha^8$. We know that the errors must have occurred in the locations corresponding to x^8 and x^{14}, which are the 9th and 15th locations in $\mathbf{w}$. Thus, flipping the 9th and 15th digits in $\mathbf{w}$, we obtain the transmitted codeword as $\mathbf{c} = 10011100000001$. Further, the codeword $\mathbf{c}$ is decoded by dividing the corresponding polynomial by the generator polynomial $g(x) = (x^4 + x + 1)(x^4 + x^3 + x^2 + x + 1)$, which gives $1 + x^3 + x^5 + x^6$ with zero remainder. Hence the original intended message symbol was 1001011. ∎

LINEAR ENCODING METHOD. Let $m(x)$ be a monic polynomial such that $s(x) = \sum_{i+0}^{r-1} s_i x^i$ is the remainder when $x^r m(x)$ is divided by $g(x)$, that is, $x^r m(x) = q(x)g(x) + s(x)$, $\deg\{s(x)\} < \deg\{g(x)\}$. Let

$$\mathbf{m} = (m_0, \dots, m_{k-1}) \mapsto \mathbf{c} = \{m_0, \dots, m_{k-1}, -s_0, -s_1, \dots, -s_{r-1}\}.$$
$$(10.3.4)$$

To ensure that this is the correct encoding, first note that $x^r m(x) - s(x) = q(x)g(x) = b(x) \in C$, which yields the corresponding codeword

$$\mathbf{c} = \{-s_0, -s_1, \dots, -s_{r-1}, m_0, \dots, m_{k-1}\} \in C.$$

In fact, the codeword $\mathbf{c}$ given in (10.3.4) is obtained after k right shifts, and $\mathbf{c} \in C$. Since every cyclic shift of this codeword is also a codeword, and C is symmetric on the first k locations, then this codeword is the only one with $\mathbf{m}$ in these locations. This gives the standard generator matrix also, so we encode the k different k-tuple messages $(0\,0\cdots0\,1\,0\cdots0)$ of weight 1 that corresponds to the message polynomials x^i, $0 \le i \le k - 1$, and these provide the rows of the generator polynomial.

Example 10.3.6. Let the generator polynomial $g(x) = x^3 + x + 1$ for the (7,4) binary cyclic code, where $r = 7 - 4 = 3$. We find that $x^3 x^2 = (x^2 + 1)(x^3 + x + 1)(x^2 + x + 1)$. Then, for example, the third row of the generator matrix corresponding to the message polynomial x^2 is

$$(m_0, m_1, m_2, -s_0, -s_1, -s_2) = (0\,0\,1\,0\,1\,1\,1).$$

Proceeding in this way we find the generator matrix as

$$\mathbf{G} = \begin{bmatrix} 1 & 0 & 0 & 0 & 1 & 1 & 0 \\ 0 & 1 & 0 & 0 & 0 & 1 & 1 \\ 0 & 0 & 1 & 0 & 1 & 1 & 1 \\ 0 & 0 & 0 & 1 & 1 & 0 & 1 \end{bmatrix}.$$

Note that this is a Hamming (7,4) code. ∎

A code that is equivalent to a cyclic code *need not* be cyclic itself, as seen from Example 10.3.5, which has 30 distinct binary Hamming (7,4) codes, but only two of them are cyclic.

The cyclic codes are basic tools for the construction of the BCH, RM, and RS codes, which are discussed in the following three chapters. We will have many occasions to either refer to a cyclic code or create more cyclic codes in these chapters.

10.4 Quadratic-Residue Codes

Let $p > 2$ be a prime and choose a primitive element $\alpha \in F_p$ (see §8.6.3). Then a nonzero element $r \in F_p$ is called a *quadratic residue modulo* p if $r = \alpha^{2i}$ for some integer i; otherwise r is called a *quadratic nonresidue mod* p. Thus, r is a quadratic nonresidue mod p iff $r = \alpha^{2j-1}$ for some integer j (see also §8.2, Property (v)). Examples of quadratic-residue (QR) codes are the binary Hamming $(7, 4, 3)$ code, the binary Golay $(23, 12, 7)$ code, and the ternary Golay $(11, 6, 5)$ code. We will, therefore, consider the finite fields F_7, F_{11}, and F_{23} to provide some examples of nonzero quadratic residues and nonresidues mod p.

Example 10.4.1. (i) Since 3 is a primitive element in F_7, the nonzero quadratic residues mod 7 are $\{3^{2i} : i = 0, 1, 2, \dots\} = \{1, 2, 4\}$, and nonzero quadratic nonresidues mod 7 are $\{3^{2i-1} : i = 1, 2, \dots\} = \{3, 6, 5\}$ (all arithmetic being performed mod 7).

(ii) Since 2 is a primitive element in F_{11}, the nonzero quadratic residues mod 11 are $\{2^{2i} : i = 0, 1, 2, \dots\} = \{1, 4, 5, 9, 3\}$, and nonzero quadratic nonresidues mod 11 are $\{2^{2i-1} : i = 1, 2, \dots\} = \{2, 8, 10, 7, 6\}$ (all arithmetic being performed mod 11).

(iii) Since 5 is a primitive element in F_{23}, the nonzero quadratic residues mod 23 are $\{5^{2i} : i = 0, 1, 2, \dots\} = \{1, 2, 4, 8, 16, 9, 18, 13, 3, 6, 12\}$, and nonzero quadratic nonresidues mod 23 are $\{5^{2i-1} : i = 1, 2, \dots\} = \{5, 10, 20, 17, 11, 22, 21, 19, 15, 7, 14\}$ (all arithmetic being performed mod 23). ∎

Some useful results are as follows:

(a) A nonzero element $r \in F_p$ is a nonzero quadratic residue mod p iff $r \equiv a^2$ (mod p) for some $a \in F_p$. For example, 6 is a quadratic residue mod 29 since $6 = 8^2$ (mod 29).

(b) The product of two quadratic residues Q_p mod p is a quadratic residue mod p.

(c) The product of two quadratic nonresidues N_p mod p is a quadratic residue mod p.

(d) The product of a nonzero quadratic residue Q_p mod p and a quadratic nonresidue N_p mod p is a quadratic nonresidue mod p.

(e) There are exactly $(p-1)/2$ nonzero quadratic residues N_p mod p and $(p-1)/2$ quadratic nonresidues N_p mod p. Thus, $F_p = \{0\} \cup Q_p \cup N_p$.

(f) For $a \in Q_p$ and $b \in N_p$, we have

$$aQ_p = \{ar : r \in Q_p\} = Q_p, \quad aN_p = \{an : n \in N_p\} = N_p,$$
$$bQ_p = \{br : r \in Q_p\} = N_p, \quad bN_p = \{bn : n \in N_p\} = Q_p.$$

Proofs of these results can be found in most books on coding theory, e.g. Ling and Xing [2004].

Example 10.4.2. (i) Consider F_7 of Example 10.4.1(i), where the set of nonzero quadratic residues mod 7 is $Q_7 = \{1, 2, 4\}$ and the set of nonzero quadratic nonresidues mod 7 is $N_7 = \{3, 6, 5\}$. Note that their size is $|Q_7| = |N_7| = \frac{7-1}{2} = 3$. Take, for example, $a = 4$ and $b = 3$; then

$$4Q_7 = \{4 \cdot 1, 4 \cdot 2, 4 \cdot 4\} = \{4, 1, 2\} = Q_7,$$
$$4N_7 = \{4 \cdot 3, 4 \cdot 6, 4 \cdot 5\} = \{5, 3, 6\} = N_7.$$
$$3Q_7 = \{3 \cdot 1, 3 \cdot 2, 3 \cdot 4\} = \{3, 6, 5\} = N_7,$$
$$3N_7 = \{3 \cdot 3, 3 \cdot 6, 3 \cdot 5\} = \{2, 4, 1\} = Q_7.$$

(ii) Consider F_{11} of Example 10.4.1(ii), where the set of nonzero quadratic residues mod 11 is $Q_{11} = \{1, 4, 5, 9, 3\}$ and the set of nonzero quadratic nonresidues mod 11 is $N_{11} = \{2, 8, 10, 7, 6\}$. Note that their size is $|Q_{11}| = |N_{11}| = \frac{11-1}{2} = 5$. Take, for example, $a = 5$ and $b = 2$; then

$$5Q_{11} = \{5 \cdot 1, 5 \cdot 4, 5 \cdot 5, 5 \cdot 9, 5 \cdot 3\} = \{5, 9, 4, 1, 4\} = Q_{11}$$
$$5N_{11} = \{5 \cdot 2, 5 \cdot 8, 5 \cdot 10, 5 \cdot 7, 5 \cdot 6\} = \{10, 7, 6, 2, 8\} = N_{11},$$
$$2Q_{11} = \{2 \cdot 1, 2 \cdot 4, 2 \cdot 5, 2 \cdot 9, 2 \cdot 3\} = \{2, 8, 10, 7, 6\} = N_{11},$$
$$2N_{11} = \{2 \cdot 2, 2 \cdot 8, 2 \cdot 10, 2 \cdot 7, 2 \cdot 6\} = \{4, 5, 9, 3, 1\} = Q_{11}.$$

(iii) Consider F_{23} of Example 10.4.1(ii), where the set of nonzero quadratic residues mod 23 is $Q_{23} = \{1, 2, 4, 8, 16, 9, 18, 13, 3, 6, 12\}$ and the set of nonzero quadratic nonresidues mod 23 is $N_{23} = \{5, 10, 20, 17, 11, 22, 21, 19, 15, 7, 14\}$. Note that their size is $|Q_{23}| = |N_{23}| = \frac{23-1}{2} = 11$. Take, efor example, $a = 2$ and $b = 5$; then

$$2Q_{23} = \{1 \cdot 2, 2 \cdot 2, 2 \cdot 4, 2 \cdot 8, 2 \cdot 16, 2 \cdot 9, 2 \cdot 18, 2 \cdot 13, 2 \cdot 3, 2 \cdot 6, 2 \cdot 12\}$$
$$= \{2, 4, 8, 16, 9, 18, 13, 3, 6, 12, 1\} = Q_{23},$$

$$2N_{23} = \{2 \cdot 5, 2 \cdot 10, 2 \cdot 20, 2 \cdot 17, 2 \cdot 11, 2 \cdot 22, 2 \cdot 19, 2 \cdot 15, 2 \cdot 7, 2 \cdot 14\}$$
$$= \{2, 4, 8, 16, 9, 18, 13, 3, 6, 12, 1\} = N_{23},$$

$$5Q_{23} = \{5 \cdot 1, 5 \cdot 2, 5 \cdot 4, 5 \cdot 8, 5 \cdot 16, 5 \cdot 9, 5 \cdot 18, 5 \cdot 13, 5 \cdot 3, 5 \cdot 6, 5 \cdot 12\}$$
$$= \{5, 10, 20, 17, 11, 22, 21, 19, 15, 7, 14\} = N_{23},$$

$$5N_{23} = \{5 \cdot 5, 5 \cdot 10, 5 \cdot 20, 5 \cdot 17, 5 \cdot 11, 5 \cdot 22, 5 \cdot 21, 5 \cdot 19, 5 \cdot 15, 5 \cdot 7, 5 \cdot 14\}$$
$$= \{2, 4, 8, 16, 9, 18, 13, 3, 6, 12, 1\} = Q_{23} \ \blacksquare.$$

10.4.1 Generator Polynomials. Let $p \neq q$ be a prime number such that q is a quadratic residue mod p. Let $m \geq 1$ be an integer such that $q^m - 1$ is divisible by p. Let β be a primitive element of F_{q^m}, and set $\alpha = \beta^{(q^m-1)/p}$. Then the order of α is p, i.e., $1 = \alpha^0 = \alpha^p, \alpha = \alpha^1, \alpha^2, \dots, \alpha^{p-1}$ are primitive distinct, and $x^p - 1 = (x - 1)(x - \alpha)(x - \alpha^2) \cdots (x - \alpha^{p-1})$. Define the polynomials

$$g_Q(x) = \prod_{r \in Q_p} (x - \alpha^r) \in F_q[x], \quad g_N(x) = \prod_{r \in N_p} (x - \alpha^n) \in F_q[x]. \quad (10.4.1)$$

Then

$$x^p - 1 = (x - 1)(x - \alpha)(x - \alpha^2) \cdots (x - \alpha^{p-1})$$
$$= (x - 1) \prod_{r \in Q_p} (x - \alpha^r) \prod_{n \in N_p} (x - \alpha^n)$$
$$= (x - 1)g_Q(x)g_N(x). \quad (10.4.2)$$

To prove that both $g_Q(x)$ and $g_N(x)$ belong to $F_q[x]$, let $g_Q(x) = a_0 + a_1 x + \cdots + a_k x^k$, where $a_i \in F_{q^m}$ and $k = \frac{p-1}{2}$. Then , by raising each a_i to the q-th power we get

$$a_0^q + a_1^q x + \cdots + a_k^q x^k = \prod_{r \in Q_p} (x - \alpha^{rq}) = \prod_{j \in qQ_p} (x - \alpha^j)$$
$$= \prod_{j \in Q_p} (x - \alpha^j) = g_Q(x), \quad \text{since } qQ_j = Q_p.$$

Thus, $a_i = a_i^q$ for all i, $0 \leq i \leq m$, are elements of $F_q[x]$. The same argument is used to show that $g_N(x) \in F_q[x]$.

Example 10.4.3. We will consider the field F_7 of Example 10.4.1. Let $p = 7$, $q = 2 \in Q_7$. Since the nonzero quadratic residues are $\{1, 2, 4\}$ and the nonzero quadratic nonresidures are $\{3, 6, 5\}$, let α be a root of the polynomial $1 + x + x^3 \in F_2[x]$. The order of α is 7, i.e., $1 = \alpha^0, \alpha, \alpha^2, \ldots, \alpha^6$ are pairwise distinct, and $x^7 - 1 = \prod_{i=0}^{6} (x - \alpha^i)$. Then by (10.4.2)

$$g_Q(x) = \prod_{r \in Q_7} (x - \alpha^r) = (x - \alpha)(x - \alpha^2)(x - \alpha^4)$$
$$= 1 + x + x^3, \quad \text{using Table 8.6.3,}$$
$$g_N(x) = \prod_{n \in N_7} (x - \alpha^n) = (x - \alpha^3)(x - \alpha^6)(x - \alpha^5)$$
$$= 1 + x^2 + x^3, \quad \text{using Table 8.6.3.}$$

Hence, by (10.4.2),

$$x^7 - 1 = (x - 1)f_Q(x)g_N(x) = (x - 1)(1 + x + x^3)(1 + x^2 + x^3).$$

A similar result can be derived for the fields F_{11} and F_{23} by using the corresponding rules for addition and multiplication in these fields similar to the Tables 8.6.3 and 8.6.4, which will not work in these cases. Remember that multiplication uses the law of exponents so that $\alpha^i * \alpha^j = \alpha^{i+j}$. ∎

Theorem 10.4.1. *For an odd prime p and an integer r such that $\gcd(r, p) = 1$, the number r is a quadratic residue mod p iff $r^{(p-1)/2} \equiv 1 \mod p$. In particular, for an odd prime p, 2 is a quadratic residue mod p iff p is of the form $p = 8m \pm 1$, and it is a quadratic residue mod p if p is of the form $p = 8m \pm 3$.*

The proof is omitted, as it can be found, e.g., in Ling and Xing [2004].

10.4.2 Quadratic-Residue Codes. The quadratic-residue (QR) codes[5] are defined as follows: Let p and q, $p \neq q$, be two primes such that q is a quadratic residue mod p. For an integer $m \geq 1$ such that $q^m - 1$ is divisible by p, let β be a primitive element of F_{q^m}, and set $\alpha = (q^m - 1)/p$. Then by (10.4.2) the divisors of $x^p - 1$ are the polynomials $g_Q(x)$ and $g_N(x)$ defined by (10.4.1) over F_q. The q-ary cyclic codes $C_Q = \langle g_Q(x) \rangle$ and $C_N = \langle g_N(x) \rangle$ of length p are called the *quadratic-residue* (QR) codes, where the dimension of both codes is $\dim(C_q) = \dim(C_N) = p - \frac{p-1}{2} = \frac{p+1}{2}$.

Example 10.4.4. The binary QR codes $C_Q = \langle 1 + x + x^3 \rangle$ and $C_N = \langle 1 + x^2 + x^3 \rangle$ of length 7 (see Example 10.4.3) are equivalent to each other and to the binary Hamming (7,4,3) code.

---

[5] Not to be confused with common Quick-Response (QR) matrix barcodes for smart phones.

Similarly, the ternary QR codes $C_Q = \langle 2 + x^2 + 2x^3 + x^4 + x^5 \rangle$ and $C_N = \langle 2 + 2x + x^2 + 2x^3 + x^5 \rangle$ of length 11 are equivalent to each other and to the ternary Golay $(11,6,5)$ code.

Also, the binary QR codes $C_Q = \langle 1 + x^2 + x^4 + x^5 + x^6 + x^{10} + x^{11} \rangle$ and $C_N = \langle 1 + x + x^5 + x^6 + x^7 + x^9 + x^{11} \rangle$ of length 23 are equivalent to each other and to the binary Golay $(23,11,7)$ code. ∎

The relationship between length, dimension and distance for QR codes is given in Table 10.4.1.

Table 10.4.1 QR Codes

Length	Dimension	Distance
7	4	3
11	6	5
17	9	5
23	12	7
31	16	7

The following results are given without proofs, as they are easily available.

Theorem 10.4.2. *The two q-ary QR codes C_Q and C_N are equivalent.*

Theorem 10.4.3. *There exists binary QR codes of lenth p iff p is a prime of the form $p = 8m \pm 1$.*

Every codeword of the binary QR code C_Q of length p is a linear combination of the rows of the $p \times p$ matrix

$$\begin{bmatrix} 1 & 1 & \cdots & 1 \\ c_0^* & c_1^* & \cdots & c_{p-1}^* \\ c_{p-1}^* & c_0^* & \cdots & c_{p-2}^* \\ \vdots & \vdots & \cdots & \vdots \\ c_1^* & c_2^* & \cdots & c_0^* \end{bmatrix}, \tag{10.4.3}$$

where c_i^*, $0 \le i \le p - 1$, are the coefficients of the idempotent code $C_Q^*(x)$ of the CR code $C_Q(x)$, defined by

$$C_Q^*(x) = \begin{cases} 1 + \sum\limits_{i \in N_p} x^i & \text{if } p \text{ is of the form } 8m - 1, \\ \sum\limits_{i \in N_p} x^i & \text{if } p \text{ is of the form } 8m + 1, \end{cases}$$

$$= \sum_{i=0}^{p-1} c_i^* x^i. \tag{10.4.4}$$

11

BCH Codes

11.1 Binary BCH Codes

The BCH codes are named after Bose-Chaudhuri-Hocquebghem (Bose and Chaudhuri [1960], and Hocquenghem [1959]). These codes are multiple error correcting codes and are a generalization of the Hamming codes. Because of their high coding rate, they have inspired a lot of study to find fast hardware encoding and decoding schemes and to develop their applications in high-speed memory.

Recall that the Hamming (n, k) codes with $m = n - k$ parity-check bits have the following characteristics: Block length: $n = 2^m - 1$; data bits: $k = 2^m - m - 1$; and correctable errors: $t = 1$. The Golay codes are the other classes of codes that are more versatile than the Hamming codes; their minimum distance is 7, so they can correct up to $t = \lfloor 7/2 \rfloor = 3$ errors. Again, both Hamming and Golay codes are perfect codes (§7.1). In terms of the space packing geometry (§3.6), the entire space of the field $\mathrm{GF}(2^m)$ is packed by spheres where each sphere contains a valid codeword at its center as well as invalid codewords that are at a distance at most equal to 3 in the case of a Golay code.

On the other hand, the BCH code has the following characteristics: Block length: $n = 2^m - 1$; parity-check bits: $n - k \le mt$; and minimum distance: $d \ge 2t + 1$. Let $\alpha \in F_{2^m}$ be a primitive element. Then the binary BCH code is defined by

$$\mathrm{BCH}(n, d) = \{(c_0, c_2, \dots, c_{n-1}) \in F_2^n : c(\mathbf{x}) = c_0 + c_1\mathbf{x} + \dots + c_{n-1}\mathbf{x}^{n-1}$$

$$\text{satisfies } c(\alpha) = c(\alpha^2) = \dots = c(\alpha^{d-1}) = 0\}. \tag{11.1.1}$$

Note that in this definition the coefficients c_i, $0 \le i \le n - 1$, take their values only from the base field F_2 instead of the extension field F_{2^m}. The BCH codes form linear vector spaces, and we use x instead of $\mathbf{x}$ as long as we understand that x is a vector. Although the restriction that $c(\alpha) = 0$ is a constraint over the field F_{2^m}, it can be regarded as a set of m linear constraints over the field

F_2. The linearity of these constraints can be justified by the following result.

Theorem 11.1.1. *The constraint $c(\alpha) = 0$ in (11.1.1) defined over the extension field F_{2^m} is equivalent to a set of m linear constraints over F_2.*

PROOF. Define a multiplication transformation over F_{2^m} by the map $\phi(\alpha) : x \mapsto \alpha x$ which is linear on the field F_2 since $\alpha(x+y) = \alpha x + \alpha y$. Next, choose a basis $\{b_1, \dots, b_m\} \subset F_{2^m}$ of F_{2^m} over F and represent each element $x \in F_{2^m}$ as the column vector $[x_1, x_2, \dots, x_m]^T \in F_2^m$, where $x = x_1 b_1 + x_2 b_2 + \cdots + x_m b_m$. Then the linear map $\phi(\alpha)$ corresponds to a linear transformation of this vector representation, thus mapping $x = [x_1, x_2, \dots, x_m]^T$ on to a column $M_\alpha x$ of an $m \times m$ matrix $M_\alpha \in F_2^{m \times m}$, and the i-th coefficient $c_i \in F_2$, $0 \le i \le n-1$, corresponds to the vector $[c_i, 0, \dots, 0]^T \in F_2^m$. This implies that the constraint $c(\alpha) = c_0 + c_1\alpha + c_2\alpha^2 + \cdots + c_{n-1}\alpha^{n-1} = 0$ is equivalent to the constraint

$$\begin{Bmatrix} c_0 \\ 0 \\ \vdots \\ 0 \end{Bmatrix} + M_\alpha \begin{Bmatrix} c_1 \\ 0 \\ \vdots \\ 0 \end{Bmatrix} + \cdots + M_a^{n-1} \begin{Bmatrix} c_{n-1} \\ 0 \\ \vdots \\ 0 \end{Bmatrix} = \begin{Bmatrix} 0 \\ 0 \\ \vdots \\ 0 \end{Bmatrix}, \qquad (11.1.2)$$

which yields m linear constraints over F_2. ∎

The block length of the BCH(n, d) code is n, and its distance $\le d$. A useful property of the BCH code is that its dimension is $\le n - (d-1)\log(n+1)$, and we have $d-1$ constraints in the definition (11.1.1) on the extension field F_{2^m} where each constraint generates $m = \log(n+1)$ in the base field F_2. The above bound is improved as given in the following result.

Theorem 11.1.2. *Given a length $n = 2^m - 1$ and a distance d, the dimension of the BCH(n, d) code is $\le n - \lceil \frac{d-1}{2} \rceil \log(n+1)$.*

PROOF. This improved bound will follow if we show the redundancy of some of the constraints in the definition (11.1.1). To do this, note that if $c(\gamma) = 0$ for any polynomial $c(x) \in F_2[x]$ and any element $\gamma \in F_2$, then we must also have $c(\gamma^2) = 0$. In fact, if $x(\gamma) = 0$, then we must also have $c(\gamma)^2 = 0$, and so, in general, we must have $\left(c_0 + c_1\gamma + c_2\gamma^2 + \cdots + c_{n-1}\gamma^{n-1}\right)^2 = 0$. Since for any two elements $\alpha, \beta \in F_{2^m}$ we have $(\alpha+\beta)^2 = \alpha^2 + \beta^2$, so $c(\gamma) = 0$ also implies that

$$c_0^2 + (c_1\gamma)^2 + (c_2\gamma^2)^2 + \cdots + (c_{n-1}\gamma^{n-1})^2 = 0,$$

which in turn implies that

$$c_0^2 + c_1^2\gamma^2 + c_2(\gamma^2)^2 + \cdots + c_{n-1}(\gamma^2)^{n-1} = 0.$$

Further, since the coefficients $c_0, c_1, \dots, c_{n-1}$ are in F_2, we have $c_i^2 = c_i$ for all $i = 0, 1, \dots, n-1$. Hence, $c(\gamma) = 0$ implies that

$$c_0 + c_1\gamma^2 + c_2(\gamma^2)^2 + \cdots + c_{n-1}(\gamma^2)^{n-1} = c(\gamma^2) = 0.$$

Next, we note that the above argument implies that the constraints $c(\gamma^{2j}) = 0$ for $j = 1, 2, \ldots, \lfloor \frac{d-1}{2} \rfloor$ are all redundant, and we can remove them from the definition (11.1.1). Thus, it leaves only $\lceil \frac{d-1}{2} \rceil m = \lceil \frac{d-1}{2} \rceil \log(n + 1)$ constraints. ∎

Note that in view of the Hamming bound, the bound on the dimension of the BCH (n, d) code in Theorem 11.1.2 is asymptotically sharp in the sense that 2 cannot be replaced by $2 - \varepsilon$ for any $\varepsilon > 0$, i.e., the dimension of the BCH (n, d) code is always $\leq n - \lceil \frac{d-1}{2} \rceil \log(n + 1)$.

Although the BCH codes, just like the Hamming codes, have a very good rate, they are useful only when the distance is small, i.e., $d \leq \frac{n}{\log n}$. These two codes enjoy a very close connection. For example, given $n = 2^m - 1$, the BCH $(n, 3)$ code is similar to the Hamming $(2^m - 1, 2^m - 1 - m, 3)$ code, up to some coordinate permutation.

11.2 Extended Finite Field

Let F_{2^m} (or $\mathrm{GF}(2^m)$) denote an extension of the field F_2 such that any polynomial with coefficients in F_2 has all its zeros in F_{2^m}. This code satisfies the following properties:

1. $(x_1 + x_2 + \cdots + x_n)^2 = x_1^2 + x_2^2 + \cdots + x_n^2$, since $1 + 1 = 0$.

2. Two irreducible polynomials in $F_2[x]$ with a common zero in F_{2^m} are equal.

3. Let $\alpha \in F_{2^m}$ be a zero of an irreducible polynomial $g(x) \in F_2[x]$ of degree r. Then for any $j \geq 0$

   (i) α^j is a zero of some irreducible polynomial $g_j(x) \in F_2[x]$ of degree $\leq r$,

   (ii) $g_j(x)$ divides $x^{2^{r-1}} + 1$, and

   (iii) $\deg\{g_j(x)\}$ divides r.

4. The *exponent* of an irreducible polynomial $g(x) \in F_2[x]$ is the least positive integer e such that $g(x)$ divides $1 + x^e$. Note that $e \leq 2^r - 1$, where $r = \deg\{g(x)\}$, and e divides $2^r - 1$.

5. For any positive integer r there exists an irreducible polynomial of degree r and exponent $2^r - 1$. Such a polynomial is called *primitive*.

6. If $g(x)$ is a primitive polynomial of degree $\leq r$ and $\alpha \in F_{2^m}$ is a zero of $g(x)$, then $\alpha^0 = 1, \alpha^1 = \alpha, \alpha^2, \ldots, \alpha^{2^r-1}$ are distinct and $\alpha^{2^r-1} = 1$.

7. $g(\alpha) =$ implies $g(\alpha^2) = g(\alpha^4) = g(\alpha^8) = \cdots = 0$ (compare 1 above).

11.3 Construction of BCH Codes

Follow the steps given below.

STEP 1. Decide a minimal distance $2d + 1$ between the codewords and an integer r such that $2^r > 2d + 1$.

STEP 2. Choose a primitive polynomial $g_1(x)$ of degree r and denote by $\alpha \in F_{2^m}$ a zero of $g_1(x)$.

STEP 3. Construct (as in Example 11.3.1. below) irreducible polynomials $g_1(x), g_2(x), g_3(x), \ldots, g_{2d}(x)$ of degree $\leq r$ with zeros $\alpha, \alpha^2, \alpha^3, \ldots, \alpha^{2d}$, respectively.

STEP 4. Let $g(x)$ of degree k, where $k \leq dr$ (always), be the least common multiple (lcm) of the polynomials $g_1(x), \ldots, g_{2d}(x)$, i.e., the product of all different of the polynomials $g_1(x), \ldots, g_{2d}(x)$.

STEP 5. The BCH code is that polynomial code $K : F_2^m \mapsto F_2^n$ generated by $g(x)$, where $n = 2^r - 1, m = n - k$, and the weight of the code is at least $2^d + 1$.

Example 11.3.1. We will take $d = 2, r = 4$; thus, $n = 15$. Choose $g_1(x) = 1 + x^3 + x^4$ (see Table 8.1.1(b) of irreducible polynomials, Degree 4, $\alpha^{15} = 1$). The construction of polynomials $g_2(x), g_3(x), g_4(x)$ is as follows: $g_1(x) = g_2(x) = g_4(x)$, as $g_1(\alpha) = g_1(\alpha^2) = g_1(\alpha^4) = 0$ because of properties 2 and 7. In fact, we can easily compute these values using $\alpha^4 = 1 + \alpha^3$, Table 8.1.1(b) and formula (8.6.2): $g_1(\alpha) = 1 + \alpha^3 + \alpha^4 = 1 + \alpha^3 + 1 + \alpha^3 = 0$, $g_1(\alpha^2) = 1 + \alpha^6 + \alpha^8 = 1 + 1 + \alpha + \alpha^2 + \alpha^3 + \alpha + \alpha^2 + \alpha^3 = 0$, and $g_1(\alpha^4) = 1 + \alpha^{12} + \alpha^{16} = 1 + \alpha + 1 + \alpha = 0$ since $\alpha^{16} = \alpha$. To construct the remaining polynomial $g_3(x)$, let $g_3(x) = 1 + Ax + Bx^2 + Cx^3 + Dx^4$. We proceed as follows (or use Tables 8.6.3 and 8.6.4):

$$\alpha^5 = \alpha\alpha^4 = \alpha(1 + \alpha^3) = \alpha + \alpha^4 = 1 + \alpha + \alpha^3,$$

$$\alpha^6 = \alpha\alpha^5 = \alpha(1 + \alpha + \alpha^3) = \alpha + \alpha^2 + \alpha^4 = 1 + \alpha + \alpha^2 + \alpha^3,$$

$$\alpha^8 = (1+\alpha^3)^2 = 1+2\alpha^3 + \alpha^6 = 1+\alpha^6 = 1+1+\alpha+\alpha^2+\alpha^3 = \alpha+\alpha^2+\alpha^3,$$

$$\alpha^9 = \alpha^3\alpha^6 = \alpha^3(1 + \alpha + \alpha^2 + \alpha^3) = \alpha^3 + \alpha^4 + \alpha^5 + \alpha^6 = 1 + \alpha^2,$$

$$\alpha^{12} = \alpha^2\alpha^9 = \alpha^3(1 + \alpha^2) = \alpha^3 + \alpha^5 = 1 + \alpha^3,$$

Then $g_3(\alpha^3) = 1+A\alpha^3+B(1+\alpha+\alpha^2+\alpha^3)+C(1+\alpha^2)+D(1+\alpha) = 1+A\alpha^3+B(1+\alpha+\alpha^2+\alpha^3)+C(1+\alpha^2)+D(1+\alpha) = 0$. Comparing the coefficients of similar powers of α, we get $1+B+C+D = 0$, $B+D = 0$, $B+C = 0$, $A+B = 0$, which gives $A = B = C = D = 1$. Thus, $g_3(x) = 1 + x + x^2 + x^3 + x^4$. Then $g(x) = g_1(x)g_3(x) = (1+x^3+x^4)(1+x+x^2+x^3+x^4) = 1+x+x^2+x^4+x^8$. Note that $\deg\{g(x)\} = 8 = k$, and $m = 7$, which yields, for example,

$$(10011) \leftrightarrow 1+x^3+x^4 \curvearrowright (1+x^3+x^4)(1+x+x^2+x^3+x^4)$$

$$= 1+x+x^2+x^4+x^8 \leftrightarrow (111010001). \blacksquare$$

Thus, the codewords in a BCH code are constructed by taking the remainder obtained after dividing a polynomial representing the given data bits by a generator polynomial that is chosen so as to provide the code its required characteristics. All codewords so constructed are multiples of the generator polynomial, as explained in detail in §11.5.2.

11.4 General Definition

Let $b \geq 0$ be an integer and let $\alpha \in F_{q^m}$ be a primitive n-th root of unity, where m is the multiplicative order of q modulo n. A *BCH code* over F_q of length n and *designated distance* $d, 1 \leq d \leq n$, is a cyclic code defined by the roots $\alpha^b, \alpha^{b+1}, \ldots, \alpha^{b+d-2}$ of the generator polynomial. If $m^{(i)}(x)$ denotes the minimal polynomial of α^i over F_q, then the generator polynomial $g(x)$ of a BCH code is of the form

$$g(x) = \mathrm{lcm}\left(m^{(b)}(x), m^{(b+1)}(x), \ldots, m^{(b+d-2)}(x)\right). \qquad (11.4.1)$$

Some special cases are as follows:

(i) If $b = 1$, the corresponding BCH codes are called *narrow-sense BCH codes*.

(ii) If $n = q^m - 1$, the BCH codes are called *primitive BCH codes*.

(iii) If $n = q - 1$, a BCH code of length n over F_q is called a *Reed–Solomon code*.

Theorem 11.4.1. *The minimum distance of a BCH code of designated distance d is at least d.*

PROOF. The BCH code is the null space of the matrix

$$\mathbf{H} = \begin{bmatrix} 1 & \alpha^b & \alpha^{2b} & \cdots & \alpha^{(n-1)b} \\ 1 & \alpha^{b+1} & \alpha^{2(b+1)} & \cdots & \alpha^{(n-1)(b+1)} \\ \vdots & \vdots & \vdots & & \vdots \\ 1 & \alpha^{b+d-2} & \alpha^{2(b+d-2)} & \cdots & \alpha^{(n-1)(b+d-2)} \end{bmatrix}. \qquad (11.4.2)$$

For the definition of null space, see §B.5. We will show that any $(d-1)$ columns of the matrix (11.4.2) are linearly independent. To do this, take the determinant of any $(d-1)$ distinct columns of $\mathbf{H}$, which gives

$$\begin{vmatrix} \alpha^{bi_1} & \alpha^{bi_2} & \cdots & \alpha^{bi_{d-1}} \\ \alpha^{(b+1)i_1} & \alpha^{(b+1)i_2} & \cdots & \alpha^{(b+1)i_{d-1}} \\ \vdots & \vdots & & \vdots \\ \alpha^{(b+d-2)i_1} & \alpha^{(b+d-2)i_2} & \cdots & \alpha^{(b+d-2)i_{d-1}} \end{vmatrix}$$

$$= \alpha^{b(i_1+i_2+\cdots+i_{d-1})} \begin{vmatrix} 1 & 1 & \cdots & 1 \\ \alpha^{i_1} & \alpha^{i_2} & \cdots & \alpha^{i_{d-1}} \\ \vdots & \vdots & & \vdots \\ \alpha^{i_1(d-2)} & \alpha^{i_2(d-2)} & \cdots & \alpha^{i_{d-1}(d-2)} \end{vmatrix}$$

$$= \alpha^{b(i_1+i_2+\cdots+i_{d-1})} \prod_{1 \leq k < j \leq d-1} \left(\alpha^{i_j} - \alpha^{i_k}\right) \neq 0,$$

which implies that the minimum distance of the code is at least d. ∎

Example 11.4.1. Let $m^{(1)}(x) = x^4 + x + 1$ be the minimal polynomial over F_2 of a primitive element $\alpha \in F_{2^4}$. The powers $\alpha^i, 0 \leq i \leq 4$, are represented as linear combinations of $1, \alpha, \alpha^2, \alpha^3$, which gives a parity-check matrix $\mathbf{H}$ of a code equivalent to the Hamming $(15,11)$ code (see Example 4.4.2):

$$\mathbf{H} = \begin{bmatrix} 1 & 0 & 0 & 0 & 1 & 0 & 0 & 1 & 1 & 0 & 1 & 0 & 1 & 1 & 1 \\ 0 & 1 & 0 & 0 & 1 & 1 & 0 & 1 & 0 & 1 & 1 & 1 & 1 & 0 & 0 \\ 0 & 0 & 1 & 0 & 0 & 1 & 1 & 0 & 1 & 0 & 1 & 1 & 1 & 1 & 0 \\ 0 & 0 & 0 & 1 & 0 & 0 & 1 & 1 & 0 & 1 & 0 & 1 & 1 & 1 & 1 \end{bmatrix}$$

$$= \begin{bmatrix} 1 & \alpha & \alpha^2 & \alpha^3 & \alpha^4 & \alpha^5 & \alpha^6 & \alpha^7 & \alpha^8 & \alpha^9 & \alpha^{10} & \alpha^{11} & \alpha^{12} & \alpha^{13} & \alpha^{14} \end{bmatrix}.$$

Note that α^2 is also a root of $m^{(1)}(x)$. This code is an example of a narrow-sense BCH code of designated distance $d = 3$ over F_2. Its minimum distance is also 3, and so it can correct a single error. The decoding of a received word $\mathbf{w} \in F_2^{15}$ is performed by finding the syndrome $\mathbf{Hw}^T$, which for this cyclic $(15,11)$ code is given by $w(\alpha)$ in the basis $\{1, \alpha, \alpha^2, \alpha^3\}$. The method for obtaining this $w(\alpha)$ is to divide $w(x)$ by $m^{(1)}(x)$, say, $w(x) = a(x)m^{(1)}(x) + r(x)$ with $\deg\{r(x)\} < 4$, because then we have $w(\alpha) = r(\alpha)$, i.e., the components of the syndrome are equal to the coefficients of $r(x)$.

In particular, let $\mathbf{w} = [0\,1\,0\,1\,1\,0\,0\,0\,1\,0\,1\,1\,1\,0\,1]$. Then $r(x) = 1 + x$, and thus $\mathbf{Hw}^T = [1\,1\,0\,0]^T = 1 + x$. The next step is to find the error vector $\mathbf{e}$ with $\text{wt}(\mathbf{e}) \leq 1$ and the same syndrome. For this, we must determine the exponent j, $0 \leq j \leq 14$, for which $\alpha^j = \mathbf{Hw}^T$. Since $j = 4$, the 5th location in the received vector $\mathbf{w}$ is in error, which when corrected gives the transmitted codeword as $\mathbf{c} = [0\,1\,0\,1\,0\,0\,0\,0\,1\,0\,1\,1\,1\,0\,1]$. ∎

Example 11.4.2. Let $q = 2, n = 15, d = 4$. Then the irreducible polynomial over F_2 is $x^4 + x + 1$ and its roots are primitive elements of F_{2^4}. If α is such a root, then α^2 is also a root, and then α^3 is a root of $x^4 + x^3 + x^2 + x + 1$. Thus, a narrow-sense BCH code with $d = 4$ is generated by

$$g(x) = (x^4 + x + 1)(x^4 + x^3 + x^2 + x + 1).$$

Note that this polynomial $g(x)$ is also a generator for a BCH code with $d = 5$, since α^4 is a root of $x^4 + x + 1$. The dimension of this code is $15 - \deg\{g(x)\} = 7$. ∎

11.5 General Algorithm

As explained above, for any positive integer d, BCH codes of minimum distance $\geq d$ can be constructed. To find a BCH code for a larger minimum distance, we must increase the length n, which in effect increases the number m, that is, the degree of F_{q^m} over F_q. A BCH code of designated distance $d \geq 2t + 1$ will correct at most t errors, but we must use codewords of larger length to achieve the desired minimum distance. A general algorithm for BCH codes is as follows: Let $c(x), w(x)$, and $e(x)$ denote the transmitted code polynomial, the received polynomial, and the error polynomial, respectively, so that $w(x) = c(x) + e(x)$.

STEP 1: Obtain the syndrome of $\mathbf{w}$, which is

$$S(\mathbf{w}) = \mathbf{H}\mathbf{w}^T = [S_b, S_{b+1}, \ldots, S_{b+d-2}]^T,$$

where

$$S_j = w(\alpha^j) = c(\alpha^j) + e(\alpha^j) = e(\alpha^j) \quad \text{for } b \le j \le b + d - 2.$$

If $r \le t$ errors occur, then

$$e(x) = \sum_{i=1}^{r} e_i x^{a_i},$$

where $a_1, \ldots, a_r$ are distinct elements of $\{0, 1, \ldots, n-1\}$. The elements $\nu_i = \alpha^{a_i} \in F_{q^m}$ are called the *error-location numbers*, and the elements $e_i \in F_q^*$ are called *error values*. This yields the syndrome of $\mathbf{w}$:

$$S_j = e(\alpha^j) = \sum_{i=1}^{r} e_i \nu_i^j \quad \text{for } b \le j \le b + d - 2. \tag{11.5.1}$$

However, using the arithmetic operations rule in F_{q^m} we have

$$S_j^q = \left(\sum_{i=1}^{r} e_i \nu_i^j \right)^q = \sum_{i=1}^{r} e_i^q \nu_i^{jq}. \tag{11.5.2}$$

Here the unknown quantities are the pairs $(\nu_i, e_i), i = 1, \ldots, r$, but the coefficients S_j of the syndrome $S(\mathbf{w})$ are known since they can be calculated from the received word $\mathbf{w}$. In the binary case any error is completely characterized by ν_i alone, because in this case $e_i = 1$.

STEP 2: Determine the coefficients σ_i defined by the polynomial identity

$$\prod_{i=1}^{r} (\nu_i - x) = \sum_{i=0}^{r} (-1)^i \sigma_{r-i} x^i = \sigma_r - \sigma_{r-1} x + \cdots + (-1)^r \sigma_0 x^r.$$

Thus, $\sigma_i = 1$ and $\sigma_1, \ldots, \sigma_r$ are the elementary symmetric polynomials in $\nu_1, \ldots, \nu_r$. Substituting ν_i for x gives

$$(-1)^r \sigma_r + (-1)^{r-1} \sigma_{r-1} \nu_i + \cdots + (-1)\sigma_1 \nu_i^{r-1} + \nu_i^r = 0 \quad \text{for } i = 1, \ldots, r.$$

Multiplying this by $e_i \nu_i^j$ and summing all these equations for $i = 0, \ldots, r$, we get

$$(-1)^r \sigma_r S_j + (-1)^{r-1} \sigma_{r-1} S_{j+1} + \cdots + (-1)\sigma_1 S_{j+r-1} + S_{j+r} = 0 \tag{11.5.3}$$

for $j = b, b+1, \ldots, b+r-1$. The following two theorems are useful for this step:

Theorem 11.5.1. *The system of equations $\sum_{i=1}^{r} e_i \nu_i^j = S_j$ in the unknowns e_i, where $j = b, b+1, \ldots, b+r-1$, is solvable if ν_i are distinct elements of $F_{q^m}^*$.*

PROOF. The discriminant of the system is

$$
\begin{vmatrix}
\nu_1^b & \nu_2^b & \cdots & \nu_r^b \\
\nu_1^{b+1} & \nu_2^{b+1} & \cdots & \nu_r^{b+1} \\
\vdots & \vdots & & \vdots \\
\nu_1^{b+r-1} & \nu_2^{b+r-1} & \cdots & \nu_r^{b+r-1}
\end{vmatrix}
= \nu_1^b \nu_2^b \cdots \nu_r^b \prod_{1 \le i < j \le r} (\nu_j - \nu_i) \ne 0. \ \blacksquare
$$

Theorem 11.5.2. *The system of equations (11.5.3) in the unknowns $(-1)^i \sigma_i$, $i = 1, 2, \ldots, r$, is solvable uniquely iff r errors occur.*

PROOF. The matrix of the system is decomposed as

$$
\begin{bmatrix}
S_b & S_{b+1} & \cdots & S_{b+r-1} \\
S_{b+1} & S_{b+2} & \cdots & S_{b+r} \\
\vdots & \vdots & & \vdots \\
S_{b+r-1} & S_{b+r} & \cdots & S_{b+2r-2}
\end{bmatrix}
= V D V^T,
\tag{11.5.3}
$$

where

$$
V = \begin{bmatrix}
1 & 1 & \cdots & 1 \\
\nu_1 & \nu_2 & \cdots & \nu_r \\
\vdots & \vdots & & \vdots \\
\nu_1^{r-1} & \nu_2^{r-1} & \cdots & \nu_r^{r-1}
\end{bmatrix}, \quad
D = \begin{bmatrix}
e_1 \nu_1^b & 0 & \cdots & 0 \\
0 & e_2 \nu_2^b & \cdots & 0 \\
\vdots & \vdots & & \vdots \\
0 & 0 & \cdots & e_r \nu_r^{r-1}
\end{bmatrix}.
$$

The matrix of the system (11.5.3) is nonsingular iff V and D are nonsingular. Since V is a Vandermonde matrix, it is nonsingular iff ν_i ($i = 1, \ldots, r$) are distinct, and D is nonsingular iff all ν_i and e_i are nonzero. Both of these conditions are satisfied iff r errors occur. $\blacksquare$

STEP 3: Solve the error-locator polynomial

$$
s(x) = \prod_{i=1}^{r} (1 - \nu_i x) = \sum_{i=0}^{r} \tau_i x^i,
\tag{11.5.4}
$$

where $\tau_i = (-1)^i \sigma_i$. Let the roots of $s(x)$ be $\nu_1^{-1}, \nu_2^{-1}, \ldots, \nu_r^{-1}$. These roots are determined by a method due to Chien [1964], as follows: First, confirm if α^{n-1} is an error-location number, i.e., if $\alpha = \alpha^{-(n-1)}$ is a root of $s(x)$. To check this, we take the expression $\tau_1 \alpha + \tau_2 \alpha^2 + \cdots + \tau_r \alpha^r$. If it is equal to -1, then α^{n-1} is an error-location number because then $s(\alpha) = 0$. In general, α^{n-m} is tested for $m = 1, 2, \ldots, n$ in the same way. In the binary case, finding errort-location numbers is equivalent to correcting errors.

STEP 4: Introduce ν_i in the first r equations in (11.5.1) of Step 1 to determine the error values e_i. Then find the transmitted codeword $\mathbf{w}$ from $w(x) = v(x) + e(x)$.

Example 11.5.1. Consider a BCH code with designated distance $d = 5$. This code can correct any single or double errors. Let $b = 1, n = 15, q = 2$. If $m^{(i)}(x)$ denotes the minimal polynomial of α^i over F_2, where the primitive element $\alpha \in F_{2^4}$ is a root of $x^4 + x + 1$, then

$$m^{(1)}(x) = m^{(2)}(x) = m^{(4)}(x) = m^{(8)}(x) = 1 + x + x^4,$$
$$m^{(3)}(x) = m^{(6)}(x) = m^{(9)}(x) = m^{(12)}(x) = 1 + x + x^2 + x^3 + x^4.$$

Thus, a generator polynomial of this BCH code is

$$g(x) = m^{(1)}(x) m^{(3)}(x) = 1 + x^4 + x^6 + x^7 + x^8.$$

This is a $(15, 7)$ code with parity-check polynomial

$$h(x) = (x^{15} - 1)/g(x) = 1 + x^4 + x^6 + x^7.$$

The basis for this $(15, 7)$ code corresponds to $\{g(x), xg(x), x^2 g(x), x^3 g(x), x^4 g(x), x^5 g(x), x^6 g(x)\}$, which yields the generator matrix

$$\mathbf{G} = \begin{bmatrix} 1 & 0 & 0 & 0 & 1 & 0 & 1 & 1 & 1 & 0 & 0 & 0 & 0 & 0 & 0 \\ 0 & 1 & 0 & 0 & 0 & 1 & 0 & 1 & 1 & 1 & 0 & 0 & 0 & 0 & 0 \\ 0 & 0 & 1 & 0 & 0 & 0 & 1 & 0 & 1 & 1 & 1 & 0 & 0 & 0 & 0 \\ 0 & 0 & 0 & 1 & 0 & 0 & 0 & 1 & 0 & 1 & 1 & 1 & 0 & 0 & 0 \\ 0 & 0 & 0 & 0 & 1 & 0 & 0 & 0 & 1 & 0 & 1 & 1 & 1 & 0 & 0 \\ 0 & 0 & 0 & 0 & 0 & 1 & 0 & 0 & 0 & 1 & 0 & 1 & 1 & 1 & 0 \\ 0 & 0 & 0 & 0 & 0 & 0 & 1 & 0 & 0 & 0 & 1 & 0 & 1 & 1 & 1 \end{bmatrix}.$$

Suppose that the received word is $\mathbf{w} = [1\,0\,0\,1\,0\,0\,1\,1\,0\,0\,0\,0\,1\,0\,0]$, which can be written as the polynomial $w(x) = 1 + x^3 + x^6 + x^7 + x^{12}$. Now, calculate the syndrome as in Step 1 using (11.5.2) as

$$S_1 = e(\alpha) = w(a) = 1, \qquad S_2 = e(\alpha^2) = w(\alpha^2) = 1,$$
$$S_3 = e(\alpha^3) = w(\alpha^3) = \alpha^4, \quad S_4 = e(\alpha^4) = w(\alpha^4) = 1.$$

Then the largest possible system of linear equations in the unknowns $\tau_i = (-1)^i \sigma_i$ from (11.5.3) is

$$S_2 \tau_1 + S_1 \tau_2 = S_3, \quad S_3 \tau_1 + S_2 \tau_2 = S_4,$$

or
$$\tau_1 + \tau_2 = \alpha^4, \quad \alpha^4 \tau_1 + \tau_2 = 1,$$

which has a nonsingular coefficient matrix. This implies that two errors have occurred, i.e., $r = 2$. We solve this system of equations and get $\tau_1 = 1, \tau_2 = \alpha$. Substituting these values into (11.5.4) and recalling $\tau_0 = 1$, we get $s(x) = 1 + x + \alpha x^2$. Now, the roots in F_{2^4} are $\nu_1^{-1} = \alpha^8, \nu_2^{-1} = \alpha^6$; thus, $\nu_1 = \alpha^7, \nu_2 = \alpha^9$, which implies that errors have occurred in locations 8 and 10 of the codeword. Correcting these errors we obtain

$$c(x) = w(x) - e(x) = (1 + x^3 + x^6 + x^7 + x^{12}) - (x^7 + x^9) = 1 + x^3 + x^6 + x^9 + x^{12}.$$

The original message is obtained from $w(x)/g(x) = 1 + x^3 + x^4$. Hence, the codeword was $\mathbf{c} = [1\,0\,0\,1\,0\,0\,1\,0\,0\,1\,0\,0\,1\,0\,0]$, and the original message word was $1\,0\,0\,1\,1\,0\,0$. $\blacksquare$

11.5.1 BCH $(31, 16)$ **Code.** Let α be a primitive element in GF(32). We will use the minimal irreducible polynomials of the first three odd powers of α, which, as mentioned in §8.1.1 (Table 8.1.3), are:

$$m_1(x) = x^5 + x^2 + 1 = 100101,$$
$$m_3(x) = x^5 + x^4 + x^3 + x^2 + 1 = 111101,$$
$$m_5(x) = x^5 + x^4 + x^2 + x + 1 = 110111.$$

Then the generator polynomial is given by

$$g(x) = \mathrm{lcm}\,[m_1(x), m_2(x), m_3(x)] = m_1(x)m_2(x)m_3(x) \text{ since they are irreducible,}$$
$$= (x^5 + x^2 + 1)(x^5 + x^4 + x^3 + x^2 + 1)(x^5 + x^4 + x^2 + x + 1)$$
$$= x^{15} + x^{11} + x^{10} + x^9 + x^8 + x^7 + x^5 + x^3 + x^2 + x + 1$$
$$= 1000111110101111.$$

For BCH encoding, consider the letter 'A' which is the first letter in the title of this book. As a binary word, the letter is represented by a 8-bit symbol 01000001 (see Appendix A), which in a 16-bit form becomes 0000000001000001. Then we append to this 16-bit symbol the same number of zeros as the degree of the above generator polynomial $g(x)$. This means that since the polynomial representation of this symbol is $x^6 + 1$, we multiply it by x^{15}, to get $f(x) = x^{15}(x^6 + 1) = x^{21} + x^{15}$. Note that this multiplication shifts the 16 data bits of the symbol 15 places to the left, so that $f(x)$ has the binary representation 0000000001000001000000000000000. Next, we divide $f(x)$ by the generator polynomial $g(x)$ using binary arithmetic. Using the result from Example 8.7.1, we obtain the remainder as 100101000100010, or $r(x) = x^{14} + x^{11} + x^9 + x^5 + x$, which are the parity bits, giving the encoded 31-bit codeword as

$$\mathbf{c}(x) = x^{21} + x^{15} + x^{14} + x^{11} + x^9 + x^5 + x, \text{ or:}$$

$$\mathbf{c} : \underbrace{0000000001000001}_{\text{16 Data Bits}}\underbrace{100101000100010}_{\text{15 Parity Bits}}$$

The accuracy of this codeword can be checked for errors by dividing it by $g(x)$, as follows:

$$1000111110101111 \overline{)000000000100000110010100010010}$$
$$\underline{1000111110101111}$$
$$0000110010000111000$$
$$\underline{1000111110101111}$$
$$1000111110101111$$
$$\underline{1000111110101111}$$
$$0$$

Next, suppose that the above codeword is received after transmission as the word

$$\mathbf{w}(x) = x^{26} + x^{21} + x^{20} + x^{15} + x^{14} + x^{11} + x^9 + x^7 + x^5 + x$$

$$= 0000\underline{1}00001\underline{1}00001100101010100010,$$

where the underlined bits are in error (they are at bit locations 26, 20, and 7). Since this code corrects up to three errors ($t = 3$), we will discuss the error correcting process of this code.

For BCH decoding, we use the method found in Lin and Costello [1983: 151], which involves three steps: (i) Compute the syndrome from the received codeword; (ii) find the error-location polynomial from a set of equations derived from the syndrome; and (iii) use the error-location polynomial to identify the erroneous bits and correct them. Now, we proceed with these steps.

STEP 1. SYNDROME COMPUTATION. The syndrome is computed as follows:

$$s_1 = \mathbf{w}(x) \bmod m_1(x), \quad s_4(x) = \mathbf{w}(x) \bmod m_4(x),$$

$$s_2 = \mathbf{w}(x) \bmod m_2(x), \quad s_5(x) = \mathbf{w}(x) \bmod m_5(x), \qquad (11.5.5)$$

$$s_3 = \mathbf{w}(x) \bmod m_3(x), \quad s_6(x) = \mathbf{w}(x) \bmod m_6(x).$$

Since in field elements several powers of the generating element have the same minimal polynomial, we select the minimal polynomials such that if $b = 2^i$, $i \geq 0$, then α^b is a root for a polynomial in GF(2). Hence, all powers of α such as $\alpha^2, \alpha^4, \dots$ are roots of the minimal polynomials of α. In GF(32), the minimal polynomials are selected in terms of the index 2^i so that in our example

$$m_1(x) = m_2(x) = m_4(x) = x^5 + x^2 + 1,$$

$$m_3 = m_6(x) = x^5 + x^4 + x^3 + x^2 + 1,$$

$$m_5(x) = x^5 + x^4 + x^2 + x + 1.$$

Then we get a system of syndrome equations

$$s_1(\alpha) = \alpha + \alpha^2 + \cdots + \alpha^n,$$
$$s_2(\alpha^2) = (\alpha)^2 + (\alpha^2)^2 + \cdots + (\alpha^n)^2,$$
$$s_3(\alpha^3) = (\alpha)^3 + (\alpha^2)^3 + \cdots + (\alpha^n)^3, \qquad (11.5.6)$$
$$\vdots$$
$$s_j(\alpha^n) = (\alpha)^j + (\alpha^2)^j + \cdots + (\alpha^n)^j.$$

Since in our case $j = 6$, the syndrome is computed from (11.5.5) as

$$s_1(x) = s_2(x) = s_4(x) = x^3 + x^2 + x + 1,$$
$$s_3(x) = s_6(x) = x^2,$$
$$s_5(x) = x^3 + x^2.$$

The computation of $s_1(x)$ is shown below; others can be computed similarly. Since $s_1(x) = \mathbf{w}(x) \bmod m_1(x)$, we divide $\mathbf{w}(x)$ by $m_1(x)$, as follows:

```
00001000011000011001010101000010
100101
_____
    100100
    100101
    _____
       100110
       100101
       _____
          110101
          100101
          _____
             100000
             100101
             _____
                101101
                100101
                _____
                   100000
                   100101
                   _____
                      101010
                      100101
                      _____
                       0011111=x³ + x² + x + 1
                      _____
```

Then using Table 8.1.3 we obtain the set of syndromes as

$$s_1(\alpha) = \alpha^3 + \alpha^2 + \alpha + 1 = \alpha^{23},$$
$$s_2(\alpha^2) = \alpha^6 + \alpha^4 + \alpha^2 + 1 = \alpha^3 + \alpha + \alpha^4 + \alpha^2 + 1 = \alpha^{15},$$
$$s_3(\alpha^3) = \alpha^6,$$
$$s_4(\alpha^4) = \alpha^{12} + \alpha^8 + \alpha^4 + 1 = \alpha^3 + \alpha^2 + \alpha + \alpha^3 + \alpha^2 + 1 + \alpha^4 + 1 = \alpha^4 + \alpha = \alpha^{30},$$

$$s_5(\alpha^5) = \alpha^{15} + \alpha^{10} = \alpha^4 + \alpha^3 + \alpha^2 V + 1 + \alpha^4 + 1 = \alpha^3 + \alpha^2 + \alpha = \alpha^{12},$$

$$s_6(\alpha) = \alpha^{12}.$$

STEP 2. ERROR LOCATION POLYNOMIAL. The determination of error-location polynomial is based on computation of the following quantities:

$$
\begin{cases}
L^{(\mu+1)}(x) = L^{(\mu)}(x) + d_\mu d_\rho^{-1} x^{2(\mu-\rho)} L^{(\rho)}(x), \\
l_{\mu+1} = \deg\{L^{(\mu+1)}(x)\}, \\
d_0 = s_1, \quad d_{\mu+1} = s_{2\mu+3} + L_1^{(\mu+1)} s_{2\mu+2} + L_2^{(\mu+1)} s_{2\mu+1} + \cdots + L_\lambda^{(\mu+1)} s_{2\mu+3-\lambda}
\end{cases}
\tag{11.5.7}
$$

where $\lambda = l_{\mu+1}$, and $L_j^{(i)}$ is the coefficient of x^j in $L^{(i)}(x)$, $i, j = 1, 2, \ldots$. The last item in the algorithm (11.5.7), which computes $d_{\mu+1}$, has a built-in stopping rule. This recursive computation stops when $\mu = t - 1$.

The error location polynomial $L^{(\mu+1)}(x)$ is computed using the algorithm that involves the recursive formulas (11.5.7) along with an initialization as follows: $\mu = -\frac{1}{2}$, $L^{(0)}(x) = 1$, $d_0 = 1$, $l_0 = 0$, $2\mu - l_\mu = -1$. Since BCH (31,16) code corrects up to 3 errors, we consider the values of $\mu = 0, 1, 2$, one at a time, as follows:

$\boxed{\mu = 0}$: Since $d_\mu \neq 0$ ($d_0 = s_1(\alpha) = \alpha^{23}$) and $L^{(-1/2)}(x) = 1, L^{(0)}(x) = 1$, choose $\rho = -\frac{1}{2}$. Then from (11.5.7),

$$L^{(1)}(x) = L^{(0)}(x) + d_0 d_{-1/2}^{-1} x^{2(0+1/2)} L^{(-1/2)}(x) = 1 + \alpha^{23}(1)x(1)$$

$$= \alpha^{23} x + 1,$$

$$l_1 = \deg\{L^{(1)}(x)\} = 1, \quad \lambda = l_1 = 1,$$

$$d_1 = s_3 + L_1^{(1)} s_2 = \alpha^6 + \alpha^{23}\alpha^{15} = \alpha^{24}.$$

$\boxed{\mu = 1}$: Since $d_1 \neq 0$ ($d_0 = s_1(\alpha) = \alpha^{23}$), choose $\rho = 0$. Then from (11.5.7),

$$L^{(2)} = L^{(1)}(x) + d_1 d_0^{-1} x^{2(1-0)} L^{(0)}(x) = \alpha^{23}(1)x + 1 + \alpha^{24}(\alpha^{23})^{-1} x^2(1)$$
$$= \alpha x^2 + \alpha^{23} x + 1,$$

$$l_2 = \deg\{L^{(2)}(x)\} = 2, \quad \lambda = l_2 = 2,$$

$$d_2 = s_5 + L_1^{(2)} s_4 + L_2^{(2)} s_3 = \alpha^{12} + \alpha^{23}\alpha^{30} + \alpha\alpha^6 = \alpha^{23}.$$

$\boxed{\mu = 2}$: Since $d_2 \neq 0$ ($d_0 = \alpha^{24}$), choose $\rho = 1$. Then from (11.5.7),

$$L^{(3)} = L^{(2)}(x) + d_2 d_1^{-1} x^{2(2-1)} L^{(1)}(x)$$
$$= \alpha x^2 + \alpha^{23}(1)x + 1 + \alpha^{23}(\alpha^{24})^{-1} x^2(\alpha^{23} + 1)$$
$$= \alpha x^2 + \alpha^{23} x + 1 + \alpha^{30} x^2(\alpha^{23} + 1) = \alpha x^2 + \alpha^{23} + 1 + \alpha^{22} x^3 + \alpha^{30} x^2$$
$$= \alpha^{22} x^3 + (\alpha + \alpha^{30}) x^2 + \alpha^{23} x + 1 = \alpha^{22} x^3 + \alpha^4 x^2 + \alpha^{23} x + 1,$$

$$l_3 = \deg\{L^{(3)}(x)\} = 3, \quad \lambda = l_3 = 3,$$

$$d_3 = s_7 + \cdots = \text{undefined. STOP.}$$

The above above are presented in Table 11.5.1.

Table 11.5.1 Error Location Polynomial Values

μ	$L^{(\mu)}(x)$	d_μ	l_μ	$2\mu - l_\mu$
$-\frac{1}{2}$	1	1	0	-1
0	1	$s_1 = \alpha^{23}$	0	0
1	$\alpha^{23} + 1$	α^{24}	1	1
2	$\alpha x^2 + \alpha^{23}x + 1$	α^{23}	2	2
$t = 3$	$\alpha^{22}x^3 + \alpha^4 x^2 + \alpha^{23}x + 1$	$-$	$-$	$-$

STEP 3. ERROR LOCATION. The error location polynomial is $L^{(3)}(x) = \alpha^{22}x^3 + \alpha^4 x^2 + \alpha^{23}x + 1$. We find the roots of this polynomial in GF(32). Since the degree of the polynomial is small, we find these roots by trial and error substitution as $\alpha^5, \alpha^{11}, \alpha^{24}$, which can be verified by substituting in $L^{(3)}(x)$ and simplified using Table 8.1.3. For example, $L^{(3)}(\alpha^{24}) = \alpha^{22}\alpha^{72} + \alpha^4\alpha^{48} + \alpha^{23}\alpha^{24} + 1 = \alpha + \alpha^{21} + \alpha^{16} + 1 = \alpha + \alpha^4 + \alpha^3 + \alpha^4 + \alpha^3 + \alpha + 1 + 1 = 0$. However, for large codes there is the Chien algorithm (Chien [1964]) discussed in §11.5. The other two roots are similarly verified. The bit locations of the errors are the inverse of these roots, i.e., $\alpha^{26}, \alpha^{20}, \alpha^7$, respectively. Thus, the erroneous bits are at locations 26, 20, and 7. The error polynomial can be written as $\mathbf{e}(x) = x^{26} + x^{20} + x^7$. The errors are corrected by computing $\mathbf{w}(x) \oplus \mathbf{e}(x) = \mathbf{c}(x)$; thus,

$$\mathbf{w}(x) = 0000100001100001100101010100010$$
$$\oplus \quad \mathbf{e}(x) = 0000100000100000000000010000000$$
$$\overline{\mathbf{c}(x) = 0000000001000001100101000100010}$$

If there are no errors, all syndromes will be zero. Although this example has been an exercise in understanding the working of this BCH code just for transmission of a single letter, all BCH encoding and decoding is performed either by hardware or software. The Berlekamp algorithm used above is very difficult to handle manually, but its use is widespread, mostly in pagers and cellphones.

The BCH codes, and codes that are derived from them, are called *symmetric* codes, because the data bits are followed by parity bits, or conversely. This is in contrast to the *non-symmetric* codes, like Hamming and Golay codes, in which the parity bits are intertwined with the data bits in a certain pattern involving the bit locations.

12

Reed–Muller Codes

Reed–Muller (RM) codes belong to a class of linear error correcting codes that are both locally testable and decodable. They are some of the oldest codes that have been useful in sending messages over long distances or through channels where error might occur during transmission. They became more prevalent as telecommunications expanded and became useful as self-correcting codes. They are used in the design of probabilistic checkable proofs in complexity theory. Named after their discoverers, Irving S. Reed and D. E. Muller (see Reed [1954] and Muller [1954]), where the latter discovered the codes while the former provided the algorithm of logic decoding, these codes have the following special cases: Hadamard code (Chapter 9), Walsh–Hadamard code (Chapter 9), and Reed–Solomon code (Chapter 13).

The RM codes are defined as $RM(r, m)$, where r is the order of the code, $0 \leq r \leq m$, and m is the parameter related to the length of the code, $n = 2^m$. These codes are related to the binary functions on the Galois field $GF(2^m)$ over the elements (alphabet) $\{0, 1\}$. Thus,

$RM(r, 0)$ codes are repetition codes of length $n = 2^m$, rate $R = 1/n$, and minimum distance $d_{\min} = n$;

$RM(r, 1)$ codes are parity-check codes of length $n = 2^m$, rate $R = (m + 1)/n$, and minimum distance $d_{\min} = n/2$;

$RM(r, r - 1)$ codes are parity codes of length $n = 2^m$; and

$RM(r, r - 2)$ codes are the class of extended Hamming codes of length $n = 2^m$, and minimum distance $d_{\min} = 4$ (see Schlegel and Perez [2004: 149]).

12.1 Boolean Polynomials

Before we discuss the construction of RM codes by providing details of RM encoding and decoding methods, we will introduce Boolean polynomials. Let $\mathbf{x}_i$ denote a set of vectors that follow the summation and multiplication rules defined in §3.5. A *Boolean polynomial* $\mathbf{b}$ is a linear combination of Boolean monomials with coefficients in F_2 and is defined by

$$\mathbf{b} = \mathbf{x}_1^{r_1} \mathbf{x}_2^{r_2} \cdots \mathbf{x}_m^{r_m}, \quad r_i \in \{0, 1, 2, \ldots\}, \quad 1 \le i \le m.$$

The *reduced form* $\mathbf{b}'$ of $\mathbf{b}$ is obtained using the formulas

$$\mathbf{x}_i \mathbf{x}_i = \mathbf{x}_j \mathbf{x}_i \text{ (commutative property)},$$
$$\mathbf{x}_i^2 = \mathbf{x}_i \quad \text{as } 0 * 0 = 0 \text{ and } 1 * 1 = 1, \quad\quad (12.1.1)$$

until the factors are distinct. Thus, a Boolean polynomial in reduced form is a linear combination of reduced-form Boolean monomials with coefficients in F_2. Note that $\deg\{\mathbf{b}'\} = \deg\{\mathbf{b}\} =$ number of variables in $\mathbf{b}'$. A Boolean polynomial is in the reduced form if each monomial is in reduced form. For example, $\mathbf{b} = \mathbf{x}_1 + \mathbf{x}_2 + \mathbf{x}_1\mathbf{x}_2 + \mathbf{x}_1\mathbf{x}_2\mathbf{x}_3$ has degree 3. Note that the zero-degree monomial is $\mathbf{1}$, and the monomials $\mathbf{x}_1, \mathbf{x}_2, \ldots, \mathbf{x}_m$ are each of degree 1.

Example 12.1.1. Suppose we have the Boolean polynomial $\mathbf{b} = 1 + \mathbf{x}_2 + \mathbf{x}_1^5\mathbf{x}_3^2 + \mathbf{x}_1\mathbf{x}_2^4\mathbf{x}_3^{121} \in \mathcal{R}_3$. After applying formulas (12.1.1) we get the reduced form $\mathbf{b}' = 1 + \mathbf{x}_2 + \mathbf{x}_1\mathbf{x}_3 + \mathbf{x}_1\mathbf{x}_2\mathbf{x}_3$. ∎

Note that $(F_2^n, \oplus, *)$ forms a commutative ring. In the vector space $F_2^{2^m}$ we consider the ring $\mathcal{R}_m = F_2[\mathbf{x}_1, \mathbf{x}_2, \ldots, \mathbf{x}_m]$. Then there exists an isomorphism of rings between $(\mathcal{R}_m, +, *)$ and $(F_2^{2^m}, \oplus, *)$. In fact, consider the mapping $\phi : \mathcal{R}_m \mapsto F_2^{2^m}$, defined by

$$\phi(\mathbf{0}) = \underbrace{00\ldots0}_{2^m},$$

$$\phi(\mathbf{1}) = \underbrace{11\ldots1}_{2^m},$$

$$\phi(\mathbf{x}_1) = \underbrace{11\ldots1}_{2^{m-1}}\underbrace{00\ldots0}_{2^{m-1}},$$

$$\phi(\mathbf{x}_2) = \underbrace{11\ldots1}_{2^{m-2}}\underbrace{00\ldots0}_{2^{m-2}}\underbrace{11\ldots1}_{2^{m-2}}\underbrace{00\ldots0}_{2^{m-2}},$$

$$\phi(\mathbf{x}_3) = \underbrace{11\ldots1}_{2^{m-3}}\underbrace{00\ldots0}_{2^{m-3}}\underbrace{11\ldots1}_{2^{m-3}}\underbrace{00\ldots0}_{2^{m-3}}\underbrace{11\ldots1}_{2^{m-3}}\underbrace{00\ldots0}_{2^{m-3}}\underbrace{11\ldots1}_{2^{m-3}}\underbrace{00\ldots0}_{2^{m-3}},$$

$$\vdots$$

$$\phi(\mathbf{x}_i) = \underbrace{11\ldots1}_{2^{m-i}}\underbrace{00\ldots0}_{2^{m-i}}\underbrace{\ldots}_{2^{m-1}}, \quad (2^i \text{ pairs}),$$

$$\vdots$$

To calculate $\phi(\mathbf{b})$ for any $\mathbf{b} \in \mathcal{R}_m$, first find the reduced form $\mathbf{b}' = \mathbf{x}_{i_1}\mathbf{x}_{i_2}\ldots\mathbf{x}_{i_r}$, where $i_j \in \mathbb{N}_n$, $i_j = i_k$ implies $j = k$, and $1 \le r \le m$. Then, $\phi(\mathbf{b}) = \phi(\mathbf{x}_{i_1}) * \phi(\mathbf{x}_{i_2}) * \cdots * \phi(\mathbf{x}_{i_r})$. On the other hand, for any polynomial $g \in \mathcal{R}_m$

we can write $\mathbf{g} = \mathbf{m}_1 + \mathbf{m}_2 + \cdots + \mathbf{m}_r$, where $\mathbf{m}_i$ is a monomial of $\mathcal{R}_m$. Then $\phi(\mathbf{g}) = \phi(\mathbf{m}_1) + \phi(\mathbf{m}_2) + \cdots + \phi(\mathbf{m}_r)$. ∎

Example 12.1.2. Let $\mathbf{b} = 1 + \mathbf{x}_2 + \mathbf{x}_1^5 \mathbf{x}_3^2 + \mathbf{x}_1 \mathbf{x}_2^4 \mathbf{x}_3^{121}$, which has the reduced form $\mathbf{b}' = 1 + \mathbf{x}_2 + \mathbf{x}_1 \mathbf{x}_3 + \mathbf{x}_1 \mathbf{x}_2 \mathbf{x}_3$. Then

$$\phi(\mathbf{b}) - \phi(\mathbf{b}')$$
$$= \phi(1) + \phi(\mathbf{x}_2) + \phi(\mathbf{x}_1 \mathbf{x}_3) + \phi(\mathbf{x}_1 \mathbf{x}_2 \mathbf{x}_3)$$
$$= \phi(1) + \phi(\mathbf{x}_2) + \phi(\mathbf{x}_1) * \phi(\mathbf{x}_3) + \phi(\mathbf{x}_1) * \phi(\mathbf{x}_2) * \phi(\mathbf{x}_3)$$
$$= 11111111 \oplus 11001100 \oplus 11110000 \oplus 101001010 \oplus 11110000 \oplus 11001100$$
$$\qquad \oplus 10101010$$
$$= 00010011. ∎$$

The function ϕ is a bijection as well as a homomorphism of rings. Hence, $\mathcal{R}_m$ and $F_2^{2^m}$ are isomorphic, so we will use the terms 'vectors' and their 'associated reduced-form polynomials' interchangeably in the same sense. Henceforth we will denote the function $\phi(\mathbf{x}_i)$ simply by $\mathbf{x}_i$.

12.2 RM Encoding

We will present three methods to help us understand RM encoding. They are:

METHOD 1: To associate a Boolean monomial in m variables to a vector with 2^m entries, follow the following step: Define the vectors associated with the zero dgree monomial 1 and degree-1 monomials $\mathbf{x}_1, \mathbf{x}_2, \ldots, \mathbf{x}_m$.

Note that the vector associated with the monomial 1 is a vector of length 2^m with every entry as 1. For example, in a space of size 2^3 the vector associated with 1 is (11111111). Also, the vector associated with the monomial $\mathbf{x}_1$ is 2^{m-1} 1s followed by 2^{m-1} 0s; the vector associated with the monomial $\mathbf{x}_2$ is 2^{m-2} 1s followed by 2^{m-2} 0s plus another 2^{m-2} 1s followed by 2^{m-2} 0s. In general, the vector associated with the monomial $\mathbf{x}_i$ is of the form of 2^{m-i} 1s followed by 2^{m-i} 0s, repeated until 2^m values are defined. For example, in the space of 2^4, the vectors associated with $\mathbf{x}_4$ is (1010101010101010).

METHOD 2. To form the vector for the monomial $\mathbf{x}_1^{r_1} \mathbf{x}_2^{r_2} \ldots$, follow the following two steps:

(1) Put the monomial in reduced form; and

(2) Multiply the vectors associated with each monomial $\mathbf{x}_i$ in the reduced form.

Example 12.2.1. In the space with $m = 3$, the vector associated with the monomial $\mathbf{x}_1 \mathbf{x}_2 \mathbf{x}_3$ is given by

$$(11110000) * (11001100) * (1010101010) = (10000000). ∎$$

METHOD 3. To reduce the vector for a polynomial, follow the following three steps:

(a) Reduce all the monomials in the polynomial;

(b) Find the vector associated with each of the monomials; then

(c) Add all the vectors associated with each of these monomials together to form the vector associated with the polynomial.

This gives a bijection between reduced polynomials and vectors.

12.2.1 Encoding Matrices. An r-th order $RM(r, m)$ code is the set of all binary strings (vectors) of length $n = 2^m$ associated with the Boolean polynomials $b(x_1, x_2, \cdots, x_m)$ of degree $\leq r$. The two extreme RM codes are $RM(0, m)$ and $RM(m, m)$, where

(i) $RM(0, m)$ code is of order 0, which consists of the binary strings associated with the constant monomials $\mathbf{0}$ and $\mathbf{1}$, i.e., $RM(0, m) = \{\mathbf{0}, \mathbf{1}\}$ which is a repetition code 2^m, and the monomials $\mathbf{0}$ and $\mathbf{1}$ are equivalent to the vectors $\{\underbrace{00\ldots00}_{2^m}\underbrace{11\ldots11}_{2^m}\}$. Thus, $RM(m, m)$ code is a repetition of either 0s or 1s of length 2^m.

(ii) $RM(m, m)$ code of order m, which consists of all binary strings of length 2^m.

(iii) $RM(m, m) = \mathcal{R}_m \simeq F_2^{2^m}$.

ENCODING METHOD. To encode for $RM(r, m)$ code and generate the encoding matrix, the method is as follows:

First row of this matrix is $\mathbf{1}$, i.e., it is a vector of length 2^m with all entries equal to 1.

If $r = 0$, then this (first) row is the only leading matrix.

If $r = 1$, then add m rows corresponding to the vectors $x_1, \ldots, x_m$ to the $RM(0, m)$ encoding matrix.

If $r > 1$, add $\binom{m}{r}$ rows to the $RM(r - 1, m)$ encoding matrix. These added rows consist of all possible reduced monomials of degree r, which can be formed using the rows $x_1, \ldots, x_m$.

Example 12.2.2. Let $m = 3$. Then the encoding matrix for $RM(1, 3)$ is

$$
\mathbf{G}_{(1,3)} = \begin{bmatrix} \mathbf{1} \\ x_1 \\ x_2 \\ x_3 \end{bmatrix} \equiv \begin{bmatrix} 1 & 1 & 1 & 1 & 1 & 1 & 1 & 1 \\ 1 & 1 & 1 & 1 & 0 & 0 & 0 & 0 \\ 1 & 1 & 0 & 0 & 1 & 1 & 0 & 0 \\ 1 & 0 & 1 & 0 & 1 & 0 & 1 & 0 \end{bmatrix}.
$$

The rows $x_1 x_2 = 11000000$, $x_1 x_3 = 10100000$, and $x_2 x_3 = 10001000$, which when added give the encoding matrix for $RM(2, 3)$:

$$
\mathbf{G}_{(2,3)} = \begin{bmatrix} 1 \\ \mathbf{x}_1 \\ \mathbf{x}_2 \\ \mathbf{x}_3 \\ \mathbf{x}_1\mathbf{x}_2 \\ \mathbf{x}_2\mathbf{x}_3 \end{bmatrix} \equiv \begin{bmatrix} 1 & 1 & 1 & 1 & 1 & 1 & 1 & 1 \\ 1 & 1 & 1 & 1 & 0 & 0 & 0 & 0 \\ 1 & 1 & 0 & 0 & 1 & 1 & 0 & 0 \\ 1 & 0 & 1 & 0 & 1 & 0 & 1 & 0 \\ 1 & 1 & 0 & 0 & 0 & 0 & 0 & 0 \\ 1 & 0 & 0 & 0 & 1 & 0 & 0 & 0 \end{bmatrix}.
$$

Finally, the row $\mathbf{x}_1\mathbf{x}_2\mathbf{x}_3 = 10000000$ is added to give the encoding matrix for RM$(3,3)$:

$$
\mathbf{G}_{(3,3)} = \begin{bmatrix} 1 \\ \mathbf{x}_1 \\ \mathbf{x}_2 \\ \mathbf{x}_3 \\ \mathbf{x}_1\mathbf{x}_2 \\ \mathbf{x}_1\mathbf{x}_3 \\ \mathbf{x}_2\mathbf{x}_3 \\ \mathbf{x}_1\mathbf{x}_2\mathbf{x}_3 \end{bmatrix} \equiv \begin{bmatrix} 1 & 1 & 1 & 1 & 1 & 1 & 1 & 1 \\ 1 & 1 & 1 & 1 & 0 & 0 & 0 & 0 \\ 1 & 1 & 0 & 0 & 1 & 1 & 0 & 0 \\ 1 & 0 & 1 & 0 & 1 & 0 & 1 & 0 \\ 1 & 1 & 0 & 0 & 0 & 0 & 0 & 0 \\ 1 & 0 & 1 & 0 & 1 & 0 & 1 & 0 \\ 1 & 0 & 0 & 0 & 1 & 0 & 0 & 0 \\ 1 & 0 & 0 & 0 & 0 & 0 & 0 & 0 \end{bmatrix}. \blacksquare
$$

Example 12.2.3. Let $m = 4$. Then the encoding matrix for RM$(2,4)$ is

$$
\mathbf{G}_{(2,4)} = \begin{bmatrix} 1 \\ \mathbf{x}_1 \\ \mathbf{x}_2 \\ \mathbf{x}_3 \\ \mathbf{x}_4 \\ \mathbf{x}_1\mathbf{x}_2 \\ \mathbf{x}_1\mathbf{x}_3 \\ \mathbf{x}_1\mathbf{x}_4 \\ \mathbf{x}_2\mathbf{x}_3 \\ \mathbf{x}_2\mathbf{x}_4 \\ \mathbf{x}_3\mathbf{x}_4 \end{bmatrix} \equiv \begin{bmatrix}
1 & 1 & 1 & 1 & 1 & 1 & 1 & 1 & 1 & 1 & 1 & 1 & 1 & 1 & 1 & 1 \\
1 & 1 & 1 & 1 & 1 & 1 & 1 & 1 & 0 & 0 & 0 & 0 & 0 & 0 & 0 & 0 \\
1 & 1 & 1 & 1 & 0 & 0 & 0 & 0 & 1 & 1 & 1 & 1 & 0 & 0 & 0 & 0 \\
1 & 1 & 0 & 0 & 1 & 1 & 0 & 0 & 1 & 1 & 0 & 0 & 1 & 1 & 0 & 0 \\
1 & 0 & 1 & 0 & 1 & 0 & 1 & 0 & 1 & 0 & 1 & 0 & 1 & 0 & 1 & 0 \\
1 & 1 & 1 & 1 & 0 & 0 & 0 & 0 & 0 & 0 & 0 & 0 & 0 & 0 & 0 & 0 \\
1 & 1 & 0 & 0 & 1 & 1 & 0 & 0 & 0 & 0 & 0 & 0 & 0 & 0 & 0 & 0 \\
1 & 0 & 1 & 0 & 1 & 0 & 1 & 0 & 0 & 0 & 0 & 0 & 0 & 0 & 0 & 0 \\
1 & 1 & 0 & 0 & 0 & 0 & 0 & 0 & 1 & 1 & 0 & 0 & 0 & 0 & 0 & 0 \\
1 & 0 & 1 & 0 & 0 & 0 & 0 & 0 & 1 & 0 & 1 & 0 & 0 & 0 & 0 & 0 \\
1 & 0 & 0 & 0 & 1 & 0 & 0 & 0 & 1 & 0 & 0 & 0 & 1 & 0 & 0 & 0
\end{bmatrix}. \blacksquare
$$

Since the dimension of this code is k, where $k = 1 + \binom{m}{1} + \binom{m}{2} + \cdots + \binom{m}{r}$, the message is sent in blocks of length k. This means that the encoding matrix has k rows. Let $\mathbf{m} = (m_1, m_2, \ldots, m_k)$ be a block. The encoded message (i.e., the codeword) is $\mathbf{c} = \sum\limits_{i=1}^{k} m_i R_i$, where R_i is the i-th row of the encoding matrix for RM(r,m).

Example 12.2.4. (a) Using RM$(1,3)$ to encode $\mathbf{m} = (0110)$ we get

$$0 * (11111111) \oplus 1 * (11110000) \oplus 1 * (11001100) \oplus 0 * (10101010)$$

$$= (00000000) \oplus (11110000) \oplus (11001100) \oplus (00000000)$$

$$= (11110000) \oplus (11001100) \quad \text{adding two terms at a time}$$

$$= (00111100).$$

Thus, the codeword[1] for the original message 0110 is 00111100.

(b) Using RM(2, 4) (Example 12.2.3) to encode $\mathbf{m} = (10101110010)$ we get

$1 * (1111111111111111) \oplus 0 * (1111111100000000) \oplus 1 * (1111000011110000)$

$\quad \oplus 0 * (1100110011001100) \oplus 1 * (1010101010101010) \oplus 1 * (1111000000000000)$

$\quad \oplus 1 * (1100110000000000) \oplus 0 * (1010101000000000) \oplus 0 * (1100000011000000)$

$\quad \oplus 1 * (1010000010100000) \oplus 0 * (1000100010001000)$

$= 0011100100000101.$

Thus, the codeword for the message 10101110010 is 0011100100000101, where the length of the codeword is 2^m, in this case $2^4 = 16$ bits. ∎

Theorem 12.2.1. *The encoding matrix* $\mathbf{G}_{(r,m)}$ *for the RM(r, m) code has dimension* $k \times n$, *where* $k = \sum_{i=0}^{r} \binom{m}{i}$ *and* $n = 2^m$. *Hence, the dimension of this code is* k.

PROOF. The rows of the encoding matrix may be partitioned by the degree of the monomial. Thus, for RM(r, m) code there are $\binom{m}{0}$ rows associated with monomials of degree 0, i.e., $\mathbf{0}$; $\binom{m}{1}$ rows associated with monomials of degree 1, i.e., $(\mathbf{x}_1, \mathbf{x}_2, \ldots, \mathbf{x}_m)$; $\binom{m}{2}$ rows associated with the monomials of degree 2, i.e., $(\mathbf{x}_1 \mathbf{x}_2, \mathbf{x}_1 \mathbf{x}_3, \ldots, \mathbf{x}_{m-1} \mathbf{x}_m)$; and so on. Thus, these rows all add up to $\binom{m}{0} + \binom{m}{1} + \cdots + \binom{m}{r} = \sum_{i=0}^{r} \binom{m}{i}$. It is obvious that the dimension of the code is k and the number of columns is 2^m. ∎

12.3 Generating Matrix for RM Codes

First we construct a generating matrix for an RM code of length $n = 2^m$. Let $X = F_2^r = \{x_1, \ldots, x_{2^r}\}$ denote a set, each member of which is a point in F_2^r, and let F_2^n denote an n-dimensional space of indicator vectors $\mathbf{I}_A$ on subsets $A \subset X$, where

$$(\mathbf{I}_A)_i = \begin{cases} 1 & \text{if } x_i \in A, \\ 0 & \text{otherwise.} \end{cases}$$

For two arbitrary points $\mathbf{w} = \{w_1, \ldots, w_n\}$ and $\mathbf{z} = \{z_1, \ldots, z_n\}$ in F_2^n, define the binary operation

$$\mathbf{w} \wedge \mathbf{z} = \{w_1 \cdot z_1, \ldots, w_n \cdot z_n\},$$

known as the *wedge product* (see §3.5.1), where the operation · is the usual multiplication in F_2^n. Since F_2^r is an r-dimensional vector space over the

---

[1] Remember that the terms 'vector', 'symbol', and 'codeword' have the same meaning in coding theory.

field F_2, we can write $(F_2)^r = \{\{y_r, \ldots, y_1\} \mid y_i \in F_2\}$. Next, we define the following vectors with length n in the n-dimensional space F_2^n:

$$\mathbf{v}_0 = \{1, 1, 1, 1, 1, 1, 1, 1\} \quad \text{and} \quad \mathbf{v}_i = \mathbf{I}_{H_i},$$

where H_i are the $(r-1)$-dimensional hyperplanes in $(F_2)^r$, defined by

$$H_i = \{y \in (F_2)^r \mid y_i = 0\}.$$

Then, the RM(r, m) code of order r and length $n = 2^m$ is defined as the code generated by $\mathbf{v}_0$ and the wedge product up to m of the $\mathbf{v}_i$, $i = 1, 2, \ldots, r$. Note that by convention, a wedge product of fewer than one vector is the identity for the operation.

Example 12.3.1. Let $r = 3$. Then $n = 8$ and
$$X = F_2^3 = \{(0,0,0), (0,0,1), (0,1,0), (0,1,1), (1,0,0), (1,0,1), (1,1,0), (1,1,1)\},$$
so that

$$\mathbf{v}_0 = \{1,1,1,1,1,1,1,1\}, \quad \mathbf{v}_1 = \{1,0,1,0,1,0,1,0\},$$
$$\mathbf{v}_2 = \{1,1,0,0,1,1,0,0\}, \quad \mathbf{v}_3 = \{1,1,1,1,0,0,0,0\}.$$

Thus, the RM(3,1) code is generated by the set $\{\mathbf{v}_0, \mathbf{v}_1, \mathbf{v}_2, \mathbf{v}_3\}$, i.e., the generator matrix of this code is

$$\mathbf{G} = \begin{bmatrix} \mathbf{v}_0 \\ \mathbf{v}_1 \\ \mathbf{v}_2 \\ \mathbf{v}_3 \end{bmatrix} = \begin{bmatrix} 1 & 1 & 1 & 1 & 1 & 1 & 1 & 1 \\ 1 & 0 & 1 & 0 & 1 & 0 & 1 & 0 \\ 1 & 1 & 0 & 0 & 1 & 1 & 0 & 0 \\ 1 & 1 & 1 & 1 & 0 & 0 & 0 & 0 \end{bmatrix}. \quad \blacksquare$$

Example 12.3.2. RM(3,2) code is generated by the set

$$\{\mathbf{v}_0, \mathbf{v}_1, \mathbf{v}_2, \mathbf{v}_3, \mathbf{v}_1 \wedge \mathbf{v}_2, \mathbf{v}_1 \wedge \mathbf{v}_3, \mathbf{v}_2 \wedge \mathbf{v}_3\}.$$

Using the values of $\mathbf{v}_i$, $i = 0, 1, 2, 3$, we find that

$\mathbf{v}_1 \wedge \mathbf{v}_2 = \{1,0,1,0,1,0,1,0\} \wedge \{1,1,0,0,1,1,0,0\} = \{1,0,0,0,1,0,0,0\}$, and similarly $\mathbf{v}_1 \wedge \mathbf{v}_3 = \{1,0,1,0,0,0,0,0\}$, and $\mathbf{v}_2 \wedge \mathbf{v}_3 = \{1,1,0,0,0,0,0,0\}$. Thus, the generator matrix for RM(3,2) is

$$\mathbf{G} = \begin{bmatrix} 1 & 1 & 1 & 1 & 1 & 1 & 1 & 1 \\ 1 & 0 & 1 & 0 & 1 & 0 & 1 & 0 \\ 1 & 1 & 0 & 0 & 1 & 1 & 0 & 0 \\ 1 & 1 & 1 & 1 & 0 & 0 & 0 & 0 \\ 1 & 0 & 0 & 0 & 1 & 0 & 0 & 0 \\ 1 & 0 & 1 & 0 & 0 & 0 & 0 & 0 \\ 1 & 1 & 0 & 0 & 0 & 0 & 0 & 0 \end{bmatrix}. \quad \blacksquare$$

12.4 Properties of RM Codes

The properties of RM codes are listed as follows:

(i) The set of all possible wedge products of up to order r of $\mathbf{v}_i$ forms a basis for F_2^n.

(ii) RM(r, m) code has rank $\sum_{j=0}^{r} \binom{m}{j}$.

(iii) RM$(r, m) = $ RM$(r, m - 1) \,|\,$ RM$(r - 1, m - 1)$, where $|$ denotes the bar product of two codes (see §3.5.2 for the definition of bar product).

(iv) RM(r, m) has minimum Hamming weight 2^{r-m}.

PROOF. The property (i) is proved as follows: Since there are $\sum_{j=0}^{r} \binom{r}{j} = 2^r = n$ and since F_2^n has dimension n, it suffices to check that the n vectors span F_2^n, or that RM$(r, r) = F_2^n$. Let $\mathbf{x}$ be an element of X and define

$$y_i = \begin{cases} v_i & \text{if } x_i = 0, \\ 1 + v_i & \text{if } x_i = 1. \end{cases}$$

Then $\mathbf{I}_{\{\mathbf{x}\}} = y_i \wedge \cdots \wedge y_r$, which when expanded using the distribution property of the wedge product, gives $\mathbf{I}_{\{\mathbf{x}\}} \in$ RM(r, r). Then since the vectors $\{\mathbf{I}_{\{\mathbf{x}\}} \,|\, \mathbf{x} \in X\}$ span F_2^n, we have RM$(r, r) = F_2^n$.

Property (i) implies that all such wedge products must be linearly independent, so the rank of RM (r, r) must be equal to the number of such vectors. This proves property (ii).

Property (iii) is obvious.

Property (iv) is proved by induction as follows: RM$(r, 0)$ code is the repetition code of length $n = 2^r$ and weight $n = 2^{r-0} = 2^r$. By property (i) RM$(r, r) = F_2^n$ and has weight $2^{r-r} = 2^0 = 1$. ∎

Note that the weight of the bar product of two codes C_1 and C_2 is given by $\min\{2\,\mathrm{wt}(C_1), \mathrm{wt}(C_2)\}$. If $0 < m < r$ and if (a) RM$(r, m - 1)$ has weight 2^{r-1-m}, and (b) RM$(r - 1, m - 1)$ has weight $2^{r-1-(m-1)} = 2^{r-m}$, then the bar product has the weight

$$\min\{2 \times 2^{r-1-m}, 2^{r-m}\} = 2^{r-m}.$$

12.5 Classification of RM Codes

First we will prove two results which are as follows:

Theorem 12.5.1. RM$(r+1, m+1) = \{(\mathbf{f}, \mathbf{f}+\mathbf{g})$ *for all* $\mathbf{f} \in$ RM$(r+1, m)$ *and* $\mathbf{g} \in$ RM$(r, m)\}$.

PROOF. The codewords of RM$(r + 1, m + 1)$ are associated with Boolean monomials in $m + 1$ variables of degree $\leq r + 1$. If $\mathbf{c}(\mathbf{x}_1, \ldots, \mathbf{x}_{m+1})$ is such a codeword, then we can write

$$\mathbf{c}(\mathbf{x}_1, \ldots, \mathbf{x}_{m+1}) = \mathbf{f}(\mathbf{x}_1, \ldots, \mathbf{x}_m) + \mathbf{x}_{m+1} + \mathbf{g}(\mathbf{x}_1, \ldots, \mathbf{x}_m).$$

Then $\mathbf{f}$ is a Boolean polynomial in m variables of degree $\leq r + 1$ and $\mathbf{g}$ has degree $\leq r$; thus, the corresponding vectors $\mathbf{f}$ and $\mathbf{g}$ are in $\mathrm{RM}(r + 1, m)$ and $\mathrm{RM}(r, m)$, respectively. ∎

Lemma 12.5.1. $v(\mathbf{x}+\mathbf{y}) \geq v(\mathbf{x}) - v(\mathbf{y})$ *for any two distinct vectors* $\mathbf{x}, \mathbf{y}$.

PROOF. Let v_{xy} denote the number of locations at which the nonzero digits of $\mathbf{x}$ and $\mathbf{y}$ overlap. Then $v(\mathbf{x} + \mathbf{y}) = [v(\mathbf{x}) - v_{xy}] + [v(\mathbf{y}) - v_{xy}]$. The result follows since $2v(\mathbf{y}) \geq 2v_{xy}$. ∎

Theorem 12.5.2. *The minimum distance of* $RM(r, m)$ *is* 2^{m-r}.

PROOF. (By induction) For $m = 1$ the $\mathrm{RM}(0, 1)$ code is the repetition code of length 2, so in this case $d_{\min} = 2$; the $\mathrm{RM}(1, 1)$ code has four codewords of length 2, so $d_{\min} = 1$. Now assume that the result is true up to m, i.e., for $d_{\min} = 2^{m-r}$ for $0 \leq r \leq m$. We will show that $d_{\min} = 2^{m-r+1}$ for $\mathrm{RM}(r, m + 1)$. Let $\mathbf{f}, \mathbf{f}' \in \mathrm{RM}(r, m)$, and let $\mathbf{g}, \mathbf{g}' \in \mathrm{RM}(r - 1, m)$. Then by Theorem 12.5.1, the vectors (codewords) $\mathbf{c}_1 = (\mathbf{f}, \mathbf{f} + \mathbf{g})$ and $\mathbf{c}_2 = (\mathbf{f}', \mathbf{f}' + \mathbf{g}')$ must be in $\mathrm{RM}(r, m + 1)$. If $\mathbf{g} = \mathbf{g}'$ then $d(\mathbf{c}_1, \mathbf{c}_2) = 2d(\mathbf{f}, \mathbf{f}') \geq 2 \cdot 2^{m-r}$. However, if $\mathbf{g} \neq \mathbf{g}'$, then $d(\mathbf{c}_1, \mathbf{c}_2) = v(\mathbf{f} - \mathbf{f}') + v(\mathbf{g} - \mathbf{g}') + v(\mathbf{g} - \mathbf{g}' + \mathbf{f} - \mathbf{f}')$. In view of Lemma 12.5.1, $d(\mathbf{c}_1, \mathbf{c}_2) \geq v(\mathbf{f} - \mathbf{f}') + v(\mathbf{g} - \mathbf{g}') - v(\mathbf{f} - \mathbf{f}') = v(\mathbf{g} - \mathbf{g}')$. Since $\mathbf{g} - \mathbf{g}' \in \mathrm{RM}(r - 1, m)$, so $v(\mathbf{g} - \mathbf{g}') \geq 2^{m-(r-1)} = 2^{m-r+1}$. ∎

The $\mathrm{RM}(r, m)$ code exists for any integers $m \geq 0$ and $0 \leq r \leq m$. The $\mathrm{RM}(m, m)$ code is defined as the *universe* $(2^m, 2^m, 1)$ code, whereas the $\mathrm{RM}(-1, m)$ is the *trivial* $(2^m, 0, \infty)$ code. In between these two limits the remaining codes are constructed from these two elementary codes using the *length-doubling construction method* defined by

$$\mathrm{RM}(r, m) = \{(\mathbf{u}, \mathbf{u} + \mathbf{v} \mid \mathbf{u} \in \mathrm{RM}(r - 1, m - 1)\}. \tag{12.5.1}$$

This formula yields $\mathrm{RM}(r, m)$ as a binary linear block code (n, k, d) with length $n = 2^m$, dimension $k(r, m) = k(r, m - 1) + k(r - 1, m - 1)$, and minimum distance $d = 2^{m-r}$ for $r \geq 0$. A complete list of $\mathrm{RM}(r, m)$ codes of length up to 32 is as follows:

UNIVERSE CODES: $\mathrm{RM}(m, m)$ as a $(2^m, 2^m, 1)$ code; an example is $\mathrm{RM}(5, 5)$ as a (32, 32,1) code.

SPC CODES: $\mathrm{RM}(m-1, m)$ as a $(2^m, 2^m - 1, 2)$ code; examples are: $\mathrm{RM}(3, 3)$ as an (8,8,1) code, $\mathrm{RM}(4, 4)$ as a (16, 16,1) code; $\mathrm{RM}(4, 5)$ as a (32, 31,2) code.

EXTENDED HAMMING CODES: $\mathrm{RM}(m - 2, m)$ as a $(2^m, 2^m - m - 1, 4)$ code; examples are $\mathrm{RM}(0, 0)$ as a (1,1,1) code; $\mathrm{RM}(1, 1)$ as a (2, 2, 1) code; $\mathrm{RM}(1, 2)$ as a (4,3,2) code; $\mathrm{RM}(2, 2)$ as a (4,4,1) code; $\mathrm{RM}(2, 3)$ as a

(8,7,2) code; RM(2, 4) as a (16,11, 4) code; RM(3, 4) as a (16, 15,2) code; and RM(3, 5) as a (32, 26,4) code

SELF-DUAL CODES: RM(−1, 0) as a (1, 0, ∞) code; RM(−1, 1) as a (4, 0, ∞) code; RM(0, 1) as a (2,1,2) code; RM(0, 2) as a (4,1,4) code; RM(0, 3) as a (8,1,8) code; RM(1, 3) as a (8,4,4) code; RM(1, 4) as a (16,5,8) code; RM(1, 5) as a (32,6,16) code; and RM(2, 5) as a (32,16,8) code;

BIORTHOGONAL CODES: RM(1, m) as a $(2^m, m+1, 2^{m-1})$ code; examples are: RM(−1, 2) as a (4, 0, ∞) code; RM(−1, 3) as a (16, 0, ∞) code; ; RM(0, 4) as a (16,1,16) code; and RM(0, 5) as a (32,1,32) code.

REPETITION CODES: RM(0, m) as a $(2^m, 1, 2^m)$ code: examples are: RM(−1, 4) as a (16, 0, ∞) code; and RM(−1, 5) as a (32, 0, ∞) code.

TRIVIAL CODES: RM(−1, m) as a $(2^m, 0, ∞)$ code.

Note that since the dual code to RM(r, m) is RM(m − r − 1, m), the repetition and SPC codes are duals, biorthogonal and extended Hamming codes are duals, and the codes with $k = n/2$ are self-dual.

12.6 Decoding of RM Codes

As suggested by Reed [1954], the RM(r, m) codes are decoded using majority logic decoding, which is based on constructing several checksums for each received codeword **w**. Since each of the different checksums must have the same value, i.e., the same value of the weight of the codeword, a majority logic decoding is used to decipher the value of the codeword. Once each order of the associated Boolean polynomial is decoded, the received word is modified by removing the corresponding codewords weighted by the decoded message contributions, up to the current stage. Thus, for an RM code of order r, we must decode iteratively r+1 times before we reach the final received codeword. The values of the message bits are computed via this scheme. Finally, we compute the codeword by multiplying the just-decoded message word with the generator matrix **G** of the RM code. A test for a successful decoding is to have an all-zero modified received word at the end of the (r + 1)-st stage.

Although decoding of RM(r, m) codes is more complex than encoding, the theory of encoding and decoding is based on the distance between vectors. Recall that the distance between two vectors is equal to the number of locations in the two vectors that have different values (§3.3). The distance between any two codewords in RM(r, m) is 2^{m-r}, provided the codewords have the same length $n = 2^m$.

The RM encoding is based on the assumption that the closest codeword in RM(r, m) for the received message is the original codeword (encoded message). Thus, if t errors are to be corrected in the received message, the distance between any two of the codewords in RM(r, m) must be greater than 2t. Although the decoding method is not very efficient, it is easy to use. It checks each row of the encoding matrix and uses *majority logic* (to be described below in Steps 1–3) to determine whether that row was used in constructing the codeword as explained in §12.2. This makes it possible to determine the error-free original message. The *algorithm* for decoding is carried out by applying the following three steps to each row of the encoding matrix starting from the last row and working upward:

STEP 1: Choose a row in the $\mathrm{RM}(r, m)$ encoding matrix. Find 2^{m-r} characteristic vectors (as described below) for each row, and then take the dot product of each of these rows with the codeword.

STEP 2: Take the majority of the values of the dot product and assign that value to the coefficient of the row.

STEP 3: For each row, except the top row of the encoding matrix, multiply each coefficient by its corresponding row and add the resulting vector to form the vector $\mathbf{c}_y$. Add this result to the received codeword. If the resulting vector has more 1s than 0s, then the top row's coefficient is 1; otherwise it is 0. Add the top row multiplied by its coefficient to $\mathbf{c}_y$. This is the original codeword, and the errors can be identified accordingly. Thus, the vector formed by the sequence of coefficients starting from the top row of the encoding matrix and ending at the bottom row is the original message.

Example 12.6.1. The message is transmitted in blocks of length k. Suppose the original message is $\mathbf{m} = (0110)$. We will use $\mathrm{RM}(1,3)$ which can correct one error. The codeword from Example 12.2.4(a) is (00111100). Now suppose that the received word after transmission with one error in the first bit is $\mathbf{w} = (10111100)$. Starting with the last row, the characteristic vectors of the last row $\mathbf{x}_3$ are $\mathbf{x}_1\mathbf{x}_2, \mathbf{x}_1\bar{\mathbf{x}}_2, \bar{\mathbf{x}}_1\mathbf{x}_2$, and $\bar{\mathbf{x}}_1\bar{\mathbf{x}}_2$.

The vector associated with $\mathbf{x}_1$ is $(11110000) \Rightarrow \bar{\mathbf{x}}_1 = (00001111)$.

The vector associated with $\mathbf{x}_2$ is $(11001100) \Rightarrow \bar{\mathbf{x}}_2 = (00110011)$.

Thus, $\mathbf{x}_1\mathbf{x}_2 = (11001100)$; $\mathbf{x}_1\bar{\mathbf{x}}_2 = (00110000)$; $\bar{\mathbf{x}}_1\mathbf{x}_2 = (00001100)$; $\bar{\mathbf{x}}_1\bar{\mathbf{x}}_2 = (00000011)$.

Taking the dot product of these four vectors with $\mathbf{w}$, we get

$(11000000) \cdot (10111100) = 1$; $(00110000) \cdot (10111100) = 0$;

$(00001100) \cdot (10111100) = 0$; $(00000011) \cdot (10111100) = 0$.

Hence, the coefficient of $\mathbf{x}_3 = 0$.

Now, we repeat the above steps for the row of $\mathbf{x}_2 = (11001100)$. The characteristic vectors are $\mathbf{x}_1\mathbf{x}_3, \mathbf{x}_1\bar{\mathbf{x}}_3$, and $\bar{\mathbf{x}}_1\bar{\mathbf{x}}_3$.

The vector associated with $\mathbf{x}_1$ is $(11110000) \Rightarrow \bar{\mathbf{x}}_1 = (00001111)$.

The vector associated with $\mathbf{x}_2$ is $(11001100) \Rightarrow \bar{\mathbf{x}}_2 = (00110011)$.

The vector associated with $\mathbf{x}_3$ is $(10101010) \Rightarrow \bar{\mathbf{x}}_3 = (010100101)$.

Thus, $x_1 x_3 = (10100000)$; $x_1 \bar{x}_3 = (01010000)$; $\bar{x}_1 x_3 = (00001010)$.

Taking the dot product of these four vectors with $\mathbf{w}$, we get

$$(10100000) \cdot (10111100) = 1; \quad (01010000) \cdot (10111100) = 1;$$
$$(00001010) \cdot (10111100) = 0; \quad (00000101) \cdot (10111100) = 1.$$

Hence, the coefficient of $x_1 = 1$.

Then compute c_y using the formula

$c_y = \{\text{coefficient of } x_3\} * x_3 \oplus \{\text{coefficient of } x_2\} * x_2 \oplus \{\text{coefficient of } x_1\} * x_1$
$= 0 * (10101010) \oplus 1 * (11001100) \oplus 1*)11110000) = (00111100).$

Thus, $\mathbf{w} + c_y = (10111100) \oplus (00111100) = (10000000)$.

This result has more 0s than 1s, so the coefficient of the first row of the encoding matrix is 0. Then the coefficients of the four rows of the encoding matrix are 0, 1, 1, 0, respectively, which yield the original message (0110). Also, the error in $\mathbf{w}$ occurred in the first bit of the codeword (00111100). ∎

12.6.1 Another Version. There are different versions of the decoding method. Before we describe this method based on majority logic decoding, recall that the distance between any two codewords in RM(r, m) code is 2^{m-r}. To correct t errors, we must have a distance greater than $2t$. Then only we can correct $\max\{0, 2^{m-r-1} - 1\}$ errors.

The basic idea behind majority logic is that, for each row of the generator matrix, we attempt to determine a majority vote whether or not that row was used in the formation of the codeword associated to the original message.

Let $\mathbf{b}$ be any Boolean monomial of degree d with its reduced form $\mathbf{b}'$. Let J denote the set of variables not in $\mathbf{b}'$ and their complements. Then the *characteristic vectors* of $\mathbf{b}$ are all vectors associated to the monomials of degree $m - d$ over the variables of J. Note that since any monomial containing a variable and its complement are associated to the vector $\mathbf{0}$, and since ϕ is a bijection and $\phi^{-1}(\mathbf{0}) = \mathbf{0}$, any monomial containing both a variable and its complement is equivalent to the monomial of degree 0. Thus, wlog, we consider only the monomials where the variables are distinct, i.e., no variable and its complement appear. For example, in RM$(3, 4)$ the characteristic vectors of $x_1 x_2 x_3$ are the vectors associated with the monomials $\{x_3, \bar{x}_3\}$, and the characteristic vectors of $x_1 x_3$ are the vectors associated with the monomials $\{x_2 x_4 \bar{x}_2 x_4, x_2 \bar{x}_4, \bar{x}_2 \bar{x}_4\}$.

This method, like the previous one, starts at the bottom of the generator matrix and works upward. The algorithm starts by examining the rows with monomials of degree r. First, we calculate 2^{m-r} characteristic vectors for the row, and then take the dot product of each of these vectors with the received message $\mathbf{w}$. If the majority of the dot products is 1, we assume that this row was used in the construction of the codeword $\mathbf{c}$. We set the location in $\mathbf{c}$ associated to this row to 1. If the majority of the dot products are 0, then we assume that this row was not used in the construction of $\mathbf{c}$, and so we set the corresponding entry in $\mathbf{c}$ to 0. After we have gone through every row associated with the monomials of degree r, we take the vector of length $\binom{m}{r}$ associated with the portion of the message we have just calculated and multiply it by the $\binom{m}{r}$ rows of the generator matrix that was just considered. This gives us a vector $\mathbf{s}$ of length n. By adding (XOR-ing) $\mathbf{s}$ to the received message $\mathbf{w}$ we proceed recursively on the rows associated with monomials of degree $r - 1$.

Example 12.6.2. To decode the message $\mathbf{w} = (00110110)$ in $\mathrm{RM}(2,3)$ we use the encoding matrix from Example 12.2.2. Starting from the bottom of this matrix up, we first consider thy rows associated to the monomials of degree 2.:

Row $\mathbf{x}_2\mathbf{x}_3$ has the characteristic vectors $\mathbf{x}_1, \bar{\mathbf{x}}_1$. So

$\mathbf{w} * \mathbf{x}_1 = 0$ and $\mathbf{w} * \bar{\mathbf{x}}_1 = 0$, which yield $\mathbf{m} = (- - - - - - 0)$.

Row $\mathbf{x}_1\mathbf{x}_2$ has the characteristic vectors $\mathbf{x}_2, \bar{\mathbf{x}}_2$. So

$\mathbf{w} * \mathbf{x}_2 = 1$ and $\mathbf{w} * \bar{\mathbf{x}}_2 = 1$, which yield $\mathbf{m} = (- - - - - 1\,0)$.

Row $\mathbf{x}_1\mathbf{x}_2$ has the characteristic vectors $\mathbf{x}_2, \bar{\mathbf{x}}_2$. So

$\mathbf{w} * \mathbf{x}_2 = 0$ and $\mathbf{w} * \bar{\mathbf{x}}_2 = 0$, which yield $\mathbf{m} = (- - - - 0\,1\,0)$.

This completes the processing of the rows associated to the monomials of degree $r = 2$. Now we compute

$$\mathbf{s} = [0\ 1\ 0] \begin{bmatrix} 1\,1\,0\,0\,0\,0\,0\,0 \\ 1\,0\,1\,0\,0\,0\,0\,0 \\ 1\,0;\,0\,0\,1\,0\,0\,0 \end{bmatrix} = [1\,0\,1\,0\,0\,0\,0\,0].$$

Then we add $\mathbf{s}$ to $\mathbf{w}$, so $\mathbf{s} \oplus \mathbf{w} = 10010110$. Next, we process the rows associated to monomials of degree $r = 1$, as follows:

Row $\mathbf{x}_2$ has the characteristic vectors $\mathbf{x}_1\mathbf{x}_2, \bar{\mathbf{x}}_1\mathbf{x}_2, \mathbf{x}_1\bar{\mathbf{x}}_2, \bar{\mathbf{x}}_1\bar{\mathbf{x}}_2$. So

$\mathbf{w} * \mathbf{x}_1 * \mathbf{x}_2 = 1, \mathbf{w} * \bar{\mathbf{x}}_1\mathbf{x}_2 = 1, \mathbf{w} * \mathbf{x}_1 * \mathbf{x}_2 = 1, \mathbf{w} * \bar{\mathbf{x}}_1\mathbf{x}_2 = 1$, which yield $\mathbf{m} = (- - - 1\,0\,1\,0)$.

Continuing in this manner, we find that the original message was $\mathbf{c} = (0111010)$. This can be done by verifying that $\mathbf{c} * \mathbf{G}_{2,3} = \mathbf{w}$. ∎

12.6.2 Decoding First-Order RM Codes. We describe an efficient method for $\mathrm{RM}(1,m)$ codes. This method is explained for $\mathrm{RM}(1,3)$ code, but can be used for any $\mathrm{RM}(1,m)$ code. For example, the $\mathrm{RM}(1,3)$ code is generated by

$$\mathbf{c} = (c_0, c_1, \ldots, c_7) = \mathbf{mG} = (m_0, m_1, m_2, m_3) \begin{bmatrix} 1 \\ \mathbf{x}_1 \\ \mathbf{x}_2 \\ \mathbf{x}_3 \end{bmatrix}$$

$$= (m_0, m_1, m_2, m_3) \begin{bmatrix} 1 & 1 & 1 & 1 & 1 & 1 & 1 & 1 \\ 0 & 1 & 0 & 1 & 0 & 1 & 0 & 1 \\ 0 & 0 & 1 & 1 & 0 & 0 & 1 & 1 \\ 0 & 0 & 0 & 0 & 1 & 1 & 1 & 1 \end{bmatrix}$$

$$\equiv \begin{bmatrix} 1 & 1 & 1 & 1 & 1 & 1 & 1 & 1 \\ 1 & 1 & 1 & 1 & 0 & 0 & 0 & 0 \\ 1 & 1 & 0 & 0 & 1 & 1 & 0 & 0 \\ 1 & 0 & 1 & 0 & 1 & 0 & 1 & 0 \end{bmatrix},$$

as given in Example 12.2.2. The matrix $\mathbf{G}$ is a 4-tuple (1000) increasing to (1111), i.e.,

$$\mathbf{G} = \begin{bmatrix} 1 & 1 & 1 & 1 & 1 & 1 & 1 & 1 \\ 0 & 0 & 0 & 0 & 1 & 1 & 1 & 1 \\ 0 & 0 & 1 & 1 & 0 & 0 & 1 & 1 \\ 0 & 1 & 0 & 1 & 0 & 1 & 0 & 1 \end{bmatrix}.$$

Let $\mathbf{w}$ be a binary received word. Translating $\mathbf{w}$ to a (± 1) sequence by the mapping $f(\mathbf{w}) = \mathbf{F} = \{(-1)^{w_1}, (-1)^{w_2}, \ldots, (-1)^{w_n}\}$, and defining the correlation function ψ between these two sequences as

$$\psi(\mathbf{F}, \mathbf{G}) = \psi((F_1, F_2, \ldots, F_n), (G_1, G_2, \ldots, G_n)) = \sum_{i=1}^{n} F_i G_i,$$

we find the codeword $\mathbf{c}$ that minimizes the distance $d(\mathbf{w}, \mathbf{c})$. This is equivalent to finding the codeword that maximizes the function $\psi(\mathbf{F}(\mathbf{w}), \mathbf{F}(\mathbf{c}))$.

The computation of the function ψ can be carried out using the Hadamard transform. Let $\hat{\mathbf{F}} = \mathbf{F}\mathfrak{H}_{2^m}$ denote the Hadamard transform of $\mathbf{F}$, where $\mathfrak{H}_{2^m}$ is the $2^m \times 2^m$ Hadamard matrix. For example, for RM(1, 3), this Hadamard matrix is

$$\mathfrak{H}_8 = \begin{bmatrix} 1 & 1 & 1 & 1 & 1 & 1 & 1 & 1 \\ 1 & - & 1 & - & 1 & - & 1 & - \\ 1 & 1 & - & - & 1 & 1 & - & - \\ 1 & - & - & 1 & 1 & - & - & 1 \\ 1 & 1 & 1 & 1 & - & - & - & - \\ 1 & - & 1 & - & - & 1 & - & 1 \\ 1 & 1 & - & - & - & - & 1 & 1 \\ 1 & - & - & 1 & - & 1 & 1 & - \end{bmatrix},$$

where $-$ is the wild card, generally replaced by -1 for a Hadamard matrix or by 0 for the binary case. Notice that the second, third, and fifth columns are $f(\mathbf{x}_3)$, $f(\mathbf{x}_2)$, and $f(\mathbf{x}_1)$, respectively. The $(i-1)$-st column of $\mathfrak{H}_8$ is the binary vector

$$f(\mathbf{c}_i) = \mathbf{F}(i_1\mathbf{x}_1 + i_2\mathbf{x}_2 + i_3\mathbf{x}_3),$$

where $i - 1 = (i_3, i_2, i_1)$. For example, the sixth column $(i - 1 = 5 = (101)_2)$ consists of the vector[2]

$$f(1 * \mathbf{x}_1 \oplus 0 * \mathbf{x}_2 \oplus 1 * \mathbf{x}_3) = 1 * (01010101) \oplus 0 * (00110011) \oplus 1 * (00001111)$$

$$= (01010101) \oplus (00000000) \oplus (00001111)$$

$$= (01010101) \oplus (00001111) = (01011010) \equiv (1 \ - \ 1 \ - \ - 1 \ - \ 1).$$

Finally, the $\mathrm{RM}(1, m)$ code has 2^{m+1} codewords, but the Hadamard transform yields only 2^m codewords. This is remedied by taking the complement of each bit by $f(\mathbf{1}+\mathbf{c}) = f(\mathbf{1}+c_1\mathbf{x}_1+c_2\mathbf{x}_2+\cdots+c_m\mathbf{x}_m)$. Then the correlation function is the negative of $f(\mathbf{c})$ (see §2.2.12). Thus, we check the sign and accordingly get the other half of the codewords.

12.6.3 One More Algorithm. The decoding algorithm for $\mathrm{RM}(2,4)$ code with the generator matrix $\mathbf{G}_{(2,4)}$ presented in Example 12.2.3, is again based on the majority logic, but it starts with the generator matrix, which is written as

$$\mathbf{G} = \begin{bmatrix} \mathbf{1} \\ \hline \mathbf{x}_1 \\ \mathbf{x}_2 \\ \mathbf{x}_3 \\ \hline \mathbf{x}_1\mathbf{x}_4 \\ \mathbf{x}_2\mathbf{x}_3 \\ \mathbf{x}_2\mathbf{x}_4 \\ \mathbf{x}_3\mathbf{x}_4 \end{bmatrix} \equiv \begin{bmatrix} \mathbf{G}_0 \\ \hline \mathbf{G}_1 \\ \hline \mathbf{G}_2 \end{bmatrix}.$$

The 11 input bits corresponding to the rows of the matrix are written as

$$\mathbf{m} = (m_0, m_1, m_2, m_3, m_4, m_{12}, m_{13}, m_{14}, m_{23}, m_{24}, m_{34}) \equiv (\mathbf{m}_0, \mathbf{m}_1, \mathbf{m}_2),$$

where the bits in $\mathbf{m}_0$ are associated to the zeroth-order term $\mathbf{1}$, the $\mathbf{m}_1$ bits to the first order terms, and the $\mathbf{m}_2$ bits to the second-order terms. Thus, the encoding operation that construct the codeword is

$$\mathbf{c} = (c_0, c_1, c_2, \dots, c_{15}) = \mathbf{mG} = [\mathbf{m}_0, \ \mathbf{m}_1, \ \mathbf{m}_2] \begin{bmatrix} \mathbf{G}_0 \\ \mathbf{G}_1 \\ \mathbf{G}_2 \end{bmatrix}.$$

The algorithm operates as follows: Given a received word $\mathbf{w}$,

---

[2] Some authors use the $+$ sign for $\oplus$, and others sometimes fail to distinguish between $*$ and $\cdot$ for vector multiplication.

(i) Obtain the estimates for the highest order block of message bits, $\mathbf{m}_2$;

(ii) subtract $\mathbf{m}_2 \mathbf{G}_2$ from $\mathbf{w}$ (this will leave lower order terms);

(iii) subtract $\mathbf{m}_1 \mathbf{G}_1$ from $\mathbf{w}$,

and so on. The main point of this algorithm is to write *multiple* equations for the same quantity and to take a majority vote. Note that $c + 0 = m + 0$, $c_1 = m_0 + m_1$, $c_2 = m_0 + m_2$, $c_3 = m + 0 + m_1 + m_2 + m_{12}$. Adding these code bits together, we get $c_0 + c_1 + c_2 + c_3 = m_{12}$. Similarly, the next four code bits give $c_4 + c_5 + c_6 + c_7 = m_{12}$, the next four code bits give $c_8 + c_9 + c_{10} + c_{11} = m_{12}$, and the final four code bits give $c_{12} + c_{13} + c_{14} + c_{15} = m_{12}$. Next, using the received word $\mathbf{w} = (w_0, w_1, \ldots, w_{15})$ and the above four equations, we obtain the four estimates of m_{12}:

$$\hat{m}_{12} = w_0 + w_1 + w_2 + w_3, \qquad \hat{m}_{12} = w_4 + w_5 + w_6 + w_7,$$
$$\hat{m}_{12} = w_8 + w_9 + w_{10} + w_{11}, \qquad \hat{m}_{12} = w_{12} + w_{13} + w_{14} + w_{15}.$$

From these four estimates, we determine the value of m_{12} by majority vote, which states that if only one is incorrect, no error has occurred; if two are incorrect, we can detect that at least one error has occurred.

In the same manner, we set up multiple equations for the other second-order bits $m_{13}, m_{14}, \ldots, m_{34}$. This step eventually leads us to an estimate of the entire second-order block:

$$\hat{m}_2 = (\hat{m}_{12}, \hat{m}_{13}, \hat{m}_{23}, \hat{m}_{14}, \hat{m}_{24}, \hat{m}_{34}).$$

Then, we determine $\mathbf{w}' = \mathbf{w} - \mathbf{m}_2 \mathbf{G}_2$.

Next, we repeat the above process for the first-order bits. There are 8 checksums on each of the first-order message bits, namely,

$$\begin{array}{ll} m_1 = c_0 + c_1, & m_1 = c_8 + c_9, \\ m_1 = c_2 + c_3, & m_1 = c_{10} + c_{11}, \\ m_1 = c_4 + c_5, & m_1 = c_{12} + c_{13}, \\ m_1 = c_6 + c_7, & m_1 = c_{14} + c_{15}. \end{array}$$

As before, from these 8 estimates obtained from $\mathbf{w}$ we estimate m_1 by majority vote. Then, after obtaining all the bits in $\mathbf{m}_1$, we remove it to get $\mathbf{w}'' = \mathbf{w}' - \mathbf{m}_1 \mathbf{G}_1$.

Next, we look for $\mathbf{m}_0$. Since $\mathbf{w}'' = m_0 \mathbf{1} + \mathbf{e}$, the parity sums are the bits $w_0, \ldots, w_{15}$.

In the general case of $\text{RM}(r, m)$ codes, where the codeword is $\mathbf{c} = (c_0, c_1, \ldots, c_{m-1})$, we associate c_i with the complement of the binary equivalent of i, which is the m-tuple denoted by P_i. For example, for $\mathrm{RM}(2,4)$,

i	Binary	P_i
0	0000	1111
1	0001	1110
2	0010	1101

Each codeword in $\mathrm{RM}(r, m)$ is an *incidence vector* that defines a subspace in $\mathrm{RM}(r, m)$. For example, the codeword $(1100110011001100) \in \mathrm{RM}(2,4)$ is an incidence vector for the subspace containing the points $\{P_0, P_1, P_4, P_5, P_8, P_9, P_{12}, P_{13}\}$. This geometry is used to find equations for orthogonal checksums as follows: First, to find a set of orthogonal checksums that provide an estimate for the k-th-order message bit $m_{I_1, i_2, \ldots, i_k}$ associated to the basis vectors $\mathbf{x}_1, \mathbf{x}_{i_2}, \ldots, \mathbf{x}_{i_k}$, let S be the subspace of points associated to the incidence vectors $\mathbf{x}_{i_1}, \mathbf{x}_{i_2}, \ldots, \mathbf{x}_{i_k}$. Then

(i) Form the 'set difference': $j_1, j_2, \ldots, j_{m-k} = \{1, 2, \ldots, m\} - \{i_1, i_2, \ldots, i_k\}$. Let T be the subspace of points associated to the incidence vectors $\mathbf{x}_{j_1}, \ldots, \mathbf{x}_{j_k}$. The space T is the *complementary subspace* to S.

(ii) The first checksum consists of the sum of the codeword coordinates specified by T.

(iii) The rest of the codewords are obtained by translating T with respect to the nonzero elements of S.

For example, the difference set is $\{j_1, j_2\} = \{1, 2, 3, 4\} - \{3, 4\} = \{1, 2\}$. So we use $\mathbf{x}_1 \mathbf{x}_2 = (0001000100010001)$, which corresponds to the set of points $T = \{P_3, P_7, P_{11}, P_{15}\} = \{(1100)(1011)(0100)(000)\}$. This yields $M_{34} = c_3 + c_7 + c_{11} + c_{15}$. The translations of T by the nonzero elements of S are

$$\text{by } P_{12} = (0011) \quad \rightarrow \quad \{(1111)(1011)(0111)(0011)\} = \{P_0, P_4, P_8, P_{12}\},$$
$$\text{by } P_{13} = (0010) \quad \rightarrow \quad \{(1110)(1010)(0101)(0010)\} = \{P_1, P_5, P_9, P_{13}\},$$
$$\text{by } P_{14} = (0001) \quad \rightarrow \quad \{(1101)(1001)(0101)(0001)\} = \{P_2, P_6, P_{10}, P_{14}\},$$

which are associated, respectively, with the checksums

$$m_{34} = c_0 + c_4 + c_8 + c_{12},$$
$$m_{34} = c_1 + c_5 + c_9 + c_{13},$$
$$m_{34} = c_2 + c_6 + c_{10} + c_{14}.$$

Other checksums $m_{12}, m_{13}, m_{23}, m_{14}, m_{24}$ can be similarly obtained.

12.7 Recursive Definition

We describe the method that can define the $\mathrm{RM}(r,m)$ recursively, although it will yield the generator matrix which is different from the one obtained in §12.2.1 using the function ϕ. If both encoding and decoding are done by this recursive definition, the final result of error correction will remain the same. The recursive definition for $\mathrm{RM}(r,m)$, as found in Hankerson et al. [1991], is as follows:

(i) $\mathrm{RM}(0,m) = \{\underbrace{00\ldots0}_{2^m}\}$;

(ii) $\mathrm{RM}(m,m) = F_2^{2^m}$;

(iii) $\mathrm{RM}(r,m) = \{(\mathbf{x}, \mathbf{x} \oplus \mathbf{y} \; : \; \mathbf{x} \in \mathrm{RM}(r, m-1), \mathbf{y} \in \mathrm{RM}(r-1, m-1); 0 \le r \le m\}$.

Example 12.7.1. To use the recursive definition and find $\mathrm{RM}(1,2)$, we have

$$\begin{aligned}
\mathrm{RM}(1,2) &= \{(\mathbf{x}, \mathbf{x} \oplus \mathbf{y} \; : \; \mathbf{x} \in \mathrm{RM}(r,1), \mathbf{y} \in \mathrm{RM}(0,1)\} \\
&= \{(\mathbf{x}, \mathbf{x} \oplus \mathbf{y} \; : \; \mathbf{x} \in \{00, 01, 10, 11\}, \mathbf{y} \in \{00, 11\}\} \\
&= \{0000, 0011, 0101, 0110, 1010, 1001, 1111, 1100\}. \;\blacksquare
\end{aligned}$$

In accordance with the above recursive definition for $\mathrm{RM}(r,m)$ codes, the recursive definition for the generator matrix is as follows:

$$\mathbf{G}_{(r,m)} = \begin{bmatrix} \mathbf{G}_{(r-1,m)} & \mathbf{G}_{r-1,m} \\ \mathbf{0} & \mathbf{G}_{r-1,m-1} \end{bmatrix},$$

where the two extreme case are

$$\mathbf{G}_{(0,m)} = \{\underbrace{11\ldots1}_{2^m}\}, \quad \text{and} \quad \mathbf{G}_{(m,m)} = \begin{bmatrix} \mathbf{G}_{m-1,m} \\ \underbrace{00\ldots0}_{2^m - 1} \end{bmatrix}.$$

Example 12.7.2. Using the recursive definition, the generator matrix for the $\mathrm{RM}(2,3)$ code is

$$\mathbf{G}_{(2,3)} = \begin{bmatrix} \mathbf{G}_{(2,2)} & \mathbf{G}_{(2,2)} \\ \mathbf{0} & \mathbf{G}_{(1,2)} \end{bmatrix} = \begin{bmatrix} \mathbf{G}_{(1,1)} & \mathbf{G}_{(1,1)} & \mathbf{G}_{(1,1)} & \mathbf{G}_{(1,1)} \\ 0\,0 & \mathbf{G}_{(1,1)} & 0\,0 & \mathbf{G}_{(1,1)} \\ 0\,0 & 0\,1 & 0\,0 & 0\,1 \\ 0\,0 & 0\,0 & \mathbf{G}_{(1,1)} & \mathbf{G}_{(1,1)} \\ 0\,0 & 0\,0 & 0\,0 & \mathbf{G}_{(1,1)} \end{bmatrix}$$

$$= \begin{bmatrix} 1 & 1 & 1 & 1 & 1 & 1 & 1 & 1 \\ 0 & 1 & 0 & 1 & 0 & 1 & 0 & 1 \\ 0 & 0 & 1 & 1 & 0 & 0 & 1 & 1 \\ 0 & 0 & 0 & 1 & 0 & 0 & 0 & 1 \\ 0 & 0 & 0 & 0 & 1 & 1 & 1 & 1 \\ 0 & 0 & 0 & 0 & 0 & 1 & 0 & 1 \\ 0 & 0 & 0 & 0 & 0 & 0 & 1 & 1 \end{bmatrix}. \;\blacksquare$$

12.8 Probability Analysis

We have thus far studied linear $RM(r, m)$ codes over the field F_2. The RM codes defined over the field F_q are denoted by $RM_q(r, m)$. Their formal definition is as follows: Given a field F_q of size $q \geq 2$, a number m of variables, and a total degree bound r, the $RM_q(r, m)$ code is a linear code over F_q defined by the encoding mapping

$$f(\mathbf{x}_1, \ldots, \mathbf{x}_m) \mapsto \langle f(\alpha) \rangle |_{\alpha \in F_q^m}$$

applied to the domain of all polynomials in $F_q[\mathbf{x}_1, \ldots, \mathbf{x}_m]$ of total degree $\deg(f) \leq r$. The total degree of the Boolean monomials $\mathbf{x}_1^{k_1}, \ldots, \mathbf{x}_m^{k_m}$ is $k_1 + k_2 + \cdots + k_m$. This leads to the total degree of the polynomial as the maximum total degree over all its monomials with nonzero coefficients. The $RM_q(r, m)$ codes are multivariate codes with the variables $\mathbf{x}_1, \ldots, \mathbf{x}_m$, whereas the Reed–Solomon codes (Chapter 13) are univariate. The block length of the $RM_q(r, m)$ code is q^m, and its dimension is the number of polynomials in $F_q[\mathbf{x}_1, \ldots, \mathbf{x}_m]$. In the case when $q = 2$, there are $\binom{m}{0} + \binom{m}{1} + \cdots + \binom{m}{r}$ polynomials. In general, for $q \geq 2$, the number of polynomials in m variables of total degree at most r is

$$\left| \left\{ (i_1, \ldots, i_m) : 0 \leq i_j \leq q - 1, \text{ where } \sum_{j=1}^{m} i_j \leq r \right\} \right|.$$

As we have seen, the distance of RM codes is very important. This parameter is computed by finding the minimum number of any nonzero polynomials, using the formula $\alpha^q = \alpha$ for $\alpha \in F_q$, which for $q = 2$ reduces to formula (12.2.1). The number of zeros of multivariate polynomials f is given below by Theorem 12.8.1, which needs the following two lemmas.

Lemma 12.8.1. (Schwarz [1980]) Let $f \in F_q[\mathbf{x}_1, \ldots, \mathbf{x}_m]$ be a nonzero Boolean polynomial of total degree at most $l < q$. Then the probability

$$\mathfrak{P}_{(\alpha_1, \ldots, \alpha_m) \in F_q^m} [f(\alpha_1, \ldots, \alpha_m) = 0] \leq \frac{l}{q}. \qquad (12.8.1)$$

PROOF. (By induction) For $m = 0$, the mapping function f is a single variable polynomial. Consider a decomposition of f as

$$f(\mathbf{x}_1, \ldots, \mathbf{x}_m) = $$
$$\mathbf{x}_m^{d_m} g_m(\mathbf{x}_1, \ldots, \mathbf{x}_{m-1}) + \cdots + \mathbf{x}_m g_1(\mathbf{x}_1, \ldots, \mathbf{x}_{m-1}) + g_0(\mathbf{x}_1, \ldots, \mathbf{x}_{m-1}), \qquad (12.8.2)$$

where $d_m = \deg(f)$ in $\mathbf{x}_m$. Then g_m is a nonzero polynomial of total degree at most $l - d_m$. Thus, by induction,

$$\mathfrak{P}_{(\alpha_1, \ldots, \alpha_m) \in F_q^{m-1}} [g_m(\alpha_1, \ldots, \alpha_{m-1} = 0] \leq \frac{l - d_m}{q}. \qquad (12.8.3)$$

If $g_m(\alpha_1, \ldots, \alpha_{m-1}) \neq 0$, then $f(\alpha_1, \ldots, \alpha_{m-1}, \mathbf{x}_m)$ is a nonzero univariate polynomial of degree at most d_m. Then

$$\mathfrak{P}_{(\alpha_1, \ldots, \alpha_m) \in F_q^m} \left[f(\alpha_1, \ldots, \alpha_m = 0 : g_m(\alpha_1, \ldots, \alpha_{m-1}) \neq 0 \right] \leq \frac{d_m}{q}. \quad (12.8.4)$$

Hence,

$$\mathfrak{P}_{(\alpha_1, \ldots, \alpha_m) \in F_q^m} \left[f(\alpha_1, \ldots, \alpha_m) = 0 \right]$$
$$\leq \mathfrak{P}_{(\alpha_1, \ldots, \alpha_{m-1}) \in F_q^{m-1}} \left[g_m(\alpha_1, \ldots, \alpha_{m-1}) = 0 \right]$$
$$+ \mathfrak{P}_{(\alpha_1, \ldots, \alpha_m) \in F_q^m} \left[f(\alpha_1, \ldots, \alpha_m) = 0 : g_m(\alpha_1, \ldots, \alpha_{m-1}) \neq 0 \right]$$
$$\leq \frac{l - d_m}{q} + \frac{d_m}{q} = \frac{l}{q}. \quad (12.8.5)$$

Lemma 12.8.2. (Zippel [1979]) *Let $f \in F_q[\mathbf{x}_1, \ldots, \mathbf{x}_m]$ be a nonzero Boolean polynomial of minimum degree $\deg_{\mathbf{x}_i}(f) \leq d_i$, $1 \leq i \leq m$. Then*

$$\mathfrak{P}_{(\alpha_1, \ldots, \alpha_m) \in F_q^m} \left[f(\alpha_1, \ldots, \alpha_m) \neq 0 \right] \geq \frac{\prod_{i=1}^m (q - d_i)}{q^m}. \quad \blacksquare \quad (12.8.6)$$

PROOF. (By induction) If $m = 0$, i.e., if f is univariate, the result is true because a polynomial of degree d has at most d zeros. Let $m \neq 0$. Then in view of the decomposition (12.8.2), the multivariate polynomial f can be regarded as a univariate polynomial in $F_q[\mathbf{x}_1, \ldots, \mathbf{x}_{m-1}][\mathbf{x}_m]$, i.e., f becomes a univariate polynomial of the variable $\mathbf{x}_m$ with its coefficient taken from the field $\mathfrak{K} = F_q[\mathbf{x}_1, \ldots, \mathbf{x}_{m-1}]$, which is the field of rational functions in the variables $\mathbf{x}_1, \ldots, \mathbf{x}_{m-1}$, and we obtain at most d_m values $\beta \in \mathfrak{K}$ for which $f(\mathbf{x}_1, \ldots, \mathbf{x}_{m-1}, \beta) = 0$ in the field $\mathfrak{K}$. This means that there exist at least $q - d_m$ values $\alpha \in F_q$ that can be assigned to $\mathbf{x}_m$ such that $f(\mathbf{x}_1, \ldots, \mathbf{x}_{m-1}, \alpha)$ is a nonzero polynomial with $(m - 1)$ variables $\mathbf{x}_1, \ldots, \mathbf{x}_{m-1}$. By applying the induction hypothesis, the proof is complete. ∎

The two lemmas 12.8.1 and 12.8.2 together are called the *Schwarz–Zippel Lemma*.

Theorem 12.8.1. *Let $f \in F_q[\mathbf{x}_1, \ldots, \mathbf{x}_m] \neq 0$ be a polynomial of total degree $r = a(q - 1) + b$, $0 \leq b < q - 1$, with the minimum degree of each monomial $\mathbf{x}_i$ bounded by $(q - 1)$. Then for all $\alpha \in F_q^M$ the probability*

$$\mathfrak{P}[f(\alpha) \neq 0] \geq \frac{1}{q^a} \left(1 - \frac{b}{q} \right). \quad (12.8.7)$$

PROOF. The result (12.8.7) follows from the Schwarz–Zippel Lemma. ∎

Note that in the case of $q = 2$, the binary $RM(r, m)$ code can be defined by $RM(r, m) = \left\{ \langle f(\alpha) \rangle_{\alpha \in F_2^m} : f \text{ has total degree } \leq r \right\}$, with the block length $n = 2^m$, the dimension $k = \binom{m}{0} + \binom{m}{1} + \cdots + \binom{m}{r} \approx m^r$, and the distance 2^{m-r}. The last statement on distance was proved in §12.5. Another proof is as follows: Consider the polynomial $f(\mathbf{x}_1, \ldots, \mathbf{x}_m) = \mathbf{x}_1, \ldots, \mathbf{x}_r$. This polynomial $f \in F_2[\mathbf{x}_1, \ldots, \mathbf{x}_m]$ is a nonzero polynomial of degree r, and $f(\alpha_1, \ldots, \alpha_m) \neq 0$ only if $\alpha_1 = \alpha_2 + \cdots + \alpha_r = 1$. Since there are 2^{m-r} choices of $\alpha \in F_2^n$, we have $\text{wt}\big(\langle f(\alpha)\rangle\big)_{\alpha \in F_2^q} = 2^{m-r}$. If we can show that the weight of any nonzero codewords in $\text{RM}(r, m)$ is at least 2^{m-r}, then the distance of this code is equal to 2^{m-r}. For this purpose, consider any nonzero polynomial $f(\mathbf{x}_1, \ldots, \mathbf{x}_m)$ of total degree at most r. Rewrite f as $f(\mathbf{x}_1, \ldots, \mathbf{x}_m) = \mathbf{x}_1 \cdots \mathbf{x}_s + g(\mathbf{x}_1, \ldots, \mathbf{x}_m)$, where $s \leq r$ and $\mathbf{x}_1 \cdots \mathbf{x}_s$ is a maximum degree term in f. Now assign any values to the variables $\mathbf{x}_1, \ldots, \mathbf{x}_m$. Then the resulting polynomial with the variable $\mathbf{x}_1 \cdots \mathbf{x}_s$ is a nonzero polynomial because the terms $\mathbf{x}_1 \cdots \mathbf{x}_s$ cannot be cancelled. Hence, the resulting polynomial is nonzero for each 2^{m-s} possible assignment of values to the variables $\mathbf{x}_1, \ldots, \mathbf{x}_m$. Because a nonzero polynomial always has at least one assignment of values to its variables such that the polynomial does not vanish, then for each assignment of values $\alpha_1, \ldots, \alpha_s$ to $\mathbf{x}_1, \ldots, \mathbf{x}_s$ there will always be at least one assignment of these values to the s variables for which $f(\alpha_1, \ldots, \alpha_m) \neq 0$. However, this means that $\text{wt}(\langle f(\alpha)\rangle_{\alpha \in F_2^q} = 2^{m-s} \geq 2^{m-r}$. This shows that the distance of $\text{RM}(r, m)$ is 2^{m-r}.

12.9 Burst Errors

Although the Hamming code corrects only 1-bit errors, it can be modified to correct burst errors as follows: Suppose k codewords of length n each are transmitted. Arrange the text in matrix form, where each row is a codeword. This matrix is $k \times n$, i.e., it has k rows and n columns. The success of the method lies in transmitting this matrix column-by-column instead of the natural manner of transmitting row-by-row. Then, if a burst error of length k occurs in the entire $k \times n$ matrix (block), and there are no other errors, then at most one bit is affected in each codeword. The Hamming code can reconstruct each codeword, and thereby one-by-one the entire block. Thus, for example, the Hamming $(11, 4)$ code will have $k = 11$ with 4 check (parity bits), and this technique will make the data bits error-free from a single burst error up to length k.

Example 12.9.1. Consider the following codewords transmitted, where the parity bits are $p_1\, p_2\, p_3\, p_4$, which are prefixed to the 7-bit ASCII codes using the method of §4.3, and the order of transmission is columnwise. The boldface bits represent the burst errors that occurred.

char	ASCII Code	Transmitted $\mathbf{c}$	Received $\mathbf{w}$
C	1000011	11100000011	11110000011
o	1101111	00111001111	00101001111
d	1100100	11101001100	11111001100
i	1101001	00111011001	00101011001
n	1101110	11101011110	11111011110
g	1100111	11101010111	11111010111
␣	0100000	11011000000	11010000000
T	1010100	11100110100	11101110100
h	1101000	11101011000	11100011000
e	1100101	00111000101	00110000101
o	1101111	00111001111	00110001111
r	1110010	00111110010	00110110010
y	1111001	11101101001	11100101001 ∎

The errors in this example can be easily corrected by treating each received word $\mathbf{w}$ individually and using the SEC Hamming decoder. This example does not represent the seriousness of burst errors that occur sequentially in time and in groups. As the length of a codeword increases, the magnitude of burst errors may increase to a point when their correction can become difficult for some codes. The RM codes can deal with burst errors more effectively. A viable method is that of interleaving, which is discussed below.

12.9.1 Interleaving. To simplify the case of burst errors, the first course of action is to use the known error correcting codes to protect against short burst errors. This method, known as *interleaving*, works by spreading the bits to be transmitted throughout the entire message (sequence of codewords). We will explain the concept of interleaving by the following example.

Example 12.9.2. Consider a (31,26) code, where 26 data bits $\{0,1\}$ are represented by 26 letters 'a–z', and five parity check bits by the numbers '1–5'. For a visual representation, let these 31 bits be arranged in a circular sequence shown in Figure 12.9.1(a). Suppose that there has occurred a transmission burst of three or more consecutive errors. Since we are using a code capable of correcting only a single error in any block of 8 bits, two or more errors in any block of 8 bits will destroy the block. A remedy to avoid this situation is to skip, say, every eight locations in laying the message in the circular sequence shown in Figure 12.9.1(b), where the bits are transmitted in a clockwise order, skipping 8 locations at each step until all 31 transmitted bits are finished. Any burst of three errors must occur in eight distinct blocks, and each block can correct its single error, provided no other errors have occurred in that block. More details on interleaving are given below in §12.9.2, and also in McEliece [1977].

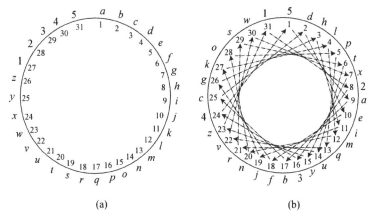

Figure 12.9.1. Interleaving for 31 transmitted bits.

It can be noticed from Figure 12.9.1(b) that in eight passes (i.e., in a full run of eight transmitted words), the errors follow the following pattern where the numbers represent the bit locations in every transmitted word as noted: $1 \to 9 \to 17 \to 25$(first word) $\to 2 \to 10 \to 18 \to 26$(second word) $\to 3 \to 11 \to 19 \to 27$(third word) $\to 4 \to 12 \to 20 \to 28$(fourth word) $\to 5 \to 13 \to 21 \to 29$(fifth word) $\to 6 \to 14 \to 22 \to 30$(sixth word) $\to 7 \to 15 \to 23 \to 31$(seventh word) $\to 8 \to 16 \to 24$(eighth word) $\to 1$, after which the process repeats with a different sequence of bits. Thus, after each eight runs the same periodic behavior of burst errors may occur. Also notice that each pass encounters four errors in each transmitted word.

If instead of eight passes, every seventh bit location is in error, we will find that there are seven runs after which the pattern starts repeating with a different sequence of bits, and each transmitted word will show four or more errors. ∎

12.9.2 Interleaving Method. The RM codes are reliably efficient in dealing with burst errors. Consider an RM (n, k) code that takes a block of k data symbols (bytes) and converts the k symbols into n encoded symbols by computing $(n - k)$ parity-check symbols, also known as check bytes, thus producing a codeword of n bytes. Such codewords are sent consecutively over a noisy channel that causes random occurrences of errors. This code has a random error correction power t, where t is a positive integer, which can correct up to t errors per block, i.e., t is the number of correctable errors per block. In the case when a channel causes burst errors in transmission either by fading or by large media or by mechanical defects, suppose that the length of burst is b, i.e., a string of b errors with the first and the bth symbol is in error, where obviously $b \le t$. However, if $b > t$, then the error correcting code will fail. In such a situation, the interleaving method becomes useful.

Assume that $b > t$ and $b \leq t \times i$, where i, known as the *interleaving depth*, is a positive integer. The $\mathrm{RM}(n, k)$ code is useful provided the burst error sequence can be spread over several code blocks such that each block has no more than t errors that are correctable. To accomplish this requirement, *block interleaving* is used as follows: Instead of encoding blocks of k symbols and then transmitting the encoded symbols consecutively, we interleave the encoded blocks and transmit the interleaved symbols. This process is explained in the following example.

Example 12.9.3. Consider the $\mathrm{RM}(255, 235)$ code, where $n = 255, k = 235$, and let $t = 10$ and $i = 5$. Then the encoded bytes output by the RM encoder would appear as

$$\boxed{a_0 \cdots a_{254} \quad | \quad b_0 \cdots b_{254} \quad | \quad c_0 \cdots c_{254} \quad | \quad d_0 \cdots d_{254} \quad | \quad e_0 \cdots e_{254}} \,,$$

where the bytes numbered 0 through 234 are the 235 data bytes, and the bytes numbered 235 through 254 are the parity check bytes. The encoded bytes are then read into the interleaver RAM row by row to construct an $i \times k$ matrix

Column # 1   2   3   ······   235   ‖   236   ······   255

$$\begin{bmatrix} a_0 & a_1 & a_2 & \cdots\cdots & a_{234} & \| & a_{235} & \cdots\cdots & a_{255} \\ b_0 & b_1 & b_2 & \cdots\cdots & b_{234} & \| & b_{235} & \cdots\cdots & b_{255} \\ c_0 & c_1 & c_2 & \cdots\cdots & c_{234} & \| & c_{235} & \cdots\cdots & c_{255} \\ d_0 & d_1 & d_2 & \cdots\cdots & d_{234} & \| & d_{235} & \cdots\cdots & d_{255} \\ e_0 & e_1 & e_2 & \cdots\cdots & e_{234} & \| & e_{235} & \cdots\cdots & e_{255} \end{bmatrix}, \qquad (12.9.1)$$

Data bytes $235 \times 5 = 1175$ Check bytes $20 \times 5 = 100$

where the rows 1 through 5 in the matrix (12.9.1) represent the codeword 1 through codeword 5, respectively. To interleave the data bytes the data is sent to an output device column by column: The data bytes output from the block interleaver is in the following form:

$$\boxed{a_0} \; \boxed{b_0} \; \boxed{c_0} \; \boxed{d_0} \; \boxed{e_0} \quad \cdots \quad \boxed{a_{254}} \; \boxed{b_{254}} \; \boxed{c_{254}} \; \boxed{d_{254}} \; \boxed{e_{254}} \,,$$

where the 1175 data bytes are followed by the final 100 check bytes.

The effect of a burst error of length $b > t$ on the received symbols in the matrix (12.9.1), where t is the number of correctable errors per block, and $b \leq \nu \times i$ for some ν, is as follows: Since the order in which the symbols are sent has already been established, a burst length $\leq \nu \times i$ will affect at most $\nu + 1$ consecutive columns of the matrix (12.9.1), depending on where the burst starts. Since a single row in the matrix (12.9.1) corresponds to a codeword, such a row will have no more than ν errors. If $\nu < t$, such errors will be corrected by the code itself. In this case, i becomes the interleaving depth. For this method to work efficiently, we would require extra buffer space to store the interleaver matrix (12.9.1), which would add some additonal cost.

The worst-case burst will always determine the size of the interleaver matrix (12.9.1) and the interleaver depth i, and these factors finally determine the additional buffer space. ■

As reported in an AHA white paper (see Paper # ANRS02_0404), the AHA4011 encoder/decoder chip was used to test the performance of RS(255,235) code. This chip is capable of correcting up to $t = 10$ bytes in error per message block (i.e., data bytes) of length up to 235 bytes. Assuming that the maximum burst size is $b = 233$ bytes, we will require $b < t \times i$ bytes, which gives the maximum interleaving depth as $i > b/t = 223/10 = 22.3$. So a value of $i = 23$ is sufficient, which gives the size of the interleaving matrix (12.9.1) as $n \times i = 255 \times 23 = 5865$ bytes. A block diagram for the performance of this chip is given in Figure 12.9.2.

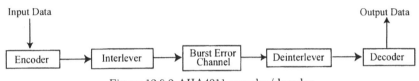

Figure 12.9.2 AHA4011 encoder/decoder.

12.9.3 Types of Interleavers. We describe three types of interleavers, as follows.

(i) **Random Interlever.** It uses a fixed random permutation and maps the input data sequence according to a predetermined permutation order. The length of the input sequence is generally assumed to be 8. A random interleaver is shown in Figure 12.9.3.

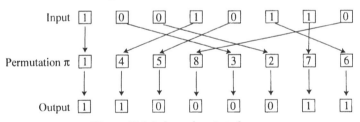

Figure 12.9.3 A random interleaver.

(ii) **Circular-Shifting Interleaver.** The permutation π of this interleaver is defined by $\pi(i) = (a\,i + s) \bmod l$, where a is the step size such that $a < l$, a being relatively prime to l, i is the index, and $s < l$ is the offset.

Figure 12.9.4 shows this type of interleaver with $a = 5$ and $s = 0$.

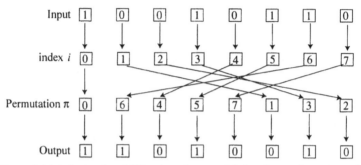

Figure 12.9.4 A circular-shifting interleaver with $l = 8, a = 5$, and $s = 0$.

(iii) Semirandom Interleaver. This interleaver is a combination of a random and a 'designed' interleaver, viz., the block and the circular-shifting interleavers. The permutation scheme for such an interleaver is as follows:

STEP 1. Choose a random index $i \in [0, l - 1]$.

STEP 2. Choose a positive integer $s < \sqrt{l/2}$.

STEP 3. Compare l to the previous values of s. For each value of s, compare the index i to see if it lies in the interval $[0, s]$. If i does lie in this interval, then go back to Step 1; otherwise, keep i.

STEP 4. Go back to Step 1 until all l positions are filled.

Figure 12.9.5 shows such an interleaver with $l = 16$ and $s = 2$.

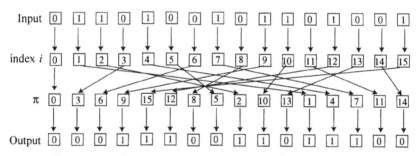

Figure 12.9.5 A semirandom interleaver with $l = 16$ and $s = 2$.

13

Reed–Solomon Codes

13.1 Definition

Reed–Solomon or $RS(n, k)$ codes are nonbinary cyclic codes composed of sequences of symbols of m-bits, where $m > 2$ is an integer, which exist for all n and k, $0 < k < n < 2^m + 2$. In this notation, n is the total number of code symbols in an encoded block (usually called *length*), and k is the number of data symbols to be encoded. Generally, $(n, k) = (2^m - 1, 2^m - 1 - 2t)$, where t is the (symbol) error correcting capability of the code. An extended RS code can have $n = 2^m$ or $n = 2^m + 1$. Among all linear codes with the same encoding input and output block lengths, RS codes have the largest possible minimum distance. Similar to the Hamming distance, the distance between two codewords in a nonbinary code is defined as the number of symbols in which the sequences differ. For RS codes, the minimum distance is given by $d_{\min} = n - k + 1$. This code can correct any combination of t or fewer errors, where $t = \left\lfloor \dfrac{d_{\min} - 1}{2} \right\rfloor = \lfloor (n - k)/2 \rfloor$. This definition of t implies that correction of t symbols using an RS code requires at most $2t$ parity symbols. In fact, the RS decoder has $(n - k)$ redundant symbols, which is equal to $2t$, i.e., twice the number of correctable errors. In the case of erasures, the correction capability of the code is $\gamma = d_{\min} - 1 = n - k$. Since RS codes are used as both error correcting and erasure codes (see §13.6), the simultaneous error-correction and erasure-correction capability of the code is expressed as $2\alpha + \gamma < d_{\min} < n - k$, where α is the number of correctable symbol-error patterns and γ the correctable number of symbol erasure patterns (see §13.6 on erasures).

The advantage of a nonbinary RS code over a binary code can be explained by the following example.

Example 13.1.1. Consider a binary $(n, k) = (7, 3)$ code. There are only $2^n = 2^7 = 128$ n-tuples, of which $2^k = 2^3 = 8$ are codewords, i.e., the number of codewords is only $1/16$ of all the n-tuples. However, in the case of a nonbinary $(n, k) = (7, 3)$ code, each symbol is composed of $m = 3$ bits, and there are $2^{mn} = 2^{21} = 2097152$ n-tuples, of which $2^{km} = 2^9 = 512$ are codewords, i.e., the number of codewords is $1/4096$ of all the n-tuples. Thus, in the nonbinary case only a small fraction of possible n-tuples are codewords, and this number decreases as m increases. By using a small fraction of n-tuples for codewords, a large $d_{\min}$ is created. ∎

A typical system takes a data source through an RS encoder that generates codewords that are transmitted through a communication channel or a storage device where noise or errors may corrupt the transmitted word. The received word is processed by an RS decoder and finally yields the data sink.

13.2 Reed–Solomon's Original Approach

Before we discuss the details of RS encoding and decoding, we will provide some details of Reed and Solomon's original approach to constructing their famous codes (Reed and Solomon [1960]). Given a set of k data symbols $\{d_0, d_1, \dots, d_{k-1}\}$ from the finite field $\mathrm{GF}(q)$, construct a polynomial $F(x) = d_0 + d_1 x + \cdots + d_{k-1} x^{k-1}$. Then an RS codeword $\mathbf{c}$ is formed by evaluating $F(x)$ at each of the q elements in the field $\mathrm{GF}(q)$:

$$\mathbf{c} = \{c_0, c_1, \dots, c_{q-1}\} = \{F(0), F(\alpha), F(\alpha^2), \dots, F(\alpha^{q-1})\}. \quad (13.2.1)$$

Thus, a complete set of codewords is constructed if the k data symbols take all the possible values. Since the data symbols belong to the field $\mathrm{GF}(q)$, they can each take q different values.

A code is said to be *linear* if the sum of any two codewords is also a codeword. Accordingly, in view of Eq (13.2.1), RS codes are linear because the sum of two polynomials of degree $(k-1)$ is again a polynomial of degree $(k-1)$.

The number of data symbols k is often referred to as the *dimension* of the code. This term alludes to the fact that RS codewords are composed of vectors (symbols) and thus form a vector space of dimension k over $\mathrm{GF}(q)$. Since each codeword has q coordinates, this quantity is called the *length* of the code; thus, $n = q$, and the RS codes of length n and dimension k are designated as $\mathrm{RS}(n, k)$ codes. Each RS codeword, defined by Eq (13.2.1), can be related to the following system of n linear equations in k variables:

$$F(0) = d_0,$$
$$F(\alpha) = d_0 + d_1 \alpha + d_2 \alpha^2 + \cdots + m_{k-1} \alpha^{k-1},$$
$$F(\alpha^2) = d_0 + d_1 \alpha^2 + d_2 \alpha^4 + \cdots + m_{k-1} \alpha^{2(k-1)},$$

$$\vdots$$

$$F(\alpha^{n-1}) = d_0 + d_1 \alpha^{n-1} + d_2 \alpha^{2(n-1)} + \cdots + m_{k-1} \alpha^{(k-1)(n-1)}. \quad (13.2.2)$$

Any k of the above equations can be used to obtain a system of k equations in k variables (unknowns). For example, the first k of the above equations form the following system, which is represented in matrix form $\mathbf{Ad} = \mathbf{F}$ as

$$
\begin{bmatrix}
1 & 0 & 0 & \cdots & 0 \\
1 & \alpha & \alpha^2 & \cdots & \alpha^{k-1} \\
F\!:\ \vdots & \vdots & \vdots & \cdots & \vdots \\
1 & \alpha^{k-1} & \alpha^{2(k-1)} & \cdots & \alpha^{(k-1)(k-1)}
\end{bmatrix}
\left\{
\begin{array}{c}
d_0 \\ d_1 \\ d_2 \\ \vdots \\ d_{k-1}
\end{array}
\right\}
=
\left\{
\begin{array}{c}
F(0) \\ F(\alpha) \\ F(\alpha^2) \\ \vdots \\ F(\alpha^{k-1})
\end{array}
\right\}.
$$

$$\tag{13.2.3}$$

This system has a unique solution for the k data symbols $d_0, d_1, \ldots, d_{k-1}$ by the *autoregressive method* (Blahut [1983]) by first finding the determinant of the coefficient matrix $\mathbf{A}$, and using the cofactor expansion across the first row of this matrix so that the determinant of this matrix reduces to that of a Vandermonde matrix, which is always nonsingular.[1]

The use of RS codes to correct errors is carried out as follows: Let t of the transmitted codeword coordinates be corrupted by noise and received with errors. Then the corresponding equations in the system (13.2.2) are incorrect and give an incorrect solution if one or more of them were used in a system of equations of the form (13.2.3). Let us assume that we are unaware of the locations and values of the errors. Then we must construct all possible distinct systems of k equations from the system (13.2.2). There are $\binom{q}{k}$ such systems of which $\binom{t+k-1}{k}$ will give incorrect data symbols. If we take the majority opinion among the solutions of all possible linear systems, we will obtain correct data bits so long as $\binom{t+k-1}{k} < \binom{q-t}{k}$. However, this condition holds iff $t + k - 1 < q - t$, which in turn holds iff $2t < q - k + 1$. Thus, a RS code of length q and dimension k can correct up to t errors, where $t = \lfloor (q - k + 1)/2 \rfloor$.

Next, suppose that some of the equations in the system (13.2.2) are missing altogether. In many digital communication systems, the demodulator can make a rough estimate as to whether a given symbol at the output of the detector is reliable or not. For example, a binary detector might consists of a simple hard limiter which makes analog signals received below a certain threshold into 0s, and those above the threshold 1s. If we know that a particular signal is close to the threshold, we may want to declare that signal being 'erased' instead of assigning a binary value that has a very low probability of being correct. When an RS codeword is erased, we have in fact deleted a cor-

---

[1] Let $A = [a_{ij}]$ be an $n \times n$ matrix. If the i-th row and j-th column of A are deleted, the remaining $(n-1)$ rows and $(n-1)$ columns can be formed into another matrix M_{ij}. If the diagonal elements of M_{ij} are the diagonal elements of A, i.e., if $i = j$, then the cofactor of A is defined as $(-1)^{i+j} \det [M_{ij}]$. Let c_{ij} denote the cofactor of a_{ij} for $i, j = 0, \ldots, n$, and let C be the matrix formed by the cofactors of A, that is, let $C = [c_{ij}]$. Then the matrix C^T is called the *adjoint* of A, written as $\mathrm{adj}(A)$, and the inverse of A is defined by
$$A^{-1} = \frac{1}{\det [A]} \, \mathrm{adj}(A).$$

responding equation from the system (13.2.2). Since we need only k correct equations in the system (13.2.2), we can erase up to $q - k$ of the codewords. Hence, combining this result with that for error correction, an RS code can correct up to t errors and v erasures as long as $2t + v < q - k + 1$.

The original Reed–Solomon approach for constructing error correcting codes was initially unsuccessful and failed to gain any support until the 'generator polynomial approach' was implemented. In fact, it was not until Tsfasman et al. [1982] that the RS codes found a proper foundation by showing that they satisfied a better bound than the Gilbert–Varshamov bound. The RS codes were then developed on the Galois field elements $\{0, \alpha, \alpha^2, \dots, \alpha^{q-1} = 1\}$, which were treated as points on a rational curve, along with the point at infinity, thus forming the one-dimensional projective line. It was also determined that the RS codes are a subset of the BCH codes.

13.3 Parity-Check Matrix

The definition in §13.1 is based on the encoding procedure. An alternative definition of $\mathrm{RS}(n, k)$ codes, based on the parity-check matrix formulation, is as follows: Let k be an integer $1 \le k < n$. Consider a primitive element α in F^*, and a set $S = \{1, \alpha, \alpha^2, \dots, \alpha^{n-1}\}$, in a field F_q^n. Then the Reed–Solomon code over F is defined by

$$\mathrm{RS}(n, k) = \big\{ (c_0, c_1, \dots, c_{n-1}) \in F^n : \mathbf{c}(x) = c_0 + c_1 x + \cdots + c_{n-1} x^{n-1}$$

$$\text{satisfies } c(\alpha) = c(\alpha^2) = \cdots = c(\alpha^{n-k}) = 0 \big\}. \qquad (13.3.1)$$

This definition implies that the codewords of $\mathrm{RS}(n, k)$ codes, which are evaluated at the points $1, \alpha, \dots, \alpha^{n-1}$, are associated with the polynomials of degree $n - 1$ and vanish at the points $\alpha, \alpha^2, \dots, \alpha^{n-k}$. This definition is found in most textbooks. In fact, a q-ary $\mathrm{RS}(n, k)$ code is a q-ary BCH code of length $q - 1$ generated by $\mathbf{g}(x) = (x - \alpha)(x - \alpha^2) \cdots (x - \alpha^{n-k})$. Note that binary $\mathrm{RS}(n, k)$ codes are never considered because their length becomes $q - 1 = 1$.

Example 13.3.1. The $\mathrm{RS}(7, 5)$ code of length 7 with generator polynomial $\mathbf{g}(x) = (x - \alpha)(x - \alpha^2) = \alpha^3 - (\alpha + \alpha^2)x + x^2 = 1 + \alpha + (\alpha^2 + \alpha)x + x^2$, where α is a root of $1 + x + x^3 \in F_2[x]$ has the generator matrix

$$\mathbf{G} = \begin{bmatrix} \alpha+1 & \alpha^2+\alpha & 1 & 0 & 0 & 0 & 0 \\ 0 & \alpha+1 & \alpha^2+\alpha & 1 & 0 & 0 & 0 \\ 0 & 0 & \alpha+1 & \alpha^2+\alpha & 1 & 0 & 0 \\ 0 & 0 & 0 & \alpha+1 & \alpha^2+\alpha & 1 & 0 \\ 0 & 0 & 0 & 0 & \alpha+1 & \alpha^2+\alpha & 1 \end{bmatrix},$$

which is obtained from $\mathbf{g}(x)$; and the parity-check matrix

$$\mathbf{H} = \begin{bmatrix} 1 & \alpha^4 & 1 & 1+\alpha^4 & 1+\alpha^4 & \alpha^4 & 0 \\ 0 & 1 & \alpha^4 & 1 & 1+\alpha^4 & 1+\alpha^4 & \alpha^4 \end{bmatrix},$$

which is obtained from $\mathbf{p}(x) = (x^7 - 1)/\mathbf{g}(x) = \alpha^4 + (1 + \alpha^4)x + (1 + \alpha^4)x^2 + x^3 + \alpha^4 x^4 + x^5$. This step can be verified directly, or by multiplying $\mathbf{g}(x)$ and $\mathbf{p}(x)$ to get $x^7 - 1$, and using formula (8.6.2) and Tables 8.6.3 and 8.6.4. ∎

13.3.1 Properties of RS Codes. The block length of the RS(n, k) code is n, and the minimum distance $n - k + 1$ (see Theorem 13.3.1 below). Thus, the encoding map (13.1.2) is injective, which implies that the code has dimension k. The following results describe certain useful properties of RS codes.

Theorem 13.3.1. *The RS(n, k) code has distance $n - k + 1$.*

The minimum distance of RS(n, k) codes satisfies the Singleton bound (§6.6). This means that although the RS(n, k) codes are simple, they need a large alphabet. Moreover, these codes are univariate, depending only on a single variable x, unlike the RM(r, m) codes, which are multivariate (see Chapter 12). Despite of all these properties, the rate of RS(n, k) codes is optimal. As mentioned in §6.6, a code for which $k + d = n + 1$ is called a *minimum distance separable code* (MDS code). Such codes, when available, are in some sense the best possible ones. Thus, RS(n, k) codes are MDS codes. The RS(7,5) code of Example 13.3.1 is an 8-array (7, 5, 3) MDS code.

Theorem 13.3.2. *The RS(n, k) codes are MDS codes.*

13.3.2 Extended RS Codes. The following theorem provides the construction of extended RS(n, k) codes which are also MDS codes:

Theorem 13.3.3. *Let C be a q-ary RS(n, k) code generated by the polynomial $p(x) = \displaystyle\prod_{i=1}^{k-1} (x - \alpha^i)$, $2 \leq a \leq q - 1$. Then the extended code $\overline{C}$ is also an MDS code.*

Example 13.3.2. The RS$(6, 3)$ code of length 6 has the generator polynomial $\mathbf{g}(x) = (x - 3)(x - 3^2)(x - 3^3) = 6 + x + 3x^2 + x^3$. The generator matrix is

$$\mathbf{G} = \begin{bmatrix} 6 & 1 & 3 & 1 & 0 & 0 \\ 0 & 6 & 1 & 3 & 1 & 0 \\ 0 & 0 & 6 & 1 & 3 & 1 \end{bmatrix},$$

and the parity-check matrix, obtained from $\mathbf{p}(x) = (x^6 - 1)/\mathbf{g}(x) = 1 + x + 4x^2 + x^3$, is

$$\mathbf{H} = \begin{bmatrix} 1 & 4 & 1 & 1 & 0 & 0 \\ 0 & 1 & 4 & 1 & 1 & 0 \\ 0 & 0 & 1 & 4 & 1 & 1 \end{bmatrix},$$

which implies that the minimum distance is 4. If the parity-check matrix of this RS$(6, 3)$ code is extended to

$$H = \begin{bmatrix} 1 & 4 & 1 & 1 & 0 & 0 & 0 \\ 0 & 1 & 4 & 1 & 1 & 0 & 0 \\ 0 & 0 & 1 & 4 & 1 & 1 & 0 \\ 1 & 1 & 1 & 1 & 1 & 1 & 1 \end{bmatrix},$$

it becomes the parity-check matrix of the extended RS code of minimum distance 5; it is also a (7,3,5)-MDS code. ∎

Example 13.3.3. The parity-check matrix of the extended RS code of Example 13.3.1 is

$$H = \begin{bmatrix} 1 & \alpha^4 & 1 & 1+\alpha^4 & 1+\alpha^4 & \alpha^4 & 0 & 0 \\ 0 & 1 & \alpha^4 & 1 & 1+\alpha^4 & 1+\alpha^4 & \alpha^4 & 0 \\ 1 & 1 & 1 & 1 & 1 & 1 & 1 & 1 \end{bmatrix},$$

which has minimum distance 4, and so it is a $(8, 5, 4)$ MDS code. ∎

13.3.3 Structure of RS Codes. The RS codes are a subset of BCH codes and are linear block codes. An RS code is specified as RS(n, k) code with s-bit symbols, i.e., the encoder takes k data symbols of s bits each and adds parity symbols to make an n symbol codeword. Thus, there are $n - k$ parity symbols of s bits each. An RS decoder can correct up to $t = (n - k)/2$ symbols that contain errors in a codeword.

A typical RS codeword, known as a *systematic code* because the data is left unchanged and the parity symbols are appended, is shown below, where one of the structures A or B can be chosen:

Total Symbols n

A:	2t Parity Symbols **p**		k Data Symbols **d**

B:	k Data Symbols **d**		2t Parity Symbols **p**

Example 13.3.4. Consider the RS$(255, 223)$ code with 8-bit symbols. Each codeword contains 255 codeword bytes, of which 223 bytes are data and 32 bytes are parity. For this code $n = 255, k = 223, s = 8, t = 16$. The decoder can correct up to 16 symbol errors, i.e., errors up to 16 bytes anywhere in the codeword can be automatically corrected. Given the symbol size s, the maximum length n for an RS code is given by $n = 2^s - 1$. For example, the maximum code length of an 8-bit code is 255 bytes. ∎

RS codes may be shortened by (conceptually) making a number of data symbols zero at the encoder, not transmitting them, and then re-inserting them at the decoder.

Example 13.3.5. The RS$(255, 223)$ code of Example 13.3.4 can be shortened to $(200, 168)$. The encoder takes a block of 168 data bytes, (conceptually) adds 55 zero bytes, creates a $(255, 223)$ codeword, and transmits only 168 data bytes and 32 parity bytes. ∎

The amount of *processing power* required to encode and decode RS codes is related to the number of parity symbols per codeword. A large value of t means that a large number of errors can be corrected but it will require more computational power.

13.3.4 Symbol Errors. One symbol error occurs when one bit in a symbol is incorrect or when all bits in a symbol are incorrect. For example, RS(255, 223) can correct up to 16 symbol errors. In the worst case, 16 errors may occur, each in a separate symbol (byte) so that the decoder corrects 16 bit errors. In the best case, 16 complete byte errors occur so that the eecoder corrects 16×8 bit errors. RS codes are particularly suited for correcting burst errors; in this situation a series of bits in the codeword is received in error.

13.3.5 Solution for Large Alphabet Size. Reed–Solomon codes were introduced by Reed and Solomon [1960] and they are still very popular as they are used in many applications, namely, the storage of digital data on hard drives and optical disks such as CDs and DVDs. In addition, RS codes correct more errors than other codes and that of burst errors (discussed below in §13.5). A basic reason for their popularity is that they are optimal codes. However, a major problem with these codes is that they require a large alphabet size. There is no way to avoid it since these codes achieve the Singleton bound, and therefore must need a large alphabet. This feature demands the solution of an important question: If the RS codes operate on bits, how should the codewords be converted over the large field in the binary alphabet $A = \{0, 1\}$. Suppose an RS code is defined over the field $F_{2^8} = F_{256}$. Then can an element in this field be written as an 8-bit vector? This question can be formally posed as follows: Suppose a message is associated with the polynomial $p(x) \in F[x]$. Then its encoding in the RS code will be a set of values $\{p(\alpha_1), p(\alpha_2), \dots, p(\alpha_n)\}$. These values can be expressed in the alphabet A with $\log |F|$ bits each using the logarithmic scale. Thus, this simple transformation yields a code over the field F_2 provided that the RS code is defined over a field that is an extension field of F_2. In fact, the bit vectors are represented precisely in this manner so that the resulting code is a binary linear code. This method is used in practice, as the following example shows, which also answers the question about the error correction capabilities of the resulting binary code.

Example 13.3.6. Consider an RS$(256, 230)$ code, with $n = 256$ and $k = 230$. By theorem 13.1.1, the distance of the code is $d = n - k + 1 = 27$, so this code can correct $\lfloor d/2 \rfloor = 13$ errors. The transformation to a binary code yields a *binary* RS (n', k') code, where $n' = 256 \times 8$-bits and $k' = 230 \times 8$-bits. This is simply a scaling transformation with the distance $d' \geq d = 27$, so it can correct at least 13 errors. ∎

This example can be generalized as follows: Suppose we have an (N, K, D) code of length N over F, where $N = 2^m$. Then the above transformation will yield an $(N \log N, K \log K, D')$ binary linear code over F_2, where $D' \geq D$. If we set $n = N \log N$, and consider the case where $K + N - D + 1$, the transformation of a RS code to a binary linear code yields an $(n, n - (D - 1) \log N, D' \geq D)$ code over F_2. The resulting code has a descent rate that is not optimal, and it can be compared to the BCH codes which are $(n, n - \lceil \frac{D-1}{2} \rceil \log(n + 1), D' \geq D)$ codes. Despite of all these shortcomings, the RS codes are still very popular because they are the best at correcting burst errors as a burst of consecutive errors affect the bits that correspond to a much smaller number of elements in the field of definition of RS codes.

13.4 RS Encoding and Decoding

Theoretical decoders were described by Reed and Solomon [1960] who established that such decoders corrected errors by finding the most popular message polynomial. The decoder for an $RS(n, k)$ code would examine all possible subsets of k symbols from the set of n symbols that were received. In general, the correctability of an RS code would require that at least k symbols be received correctly, because the k symbols are needed to interpolate the message polynomial $\mathbf{c}(x)$. The decoder would interpolate a message polynomial for each subset and keep track of the resulting polynomial. The most popular message is the corrected result. However, there are a large number of subsets. For example, the number of subsets is the binomial coefficient $\binom{n}{k} = \dfrac{n!}{k!\,(n-k)!}$, which becomes huge for a modest code; e.g., for the simple $RS(7, 3)$ code that can correct up to two errors, the theoretical decoder would examine $\binom{7}{3} = 35$ subsets, and for another simple $RS(255, 249)$ code that can correct up to $(255 - 249)/2 = 3$ errors, the theoretical decoder would examine $\binom{255}{249} > 359 \times 10^9$ subsets. So the Reed–Solomon algorithm became impractical and needed a practical decoder. Finally, the Peterson algorithm (Peterson [1960]) was developed as a practical decoder based on syndrome decoding, and subsequently improved upon by Berlekamp [1968].

The RS encoder takes a block of digital data and adds extra 'redundant' bits. Error occurs during transmission or storage for a number of reasons, e.g., noise or interference, scratches on a CD, and so on. The RS decoder processes each block and attempts to correct errors and recover the original data. The number and type of errors that can be corrected depend on the characteristics of the RS code used.

For decoding RS codes, the Berlekamp–Massey algorithm works as follows. The codeword polynomial $\mathbf{c}(x)$ is divisible by the generator polynomial $\mathbf{g}(x)$. After transmission, the received word $\mathbf{w}(x)$ has errors, so $\mathbf{w}(x) = \mathbf{c}(x) +$

$e(x)$. The polynomial $w(x)$ is evaluated at the roots of $g(x)$. Since $g(x)$ divides $c(x)$, the polynomial $c(x)$ will have no effect on the syndrome, and therefore is eliminated. Only $e(x)$ affects the syndrome, so use the syndrome to determine $e(x)$, which in turn will be used to reconstruct $c(x) = w(x) - e(x)$. This algorithm is useful in finding the error-location polynomial $L(x)$. The roots of $L(x)$ identify the nonzero terms of $e(x)$. The roots are found by the Chien search algorithm. Once the error locations are known, the error values are found by the Forney algorithm. Thus, knowing the error locations and their values, the original message is reconstructed. Details can be found in Berlekamp [1968] and Massey [1969]. The latest practical decoder is the modified Peterson algorithm which is described below.

13.4.1 Modified Peterson Algorithm. We follow the procedure developed by Peterson [1961], Berlekamp [1968], Massey [1969], Chien [1964] and Forney [1965] for encoding and decoding of RS codes for correcting transmission errors, first by constructing the generator polynomial and the parity polynomial, which together will provide the codeword polynomial, and then providing the details of the decoding scheme. Consider $RS(n, k)$ codes in which the parameters n, k, t and any positive integer $m > 2$ are defined as $(n, k) = (2^m - 1, 2^m - 1 - 2t)$, where $n - k = 2t$ are the parity-check symbols, and t denotes the error correction capability of the code. The generating polynomial of an $RS(n, k)$ code is given by

$$g(x) = g_0 + g_1 x + g_2 x^2 + \cdots + g_{2t-1} x^{2t-1} + x^{2t}. \tag{13.4.1}$$

Obviously, the degree of the polynomial $g(x)$ is equal to the number $2t = n - k$ of the parity symbols. Hence, there must be exactly $2t$ successive powers of α that are the roots of the polynomial $g(x)$. Let these roots be denoted by $\alpha, \alpha^2, \ldots, \alpha^{2t}$, where the order of these roots in not important. Thus, the generator polynomial $g(x)$ can be written as

$$g(x) = (x - \alpha)(x - \alpha^2) \cdots (x - \alpha^{2t}), \tag{13.4.2}$$

which must be simplified using the addition and multiplication lookup tables for $GF(2^m)$. The simplified form of $g(x)$ must be checked to ensure that $g(\alpha^i) = 0$ for all $i = 1, \ldots, 2t$.

Next, we consider the data (or message) polynomial $d(x)$ composed of k symbols that are appended to the $(n - k)$ parity-symbol polynomial $p(x)$. This is in conformity with the codeword structure A presented in §13.3.3.[2] Next, to right-shift the data polynomial $d(x)$ by $n - k$ positions, we multiply (upshift) $d(x)$ by x^{n-k}, and divide $x^{n-k}d(x)$ by the generator polynomial $g(x)$, which gives

$$x^{n-k}d(x) = q(x)g(x) + p(x), \tag{13.4.3}$$

---

[2] The structure B can also be chosen, but then the entire discussion must conform to the succession of parity symbols following the data symbols.

where $\mathbf{q}(x)$ and $\mathbf{p}(x)$ are the quotient and remainder polynomials, respectively. Note that the remainder is in parity, as in the binary case. Eq (13.4.3) can be expressed as $\mathbf{p}(x) = x^{n-k}\mathbf{d}(x)$ (mod $\mathbf{g}(x)$), so that the codeword can be written as

$$\mathbf{c}(x) = \mathbf{p}(x) + x^{n-k}\mathbf{d}(x). \tag{13.4.4}$$

Then, $x^{n-k}\mathbf{d}(x)$ is divided by $\mathbf{g}(x)$, to obtain the quotient $\mathbf{q}(c)$ and the remainder $\mathbf{p}(x)$, which is the parity polynomial. Hence, using (13.4.4), we obtain the codeword polynomial $\mathbf{c}(x)$. At this point it is advisable to check the accuracy of the encoded codeword by verifying that $\mathbf{c}(\alpha) = \mathbf{c}(\alpha^2) = \cdots = \mathbf{c}(\alpha^{2t}) = 0$. This completes the encoding part.

Now we discuss RS decoding, by assuming that during transmission the n-symbol codeword gets corrupted with j errors of unknown type and location. Let the error polynomial $\mathbf{e}(x)$ have the form $\mathbf{e}(x) = \sum\limits_{i=0}^{j} e_i x^i$. Thus, the received word polynomial $\mathbf{w}(x)$ is represented as the sum of transmitted codeword polynomial $\mathbf{c}(x)$ and the error polynomial $\mathbf{e}(x)$: $\mathbf{w}(x) = \mathbf{c}(x) + \mathbf{e}(x)$.

To determine the errors, the *syndrome computation* is needed, which is accomplished using the Chien search as follows (see Chien [1964]). As mentioned in §3.4.13, the syndrome results from a parity check performed on the received word $\mathbf{w}(x)$, and determines whether $\mathbf{w}(x)$ belongs to of the set of codewords. If $\mathbf{w}$ is a valid member of this set, then the syndrome vector $\mathbf{s}$ has value $\mathbf{0}$, and any nonzero value of $\mathbf{s}$ means that an error has occurred. The syndrome $\mathbf{s}$ consists of $(n-k)$ symbols $\{s_i\}$, $i = 1, 2, \ldots, n-k$. Thus, a syndrome symbol can be computed from the formula

$$s_i = \mathbf{w}(x)\Big|_{x=\alpha^i} = \mathbf{w}(\alpha^i), \quad i = 1, 2, \ldots, n-k, \tag{13.4.5}$$

where a lookup addition table can be used. One nonzero value of s_i will signal the presence of an error. Alternatively, this information can be obtained using the fact that each element in a coset (row) in the standard array has the same syndrome (Sklar [2001]). Thus, since $s_i = \big[\mathbf{c}(x) + \mathbf{e}(x)\big]_{x=\alpha^i}$, $i = 0, 1, 2, \ldots, n-k = 4$, we find from (13.4.5) that $s_i = \mathbf{w}(\alpha^i) = \mathbf{c}(\alpha^i) = \mathbf{e}(\alpha^i) = 0 + \mathbf{e}(\alpha^i)$.

To determine the *error locations*, suppose that there are ν errors in the received codeword at the locations $x^{j_1}, x^{j_2}, \ldots, x^{j_\nu}$. Then

$$\mathbf{e}(x) = e_{j_1} x^{j_1} + e_{j_2} x^{j_2} + \cdots + e_{j_\nu} x^{j_\nu}, \tag{13.4.6}$$

where the indices $j = 1, 2, \ldots, \nu$ refer to the first, second, $\ldots$, ν-th location, respectively. We will determine each e_{j_l} and x^{j_l}, $l = 1, 2, \ldots, \nu$. Let $\beta_l = \alpha_{j_l}$ define an error location number. Then obtain $n - k = 2t$ syndrome symbols from (13.4.5) for $i = 1, 2, \ldots, 2t$, that is,

$$s_1 = \mathbf{w}(\alpha) = e_{j_1}\beta_1 + e_{j_2}\beta_2 + \cdots + e_{j_\nu}\beta_\nu,$$
$$s_2 = \mathbf{w}(\alpha^2) = e_{j_1}\beta_1^2 + e_{j_2}\beta_2^2 + \cdots + e_{j_\nu}\beta_\nu^2,$$
$$\vdots$$
$$s_{2t} = \mathbf{w}(\alpha^{2t}) = e_{j_1}\beta_1^{2t} + e_{j_2}\beta_2^{2t} + \cdots + e_{j_\nu}\beta_\nu^{2t}. \tag{13.4.7}$$

In this system of equations there are $2t$ unknowns (t errors and t error locations). Since these $2t$ equations are nonlinear, we will use an RS *decoding algorithm*, as follows: Define an error-location polynomial as

$$\mathbf{L}(x) = \prod_{i=1}^{\nu}(1 + \beta_i x) = 1 + L_1 x + L_2 x^2 + \cdots + L_\nu x^\nu. \tag{13.4.8}$$

The roots of $\mathbf{L}(x)$ are $1/\beta_i$, $i = 1, 2, \ldots, \nu$. The reciprocal of these are the error-location numbers of the error polynomial $\mathbf{e}(x)$. Thus, we use the autoregressive method, defined above in §13.1 (see Footnote 1), and form a syndrome matrix, where the first t syndromes are used to predict the next syndrome:

$$\begin{bmatrix} s_1 & s_2 & s_3 & \cdots & s_{t-1} & s_t \\ s_2 & s_3 & s_4 & \cdots & s_t & s_{t+1} \\ \vdots & & & \cdots & & \vdots \\ s_{t-1} & s_t & s_{t+1} & \cdots & s_{2t-3} & s_{2t-2} \\ s_t & s_{t+1} & s_{t+2} & \cdots & s_{2t-2} & s_{2t-1} \end{bmatrix} \begin{Bmatrix} L_t \\ L_{t-1} \\ \vdots \\ L_2 \\ L_1 \end{Bmatrix} = \begin{Bmatrix} -s_{t+1} \\ -s_{t+2} \\ \vdots \\ -s_{2t-1} \\ -s_{2t} \end{Bmatrix}. \tag{13.4.9}$$

This matrix system is then applied using the largest dimensional matrix that is nonsingular (i.e., has a nonzero determinant). To solve for the coefficients $L_1, \ldots, L_t$, we first determine $M^{-1} = \dfrac{\mathrm{adj}(M)}{\det(M)}$. It is advisable to ensure the accuracy of M^{-1} by checking that $MM^{-1} = I$.

Once M^{-1} is determined, Eq (13.4.9) is solved to obtain the values of $L_1, \ldots, L_\nu$. The error-locator polynomial $\mathbf{L}(x)$ is constructed as a monic polynomial $\mathbf{L}(x)$ defined by (13.4.8). Since the roots of $\mathbf{L}(x)$ are reciprocals of the error locations, we test $\mathbf{L}(x)$ with each field element $1, \alpha, \alpha^2, \ldots, \alpha^{2t}$. Any one of these elements that yields $\mathbf{L}(x) = 0$ is a root and therefore provides the location of an error in that particular element. At the error locations we take $\beta_{l_\nu} = \alpha^{j_\nu}$.

Finally, to determine the *error values* at the known error locations, we use the Forney algorithm: Let us denote e_{j_ν} by $e_1\nu$, and the error locations by $\beta_\nu = \alpha^\nu$. Then use any ν of the $2t$ syndrome equations (13.4.5) and compute

s_ν, which can be written in matrix form as

$$
\begin{bmatrix}
\beta_1 & \beta_2 & \cdots & \beta_\nu \\
\beta_1^2 & \beta_2^2 & \cdots & \beta_\nu^2 \\
\cdots & \cdots & \vdots & \cdots \\
\beta_1^\nu & beta_2^\nu & \cdots & \beta_{n}u^\nu
\end{bmatrix}
\begin{Bmatrix}
e_1 \\ e_2 \\ \vdots \\ e^\nu
\end{Bmatrix}
=
\begin{Bmatrix}
s_1 \\ s_2 \\ \vdots \\ s_\nu
\end{Bmatrix} .
$$

This matrix can be solved for e_i by again using the autoregressive method, as follows: Let Q denote the square matrix on the left side. Then $Q^{-1} = \dfrac{\mathrm{adj}(Q)}{\det(Q)}$ and the error values e_i are determined. Thus, we obtain the estimated error polynomial

$$
\hat{e}(x) = e_1 x^{j_1} + e_2 x^{j_2} + \cdots + e_\nu x^{j_\nu} . \tag{13.4.10}
$$

The received codeword polynomial is then estimated to be

$$
\hat{c}(x) = c(x) + \hat{e}(x) = c(x) + e(x) + \hat{e}(x), \tag{13.4.11}
$$

which gives all the corrected decoded data symbols.

The above procedure is explained by means of an example.

Example 13.4.1. Consider the RS(7, 3) code, for which $2t = 4$, so it is a double-error correcting code. Also, it has 4 roots, say, $\alpha, \alpha^2, \alpha^3, \alpha^4$. Thus, the generator polynomial (13.4.1) for this code is

$$
\begin{aligned}
\mathbf{g}(x) &= (x - \alpha)(x - \alpha^2)(x - \alpha^3)(x - \alpha^4) \\
&= \left[x^2 - (\alpha + \alpha^2)x + \alpha^3 \right] \left[x^2 - (\alpha^3 + \alpha^4)x + \alpha^7 \right] \\
&= \left[x^2 - \alpha^4 x + \alpha^3 \right] \left[x^2 - \alpha^6 x + \alpha^7 \right] \\
&= x^4 - (\alpha^4 + \alpha^6)x^3 + (\alpha^3 + \alpha^{10} + \alpha^0)x^- (\alpha^4 + \alpha^9)x + \alpha^3 \\
&= x^4 - \alpha^3 x^3 + \alpha^0 x^2 - \alpha^1 x + \alpha^3 \\
&= \alpha^3 + \alpha x + x^2 + \alpha^3 x^3 + x^4, \tag{13.4.12}
\end{aligned}
$$

where we have used formulas (8.6.2) and Table 8.6.3.

For the purpose of discussing the encoding we take specifically a non-binary 3-symbol data $\underbrace{010}_{\alpha} \ \underbrace{110}_{\alpha^3} \ \underbrace{111}_{\alpha^5}$, and we multiply the data polynomial $\mathbf{d}(x)$ by $x^{n-k} = x^4$, which gives

$$
x^4 \mathbf{d}(x) = x^4 \left(\alpha + \alpha^3 x + \alpha^5 x^2 \right) = \alpha x^4 + \alpha^3 x^5 + \alpha^5 x^6 .
$$

Then, by dividing $x^4 \mathbf{d}(x)$ by $\mathbf{g}(x)$ we get

$$\alpha^5 x^2 + x + \alpha^4$$

$$x^4 + \alpha^3 x^3 + x^2 + \alpha x + \alpha^3 \,\big|\, \alpha^5 x^6 + \alpha^3 x^5 + \alpha x^4$$

$$\alpha^5 x^6 + \alpha x^5 + \alpha^5 x^4 + \alpha^6 x^3 + \alpha x^2$$

$$x^5 + \alpha^6 x^4 + \alpha^6 x^3 + \alpha x^2$$
$$x^5 + \alpha^3 x^4 + x^3 + \alpha x^2 + \alpha^3 x$$

$$\alpha^4 x^4 + \alpha^2 x^3 + \quad + \alpha^3 x$$
$$\alpha^4 x^4 + x^3 + \alpha^4 x^2 + \alpha^5 x + 1$$

$$\alpha^6 x^3 + \alpha^4 x^2 + \alpha^2 x + 1$$

which gives $\mathbf{q}(x) = \alpha^4 + x + \alpha^5 x^2$, and $\mathbf{p}(x) = 1 + \alpha^2 x + \alpha^4 x^2 + \alpha^6 x^3$. Hence, in view of (13.4.4), the codeword polynomial is

$$\mathbf{c}(x) = \sum_{i=0}^{6} c_i x^i = 1 + \alpha^2 x + \alpha^4 x^2 + \alpha^6 x^3 + \alpha x^4 + \alpha^3 x^5 + \alpha^5 x^6$$

$$= (100) + (001)x + (011)x^2 + (101)x^3 + (10)x^4 + (110)x^5 + (111)x^6.$$
$$(13.4.13)$$

To ensure that the encoding is correct, a check on the accuracy of the encoded codeword can be implemented by using the following criterion: The roots of the generator polynomial $\mathbf{g}(x)$ must be the same as those of codeword $\mathbf{c}(x)$, since $\mathbf{c}(x) = \mathbf{d}(x)\mathbf{g}(x)$. We must verify that $\mathbf{c}(\alpha) = \mathbf{c}(\alpha^2) = \mathbf{c}(\alpha^3) = \mathbf{c}(\alpha^4) = 0$. By using Tables 8.6.3 and 8.6.4, for example, $\mathbf{c}(\alpha) = 1 + \alpha^3 + \alpha^6 + \alpha^9 + \alpha^5 + \alpha^8 + \alpha^{11} = 1 + \alpha^3 + \alpha^6 + \alpha^2 + \alpha^5 + \alpha + \alpha^4 = \alpha + 1 + \alpha^6 + \alpha^4 = \alpha^3 + \alpha^3 = 0$, and so for $\mathbf{c}(\alpha^2), \mathbf{c}(\alpha^3)$, and $\mathbf{c}(\alpha^4)$. This verification is left as an exercise.

For RS decoding, using the above example of an RS$(7, 3)$ code, we will assume that during transmission the 7-symbol codeword gets corrupted with two-symbol errors. Let the error polynomial $\mathbf{e}(x)$ have the form $\mathbf{e}(x) = \sum_{i=0}^{6} e_i x^i$, and suppose that these two errors consist of one parity error, say a 1-bit error in the coefficient of x^3 such that α^6 (101) changes to α^2 (001), and one data-symbol error, say a 2-bit error in the coefficient of x^4 such that α (010) changes to α^5 (111), i.e., the two-symbol error can be represented as

$$\mathbf{e}(x) = 0 + 0x + 0x^2 + \alpha^2 x^3 + \alpha^5 x^4 + 0x^5 + 0x^6$$

$$= (000) + (000)x + (000)x^2 + (001)x^3 + (111)x^4 + (000)x^5 + (000)x^6.$$
$$(13.4.14)$$

Thus, the received word polynomial $\mathbf{w}(x)$ is represented as the sum of the transmitted codeword polynomial $\mathbf{c}(x)$ and the error polynomial $\mathbf{e}(x)$: $\mathbf{w}(x) = \mathbf{c}(x) + \mathbf{e}(x)$, which, in this example, from (13.4.13) and (13.4.14) becomes

$$\mathbf{w}(x) = (100) + (001)x + (011)x^2 + (100)x^3 + (101)x^4 + (110)x^5 + (111)x^6$$

$$= 1 + \alpha^2 x + \alpha^4 x^2 + x^3 + \alpha^6 x^4 + \alpha^3 x^5 + \alpha^5 x^6.$$
$$(13.4.15)$$

There are four unknowns: 2 error locations and 2 error values, so four equations are needed. To determine the error locations, the *syndrome computation* is needed. As mentioned above, the syndrome results from a parity check performed on the received word $\mathbf{w}(x)$ and determines whether $\mathbf{w}(x)$ belongs to of the set of codewords. If $\mathbf{w}$ is a valid member of this set, then the syndrome vector $\mathbf{s}$ has value $\mathbf{0}$, and any nonzero value of $\mathbf{s}$ means that an error has occurred. The syndrome $\mathbf{s}$ consists of $(n-k)$ symbols $\{s_i\}$, $i = 1, 2, \ldots, n-k = 4$ and is computed from

$$s_i = \mathbf{w}(x)\Big|_{x=\alpha^i} = \mathbf{w}(\alpha^i), \quad i = 1, 2, \ldots, 4, \tag{13.4.16}$$

which yields (using Table 8.6.3)

$$s_1 = \mathbf{w}(\alpha) = 1 + \alpha^3 + \alpha^6 + \alpha^3 + \alpha^{10} + \alpha^8 + \alpha^{11} = \alpha^3,$$
$$s_2 = \mathbf{w}(\alpha^2) = 1 + \alpha^4 + \alpha^8 + \alpha^6 + \alpha^{14} + \alpha^{13} + \alpha^{17} = \alpha^5,$$
$$s_3 = \mathbf{w}(\alpha^3) = 1 + \alpha^5 + \alpha^{10} + \alpha^9 + \alpha^{18} + \alpha^{23} = \alpha^6,$$
$$s_4 = \mathbf{w}(\alpha^4) = 1 + \alpha^6 + \alpha^{12} + \alpha^{12} + \alpha^{22} + \alpha^{23} + \alpha^{29} = 0.$$

Alternatively, since $s_i = \big[\mathbf{c}(x) + \mathbf{e}(x)\big]_{x=\alpha^i}$, $i = 0, 1, 2, \ldots, 4$, and since $\mathbf{e}(x) = \alpha^2 x^3 + \alpha^5 x^4$, we find from (13.4.16) that

$$s_1 = \mathbf{e}(\alpha) = \alpha^5 + \alpha^9 = \alpha^5 + \alpha^2 = \alpha^3,$$
$$s_2 = \mathbf{e}(\alpha^2) = \alpha^8 + \alpha^{13} = \alpha + \alpha^6 = \alpha^5,$$
$$s_3 = \mathbf{e}(\alpha^3) = \alpha^{11} + \alpha^{17} = \alpha^4 + \alpha^3 = \alpha^6,$$
$$s_4 = \mathbf{e}(\alpha^4) = \alpha^{14} + \alpha^{21} = 1 + 1 = 0.$$

To determine the *error locations*, since there are 2 errors in the received codeword at the locations x^3 and x^4, we find from (13.4.6) that $\mathbf{e}(x) = e_1 x^3 + e_2 x^4$. To determine the error locations L_1 and L_2, we notice that the matrix equation (13.4.9) reduces to two 2×2 matrices

$$\begin{bmatrix} s_1 & s_2 \\ s_2 & s_3 \end{bmatrix} \begin{Bmatrix} L_2 \\ L_1 \end{Bmatrix} = \begin{Bmatrix} s_3 \\ s_4 \end{Bmatrix}, \tag{13.4.17}$$

$$\begin{bmatrix} \alpha^3 & \alpha^5 \\ \alpha^5 & \alpha^6 \end{bmatrix} \begin{Bmatrix} L_2 \\ L_1 \end{Bmatrix} = \begin{Bmatrix} \alpha^6 \\ 0 \end{Bmatrix}. \tag{13.4.18}$$

To solve for the coefficients L_1 and L_2, we use the autoregressive method: Let M denote the 2×2 matrix on the left side of (13.4.18). Then

$$\det(M) = \begin{vmatrix} \alpha^3 & \alpha^5 \\ \alpha^5 & \alpha^6 \end{vmatrix} = \alpha^3\alpha^6 - \alpha^5\alpha^5 = \alpha^5 + \alpha^{10} = \alpha^2 + \alpha^3 = \alpha^5,$$

Since $\text{adj}(M) = \begin{bmatrix} \alpha^6 & \alpha^5 \\ \alpha^5 & \alpha^3 \end{bmatrix}$, we get

$$M^{-1} = \frac{\text{adj}(M)}{\det(M)} = \frac{\begin{bmatrix} \alpha^6 & \alpha^5 \\ \alpha^5 & \alpha^3 \end{bmatrix}}{\alpha^5} = \alpha^{-5} \begin{bmatrix} \alpha^6 & \alpha^5 \\ \alpha^5 & \alpha^3 \end{bmatrix}$$

$$= \alpha^2 \begin{bmatrix} \alpha^6 & \alpha^5 \\ \alpha^5 & \alpha^3 \end{bmatrix} = \begin{bmatrix} \alpha^8 & \alpha^7 \\ \alpha^7 & \alpha^5 \end{bmatrix} = \begin{bmatrix} \alpha & 1 \\ 1 & \alpha^5 \end{bmatrix} \text{ since } \alpha^5 \alpha^2 = 1 \text{ (Table 8.6.4).}$$

Once M^{-1} is determined,[3] we can solve (13.4.18) to get

$$\left\{ \begin{matrix} L_2 \\ L_1 \end{matrix} \right\} = \begin{bmatrix} \alpha & 1 \\ 1 & \alpha^5 \end{bmatrix} \left\{ \begin{matrix} \alpha^6 \\ 0 \end{matrix} \right\} = \left\{ \begin{matrix} \alpha^7 \\ \alpha^6 \end{matrix} \right\} = \left\{ \begin{matrix} 1 \\ \alpha^6 \end{matrix} \right\},$$

which determines the error-locator polynomial (13.4.8) as

$$\mathbf{L}(x) = 1 + L_1 x + L_2 x^2 = 1 + \alpha^6 x + x^2. \tag{13.4.19}$$

Since the roots of $\mathbf{L}(x)$ are reciprocals of the error locations, we test $\mathbf{L}(x)$ with each field element $1, \alpha, \alpha^2, \dots, \alpha^6$. Any one of these elements that yields $\mathbf{L}(x) = 0$ is a root and therefore provides the location of an error. Thus,

$\mathbf{L}(1) = 1 + \alpha^6 + (1)^2 = \alpha^6 \neq 0,$

$\mathbf{L}(\alpha) = 1 + \alpha^6 \alpha + (\alpha)^2 = 1 + \alpha^7 + \alpha^2 = 1 + 1 + \alpha^2 = \alpha^2 \neq 0,$

$\mathbf{L}(\alpha^2) = 1 + \alpha^6 \alpha^2 + (\alpha^2)^2 = 1 + \alpha^8 + \alpha^4 = 1 + \alpha + \alpha^4 = \alpha^6 \neq 0,$

$\mathbf{L}(\alpha^3) = 1 + \alpha^6 \alpha^3 + (\alpha^3)^2 = 1 + \alpha^9 + \alpha^6 = 1 + \alpha^2 + \alpha^6 = \alpha^6 + \alpha^6 = 0, \leftarrow \text{Error},$

$\mathbf{L}(\alpha^4) = 1 + \alpha^6 \alpha^4 + (\alpha^4)^2 = 1 + \alpha^{10} + \alpha^8 = 1 + \alpha^3 + \alpha^6 = \alpha + \alpha = 0, \leftarrow \text{Error},$

$\mathbf{L}(\alpha^5) = 1 + \alpha^6 \alpha^5 + (\alpha^5)^2 = 1 + \alpha^{11} + \alpha^{10} = 1 + \alpha^4 + \alpha^3 = \alpha^5 + \alpha^3 = \alpha^2 \neq 0,$

$\mathbf{L}(\alpha^6) = 1 + \alpha^6 \alpha^6 + (\alpha^6)^2 = 1 + \alpha^{12} + \alpha^{12} = 1 \neq 0.$

Hence, $\mathbf{L}(\alpha^3) = 0$ means that one root exists at $1/\beta_1 = \alpha^3$, which gives $\beta_1 = \alpha^{-3} = \alpha^4$ (since $\alpha^3 \alpha^4 = 1$). Similarly, $\mathbf{L}(\alpha^4) = 0$ means one error is located at $1/\beta_2 = \alpha^4$, which gives $\beta_2 = \alpha^{-4} = \alpha^3$. So we have found two-symbols errors with the error polynomial

$$\mathbf{e}(x) = e_1 x^3 + e_2 x^4. \tag{13.4.20}$$

---

[3] It is advisable to ensure the accuracy of M^{-1} by checking that $MM^{-1} = I$: thus,

$$\begin{bmatrix} \alpha^3 & \alpha^5 \\ \alpha^5 & \alpha^6 \end{bmatrix} \begin{bmatrix} \alpha & 1 \\ 1 & \alpha^5 \end{bmatrix} = \begin{bmatrix} \alpha^4 + \alpha^5 & \alpha^3 + \alpha^{10} \\ \alpha^6 + \alpha^6 & \alpha^1 1 \end{bmatrix} = \begin{bmatrix} 1 & 0 \\ 0 & 1 \end{bmatrix}.$$

Thus, we find that the errors are located at locations α^3 and α^4.

Finally, to determine the *error values* at these two error locations $\beta_1 = \alpha^3$ and $\beta_2 = \alpha^4$, we use any two of the four syndrome equations (13.4.16), say s_1 and s_2:

$$s_1 = \mathbf{w}(a) = e_1\beta_1 + e_2\beta_2, \quad s_2 = \mathbf{w}(\alpha^2) = e_1\beta^2 + e_2\beta_2^2,$$

which in matrix form is written as

$$\begin{bmatrix} \beta_1 & \beta_2 \\ \beta_1^2 & \beta_2^2 \end{bmatrix} \begin{Bmatrix} e_1 \\ e_2 \end{Bmatrix} = \begin{Bmatrix} s_1 \\ s_2 \end{Bmatrix}, \quad \text{or} \quad \begin{bmatrix} \alpha^3 & \alpha^4 \\ \alpha^6 & \alpha^8 = \alpha \end{bmatrix} \begin{Bmatrix} e_1 \\ e_2 \end{Bmatrix} = \begin{Bmatrix} \alpha^3 \\ \alpha^5 \end{Bmatrix}.$$

$$(13.4.21)$$

Let Q denote the square matrix on the left side, i.e., $Q = \begin{bmatrix} \alpha^3 & \alpha^4 \\ \alpha^6 & \alpha \end{bmatrix}$. Then

$$Q^{-1} = \frac{\text{adj}(Q)}{\det(Q)} = \frac{\begin{bmatrix} \alpha & \alpha^4 \\ \alpha^6 & \alpha^3 \end{bmatrix}}{\alpha^3\alpha - \alpha^6\alpha^4} = \frac{\begin{bmatrix} \alpha & \alpha^4 \\ \alpha^6 & \alpha^3 \end{bmatrix}}{\alpha^4 + \alpha^{10}} = \frac{\begin{bmatrix} \alpha & \alpha^4 \\ \alpha^6 & \alpha^3 \end{bmatrix}}{\alpha^4\alpha + \alpha^3} = \frac{\begin{bmatrix} \alpha & \alpha^4 \\ \alpha^6 & \alpha^3 \end{bmatrix}}{\alpha^6}$$

$$= \alpha \begin{bmatrix} \alpha & \alpha^4 \\ \alpha^6 & \alpha^3 \end{bmatrix} = \begin{bmatrix} \alpha^2 & \alpha^5 \\ 1 & \alpha^4 \end{bmatrix} \quad (\text{since } \alpha\alpha^6 = 1).$$

Then from (13.4.21) the error values are

$$\begin{Bmatrix} e_1 \\ e_2 \end{Bmatrix} = \begin{bmatrix} \alpha^2 & \alpha^5 \\ 1 & \alpha^4 \end{bmatrix} \begin{Bmatrix} \alpha^3 \\ \alpha^5 \end{Bmatrix} = \begin{Bmatrix} \alpha^5 + \alpha^{10} \\ \alpha^3 + \alpha^9 \end{Bmatrix} = \begin{Bmatrix} \alpha^5 + \alpha^3 \\ \alpha^3 + \alpha^2 \end{Bmatrix} = \begin{Bmatrix} \alpha^2 \\ \alpha^5 \end{Bmatrix}.$$

$$(13.4.22)$$

From (13.4.20) and (13.4.22) we obtain the estimated error polynomial $\hat{\mathbf{e}}(x) = \alpha^2 x^3 + \alpha^5 x^4$. The received codeword polynomial is estimated to be

$$\hat{\mathbf{c}}(x) = \mathbf{c}(x) + \hat{\mathbf{e}}(x) = \mathbf{c}(x) + \mathbf{e}(x) + \hat{\mathbf{e}}(x), \qquad (13.4.23)$$

where

$$\mathbf{w}(x) = 1 + \alpha^2 x + \alpha^4 x^2 + x^3 + \alpha^6 x^4 + \alpha^3 x^5 + \alpha^5 x^6,$$

$$\hat{\mathbf{e}}(x) = \alpha^2 x^3 + \alpha^5 x^4,$$

$$\hat{\mathbf{c}}(x) = 1 + \alpha^2 x + \alpha^4 x^2 + (1 + \alpha^2)x^3 + (\alpha^6 + \alpha^5)x^4 + \alpha^3 x^5 + \alpha^5 x^6$$

$$= 1 + \alpha^2 x + \alpha^4 x^2 + \alpha^6 x^3 + \alpha x^4 + \alpha^3 x^5 + \alpha^5 x^6$$

$$= \underbrace{(100) + (001)x + (011)x^2 + (101)x^3}_{\text{4 parity symbols}} + \underbrace{(010)x^4 + (110)x^5 + (111)x^6}_{\text{3 data symbols}}.$$

This gives the corrected decoded data symbols as $\underbrace{010}_{\alpha}\ \underbrace{110}_{\alpha^3}\ \underbrace{111}_{\alpha^5}$. ∎

13.4.2 LSFR Encoder. The parity-symbol polynomial, which has been generated above by purely algebraic rules in the Galois field, is electronically determined easily by the built-in circuitry of an $(n-k)$-stage shift register is shown in Figure 13.4.1 (Sklar [2001]). Since we are considering an RS $(7,3)$ code, it requires a linear feedback shift register (LFSR) circuit. The multiplier terms taken from left to right correspond to the coefficients of the generator polynomial in Eq (13.4.12) which is written in low order to high order.

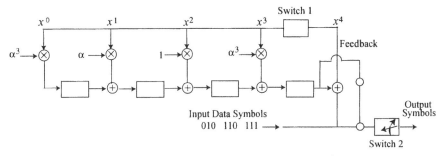

Figure 13.4.1 Linear feedback shift register circuit for RS$(7,3)$ code.

In this example, the RS$(7,3)$ codeword $\mathbf{c}(x)$, defined by (13.4.13), has $2^m - 1 = 2^3 - 1 = 7$ symbols (bytes), where each symbol is composed of $m = 3$ bits. Each stage in the shift register of Figure 13.4.1 will hold a 3-bit symbol; and since each coefficient is specified by 3 bits, it can take one of $2^3 = 8$ values. The operation of the encoder controlled by the LFSR circuit will generate codewords in a systematic manner according to the following steps:

STEP 1: Switch 1 is closed during the first $k = 3$ clock cycles; this allows shifting the data symbols into the $n - k = 4$-stage shift register.

STEP 2: Switch 2 is in the drawn position during the first $k = 3$ clock cycles; this allows simultaneous transfer of data symbols directly to an output register.

STEP 3: After transfer of the k-th (third) data symbol to the output register, switch 1 is opened and switch 2 is moved to the up position.

STEP 4: The remaining $n - k = 4$ clock cycles clear the parity symbols contained in the shift register by moving them to the output register.

STEP 5: The total number of clock cycles is equal to $n = 7$, and the output register contains the codeword polynomial $\mathbf{c}$ defined by Eq (13.4.13).

The data symbols $\underbrace{010}_{\alpha}\ \underbrace{110}_{\alpha^3}\ \underbrace{111}_{\alpha^5}$ move to the LFSR from right to left, where the rightmost symbol is first symbol and the rightmost bit is the first bit. The operation of the first $k = 3$ shift of the encoding circuit of Figure

13.4.1 are given below, where Tables 8.6.3 and 8.6.4 are used.

Clock Cycle	Input Queue	Register Contents	Feedback
0	$\alpha\ \alpha^3\ \alpha^5$	0, 0 0 0	α^5
1	$\alpha\ \alpha^3$	$\alpha\ \alpha^6\ \alpha^5\ \alpha$	1
2	α	$\alpha^3\ 0\ \alpha^2\ \alpha^2$	α^4
3	–	$1\ \alpha^2\ \alpha^4\ \alpha^6$	–

At the end of the third cycle, the register contents are the four parity symbols $1, \alpha^2, \alpha^4, \alpha^6$, as shown in the above table. At this point, switch 1 is opened, switch 2 is toggled to up position, and the 4 parity symbols are shifted from the register to the output. This suggests that the output (encoded) codeword can be written as

$$\mathbf{c}(x) = \sum_{i=0}^{6} c_i x^i = 1 + \alpha^2 x + \alpha^4 x^2 + \alpha^6 x^3 + \alpha x^4 + \alpha^3 x^5 + \alpha^5 x^6,$$

which is the same as (13.4.13), where the first four terms correspond to register contents after cycle 3, and the last three terms to the data input. ∎

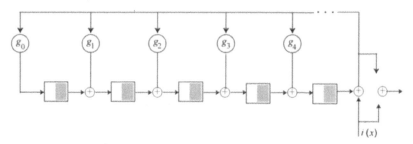

Figure 13.4.2 Systematic RS$(255, 249)$ encoder .

13.4.3 Encoding and Decoding Architecture. The RS encoding and decoding can be carried out in software or in special-purpose hardware. The RS codes are based on the Galois fields, and the RS encoder or decoder needs to carry out the arithmetic operations on finite field elements. These operations require special hardware or software functions to implement. The RS codeword is generated using a special polynomial. All valid codewords are exactly divisible by the generator polynomial, which has the general form of (13.4.1), and the codeword $\mathbf{c}(x)$ is constructed. For example, the generator polynomial for RS(255, 223) code is $\mathbf{g}(x) = (x - \alpha^0)(x - \alpha^1)(x - \alpha^2)(x - \alpha^3)(x - \alpha^4)(x - \alpha^5)$.

The $2t$ parity symbols in a systematic RS codeword are given by $\mathbf{p}(x)$. Figure 13.4.2 shows an encoder architecture for a systematic RS(255, 249) encoder in which each of the six registers hold a symbol (8 bits). The arithmetic operators $\oplus$ carry out finite field addition or multiplication on a complete symbol.

A general architecture for decoding RS codes is shown in Figure 13.4.3.

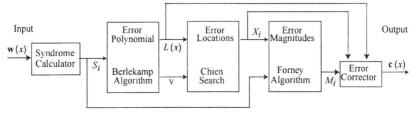

Figure 13.4.3 RS decoder.

LEGEND : $\mathbf{w}(x)$: transmitted codeword; S_i: syndrome; $L(x)$: error locator polynomial; X_i: error locations; M_i: error magnitudes; $\mathbf{c}(x)$: transmitted codeword; and ν: number of errors.

The received word $\mathbf{w}(x)$ is the original codeword $\mathbf{c}(x)$ plus errors, i.e., $\mathbf{w}(x) = \mathbf{c}(x) + \mathbf{e}(x)$. The RS decoder attempts to identify the location and magnitude of up to t errors (or $2t$ erasures) and correct the errors or erasures.

The syndrome computation is similar to the parity calculation. An RS codeword has $2t$ syndromes that depend only on errors and not on the transmitted codeword $\mathbf{c}(x)$. The syndromes can be calculated by substituting the $2t$ roots of the generator polynomial $g(x)$ into $\mathbf{w}(x)$.

Symbol-error locations can be found by solving simultaneous equations with t unknowns. Several fast algorithms are available and take advantage of the special matrix structure of RS codes and greatly reduce the computational effort required. In general, two steps are involved: (i) Find an error locator polynomial. Use the Berlekamp–Massey algorithm or Euclid's algorithm; the latter algorithm tends to be more widely used in practice because it is easier to implement, but the Berlekamp–Massey algorithm leads to more efficient hardware and software implementations. (ii) Find the roots of the error locator polynomial. This is done using the Chien search algorithm, which is described in §13.4.1. Symbol-error values are found by again solving simultaneous equations with t unknowns. A widely used fast algorithm is the Forney algorithm, which is also described in §13.4.1.

13.4.4 Hardware Implementation. A number of commercial hardware implementations exist. Many of the existing systems use 'off-the-shelf' integrated circuits that encode and decode RS codes. These ICs tend to support a certain amount of programability, e.g., RS$(255, k)$, where $t = 1$ to 16 symbols. A recent trend is toward VHDL or Verilog designs (logic cores or intellectual property cores). These have a number of advantages over standard ICs. A logic core can be integrated with other VHDL or Verlog components and synthesized to an FGPA (Field Programmable Gate Array) or ASIC (Application Specific Integrated Circuit); this enables the so-called 'System on Ship' designs where multiple modules can be combined in a single IC. Depending on production volumes, logic cores can often give significantly lower system costs than 'standard' ICs.

13.4.5 Software Implementation. Until recently, software implementation in 'real-time' required too much computational power for all but the simplest of RS codes, i.e., codes with small values of t. The major difficulty in implementing RS codes in software is that general-purpose processors do not support Galois field arithmetic operations. For example, to implement a Galois field multiplication in software requires a test for 0, two log table look-ups, a modulo add and an anti-log table lookup. However, careful design together with an increase in processor performance means that software implementations can operate at relatively high data rates. The following are some example benchmark figures on a 166 MHz Pentium PC:

Code	Data Rate
RS(255, 251)	12 Mbps
RS(255, 239)	2.7 Mbps
RS(255, 223)	1.1 Mbps

These data rates are for decoding only; encoding is considerably faster since it requires less computation.

13.5 Burst Errors

RS codes perform very well in the case of burst errors. Consider the RS(255, 247) code, with $n = 255, k = 247$, so that each symbol contains $m = 8$ bits (i.e., a byte). Since $t = \left\lfloor \dfrac{n-k}{2} \right\rfloor = 4$, this code can correct any four symbol errors in a block of 255 bits. Now, suppose that there is a noise burst that affects 27 bits in one block of data during transmission; this can be represented as follows:

symbol 1	symbol 2	symbol 3	symbol 4	symbol 5	symbol 6	$\cdots$

    ok   error   error   error   error   ok

27-Bit Noise Burst

The burst lasts for a duration of 27 contiguous bits, which affects four bytes (symbol 2 through symbol 5). This RS decoder will correct any four-symbol errors no matter what type of errors occurs in the symbols: It will replace the incorrect byte with the correct one, irrespective of the occurrence of a single-bit error or all eight-bit error in a byte. This error-correcting feature is not available in binary codes, and this is what makes the RS codes more efficient and popular. Moreover, if instead of the error in 27 contiguous bits, the 27 bits were disturbed at random, the RS(255, 247) code will not be any more useful, and to remedy this situation another RS code will be needed depending on the type and location of such errors. Error correcting codes become more efficient as the block size increases. For example, if the code rate is fixed at $7/8$, and if the block size of an RS code increases from $n = 32$ symbols (with $m = 5$ bits per symbol) to $n = 256$ symbols (with $m = 8$ bits per symbol), the block size will increase from 160 bits to 2048 bits.

RS codes are very effective for channels that have memory and those where the input is large. An important feature of RS codes of length n is that two data symbols can be added without reducing the minimum distance, which produces the extended RS code of length $n + 2$ with the same number of parity-check symbols as the original code of length n with m bits each. The relation between the RS decoded symbol-error probability $\mathfrak{p}_E$ and the channel symbol-error probability $\mathfrak{p}$ is (Odenwalder [1976])

$$\mathfrak{p}_E \approx \frac{1}{2^m - 1} \sum_{j=t+1}^{2^m - 1} j \binom{2^m - 1}{j} \mathfrak{p}^j (1 - \mathfrak{p})^{2^m - 1 - j}. \qquad (13.5.1)$$

Let $\mathfrak{p}_B$ denote the bit-error probability. For specific modulation of the MFSK type, we have the ratio

$$\frac{\mathfrak{p}_B}{\mathfrak{p}_E} = \frac{2^{m-1}}{2^m - 1}, \qquad (13.5.2)$$

that is, the bit-error probability is bounded above by the symbol-error probability. Gallager [1968] established that for RS codes, error-probability is an exponentially decreasing function of block length n, while decoding complexity is proportional to a small power of the block length.

13.6 Erasures

An erasure is defined as a failure of any device that shuts down if it fails, and the system signals the shutdown (Peterson and Weldon [1972]). A mathematical definition is based on the fact that the computation of each checksum (parity) device P_i is a function F_i of all the data devices D_j, $j = 1, \ldots, k$, where a device means a set of symbols (codewords). For the sake of simplicity, assume that each device holds just one codeword. Let $d_1, \ldots, d_k$ denote the datawords and $m = n - k$ parity words $p_1, \ldots, p_m$, which are computed from the datawords in such a way that the loss of any l words is tolerable. The parity words p_i are computed for the checksum device P_i by

$$p_i = F_i(d_1, \ldots, d_k), \qquad (13.6.1)$$

where the dataword is on the device D_j. Figure 13.6.1 illustrates the configuration of the technique for $k = 2$ and $m = 2$.

$$\boxed{D_1}\;\boxed{D_2}\;\boxed{D_3}\;\boxed{D_4}\;\boxed{D_5}\;\boxed{D_6}\;\boxed{D_7}\;\boxed{D_8}$$

$$\boxed{P_1} = F_1(D_1, D_2, D_3, D_4, D_5, D_6, D_7, D_8),$$

$$\boxed{P_2} = F_2(D_1, D_2, D_3, D_4, D_5, D_6, D_7, D_8).$$

$\boxed{D_1}$	$\boxed{D_2}$	$\boxed{P_1}$	$\boxed{P_2}$
$d_{1,1}$	$d_{2,1}$	$p_{1,1} = F_1(d_{1,1}, d_{2,1})$	$p_{2,1} = F_2(d_{1,1}, d_{2,1})$
$d_{1,2}$	$d_{2,2}$	$p_{1,2} = F_1(d_{1,2}, d_{2,2})$	$p_{2,2} = F_2(d_{1,2}, d_{2,2})$
$d_{1,3}$	$d_{2,3}$	$p_{1,3} = F_1(d_{1,3}, d_{2,3})$	$p_{2,3} = F_2(d_{1,3}, d_{2,3})$
$\vdots$	$\vdots$	$\vdots$	$\vdots$
$d_{1,n}$	$d_{2,n}$	$p_{1,n} = F_1(d_{1,n}, d_{2,n})$	$p_{2,n} = F_2(d_{1,n}, d_{2,n})$

Figure 13.6.1 Data-parity configuration.

In this figure, the contents of the parity (checksum) devices P_1 and P_2 are computed by (13.6.1) applied to the functions F_1 and F_2, respectively, where each device contains $n = (k \text{ bytes})\left(\dfrac{8\,\text{bits}}{\text{byte}}\right)\left(\dfrac{1\,\text{word}}{n\,\text{bits}}\right) = \dfrac{8k}{s}$ words, where s is the size of a word, and the contents of each device is denoted with a double index i, j.

Suppose that a dataword d_j on device D_j is erased. The system knows the location of the erased word. To reconstruct the erased word, the device D_j chooses at random a word d_j' at the location d_j from the set of words in non-failed devices. Then each checksum word p_i, $i = 1, \dots, m$, is recomputed using a function $G_{i,j}$ such that

$$p_i' = G_{i,j}(d_j, d_j', p_i). \tag{13.6.2}$$

If up to m devices fail, the system is reconstructed as follows: First, for each failed data device D_j, construct a function to restore the words in D_j from the words in the active (non-failed) devices. After it is completed, recompute any failed checksum device P_i with F_i using (13.6.1).

Example 13.6.1. In the case when $m = 1$, define a $(k + 1)$-parity in the devices P_i. Because there is only one parity device P_1 (because $m = 1$) and words consist of one bit ($s = 1$), the checksum word p_1 is computed by XOR-ing the datawords, that is,

$$p_1 = F_1(d_1, \dots, d_k) = d_1 \oplus d_2 \oplus \cdots \oplus d_k.$$

If a word d_j on the data device D_j is erased and d'_j is taken in its place, then p_1 is recomputed from the parity of its previous value and the two datawords, that is,

$$p'_1 = G_{1,j}(d_j, d'_j, p_1) = p_1 \oplus d_j \oplus d'_j.$$

If a data device D_j fails, then each word is restored as the parity of the corresponding words on the remaining devices, that is,

$$d_j = d_1 \oplus d_2 \oplus \cdots \oplus d_{j-1} \oplus d_{j+1} \oplus \cdots \oplus d_n \oplus p_1.$$

This computation will restore the system in the case of any single device failure. ∎

Hence, given k datawords $d_1, d_2, \ldots, d_k$ all of size s, define the functions F and G to calculate and maintain the parity words $p_1, p_2, \ldots, p_m$, and then find a method to reconstruct the words of any erased data device D_j when up to m devices fail. Once the datawords are restored, the checksum words are recomputed from the datawords and the function F. Thus, the entire system is reconstructed by the reconstruction algorithm defined below.

13.6.1 Reconstruction Algorithm. This algorithm has three distinct procedures:

(a) Use Vandermonde matrix to calculate and maintain parity (checksum) words;

(b) Use Gauss elimination method to recover from erasures; and

(c) Use Galois field arithmetic, explained in §8.6, and Tables 8.6.3–8.6.5 for computation.

The procedures (a) and (b) are described as follows.

Procedure (a): Define each function F_i as a linear combination of datawords, as in (13.6.1):

$$p_i = F_i(d_1, d_2, \ldots, d_k) = \sum_{j=1}^{k} d_j f_{i,j}, \qquad (13.6.3)$$

where $f_{i,j} = j^{i-1}$. In matrix form, Eq (13.6.3) can be written as $\mathbf{FD} = \mathbf{P}$, where $\mathbf{F}$ is an $m \times k$ Vandermonde matrix:

$$\mathbf{F} = \begin{bmatrix} f_{1,1} & f_{1,2} & \cdots & f_{1,k} \\ f_{2,1} & f_{2,2} & \cdots & f_{2,k} \\ \vdots & \vdots & \cdots & \vdots \\ f_{m,1} & f_{m,2} & \cdots & f_{m,k} \end{bmatrix} = \begin{bmatrix} 1 & 1 & 1 & \cdots & 1 \\ 1 & 2 & 3 & \cdots & k \\ \vdots & \vdots & \vdots & \cdots & \vdots \\ 1 & 2^{m-1} & 3^{m-1} & \cdots & k^{m-1} \end{bmatrix}.$$

Thus, the system $\mathbf{FD} = \mathbf{P}$ becomes

$$
\begin{bmatrix}
1 & 1 & 1 & \cdots & 1 \\
1 & 2 & 3 & \cdots & k \\
\vdots & \vdots & \vdots & \cdots & \vdots \\
1 & 2^{m-1} & 3^{m-1} & \cdots & k^{m-1}
\end{bmatrix}
\left\{
\begin{array}{c}
d_1 \\ d_2 \\ \vdots \\ d_k
\end{array}
\right\}
=
\left\{
\begin{array}{c}
p_1 \\ p_2 \\ \vdots \\ p_m
\end{array}
\right\}. \tag{13.6.4}
$$

When one of the datawords d_j is erased, replace it by d_j'; then each of the checksum words must be changed accordingly, by subtracting the portion of the checksum words corresponding to d_j and adding the required amount from d_j'. Thus, $G_{i,j}$ is defined as

$$
p_i' = G_{i,j}(d_j, d_j', p_i) = p_i + f_{i,j}(d_j' - d_j). \tag{13.6.5}
$$

(b) **Recovery Procedure.** Define a matrix $\mathbf{A}$ and a vector $\mathbf{P}$ by $\mathbf{A}' = \begin{bmatrix} \mathbf{I} \\ \mathbf{F} \end{bmatrix}$ and $\mathbf{E} = \left\{ \begin{array}{c} \mathbf{D} \\ \mathbf{P} \end{array} \right\}$, where $\mathbf{I}$ is an $k \times k$ identity matrix. Then Eq (13.6.5) becomes

$$
\begin{bmatrix}
1 & 0 & 0 & \cdots & 0 \\
0 & 1 & 0 & \cdots & 0 \\
\vdots & \vdots & \vdots & \cdots & \vdots \\
0 & 0 & 0 & \cdots & 1 \\
1 & 1 & 1 & \cdots & 1 \\
1 & 2 & 3 & \cdots & k \\
\vdots & \vdots & \vdots & \cdots & \vdots \\
1 & 2^{m-1} & 3^{m-1} & \cdots & k^{m-1}
\end{bmatrix}
\left\{
\begin{array}{c}
d_1 \\ d_2 \\ \vdots \\ d_k
\end{array}
\right\}
=
\left\{
\begin{array}{c}
d_1 \\ d_2 \\ \vdots \\ d_k \\ p_1 \\ p_2 \\ \vdots \\ p_m
\end{array}
\right\}, \tag{13.6.6}
$$

which can be represented as $\mathbf{A'D} = \mathbf{P}'$. The matrix $\mathbf{F}$ is an $k \times k$ Vandermonde matrix, and every subset of k rows of matrix $\mathbf{A}'$ is linearly independent, and so it is nonsingular. Thus, the column vector $\mathbf{D}$ can be computed from Eq (13.6.5) by the Gauss elimination method, using the arithmetic in $GF(2^m)$ defined in §8.6.4. Once the values of $\mathbf{D}$ are obtained, the values of any failed P_i is computed. In case fewer than m devices fail, the system is recovered by the same method. The method fails to work if more than m devices fail.

Example 13.6.2. Let $k = 3, m = 3$, and $l = 4$. Since $k + m < 2^4$, we will use the binary arithmetic in $GF(16)$ (see §8.1.1). Suppose the first words in devices D_1, D_2, D_3 are $d_1 = 4, d_2 = 11, d_3 = 8$, respectively. The Vandermonde matrix $\mathbf{F}$ is

$$
\mathbf{F} = \begin{bmatrix}
1 & 1 & 1 \\
1 & 2 & 3 \\
1 & 2^2 & 3^2
\end{bmatrix} = \begin{bmatrix}
1 & 1 & 1 \\
1 & 2 & 3 \\
1 & 4 & 9
\end{bmatrix}.
$$

Next, we calculate the first word of each of the three parity devices P_1, P_2, and P_3 using the matrix equation $\mathbf{F}\mathbf{D} = \mathbf{P}$:

$$p_1 = (13)(4) + (1)(11) + (1)(8) = 4 + 11 + 8 = 0100 \oplus 1011 \oplus 1000 = 0111 = 7,$$
$$p_2 = (1)(4) + (2)(11) + (3)(8) = 4 + 5 + 11 = 0100 \oplus 0101 \oplus 1011 = 1010 = 10,$$
$$p_3 = (1)(4) + (4)(11) + (9)(8) = 4 + 10 + 4 = 10, \text{ since } 4 + 4 = 0.$$

Now suppose that d_2 is erased, and we replace it by $d_2' = 1$. Then d_2 will have the value $(1 - 11) = 0001 \oplus 1011 = 1010 = 10$, which is sent to each parity device. Thus, the new values of their first word are

$$p_1 = 7 + (1)(10) = 7 + 10 = 0111 \oplus 1010 = 1101 = 13,$$
$$p_2 = 10 + (2)(10) = 10 + 7 = 1010 \oplus 0111 \oplus 1101 = 13,$$
$$p_3 = 10 + (4)(10) = 10 + 14 = 1010 \oplus 1110 = 0100 = 4.$$

Suppose d_2, d_3, and p_3 are lost (erased). This means that we delete the rows of $\mathbf{A}$ and $\mathbf{P}$ corresponding to D_1, P_2, and P_3, which yield the matrix equation $\mathbf{A}'\mathbf{D} = \mathbf{P}'$, i.e.,

$$\begin{bmatrix} 1 & 0 & 0 \\ 1 & 1 & 1 \\ 1 & 2 & 3 \end{bmatrix} \mathbf{D} = \left\{ \begin{array}{l} d_1 = 4 \\ p_1 = 13 \\ p_3 = 13 \end{array} \right\},$$

which gives

$$\mathbf{D} = (\mathbf{A})^{-1}\mathbf{P}' = \begin{bmatrix} 1 & 0 & 0 \\ 2 & 3 & 1 \\ 3 & 2 & 1 \end{bmatrix} \left\{ \begin{array}{l} 4 \\ 13 \\ 13 \end{array} \right\}.$$

From this equation we find that

$$d_2 = (2)(4) + (3)(13) + (1)(13) = 8 + 2 + 13 = 1000 \oplus 0010 \oplus 1101 = 0111 = 7,$$
$$d_3 = (3)(4) + (2)(13) + (1)(13) = 12 + 13 + 13 = 12,$$

so that

$$p_3 = [1\ 4\ 9] \left\{ \begin{array}{l} d_1 = 1 \\ d_2 = 7 \\ d_3 = 12 \end{array} \right\} = (1)(1) + (4)(7) + (9)(12) = 1 + 15 + 5$$
$$= 0001 \oplus 1111 \oplus 0101 = 1011 = 11,$$

and the system is restored by taking $p_3 = 11$ as the correct value of d_2. ∎

13.7 Concatenated Systems

The RS codes are sometimes used as a concatenated system, in which an inner convolutional decoder first provides some error control by operating on soft-decision demodulator outputs, and then the convolutional decoder provides hard-decision data to the outer RS decoder to further reduce the error probability. Concatenated codes were first introduced by Forney [1966]. In order to obtain a binary code with good distance from the RS code that satisfies the Singleton bound (§6.6), we will introduce a new idea to the above transformation, by looking at the step where the values are taken from the field F_{2^m} and encoded with m bits in the binary alphabet. Instead of using the minimum number of bits to encode those elements in the binary alphabet, we will use both more bits, say $2m$ bits, and an encoder that adds more distance to the final code. This is, in fact, the idea behind concatenated codes.

We use the simple transformation, discussed in §13.1.3, to convert RS codes to binary codes. Recall that this transformation starts with a polynomial f of degree $k - 1$ and evaluates it over $\alpha_1, \alpha_2, \ldots, \alpha_m$ to obtain the values $f(\alpha_1), f(\alpha_2), \ldots, f(\alpha_m) \in F_{2^m}^n$. Then each of the values $f(\alpha_i)$, $1 \leq i \leq m$, are encoded in the binary alphabet with m bits. The binary code so obtained has block length nm and distance at least $d = n - k + 1$. This distance is not very good, since the lower bound on the relative distance is not robust.

In general, a concatenated code C is composed of two codes: $C = C_{\text{out}} \diamond C_{\text{in}}$, where the *outer code* $C_{\text{out}} \subset A_1^{n_1}$ converts the input message into a codeword over a large alphabet A_1, while the *inner code* $C_{\text{in}} \subset A_2^{n_2}$ is a much smaller code that converts symbols from A_1 to codewords over A_2. In particular, when $A_2 = \{0, 1\}$, the code C is a binary concatenated code, which is defined as follows:

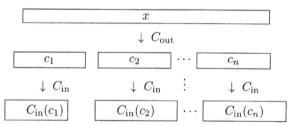

Note that in the above definition, the inner code C_{in} is a small code in the sense that it only requires one codeword for each symbol in the alphabet A_1. This alphabet, in general, is of much smaller size than the total number of codewords encoded by C.

The rate of the concatenated code $C = C_{\text{out}} \diamond C_{\text{in}}$ is

$$R(C) = \frac{\log |C_{\text{out}}|}{n_1 n_2 \log |A_2|} = \frac{\log |C_{\text{out}}|}{n_1 \log |A_1|} \cdot \frac{\log |A_1|}{n_2 \log |A_2|} = R(C_{\text{out}}) \cdot R(C_{\text{in}}), \quad (13.7.1)$$

where in the last quantity we have used $|C_{\text{in}}| = |A_1|$. Eq (13.7.1) suggests that we can replace the inner code with any other code and get a cost proportional to C_{in}.

The distance of a concatenated code $C = C_{\text{out}} \diamond C_{\text{in}}$ satisfies the inequality

$$d(C) \geq d(C_{\text{out}}) \cdot d(C_{\text{in}}). \tag{13.7.2}$$

This result is proved by considering two distinct messages x and y, where the distance property of the outer code implies that the encodings $C_{\text{out}}(x)$ and $C_{\text{out}}(y)$ differ in at least $d(C_{\text{out}})$ symbols. For each of the symbols where they differ, the inner code will encode the symbols into codewords that differ in at least $d(C_{\text{in}})$ locations.

Note that since the lower bound in (13.7.2) is not rigid, the distance of concatenated codes may generally be much larger. Such an inference seems to be counter-intuitive, as we can have two codewords at the outer level that differ in only $d(C_{\text{out}})$ locations, whereas at the inner level we may also have two different symbols in A_1 whose encodings under C_{in} differ in only $d(C_{\text{in}})$ locations. These two situations are not necessarily independent, as it could happen that there are two codewords at the outer level that differ at only $d(C_{\text{out}})$ symbols, but then they must differ in such a manner that the inner code does much better than its worst situation. In fact, a probabilistic argument shows that when the outer code is an RS code and the inner codes are 'random projections' obtained by mapping the symbols of the alphabet A_1 onto the codewords in the alphabet A_2 with independently chosen random bases, then the resulting concatenated code achieves the Gilbert–Varshamov bound with high probability (for this bound, see §6.4). This guarantees that such a concatenated code has distance much larger than the lower bound in (13.7.2).

Example 13.7.1. (Construction of a family of binary concatenated codes with good distance.) Let $0 < R < 1$ be a fixed rate that we would like to achieve. To build a code with rate R and distance as large as possible, let $C_{\text{out}} = (n, k, n-k+1) \in F_{2^m}$ be the RS code with block length $n = 2^m$. The rate of this outer code is $R_{\text{out}} = k/n$ and relative distance $d_{\text{out}} = \dfrac{n-k+1}{n} \geq 1 - R_{\text{out}}$. Let C_{in} be a binary linear code with parameters $(m/r, m, d) \in F_2$, with rate $r = R(C_{\text{in}})$. Then the rate R of the concatenated code $C = C_{\text{out}} \diamond C_{\text{in}}$ is $R = R_{\text{out}}\, r$, which gives $R_{\text{out}} = R/r$. The outer code which is an RS code and therefore optimal, has been constructed thus far. It remains to construct the inner code, which must be defined as a linear binary code with rate r, and a distance as large as possible. According to the Gilbert–Varshamov bound, there exists a linear code C_{in} with rate $r \geq 1 - h(d_{\text{in}})$. This means that there exists a code with rate r and distance $d_{\text{in}} \geq (1-r)/h$. With this inner code, we obtain the concatenated code C with distance

$$d(C) \geq (d_{\text{out}})(d_{\text{in}}) \geq \frac{(1 - R_{\text{out}})(1-r)}{h} = \frac{1}{h}\left(1 - \frac{R}{r}\right)(1-r). \tag{13.7.3}$$

Now, to find an inner code C_{in} with minimum distance $d_{\text{in}} \geq (1-r)/h$, we must use the *greedy algorithm* because any other algorithm would generate all possible generator matrices $\mathbf{G}$ for C_{in}, which are $2^{m^2/r} = n^{\log(n)/r}$ in number, and this number may not be run in time polynomial in the given block length. This algorithm constructs the parity-check matrix $\mathbf{H}$ such that every set of $(d-1)$ columns of $\mathbf{H}$ is linearly independent. So enumerate all possible columns, and if the current column is not contained in the linear span of any $(d-2)$ columns already in $\mathbf{H}$, add it to $\mathbf{H}$. Thus, the greedy algorithm examines $2^{m/r-m} = n^{1/r-1}$ columns, and the process will continue to generate columns of $\mathbf{H}$ as long as $2^{m/r-m} > \sum_{i=0}^{d-2} \binom{n-1}{i}$. This completes the process of constructing the concatenated code C. ∎

13.7.1 Types of Concatenation Codes. There are two types of concatenation codes: one serial and the other parallel. The concatenation scheme for transmission of the former is shown in Figure 13.7.1(a), and for the latter in Figure 13.7.1(b).

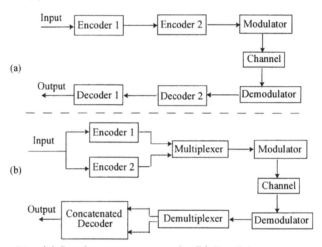

Figure 13.7.1 (a) Serial concatenation code; (b) Parallel concatenation code.

13.8 Applications

The RS codes are used most frequently, either independently or in combination with other recently discovered algorithms and new codes. These codes have a big business appeal, and there are technological enterprises that specialize in creating RS codes of different lengths to suit the applications in various electronic equipments and machines. Typical applications include the following areas.

(i) DATA STORAGE (OPTICAL MEDIA). RS codes are widely used to correct burst errors due to media defects in CDs and DVDs. For example, CDs use two layers of cross-interleaved RS codes (CIRC). The first part of the CIRC decoder is an RS(32,28) code, shortened from a (255,251) code with 8-bit symbols, which can correct up to two byte errors per 32-byte block. It flags as erasures any uncorrectable blocks that have more than two byte errors. The decoded 28-byte blocks that have erasure indicators are then spread by the deinterlever to different blocks of the (28, 24) outer code, thus making an erased 28-byte block from the inner code a single erased byte in each of 28 outer code blocks. This erasure is then corrected by the outer code, which can correct up to four erasures per block.

(ii) DATA STORAGE (DISK ARRAYS). RAID (Redundant Array of Independent Disks, originally called Redundant Array of Inexpensive Disks) is a data storage technology that provides increased storage and reliability through redundancy by combining multiple disk drive components into a logical unit and by having data distributed across the drives in various "RAID levels". RAID level 6 can continue to provide access to data even if two of the disks fail by using parity checks and Reed–Solomon error correction.

(iii) DATA TRANSMISSION. The unreliable aspect of data transmission over erasure channels (for example, over DSL lines) is overcome by certain specialized RS codes, like the Caucy-RS and Vandermonde-RS, in which encoding is performed by an $RS(n,k)$ code. Any combination of $k < n$ codewords received by the decoder is sufficient to reconstruct all the n codewords.

(iv) DEEP SPACE COMMUNICATION. RS codes have been used in several of NASA and ESA's planetary explorations. Initially it was realized that RS codes would not be suitable for deep space because deep space communication does not induce burst errors during transmission. However, once it was found that a combination of binary codes and RS codes is used in a concatenated form, where the binary code is used as the 'inner code' while the RS code is used as an error correcting code, such concatenated codes become very powerful. A concatenated conventional binary code with an RS code was used in the Voyager expeditions to Uranus and Neptune, and the close-up photographs transmitted from those distant planets were received with the expected resolution. The Galileo mission to Jupiter is another example of the use of RS codes to successfully solve the erasures by a concatenated error correcting code.

(v) FEEDBACK SYSTEMS. Mobile data transmission systems and highly reliable military communication systems use RS codes with simultaneous error detection and correction. These systems distinguish between a decoder error and a decoder failure.

(vi) BAR CODES. Some paper bar codes use Reed–Solomon error correction to allow reading even if a portion of the bar code is damaged. Unrecognizable symbols are treated as erasures. Examples of paper bar codes using RS codes include the PDF417 stacked linear barcode used by the US Postal Service to print postage, or the QR (Quick Response) matrix barcode (or two-dimensional code) designed to be read by smartphones.

(vii) FAMILY OF LT CODES. RS codes are used in the family of LT and Raptor codes as a pre-code to first correct burst errors. For more details, see Chapters 19 and 20.

Finally, a comprehensive treatise on RS codes is by Wicker and Bhargava [1999].

14

Belief Propagation

14.1 Rational Belief

Bayes' theorem (Bayes [1764]) is a mathematical statement based on the subjectivists' notion of belief governed by probability, and it is used to calculate conditional probabilities. In simplest terms, this theorem states that a hypothesis is confirmed by any body of data that its truth renders probable. Thus, the probability of a hypothesis h conditional on some given data d is defined as the ratio of the unconditional probability of the conjunction of the hypothesis with the data to the unconditional probability of the data alone. In other words,

$$\mathfrak{p}_d(h) = \frac{\mathfrak{p}(h \wedge d)}{\mathfrak{p}(d)}, \qquad (14.1.1)$$

provided that both terms of this ratio exist and $\mathfrak{p}(d) > 0$, where $\mathfrak{p}_d$ is a probability function (Birnbaum [1962]), and $\wedge$ denotes the logical AND. A consequence of (14.1.1) is that if d entails h, then $\mathfrak{p}_d(h) = 1$. Further, if $\mathfrak{p}(h) = 1$, then $\mathfrak{p}_d(h) = 1$. Thus, combining these results, we have $\mathfrak{p}(h) = \mathfrak{p}(d)\mathfrak{p}_d(h) + \mathfrak{p}(\neg d)\mathfrak{p}_{\neg d}(h)$, where $\neg$ denotes negation (Carnap [1962]). Hence,

Theorem 14.1.1. (Bayes' Theorem): $\mathfrak{p}_d(h) = \dfrac{\mathfrak{p}(h)}{\mathfrak{p}(d)}\mathfrak{p}_h(d), \qquad (14.1.2)$

where $\mathfrak{p}_h(d)$ is known as the 'prediction term'. Note that according to a slightly different statistical terminology the inverse probability $\mathfrak{p}_h(d)$ is called the *likelihood* of h on d.

Example 14.1.1. According to the Center for Disease Control, approximately 2.75 m (m short for million) of the 309 m Americans living on January 1, 2011, died during the calendar year 2010. A randomly chosen Mr. X was alive on this date. Of the roughly 21.6 m seniors of age at least 75, about 1.85 m died. The unconditional probability of the hypothesis that Mr. X died during 2010 is given by the population-wide mortality rate $\mathfrak{p}(h) = 2.75/309 \approx 0.00896$. To determine X's death conditional on the information that he is a senior citizen, we divide the probability that he was a senior citizen who died by the probability of his being a senior citizen, i.e., we divide $\mathfrak{p}(h \wedge d)$ by $\mathfrak{p}(d)$, where $\mathfrak{p}(h \wedge d) = 1.85/309 \approx 0.005987$, $\mathfrak{p}(d) = 21.6/309 \approx 0.0699$, and find from (14.1.1) that $\mathfrak{p}_d(h) = 0.085$, which is just the proportion of seniors who died. On the other hand, $\mathfrak{p}_h(d) = \mathfrak{p}(h \wedge d)/\mathfrak{p}(h) = 0.005987/0.008967 \approx 0.67$, which shows that about 67% of the total deaths occurred among the senior citizens. Again, using (14.1.2), we can directly compute the probability of X's dying given that he was a senior citizen, so that $\mathfrak{p}_d(h) = [\mathfrak{p}(h)/\mathfrak{p}(d)]\,\mathfrak{p}_h(d) = 0.1273 \times 0.667 \approx 0.085$, as expected. ■

A second form of Bayes' theorem is

$$\mathfrak{p}_d(h) = \frac{\mathfrak{p}(h)\mathfrak{p}_h(d)}{\mathfrak{p}(h)\mathfrak{p}_h(d) + \mathfrak{p}(\neg h)\mathfrak{p}_{\neg h}(d)}. \tag{14.1.3}$$

If both $\mathfrak{p}_h(d)$ and $\mathfrak{p}_{\neg h}(d)$ are known, the value of $\mathfrak{p}_d(h)$ is determined from (14.1.2) or (14.1.3). Note that Bayes' theorem in the form of (14.1.3) is useful for inferring causes from their effects since, given the presence or absence of a supposedly known cause, the probability of an effect can be easily discerned. Some special forms of Bayes' theorem are as follows:

(i) PROBABILITY RATIO RULE, denoted by PR, states that $\mathrm{PR}(h, d) = \mathrm{PR}(d, h)$, where the term on the right provides a measure of the degree to which h predicts d. Note that $\mathrm{PR}(d, h) = 0$ means that h makes d more or less predictable relative to the 'baseline' $\mathfrak{p}(d)$; thus, $\mathrm{PR}(d, h) = 0$ means that h categorically predicts $\neg d$. Again, $\mathrm{PR}(d, h) = 1$ means that adding h does not alter the baseline prediction at all. Finally, $\mathrm{PR}(d, h) = 1/\mathfrak{p}(d)$ means that h categorically predicts d.

(ii) ODDS RATIO RULE, denoted by OR, states that $\mathrm{OR}(h, d) = \dfrac{\mathfrak{p}_h(d)}{\mathfrak{p}_{\neg h}(d)}$. Note that while both OR and PR use a different way of expressing probabilities, each shows how its expression for h's probability conditional on d can be obtained by multiplying its expression for h's unconditional probability by a factor involving inverse probabilities.

(iii) LIKELIHOOD RATIO of h given d, denoted by LR, states that $\mathrm{LR}(h, d) = \dfrac{\mathfrak{p}_h(d)}{\mathfrak{p}_{\neg h}(d)}$ (see $\mathrm{OR}(h, d)$ above). The likelihood ratio is a measure of the degree to which h predicts d. Thus, $\mathrm{LR}(h, d)$ is the degree to which the hypothesis surpasses its negation as a predictor of the data. Further developments on log-likelihood ratio and its algebra are given in §14.5.

The major trend of the subjectivist concept is that beliefs come in varying gradations of strength, and that an ideally rational person's beliefs can be represented by a subjective probability function $\mathfrak{p}$. For each hypothesis h, about which a person has firm opinion, $\mathfrak{p}(h)$ measures the level of confidence (or degree of belief) in h's truth. Conditional beliefs are measured by conditional probabilities, so that $\mathfrak{p}_d(h)$ measures the confidence one has in h under the assumption that d is a fact. A significant feature of the belief philosophy is its account of evidential support, known as the Bayesian confirmation theory (Dale [1989]) which is comprised of

1. Confirmational relativity, so that evidential relationships are relativized to individuals and their degree of belief.

2. Evidence proportionism, by which a rational believer proportions the confidence in a hypothesis h to the total evidence for h so that the subjective probability for h reflects the overall balance of reasons for or against its truth.

3. Incremental confirmation, by which a body of data provides incremental evidence for h to the extent that any conditioning on the data raises h's probability.

There are two subordinate concepts related to total evidence: (a) The net evidence in favor of h is the degree to which a subject's total evidence in favor of h exceeds the total evidence in favor of $\neg h$ (negation of h); and (b) the balance of total evidence for h over the total evidence in favor of some other hypothesis h^*. For more details on the concept of belief and inferences thereof, see Birnbaum [1962], Carnap [1962], Dale [1989], Edwards [1972], Glymour [1980], Chihara [1987], and Christensen [1999].

14.2 Belief Propagation

If belief is regarded as a probabilistic phenomenon, then belief propagation becomes a stochastic process. The belief propagation algorithm is known as the *sum-product formula* (MacKay and Neal [1996], McEliece et al. [1998]), and also as the *forward/backward* algorithm (Rabiner [1989]), or as the *BCJR* algorithm[1] (Bahl et al. [1974]). It is an efficient method to solve inference problems through logical means. The problem is often pointed to the possibility of constructing a smart machine that is provided with possibly complete information about a sequence of events in Bayesian networks, and it is then asked to *infer* the probability of the occurrence of a given event. This algorithm was invented by Pearl [1982] to calculate marginals in Bayesian networks. It also works with Markov random fields (MRFs), graphical models, and Tanner graphs. The results are exact in some cases, but approximate for most problems.

MRFs, graphical models, and Tanner graphs have the common property that they all have joint probability of many variables. Thus, the probability domain

$$\mathfrak{p}(x_1, x_2, \ldots, x_N) = \frac{1}{Z} f(x_1, x_2) g(x_2) h(x_2, x_3, x_4)$$

is equivalent to the energy (log probability) domain

---

[1] The term 'BCJR' is coined from the first letters of the last names of the authors of the publication.

$$p(x_1, x_2, \ldots, x_N) = \frac{1}{Z} e^{F(x_1,x_2)+G(x_2)+H(x_2,x_3,x_4)},$$

where the factors (f, F, g, G, h, H) are called *potentials*. In general, the factors themselves are not probabilities; they are functions that determine probabilities. However, in particular cases, like Markov chains and Bayesian networks, they can be interpreted as conditional probabilities. They are non-negative (except in the log domain), but they need not be normalized to 1. An example of Bayesian network is given in Figure 14.2.1(a), which implies that $p(w, x, y, z) = p(w)p(x)p(y|w)p(z|w, x)$. An example of an MRF is given in Figure 14.2.1(b), for which $p(w, x, y, z) = \frac{1}{Z} f_{wx}(w, x) f_{xz}(x, z) f_{yz}(y, z) f_{wy}(w, y)$. An example of a factor graph is given in Figure 14.2.1(c), where each box denotes an interaction among the variables it connects to; it implies that $p(w, x, y, z) = \frac{1}{Z} f(w, x, y) g(y, z) h(z)$ (see Coughlan [2009]).

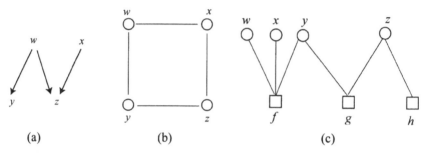

(a)   (b)   (c)

Figure 14.2.1 Examples of some networks.

The factor graph in Figure 14.1.(c) represents the parity-check matrix of the form

$$\mathbf{H} = \begin{bmatrix} w & x & y & 0 \\ 0 & 0 & y & z \\ 0 & 0 & 0 & z \end{bmatrix}.$$

In the case of marginals, we determine: $p(x_1), p(x_2), \ldots, p(x_N)$, but in the case of a maximizer, we determine $\arg\{ \max_{x_1,x_2,x\ldots,x_N} p(x_1, x_2, \ldots, x_N) \}$. Both quantities are difficult to compute, since if the state space of all x_i has S possible states, then $O(S^N)$ calculations are required, which involve a vast number of additions and search. One method, known as the *message passing algorithm* (MPA), works as follows: The neighboring variables 'talk' to each other and pass messages such as 'I (variable x_3) think that you (variable x_2) belong in these states with various likelihoods', and so on. After enough iterations, this series of conversations is likely to converge to a consensus that determines the marginal probabilities of all the variables. Estimated marginal probabilities are called beliefs. The *belief propagation algorithm* in a nutshell is: Update messages until convergence occurs, then calculate beliefs. For details about MPA, see §14.5.2.

14.2.1 Pairwise MRF. Pairwise MRF in the graphical model has only binary and pairwise factors:

$$\mathfrak{p}(x_1, x_2, \ldots, x_N) = \frac{1}{Z} \prod_{i=1}^{N} g_i(x_i) \prod_{i,j} f_{ij}(x_i, x_j),$$

where g_i are unary factors and f_{ij} the pairwise factors, and the second product is over a neighboring pair of nodes (variables) such that $i < j$, that is, it ignores a pair of variables twice. In the case of a Tanner graph, the formulation of belief propagation may, in general, have interactions of higher order than pairwise (for details see Kschischang et al. [2001]). We will only discuss pairwise MRF.

14.2.2 Messages. Let a message from node i to node j be denoted by $m_{ij}(x_j)$. Messages are similar to likelihoods; they are non-negative and do not sum to 1. A high value of $m_{ij}(x_j)$ means that the node i 'believes' the marginal value $\mathfrak{p}(x_j)$ is high. Usually all messages are initialized to 1 (uniform), or a random positive value. To update a message from i to j, consider all messages flowing into i (except for the message from j). This is presented in Figure 14.2.2.

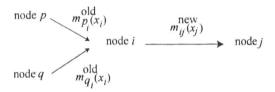

Figure 14.2.2. Message from node i to node j.

In this example, the equation is

$$m_{ij}^{\text{new}}(x_j) = \sum_{x_i} f_{ij}(x_i, x_j)\, g_i(x_i) \underbrace{\prod_{k \in \text{nbd}(i) \backslash j} m_{ki}^{\text{old}}(x_i)}_{h(x_i)},$$

where the messages (and unary factors) on the right side multiply like independent likelihoods. This yields the *update equation* of the form

$$m_{ij}^{\text{new}}(x_j) = \sum_{x_i} f_{ij}(x_i, x_j) h(x_i). \qquad (14.2.1)$$

Note that in a message update, given a pair of neighboring nodes, there is only one pairwise interaction, but messages flow in *both* directions. Hence, we define *pairwise potential* so that we can use the message update equation in both directions (from node i to j and from j to i), i.e., $f_{ij}(x_i, x_j) = f_{ji}(x_j, x_i)$. However, this condition is not the same as assuming a symmetric potential which is defined slightly differently as $f_{ij}(x_i, x_j) = f_{ij}(x_j, x_i)$.

In practice, messages are normalized to sum to 1, so that $\sum_{x_j} m_{ij}(x_j) = 1$. It is useful for numerical stability, otherwise overflow or underflow is likely to occur after a certain number of message updates. The message update schedule is either *synchronous*, i.e., all messages are updated in parallel, or *asynchronous*, i.e., all messages are updated at different times.

It is likely that messages will converge after enough updates, but the question is: Which of these two schedules must be chosen? For a chain, asynchronous is most efficient for a serial computer (an up-and-down chain once guarantees convergence). We can use a similar procedure for a tree. For a grid (e.g., stereo on a pixel lattice), we must sweep in an 'up-down-left-right' fashion. The choice of a good schedule always requires some experimentation.

14.2.3 Belief Read-Out Equation. Once the messages have converged, use the belief read-out equation

$$b_i(x_i) \approx g_i(x_i) \prod_{k \in \text{nbd}(i)} m_{ki}(x_i), \qquad (14.2.2)$$

where $\text{nbd}(i)$ denotes a neighborhood of i. If belief has been normalized, this equation approximates the marginal probability, and if there are no loops, the approximation in Eq (14.2.2) is exact. Another belief equation for pairwise beliefs that estimates pairwise marginal distributions is available in Bishop [2006]. The main cost is the message update equation (14.2.1), which is $O(S^2)$ for each pair of variables in S possible states.

14.2.4 Sum-Product versus Max-Product. The standard belief propagation described above is called the *sum-product*, because of the message update equation (14.2.1), and it is used to estimate marginals. A simple variant, called the *max-product* (or *max-sum* in log domain), is used to estimate the state configuration with maximum probability. In max-product the message update equation is the same as Eq (14.2.1) except that the sum is replaced by max, that is,

$$m_{ij}^{\text{new}}(x_j) = \max_{x_i} f_{ij}(x_i, x_j) g_i(x_i) \prod_{k \in \text{nbd}(i) \backslash j} m_{ki}^{\text{old}}(x_i). \qquad (14.2.3)$$

In this case, belief propagation is the same as before, but beliefs no longer estimate marginals. Instead, they are scoring functions whose maxima point to the most likely states.

As noted above, max-products become max-sums in the log domain, and messages can be positive or negative. However, in practice beliefs are often 'normalized' to avoid underflow or overflow, e.g., by shifting them so that the lowest belief value is zero.

Example 14.2.1. (MRF stereo) Let r and s denote distinct 2D image coordinates (pixels) (x, y). Denote the unknown disparity field $D(x, y) = D(r)$, and the prior smoothness by $\mathfrak{p}(D) = \frac{1}{Z} e^{-\beta V(D)}$, $V(D) = \sum_{<rs>} |D_r - D_s|$, where summation is taken over all neighboring pixels. Often, for greater robustness the penalty $\min(|D_r - D_s|, \tau)$ is used instead of $|D_r - D_s|$. Let m denote the marching error across the entire left and right images, i.e., $m_r(D_r) = |L(x + d_r, y) - R(x, y)|$. Then the simple likelihood function is

$$\mathfrak{p}(m_r(D_r)|D_r) = \frac{1}{Z'} e^{-\mu m_r(D_r)}. \tag{14.2.4}$$

Assume that $\mathfrak{p}(m|D) = \prod_r \mathfrak{p}(m_r(D_r)|D_r)$, which implies that there exists a conditional independence across the images. The posterior, which has both unary and pairwise factors, is defined by $\mathfrak{p}(D|m) = \mathfrak{p}(D)\frac{\mathfrak{p}(m|D)}{\mathfrak{p}(m)}$. Then the sum-product belief propagation estimates the quantity $\mathfrak{p}(D_r|m)$ at each pixel r. The message update schedule is 'left-right-up-down', where 'left' means an entire sweep that updates messages from all pixels to their left neighbors, and so on. One iteration consists of a sweep left, then right, then up, then down. The results obtained are as follows: little change after a few sweeps, but later improvements gained give better results. ∎

14.2.5 Complications and Pitfalls. They are as follows:

1. Suppose there are two state configurations that are equally probable. Belief propagation will show ties between the two solutions. To recover both globally consistent solutions Bishop's back-tracking algorithm [2006] is used.

2. If messages oscillate instead of converging, damp them with 'momentum'; for details, see Murphy et al. [1999].

Belief propagation algorithms can be applied to any graphical model with any form of potentials, even higher-order than pairwise. It is useful for a marginal or maximum probability solution; it is exact when there are no loops; and it is easy to program and easy to parallelize. However, other models may be more accurate, faster, or less memory intensive in some domains (see Szeliski et a. [2008]). For example, graph cuts are faster than belief propagation for stereo and give slightly better results.

14.3 Stopping Time

The Markov property states: "Given the present state B_s, any other information about what happened before time s is irrelevant for predicting what happens after time s." In this statement, the phrase 'what happened before time s' defines a *filtration*, which is an increasing set of fields F_s, that is, if $s \le t$, then $F_s \subset F_t$. The strong Markov property is related to the stopping rule: A random variable s with domain of definition $[0, \infty)$ is a *stopping time* for all times $t \ge 0$, where $\{s, t\} \in F_t$. For example, if we regard B_t as the sequence of encoded symbols and s as the time it takes these symbols to be decoded, then the decision to terminate the decoding before time t should be measurable with respect to the information known at time t. In problems on stopping time, the choice between $\{s < t\}$ and $s \le t$ is important in the sense of discrete time, but unimportant in continuous time $(t > 0)$. The following two results are obvious:

(i) If $\{s \le t\} \in F_t$, then $\{s < t\} = \cup_n \{s \le t - 1/n\} \in F_t$;

(ii) If $\{s < t\} \in F_t$, then $\{s \le t\} = \cap_n \{s < t - 1/n\} \in F_t$, where $n \in \mathbb{N}$.

The first result requires that $t \to F_t$ is increasing, while the second is related to the fact that $t \to F_t$ is continuous. Thus, the two results are equivalent while checking that something is a stopping time. In the case when F_s is right continuous, the definition of F_s remains unchanged even if we replace $\{s \le t\}$ by $\{s < t\}$.

Example 14.3.1. Let A be a countable union of closed sets. Then $T_A = \inf\{t : B_t \in A\}$ is a stopping time. ∎

Example 14.3.2. Let s be a stopping time and let $s_n = \dfrac{\lceil 2^n s \rceil + 1}{2^n}$. Then $s_n = \dfrac{m+1}{2^n}$ if $\dfrac{m}{2^n} \le s < \dfrac{m+1}{2^n}$. These inequalities imply that we stop at the first time when $\dfrac{k}{2^n} > s$. The s_n is a stopping time. ∎

Example 14.3.3. If s and t are stopping times, so are $\min\{s, t\}$, $\max\{s, t\}$, and $s + t$. ∎

Example 14.3.4. For a sequence t_n of stopping times, $\sup_n t_n$, $\inf_n t_n$, $\limsup_n T_n$, and $\liminf_n t_n$ are stopping times. ∎

Example 14.3.5. If $t_n \downarrow t$ are stopping times, then $F_t = \cap_n F(t_n)$. ∎

14.3.1 Continuous Local Martingales. Let X_n, $n \ge 0$, be a martingale with respect to F_n. If H_n, $n \ge 1$, is any process, we define

$$(H \cdot X)_n = \sum_{m=1}^{n} H_m \left(X_m - X_{m-1}\right).$$

Example 14.3.6. Let $\xi_1, \xi_2, \dots$ be independent processes with $\mathfrak{p}\,(\xi_i = 1) = \mathfrak{p}\,(\xi_i = -1) = \frac{1}{2}$, and let $X_n = \xi_1 + \xi_2 + \cdots + \xi_n$ be the *symmetric simple random walk*, then X_n is a martingale with respect to $F_n = \sigma\,(\xi_1, \dots, \xi_n)$. ∎

Theorem 14.3.1. *Let X_n be a martingale. If H_n is predictable and each H_n is bounded, then $(H \cdot X_n)$ is a martingale.*

In continuous time, we have the dictum "You can't make money gambling on a fair game." This means our integral must be martingales. However, since the present (t) and the past $(< t)$ are contiguous and thus not separated, the definition of the class of allowable integrable integrals becomes very subtle.

Example 14.3.7. Let $(\Omega, F, \mathfrak{p})$ be a probability space on which we define a random variable T with $\mathfrak{p}(T \le t) = t$ for $0 \le t \le 1$, and an independent random variable ξ with $\mathfrak{p}(\xi = 1) = \mathfrak{p}(\xi = -1) = \frac{1}{2}$. Let

$$X_t = \begin{cases} 0, & t < T, \\ \xi, & t \ge T, \end{cases}$$

and let $F_t = \sigma(X_s : s \le t)$. This means that we wait until time T and then flip a coin. Notice that X_t is a martingale with respect to F_t. Notice that in the two-signal case for transmission over an additive white Gaussian noise (AWGN) channel, the variable $\xi = \pm 1$ represents voltages associated with the bits 1 and 0 such that the binary 0 (or the voltage value -1) can be regarded as the null element under addition.

However, if we define the stochastic integral $I_t = \int_0^1 X_t \, dX_t$ as the Lebesgue-Stieltjes integral, then I_t is *not* a martingale. To see this, note that the measure dX_s corresponds to a mass of the size ξ at t and hence the integral is ξ times the value there. Define $Y_t = \int_0^t X_s \, dX_s$. Then $Y_1 = \int_0^1 X_s \, dX_s = X_t \cdot \xi = \xi^2 = 1$, and $Y_0 = 0$.

So the question arises, why local martingales? The reasons are two-old:

(i) since we often consider processes that depend on a random time interval $[0, \tau)$, where $\tau < \infty$, the concept of martingales becomes meaningless, as shown by Example 14.3.7, simply because X_t is not defined for large t on the whole space; and

(ii) since most results are proved by introducing stopping times t_n, the problems are reduced to the existence of martingales and proving these results becomes much easier for local martingales defined on a random time interval than for ordinary martingales.

Some useful results are:

(i) $\{X_{\gamma(t)}, F_{\gamma(t)}, t \ge 0\}$ is a martingale.

(ii) If X_t is a local martingale and $E\left(\sup_{0 \le s \le 1} |X_s|\right) < \infty$ for each t, then X_t is a martingale.

(iii) A bounded local martingale is a martingale.

Proofs of these results are available in Durrett [1996]. A martingale is like a fair game. It defines a stochastic process, a sequence of random variables $x_0, x_1, \ldots$ with finite means such that the conditional expected value of an observation X_{n+1} at time t, given all the observations $x_0, x_1, \ldots, x_n$ up to time s, is equal to an observation x_n at an earlier time s, i.e., $(x_{n+1}|x_0, \ldots, x_n) = x_n$ (Doob [1953]). The term 'martingale' was first used to describe a class of betting strategies in which the bet was doubled or halved after a loss or win, respectively. The concept of martingales in probability theory was introduced by Lévy [1925], which was extended by Doob. A simple example of a martingale is a linear random walk with steps equally likely in either direction ($\mathfrak{p} = \mathfrak{q} = 1 - \mathfrak{p} = 1/2$). Other examples are:

(i) x_n being a gambler's fortune after n tosses of a coin, where he wins for heads and loses for tails; the gambler's expected fortune after the next trial, given his history, is the same as his present fortune, and so this sequence is a martingale.

(ii) de Moivre's martingale: suppose a biased coin with probability $\mathfrak{p}$ of 'heads' and $\mathfrak{q} = 1 - \mathfrak{p}$ of 'tails'. Let $x_{n+1} = x_n \pm 1$, where the plus or minus sign is chosen according as a 'head' or a 'tail' is thrown. Let $y_n = (\mathfrak{q}/\mathfrak{p})^{x_n}$. Then the sequence $\{y_n : n = 1, 2, \ldots\}$ is a martingale with respect to $\{x_n : n = 1, 2, \ldots\}$.

(ii) Polya's urn: The urn initially contains r red and b blue marbles, out of which one is chosen at random and put back into the urn along with another marble of the same color. Let x_n denote the number of red marbles in the urn after n iterations of this process, and let $y_n = x_n/(n + r + b)$. Then the sequence $\{y_n : n = 1, 2, \ldots\}$ is a martingale.

(iv) Likelihood-ratio (LR) testing, which is discussed in detail in §14.5 and §16.5.

14.4 Probability Density Function

Given a sample space S, a mathematical description of samples (events) can be given in terms of a coordinate system that is imposed on the space S. Let X denote the Cartesian (x, y) coordinate space, which describes the locations of various samples in S in terms of its coordinates.

Example 14.4.1. A sample space S and the corresponding coordinate space X are presented in Figure 14.4.1, in which individual samples on the left are labeled $1, 2, \ldots, 35$, whereas in the coordinate system X on the right the location of each sample is specified by its (x, y) coordinates. The function that defines the location of each sample is called a *density function*, since it shows how densely packed the sample is. Assuming that the odds of selecting any one sample in S are the same as for any other, the probability density function $\mathrm{pdf}(x, y)$ is

$$\mathrm{pdf}(x, y) = \begin{cases} 1/35 \approx 0.028571428 & \text{for } -3 \leq x \leq 3, -4 \leq y \leq 4, \\ 0 & \text{otherwise.} \end{cases}$$

This means that the value of $\mathfrak{p}(x, y)$ is zero everywhere in X except where the coordinates are positive integers from 1 to 35, in which case $\mathfrak{p}(x, y) = 1/35$.

The samples are said to be a discrete distribution since $\text{pdf}(x, y) \neq 0$ only at the above 35 discrete values. Let $A \subset S$ denote an event (Figure 14.4.1). Then the probability of the event A is the sum of all probabilities of all samples within the circular region, that is, $\mathfrak{p}(A) = \sum_{\mathbf{x} \in A} \mathfrak{p}(\mathbf{x})$, where $\mathbf{x}$ is the vector representing the discrete points (x, y). Since each probability is equal to $1/35$, we obtain $\mathfrak{p}(A) = \sum_{i=1}^{9}(1/35) = 9/35 \approx 0.257$, i.e., $\mathfrak{p}(A)$ is about 25.7%. Notice that the value of $\mathfrak{p}(S) = 1$, so there is a 100% chance of choosing at least one sample from all possible samples in S.

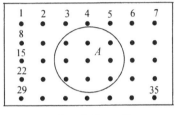
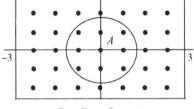

Sample Space S Coordinate Space X

Figure 14.4.1 Sample space and coordinate space.

The function $\text{pdf}(x, y)$ need not be constant for all values of x and y. If the probability increases downward with size 3 and if the probability also increases to the right with size 4, then the probability density function is given by

$$\text{pdf}(x, y) = \frac{|x + 4||y - 3|}{(35)(4)(3)} \quad \text{for } -3 \leq x \leq 3, \ -4 \leq y \leq 4.$$

Thus, for example, $\text{pdf}(-3, 2) = 2.38\%$ for the sample at $(-3, 2)$, $\text{pdf}(-3, -2) = 1.19\%$ for the sample at $(-3, -2)$, $\text{pdf}(3, -2) = 8.33\%$ for the sample at $(3, -2)$, and $\text{pdf}(0, 0) = 2.85\%$ for the sample at $(0, 0)$. Recalculating $\mathfrak{p}(A)$, we get

$$\mathfrak{p}(A) = \text{pdf}(-1, -1) + \text{pdf}(0, -1) + \text{pdf}(1, -1) + \text{pdf}(-1, 0) + \text{pdf}(0, 0)$$
$$+ \text{pdf}(-1, 1) + \text{pdf}(0, -1) + \text{pdf}(1, 1) + \text{pdf}(1, 0)$$
$$= \big[2.857 + 3.81 + 4.762 + 2.142 + 2.857 + 1.428 + 3.81 + 2.381 + 3.57\big]\%$$
$$= 27.62\%.$$

However, if the probability increases downward by 4 and to the right by 3, we have $\text{pdf}(x, y) = \dfrac{|x + 3||y - 4|}{420}$, $-3 \leq x \leq 3, \ -4 \leq y \leq 4$, which yields $\mathfrak{p}(A) = 26.19\%$. ∎

The probability density function for a discrete distribution can be defined by other mathematical functions. A probability density function must have the following properties:

(i) $\mathrm{pdf}(\mathbf{x}) \geq 0$ for all $\mathbf{x} \in X$;

(ii) $\sum_{\mathbf{x} \in X} \mathrm{pdf}(\mathbf{x}) = 1$, where $\mathbf{x} \in X$ means to include the probability calculated
at each $(x, y) \in X$; and

(iii) $\mathrm{pdf}(A) = \sum_{\mathbf{x} \in A} p(\mathbf{x})$ for an event $A \subset S$

In the case of continuous probability distributions, the probability density function would not have nonzero values only at specific coordinates; instead, it would have continuous changes in value as the coordinates (x, y) change. The probability density function, or the *probability density*, of a *continuous* random variable, denoted by f, describes the occurrence of this variable at a given point. The probability that a random variable will fall within a pre-assigned region is given by the integral of that variable's probability density function over the region. By properties (i) and (ii) the probability density function is nonnegative everywhere, and its integral over the entire space is equal to 1. Analytically, a continuous probability density function is generally associated with absolutely univariate distributions. A random variable x has density f if

$$\mathfrak{p}[a \leq x \leq b] = \int_a^b f(x)\,dx,$$

where f is a non-negative Lebesgue integrable function. Thus, if $F(x)$ denotes the cumulative distribution function of x, then $F(x) = \int_{-\infty}^x f(t)\,dt$, and if f is continuous at x, then $f(x) = \dfrac{dF(x)}{dx}$.

Example 14.4.2. The uniform distribution on the interval $[0, 1]$ has the probability density function $f(x) = \begin{cases} 1 \text{ for } 0 \leq x \leq 1, \\ 0 \text{ elsewhere.} \end{cases}$

The standard normal distribution has the probability density function $f(x) = \dfrac{1}{\sqrt{2\pi}}\, e^{-x^2/2}$.

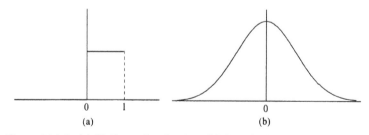

$$\begin{array}{cc} 0 \quad 1 & 0 \\ \text{(a)} & \text{(b)} \end{array}$$

Figure 14.4.2. (a) Uniform distribution (b) Standard normal distribution.

If a random variable s is given and its distribution admits a probability density function f, then the expected value of s, if it exists, is given by $E[s] = \int_{-\infty}^{\infty} x f(x)\,dx$. Note that not every probability distribution has a probability density function; for example, the distributions of discrete random variables do not. Another example is the Cantor distribution, which has no discrete component, i.e., it does not assign positive probability to any individual point. ∎

A distribution has a probability density iff its cumulative distribution function $F(x)$ is absolutely continuous, because then F is differentiable almost everywhere, and its derivative can be used as a probability density since $F'(x) = f(x)$. If the probability distribution admits a continuous probability density function f, then the probability of every one-point set $\{a\}$ is zero, as it is for finite and countable sets. Two probability density functions f and g represent the same probability distribution if they differ only on a set of Lebesgue measure zero.

The Dirac delta function[2] is used to represent certain discrete random variables as well as random variables involving both continuous and a discrete part with a generalized probability density. For example, for a binary discrete random variable taking values -1 or 1 with probability $1/2$ each, the probability density is defined by $f(t) = \frac{1}{2}[\delta(t+1) + \delta(t-1)]$. In general, if a discrete variable can take n different real values, then the associated probability density is $f(t) = \sum_{i=1}^{n} \mathfrak{p}_i \delta(t - \tau_i)$, where $\tau_1, \dots, \tau_n$ are the discrete values taken by the variable and $\mathfrak{p}_1, \dots, \mathfrak{p}_n$ are the probabilities associated with these values. This relation is the basis for unifying the discrete and continuous probability distributions.

In the case of multiple random variables $s_1, \dots, s_n$, the probability density associated with the set as a whole is called the *joint probability density*, defined as a function of the n variables such that for any domain D in the n-dimensional space of the values of n variables, the probability is defined as

$$\mathfrak{p}(s_1, \dots, s_n \in D) = \int_D f(s_1, \dots, s_n(x_1, \dots, x_n)\,dx_1 \dots dx_n.$$

If $F(x_1, \dots, x_n) = \mathfrak{p}(s_1 \le x_1, \dots, s_n \le x_n)$ is the cumulative distribution function of the vector $(s_1, \dots, s_n)$, then the joint probability density is given

---

[2] This function represents a point source and belongs to a class of generalized functions (or distributions, as they are often called, although care must be taken to distinguish them from ordinary distributions discussed in Chapter 17). In the one-dimensional case, it is defined as $\delta(x - x') = \lim_{x_0 \to 0} F(x)$, where $F(x) = \begin{cases} 0 & \text{if } x_0 < |x - x'| \\ 1/(2x_0) & \text{if } x_0 \ge |x - x'| \end{cases}$. Some important properties are: (i) $\delta(x - x') = 0$ for $x \ne x'$; (ii) $\int_{-\infty}^{\infty} \delta(x - x')\,dx = 1$; and (iii) $\int_{-\infty}^{\infty} f(x)\delta(x - x')\,dx = f(x')$. For detailed information and its applications, see Kythe [1996; 2011].

by

$$f(x) = \left. \frac{\partial^n F}{\partial x_1 \cdots \partial x_n} \right|_x .$$

Continuous random variables $s_1, \dots, s_n$ admitting a joint probability density are all independent iff $f_{s_1, \dots, s_n}(x_1, \dots, x_n) = f_{s_1}(x_1) \cdots f_{s_n}(x_n)$.

Example 14.4.3. Let $\mathbf{s}$ be a 2-dimensional random vector of coordinates (X, Y). Then the probability of obtaining $\mathbf{s}$ in the quarter plane $X > 0, Y > 0$ is given by $\mathfrak{p}(X > 0, Y > 0) = \int_0^\infty \int_0^\infty f_{X,Y}(x, y) \, dx \, dy$. ∎

The probability density of the sum of two independent random variables u and v, each of which has a probability density, is the convolution of their separate probability densities:

$$f_{u+v}(x) = \int_{-\infty}^{\infty} f_u(y) f_v(x - y) \, dy = (f_u \star f_v)(x). \tag{14.4.1}$$

This result is generalized to a sum of n independent random variables $u_1, \dots, u_n$:

$$f_{u_1 + \cdots + u_n}(x) = (f_{u_1} \star \cdots \star f_{u_n})(x). \tag{14.4.2}$$

The 'change of variable' rule for the probability density is used to calculate the probability density function of some variable $t = g(s)$, and is defined by

$$f_t(y) = \left| \frac{1}{g'(g^{-1}(y))} \right| f_s\left(g^{-1}(y)\right), \tag{14.4.3}$$

where g^{-1} denotes the inverse function and g' its derivative with respect to x. Since the probability density in a different area must be invariant under the change variable, we have

$$\left| f_t(y) \, dy \right| = \left| f_s(x) \, dx \right|,$$

or

$$f_t(y) = \left| \frac{dx}{dy} \right| f_s(x) = \left| \frac{1}{g'(x)} \right| f_s(x) = \left| \frac{1}{g'(g^{-1}(y))} \right| f_s\left(g^{-1}(y)\right). \tag{14.4.4}$$

For a non-monotonic function, the probability density for y is given by

$$f(y) = \sum_{k=1}^{n(y)} \left| \frac{1}{g'(x)} \right| f_s\left(g^{-1}(y)\right), \tag{14.4.5}$$

where $n(y)$ is the number of solutions in x for the equation $g(y) = y$, and $g_k^{-1}(y)$ are these solutions. The details of these definitions can be found in Ord [1972] and Ushakov [2001].

14.5 Log-Likelihood Ratios

A likelihood test is used in statistical analysis to compare the fit of two models, one of which, called the *null model*, is a special case of the other, called the *alternative model*. This test, based on the likelihood ratio, or equivalently its logarithm, is often used to compute its pdf that decides whether to reject one model against the other. Analytically, each of the two models is separately fitted to data and the log-likelihood recorded. The test statistic D is then defined as twice the difference in these log-likelihoods:

$$D = -2\log\left\{\frac{\text{likelihood for null model}}{\text{likelihood for alternative model}}\right\}$$

$$= 2\log(\text{likelihood for alternative model}) - 2\log(\text{likelihood for null model}).$$

In most cases the probability of D is computed, usually by approximating it using a chi-square distribution of the two models with their respective degree of freedom (dof).

Example 14.5.1. Consider a model A with 1 dof and a log-likelihood of -7064 and a model B with 3 dof and a log-likelihood of -7058. Then the probability of this difference is equal to the chi-square value of $2(7064 - 7058) = 12$ with $3 - 1 = 2$ dof. ∎

In general, let $H_0 : \theta = \theta_0$ and $H_1 : \theta = \theta_1$ be two simple hypotheses, and let $L(\theta_0|x)$ and $L(\theta_1|x)$ denote the likelihood function for the null and alternative models, respectively. Then the log-likelihood ratio (LLR) is defined by

$$\Lambda(x) = \frac{L(\theta_0|x)}{L(\theta_1|x)} = \frac{L(\theta_0|x)}{\sup\{L(\theta|x) : \theta \in (\theta_0, \theta_1)\}}, \tag{14.5.1}$$

where the decision rule is as follows: Let a and $\mathfrak{p}$ be two values chosen to obtain a specified significance level α according to the Neyman-Pearson lemma[3] (see Neyman and Pearson 1933] or Hoel et al. [1971]), i.e., $\mathfrak{p}(\Lambda = a|H_0) + \mathfrak{p}(\Lambda < a|H_0) = \alpha$, where α is the tolerance limit for the rejection of the null hypothesis. Then, if $\Lambda > a$, do not reject H_0; but if $\Lambda < a$, reject H_0; and if $\Lambda = a$, reject both with probability $\mathfrak{p}$. The likelihood ratio (LR) test rejects the null hypothesis if the value of this statistic is too small, but the smallness depends on the value of α.

If the distribution of the LR associated with a particular null and alternative hypothesis can be explicitly determined, it can be definitely used to form decision regions with regard to the acceptance or rejection of the null hypothesis. In practical cases it is very difficult to determine the exact distribution of LRs. However, according to Wilks [1938], as the sample size $n \to \infty$, the test

---

[3] This lemma states that in testing two hypotheses $H_0 : \theta + \theta_0$ and $H_1 : \theta = \theta_1$, the likelihood-ratio (LR) test to reject H_0 in favor of H_1 is $\Lambda(x) = \dfrac{L(\theta_0|x)}{L(\theta_1|x)} \leq \alpha$, where $\mathfrak{P}(\Lambda(x) \leq \eta|H_0) = \eta$ is of size η for a threshold (tolerance) α.

statistic $-2\log(\Lambda)$ for a model is asymptotically chi-square distributed with degrees of freedom equal to the difference in dimensions of θ and θ_0. From a practical point of view, this means that for a number of hypotheses, the computation of LR Λ for the data and a comparison of $-2\log(\Lambda)$ to the chi-squared value will be sufficient as an approximate statistical test. Although there are some criticisms of the above LLR, they do not seem to be significant against the strong argument of their practicality to statistical inference.

Reliability information is used from soft-decision channel decoders for different kinds of control techniques. Almost all wireless systems have some form of channel coding, such as low-density parity-check codes (see detail in Chapters 15 and 16), whereas the soft-decision decoders are invariably used to decode channel codes. Thus, soft-output reliability becomes important in designing the decoders. To estimate error rates from the log-likelihood ratio (LLR), one of the methods is to map the LLR to a bit-error probability (BEP), and then use time-averages of BEP for control. Let the BEP and LLR be regarded as random variables (or stochastic time-series). Then the probability distribution function of each metric can be determined. The LLR for the i-th transmitted bit c_i is defined as

$$\Lambda_i(c_i) \equiv \Lambda(c_i|w_i) = \log \frac{\mathfrak{p}(c_i = 1|w_i)}{\mathfrak{p}(c_i = 0|w_i)}, \qquad (14.5.2)$$

where w_i is the received bit (or signal sample). Using Bayes' rule (14.1.2), and assuming that $\mathfrak{p}(c_i = 1) = \mathfrak{p}(c_i = 0) = 0$ which refers to the two-signal case (see Example 14.3.7), Eq (14.5.2) becomes

$$\Lambda_i(w_i) = \log \frac{\mathfrak{p}(w_i|c_i = 1)\mathfrak{p}(c_i = 1)/\mathfrak{p}(w_i)}{\mathfrak{p}(w_i|c_i = 0)\mathfrak{p}(c_i = 0)/\mathfrak{p}(w_i)} = \log \frac{\mathfrak{p}(w_i|c_i = 1)}{\mathfrak{p}(w_i|c_i = 0)} + \log \frac{\mathfrak{p}(c_i = 1)}{\mathfrak{p}(c_i = 0)}$$

$$= \Lambda(w_i|c_i) + \Lambda(c_i). \qquad (14.5.3)$$

In the case when the noise pdf is a Gaussian of power σ, i.e., when $\mathfrak{p}(w_i|c_i = 1) = \frac{1}{2\pi\sigma} \exp\{-(w_i - \sqrt{E_b})^2/2\sigma^2\}$, Eq (14.5.3) reduces to

$$\Lambda_i(w_i) = \ln \frac{\frac{1}{2\pi\sigma} \exp\{-(w_i - \sqrt{E_b})^2/2\sigma^2\}}{\frac{1}{2\pi\sigma} \exp\{-(w_i + \sqrt{E_b})^2/2\sigma^2\}} = \frac{4\sqrt{E_b}}{N_0} w_i, \qquad (14.5.4)$$

where E_b is the energy per (information) bit, N_0 the variance, and σ^2 the noise variance. The quantity E_b/N_0 is measured in dB units. Then the pdf of LLR, denoted by $\text{pdf}(\Lambda_i)$, subject to the transmitted signal c_i, is given by

$$\text{pdf}(\Lambda_i) = \frac{1}{\sqrt{2\pi} \left[2\sqrt{2E_b/N_0}\right]} \exp\left\{-\frac{\left(\Lambda_i - (4\sqrt{E_b}/N_0) c_i\right)^2}{16E_b/N_0}\right\}, \qquad (14.5.5)$$

which shows that

$$\text{pdf}(\Lambda_i) \sim N\left\{\frac{4E_b}{N_0}\,\text{sgn}(c_i), \frac{8E_b}{N_0}\right\} = N\{4\gamma\,\text{sgn}(c_i), 8\gamma\}, \qquad (14.5.6)$$

where $\gamma = E_b/N_0$, and sgn is ± 1 depending on the voltage value (i.e., sign of the polarity). Thus, the pdf for LLR is a normally distributed random variable with the above mean and variance, and is obtained from Eq (14.5.6). A graph of $\text{pdf}(\Lambda_i)$ is presented in Figure 14.5.1.

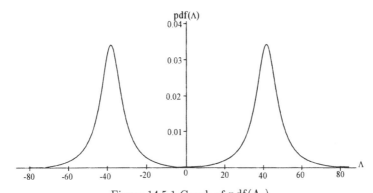

Figure 14.5.1 Graph of $\text{pdf}(\Lambda_i)$.

The bit-error probability (BEP) $\mathfrak{p}_b$ is then defined by

$$\mathfrak{p}_b = \mathfrak{p}(\Lambda_i < 0|c_i = 1)\mathfrak{p}(c_i = 1) + \mathfrak{p}(\Lambda_i > 0|c_i = 0)\mathfrak{p}(c_i = 0)$$

$$= \tfrac{1}{2}\mathfrak{p}\left(\Lambda_i < 0|u_i = +\sqrt{E_b}\right) + \tfrac{1}{2}\mathfrak{p}\left(\Lambda_i > 0|u_i = -\sqrt{E_b}\right). \qquad (14.5.7)$$

Then, given the observed value of LLR, the BEP, denoted by ε_i for the i-th received bit w_i, is given by $\varepsilon_i = \min\{\text{pdf}_0, \text{pdf}_1\}$, where $\text{pdf}_0 = \mathfrak{p}(w_i = 0)$ and $\text{pdf}_1 = \mathfrak{p}(w_i = 1)$. Since $\text{pdf}_0 = 1 - \text{pdf}_1$ and $\text{pdf}_i = \dfrac{e^{\Lambda_i}}{1 + e^{\Lambda_i}}$, we get $\varepsilon_i = \dfrac{1}{1 + e^{|\Lambda_i|}}$. As $\text{pdf}(\Lambda_i)$ is known, the pdf of BEP, $\text{pdf}(\varepsilon_i)$, is then given by

$$\text{pdf}(\varepsilon_i) = \frac{1}{\sqrt{2\pi}\,\sigma\,\varepsilon_i(1 - \varepsilon_i)}\left\{\exp\{-\left(\ln(1/\varepsilon_i - 1) + \mu\right)^2/2\sigma^2\}\right.$$

$$\left. + \exp\{-\left(\ln(1/\varepsilon_i - 1) - \mu\right)^2/2\sigma^2\}\right\} \qquad (14.5.8)$$

$$= \frac{1}{\varepsilon_i}\left[N(-\mu, \sigma^2) + N(\mu, \sigma^2)\right] \quad \text{for } 0 \leq \varepsilon_i \leq \tfrac{1}{2},$$

where $N(\pm\mu, \sigma^2)$ are two normal distributions, and μ and σ are functions of γ defined in (14.5.6). If instead of the LLR of the i-th transmitted bit c_i, the LLR of the received i-th bit w_i needs be computed, just interchange c_i

and w_i in the above discussion from (14.5.2) through (14.5.8). Note that the log-likelihood of a symbol x_i is defined as

$$\Lambda_i = \log \mathfrak{p}(x_i|y) = \log \frac{\mathfrak{p}(x_i|y)}{\sum\limits_{x_j \in S} \mathfrak{p}(x_i|y)} = \log \frac{f(y|x_i)}{\sum\limits_{x_j \in S} f(y|x_j)}$$

$$= \log f(y|x_i) - \log \sum\limits_{x_j \in S} f(y|x_j) = \log f(y|x_i) - \log \sum\limits_{x_j \in S} e^{\log f(y|x_j)}$$

$$= \log f(y|x_i) - \max{}^*{}_{x_j \in S}\Big\{ \log f(y|x_j) \Big\},$$

where the max* function is defined as

$$\max{}^*(x, y) = \log\left[e^x + e^y\right] = \max(x, y) + \log\left\{1 + e^{-|y-x|}\right\}$$
$$= \max(x, y) + f_c(|y - x|), \quad f_c(z) = \log\{1 + e^{-z}\}.$$

A graph of $f_c(z)$, $z + |y - x|$, is shown in Figure 14.5.2.

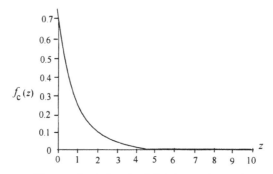

Figure 14.5.2 Graph of $f_c(z)$, $z = |y - x|$.

14.5.1 LLR Algebra. (Hagenauer [1996]) For the sake of simplicity, denote c_i and w_i by θ and x, respectively. Then formula (14.5.3) is written as

$$\Lambda(\theta|x) = \Lambda(x|\theta) + \Lambda(\theta), \tag{14.5.9}$$

where the statistic $\Lambda(x|\theta)$ is obtained by computing the channel output x at the decoder under the alternate conditions $\theta = +1$ or $\theta = -1$, and $\Lambda(\theta)$ is the a priori LLR of the bit θ. For statistically independent data θ, let the sum of two LLRs $\Lambda(\theta_1)$ and $\Lambda(\theta_2)$ be defined by

$$\Lambda(\theta_1) \boxplus \Lambda(\theta_2) \stackrel{\text{def}}{=} \Lambda(\theta_1 \oplus \theta_2) = \ln \frac{e^{\Lambda(\theta_1)} + e^{\Lambda(\theta_2)}}{1 + e^{\Lambda(\theta_1)}e^{\Lambda(\theta_2)}}, \tag{14.5.10}$$

$$\approx (-1)\,\text{sgn}\{\Lambda(\theta_1)\}\,\text{sgn}\{\Lambda(\theta_2)\}\cdot\min\{|\Lambda(\theta_1)|, |\Lambda(\theta_2)|\}, \tag{14.5.11}$$

where sgn denotes the sign of polarity, as above, $\oplus$ the bitwise mod-2 sum, and $\boxplus$ the LLR addition defined by (14.5.10), which is the LLR of the mod-2 sum of the statistically independent data bits (summands). Formula (14.5.10) is a consequence of (14.5.4); it is an example of the case when the probability domain is equivalent to the energy (log probability) domain, as mentioned in §14.3. Formula (14.5.11) gives an approximate value of formula (14.5.10) and will be useful in some computations later on. In the case when one of the LLRs $\Lambda(\theta_1)$ or $\Lambda(\theta_2)$ is very small (or very large), formula (14.5.10) provides the the following two special formulas:

$$\Lambda(\theta) \boxplus 0 = 0, \quad \text{and} \quad \Lambda(\theta) \boxplus \infty = -\Lambda(\theta), \qquad (14.5.12)$$

where 0 is referred to as the null element in GF(2). For an AWGN channel at $\sqrt{E_b}/N_0 = 0.5$ dB, Eq (15.5.4) becomes

$$\Lambda(w_i) = 2w_i. \qquad (15.5.13)$$

14.5.2 Message Passing Algorithm. (MPA) Bayes' theorem (14.1.2) becomes very complicated for large networks, and so a message passing algorithm is needed. This is defined as follows: The first messages that are simply the identity messages (bits) are passed from some set of singly-connected function nodes (white circles) to their dependent variable nodes (gray circles). Only *a priori* knowledge of the messages is required. The variable nodes then pass the same messages along to the next function node. An 8-stage example is presented in Figure 14.5.3.

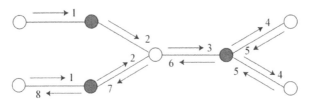

Figure 14.5.3 An 8-stage message passing.

Assuming that the function node is connected to more than one variable node, it must wait for all but one neighbors to send it a message. Once the messages are received, it creates a message of its own and passes that to the variable node that did not send it. This message is created from the message passed to it with its *a priori* knowledge, using the formula

$$\sum_{\{x\}} \left(f(x) \prod_{y \in n(f) \setminus \{x\}} w_{h \to x}(y) \right), \qquad (14.5.14)$$

where $x = n(f)$ is the set of arguments to the function f, and $w_{h \to x}(y)$ is a message from some neighbor y to node 1. This formula simply states that for each possible argument of f, compute the value of f, which is a part of the *a priori* knowledge, then multiply it by each message received according to that same argument. Then the sum of these computations is the message passed to the next variable node (marked stage 3 in Figure 14.5.3); the stage 4 is the same as stage 2; at stage 5, the functions at the end of graph reverse the passing of messages and perform the same marginals as at stage 3; at stage 6, a different type of formula for marginals is used, which is defined as $\prod\limits_{h \in n(x) \backslash \{f\}} w_{h \to x}(x)$, where x is the variable creating the message, and $n(x) \backslash \{f\}$ is the set of functions that have sent x a message without including the message received while the messages were passing in the other direction. The final stages 7 and 8 are computed as above and the algorithm terminates where it started.

14.5.3 Sum-Product Algorithm. Let $q_0 = p(c_i = 0|s_i)$, $q_1 = p(c_i = 1|s_i)$, where s_i is the event that bits in a codeword $\mathbf{c} = \{c_i\}$ satisfy the parity-check equations involving c_i; $q_{ij}(b)$ is the extrinsic information to be passed from p-node j to c-node i (this is the probability that $c_i = b$ given extrinsic information from check nodes and channel sample y_i); $\mathfrak{r}_{ij}(b)$ is the extrinsic information to be passed from c-node i to p-node j (it is the probability that the j-th check equation is satisfied given that $c_i = b$); $C_i = \{i : h_{ij} = 1\}$, where the set of row locations of the 1s in the i-th column, excluding the location j); $R_i = \{i : h_{ij} = 1\}$ (This is the set of column locations of the 1s in the j-th row.) and $R_{ij \backslash j'} = \{i' : h_{ij'} = 1\} \backslash \{j\}$, which is the set of column locations of the 1s in the j-th row, excluding location i. Then the sum-product algorithm is defined by the following four steps:

STEP 1. Initialize $q_{ij}(0) = 1 - p_i = \dfrac{1}{1 + e^{-2yi/\sigma^2}}$. Then $q_{ij}(1) = p_i = \dfrac{1}{1 + e^{2yi/\sigma^2}}$, where $q_{ij}(b)$ is the probability that $c_i = b$, given the channel sample.

STEP 2. At each c-node, update the messages using

$\mathfrak{r}_{ij}(0) = \frac{1}{2} + \frac{1}{2} \prod\limits_{j' \in R_{j \backslash j}} (1 - 2q_{ij'}(1))$, and $\mathfrak{r}_{ji}(1) = 1 - \mathfrak{r}_{ji}(0)$, where $\mathfrak{r}_{ij}(b)$ is the probability that the i-th check equation is satisfied, given $c_i = b$.

STEP 3. Update $q_{ij}(0)$ and $q_{ij}(1)$ using

$q_{ij}(0) = k_{ij}(1 - p_i) \prod\limits_{j' \in C_{i \backslash j}} (\mathfrak{r}_{ij'}(0))$, and $q_{ij}(1) = k_{ij}(p_i) \prod\limits_{j' \in C_{i \backslash j}} (\mathfrak{r}_{ij'}(1))$.

STEP 4. Make hard decisions by computing: $q_i(0) = k_{ij}(1 - p_i) \prod\limits_{j \in C_i} (\mathfrak{r}_{ij}(0))$,

and $q_i(1) = k_{ij}(1 - p_i) \prod\limits_{j \in C_i} (\mathfrak{r}_{ij}(1))$, where $c_i = \begin{cases} 1 & \text{if } Q_i(1) \geq \frac{1}{2}, \\ 0 & \text{otherwise.} \end{cases}$

An important application of the belief propagation algorithm lies in the BP decoding of low-density parity-check codes and pre-codes, which is discussed in §16.6 and used thereafter whenever needed.

15

LDPC Codes

15.1 Tanner Graphs

A Tanner graph is a pictorial representation for the parity-check constraints of an ECC (Tanner [1981]). In Tanner graphs each square represents a parity-check bit and each circle connected to a square represents a bit that participates in that parity check. Thus, the nodes of the graph are separated into two distinct sets: c-nodes (top nodes, circles) and p-nodes (bottom nodes, squares). For an (n, k) code, the c-nodes c_i, $i = 1, \ldots, n$, and the p-nodes p_j, $j = 1, \ldots, m$, $m = n - k$, represent the message bits (symbols) and the parity bits, respectively, for a transmitted word c_i (or received word denoted by w_i). These nodes are connected to each other by the following formula:

$$\left\{ \begin{array}{c} \text{p-node } p_j \text{ is connected to c-node } c_i \text{ if the element } h_{ji} \text{ of } \mathbf{H} \text{ is } 1 \\ \text{for } 1 \leq i \leq n \text{ and } 1 \leq j \leq k. \end{array} \right\} . \quad (15.1.1)$$

A Tanner graph represents a linear code C if there exists a parity-check matrix $\mathbf{H}$ for C associated with the Tanner graph. All variable c-nodes connected to a particular p-node must sum, mod 2, to zero, which is same as XOR-ing them to zero. This provides the constraints. Tanner graphs are bipartite graphs, which means that nodes of the same type cannot be connected, i.e., a c-node cannot be connected to another c-node; the same rule applies to the p-nodes. For example, in Figure 15.1.1(a) or (b) the c-nodes for a (6,3) code are denoted by $c_1, \ldots, c_6$, and the parity-check bits denoted by p_1, p_2, p_3 are the p-nodes. Tanner graphs can also be presented in the vertical form in which the c-nodes are the left nodes and p-nodes the right nodes, and c_i and p_j start at the top and move downward (see §17.5). Sometimes a general notation is used in which the c-nodes and p-nodes are denoted as x-nodes x_i and y-nodes y_j, respectively.

The constraints for the example in Figure 15.1.1 are presented in two ways, both implying the same constraints equations, but the representation (b) is generally easier in the case when the number of c-nodes and p-nodes is large.

In this example the first parity-check bit p_1 forces the sum of the bits c_1, c_2 and c_4 to be even, the second parity-check bit p_2 forces the sum of the code bits c_1, c_3 and c_5 to be even, and the third parity-check bit p_3 forces the sum of the code bits c_2, c_3 and c_6 to be even. Since the only even binary number is 0, these constraints can be written as $c_1 \oplus c_2 \oplus c_4 = 0$, $c_1 \oplus c_3 \oplus c_5 = 0$, $c_2 \oplus c_3 \oplus c_6 = 0$. The only 8 codewords that satisfy these three parity-check constraints are $\{000000, 001011, 010101, 011110, 100110, 101101, 110011, 111000\}$. In this code (Figure 15.1.1) the first three bits are the data bits and the last three bits are then uniquely determined from the constraints; for example, if the data bits are 010, then the codeword to be transmitted is 010101 as determined from the above list of 8 codewords. This construction of codewords is different from the one discussed in Chapter 4. The parity-check matrix $\mathbf{H}$ so obtained must be transformed first into the form $\left[-P^T \mid I_{n-k}\right]$; then the generator matrix $\mathbf{G}$ is obtained by transforming it into the form $[I_k \mid P]$. This method is used in Example 15.6.1, given below.

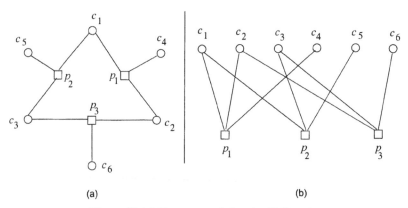

(a)   (b)

Figure 15.1.1 Tanner graph for the (6,3) code.

The graph in Figure 15.1.1 is said to be *regular* since there are the same number of constraints at each parity-check bit, as seen by the same number of lines connecting each square (p-node) to circles (c-nodes). If this number is not the same at each p-node, then the graph is said to be *irregular*, an example of which is given in Figure 15.6.1. However, another feature not present in Figure 15.1.1 but shown in Figure 15.1.2(a) is the situation of a short cycle $p_1 \rightarrow c_1 \rightarrow p_2 \rightarrow c_3 \rightarrow p_1$, which creates the so-called *4-cycle* that should be avoided, if possible, in a circuit design based on a Tanner graph. The graph of Figure 15.1.1 (a) or (b) is equivalent to the parity-check matrix

$$\mathbf{H} = \begin{bmatrix} 1 & 1 & 0 & 1 & 0 & 0 \\ 1 & 0 & 1 & 0 & 1 & 0 \\ 0 & 1 & 1 & 0 & 0 & 1 \end{bmatrix}.$$

Example 15.1.1. Consider two parity-check matrices $\mathbf{H}_1$ and $\mathbf{H}_2$ for the linear binary (5,2,3) code C, given by

$$\mathbf{H}_1 = \begin{bmatrix} 1 & 0 & 1 & 0 & 1 \\ 1 & 1 & 1 & 0 & 0 \\ 1 & 0 & 0 & 1 & 0 \end{bmatrix}, \quad \mathbf{H}_2 = \begin{bmatrix} 1 & 0 & 1 & 0 & 1 \\ 0 & 1 & 0 & 0 & 1 \\ 1 & 0 & 0 & 1 & 0 \end{bmatrix}.$$

The Tanner graph corresponding to $\mathbf{H}_1$ (Figure 15.1.2(a)) shows that $\mathbf{H}_1$ is not cycle-free, since the edges $(c_1, p_1), (p_1, c_3), (c_3, p_2), (p_2, c_1)$ constitute a 4-cycle. However, a cycle-free Tanner graph is shown in Figure 15.1.2(b), which corresponds to the parity-check matrix $\mathbf{H}_2$. ∎

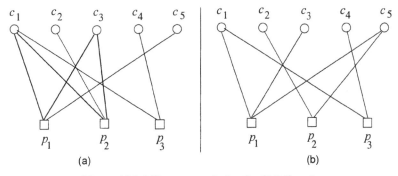

Figure 15.1.2 Tanner graph for the (5,2,3) code.

A formal definition of a Tanner graph for a code C of length n over an alphabet A is a pair $(\mathcal{G}, \mathcal{L})$, where $\mathcal{G} = (V, E)$ is a bipartite graph and $\mathcal{L} = \{C_1, \dots, C_m\}$ is a set of codes over A, called the *constraints*. If we denote the two classes of nodes of $\mathcal{G}$ by $\mathbf{c}$ and $\mathbf{p}$, so that $V = \mathbf{c} \cup \mathbf{p}$, then the nodes of $\mathbf{c}$ are called *symbol nodes* and $|\mathbf{c}| = n$, while the vertices of $\mathbf{p}$ are called the *parity-check nodes* and $|\mathbf{p}| = j$. There is a one-to-one correspondence between the constraints $C_1, \dots, C_m \in \mathcal{L}$ and the parity-check vertices $p_1, \dots, p_j \in \mathbf{p}$, so that the length of the code $C_i \in \mathcal{L}$ is equal to the degree of the nodes $p_j \in \mathbf{p}$ for all $j = 1, 2, \dots, k$. An assignment of a value from A to each symbol vertex $\mathbf{c} = \{c_1, \dots, c_n\}$ is called a *configuration*, which is said to be valid if all the constraints are satisfied. The code C represented by the Tanner graph $(\mathcal{G}, \mathcal{L})$ is then the set of all valid configurations.

In practice, however, we will be considering simple special cases where all the constraints are single parity-check codes over F_2, and the p-nodes are connected to c-nodes by formula (15.1.1). Moreover, it is found that the number of edges in any Tanner graph for a linear code C of length n is $O(n^2)$, which implies that if such a linear code C is represented by a cycle-free Tanner graph, the maximum-likelihood decoding of C (§3.4.12.4) is achievable in time $O(n^2)$ by using, for example, the min-sum algorithm. It is suggested in MacKay [1999] that powerful codes cannot be represented by cycle-free

Tanner graphs, and that such graphs can only support weak codes. So the question arises: which codes can have cycle-free Tanner graphs? Etzion et al [1999] have provided the following answers: If a linear (n, k, d) code C over F_q can be represented by a cycle-free Tanner graph and has the rate $R = k/n \geq \frac{1}{2}$, then $d \leq 2$. Also, if $R < \frac{1}{2}$, then C is necessarily obtained from a code rate $R \geq \frac{1}{2}$ and minimum distance ≤ 2 by simply repeating certain bits in each codeword. In fact, the minimum distance of cycle-free codes is given by (see Etzion et al. [1999, Theorem 5])

$$d \leq \left\lfloor \frac{n}{k+1} \right\rfloor + \left\lfloor \frac{n+1}{k+1} \right\rfloor. \qquad (15.1.2)$$

The bound in (15.1.2) reduces to $d \leq 2$ for $R = k/n = 1/2$. This result shows that linear codes with cycle-free Tanner graphs provide extremely poor exchange between rate and distance for each fixed length n. Thus, at very high signal-to-noise ratios, these codes will perform very poorly, and in general, the minimum distance of a code does not necessarily determine its performance at signal-to-noise ratios of practical interest, although there exist codes, for example Turbo codes discussed in Berrou and Glavieux [1996], that have low minimum distance yet perform extremely well at low signal-to-noise ratios.

15.1.1 Row-Graph of a Tanner Graph. For a linear binary cycle-free (n, k, d) code C, the parity-check $r \times k$ matrix H, where $m = n - k$, is said to be in *s-canonical form* if H can be represented as

$$H = \begin{bmatrix} Q & 0 \\ P & I_s \end{bmatrix} \equiv Q\|_s P, \qquad (15.1.3)$$

where all rows of P have weight ≤ 1, I_s is the $s \times s$ identity matrix, $0 \leq s \leq m$, and the right-side expression in (15.1.3) is the notation to mean that H is in the s-canonical form. For $s = 0$, the matrix $H = Q$, i.e., H is in the 0-canonical form; if $s = m$, the the canonical form is $H = [P \,|\, I_m]$. As mentioned above, the generator matrix $\mathbf{G}$ is obtained by transforming it into the form $[I_m \,|\, P]$. The following result on the structure of the matrix Q is noteworthy: If $H = Q\|sP$ is a cycle-free binary matrix in s-canonical form, then at least one of the following statement is true:

(i) The matrix Q contains a row of weight ≤ 2;

(ii) The matrix Q contains 3 identical columns of weight 1; or

(iii) The matrix Q contains 2 identical columns of weight 1, and the row of Q that contains the nonzero entries of these 2 columns has weight 3. For proof see Etzion et al. [1999, Lemma 6].

The *row-graph* for the Tanner graph $T(Q)$ of Q is constructed as follows: Since $T(Q)$ is a subgraph of $T(H)$, it is obtained by keeping only the first $(n - s)$ c-nodes $c_1, \dots, c_{n-s}$; the first $(k - s)$ p-nodes $p_1, \dots, p_{m-s}$; and all edges between these nodes. As assumed, since $T(H)$ is cycle-free, so is $T(Q)$. Now construct another graph, called the row-graph of Q, such that its p-nodes $p_1, \ldots, p_{m-s}$ correspond to the rows of Q. The set of edges of the row-graph is obtained from the columns of Q of weight ≥ 2, such that a column of weight w in Q contributes $w - 1$ edges to the row-graph.

Example 15.1.2. The construction of the row-graph for a 6×8 cycle-free matrix H is presented in Figure 15.1.3.

$$H = \begin{array}{c} \begin{array}{cccccccc} c_1 & c_2 & c_3 & c_4 & c_5 & c_6 & c_7 & c_8 \end{array} \\ \left[\begin{array}{cccccccc} 0 & 1 & 0 & 1 & 0 & 0 & 0 & 0 \\ 1 & 0 & 0 & 0 & 1 & 0 & 1 & 0 \\ 0 & 1 & 1 & 0 & 0 & 0 & 0 & 0 \\ 0 & 0 & 0 & 0 & 1 & 0 & 0 & 1 \\ 0 & 0 & 0 & 1 & 0 & 1 & 0 & 0 \\ 0 & 0 & 0 & 1 & 0 & 0 & 1 & 0 \end{array}\right] \begin{array}{c} p_1 \\ p_2 \\ p_3 \\ p_4 \\ p_5 \\ p_6 \end{array} \end{array}.$$

The Tanner graph for this matrix is shown in Figure 15.1.3(a). Notice that in this matrix the c-nodes connecting more than one p-node are as follows: c_2 connects p_1 and p_3, c_4 connects p_1 and p_5 and p_6, c_5 connects p_2 and p_4, and c_7 connects p_2 and p_6. These connections generate the row-graph for the matrix H, which are presented in Figure 15.1.3(b).

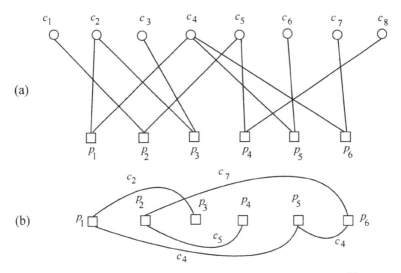

Figure 15.1.3 (a) Tanner graph and (b) row-graph for matrix H. ∎

A binary linear cycle-free (n, k, d) code is said to be in reduced canonical form if $H = Q\|_s P$ and either $s = m$ or all the rows of Q have weight ≥ 3. There exists a cycle-free parity-check matrix for C that is in reduced canonical form. A proof for this statement is given in Etzion et al. [1999]. However, the notion of the row-graph for the Tanner graph has not found much use in coding theory.

15.2 Optimal Cycle-Free Codes

To construct a family of cycle-free codes such that the bound in (15.1.2) is attained for all values of n and k, start with a single parity-check code C_{k+1} of dimension k, and repeat the symbols (entries) of C_{k+1} in each codeword for $k + 1 \leq n$ until a code C of length n is obtained. Obviously, $\dim(C) = k$ and the code C is cycle-free; however, the minimum distance of C depends on the sequence of symbol repetitions, which must be done equally often as far as possible with each symbol in C_{k+1}.

Example 15.2.1. For $n = 13, k = 3$, the parity-check matrix $\mathbf{H}$ is obtained in reduced canonical form:

$$
\mathbf{H} = \left[\begin{array}{cccc|ccccccccc}
1&1&1&1&0&0&0&0&0&0&0&0&0\\
\hline
1&0&0&0&1&0&0&0&0&0&0&0&0\\
1&0&0&0&0&1&0&0&0&0&0&0&0\\
1&0&0&0&0&0&1&0&0&0&0&0&0\\
0&1&0&0&0&0&0&1&0&0&0&0&0\\
0&1&0&0&0&0&0&0&1&0&0&0&0\\
0&0&1&0&0&0&0&0&0&1&0&0&0\\
0&0&1&0&0&0&0&0&0&0&1&0&0\\
0&0&0&1&0&0&0&0&0&0&0&1&0\\
0&0&0&1&0&0&0&0&0&0&0&0&1
\end{array}\right], \quad
\mathbf{H}' = \left[\begin{array}{ccccc|cccccccc}
1&1&1&0&0&0&0&0&0&0&0&0&0\\
\hline
0&1&0&0&0&1&0&0&0&0&0&0&0\\
0&1&0&0&0&0&1&0&0&0&0&0&0\\
0&0&1&0&0&0&0&1&0&0&0&0&0\\
0&0&1&0&0&0&0&0&1&0&0&0&0\\
0&0&0&1&0&0&0&0&0&1&0&0&0\\
0&0&0&1&0&0&0&0&0&0&1&0&0\\
0&0&0&0&1&0&0&0&0&0&0&1&0\\
0&0&0&0&1&0&0&0&0&0&0&0&1
\end{array}\right].
$$

The matrix $\mathbf{H}$ defines a (13,3,6) cycle-free code C. Another cycle-free code C' is defined with the parity-check matrix $\mathbf{H}'$ in reduced canonical form; this code was obtained by repeating symbols (entries) in a (5,3,2) code. It is easy to verify that the code C' is not equivalent to C, although both are (13,3,6) cycle-free codes. ∎

15.2.1 Codes without Cycle-Free Tanner Graphs. A binary linear (n, k, d) code C with rate $R = k/n$ can be represented by a 4-cycle-free Tanner graph iff

$$ m\, d^{\perp} \leq \left\lfloor \sqrt{nm(m-1) + \frac{n^2}{4}} + \frac{n}{2} \right\rfloor, \tag{15.2.1} $$

where $m = n - k$, and $d^{\perp}$ is the minimum distance of the dual code $C^{\perp}$.

Example 15.2.2. Consider the Hamming (7,4,3) code C with $d^{\perp} = 4$. Any parity-check matrix to this code contains 7 nonzero binary vectors of length 3 and thus is isomorphic to the matrix

$$
\mathbf{H} = \begin{bmatrix}
0&0&0&1&1&1&1\\
0&1&1&0&0&1&1\\
1&0&1&0&1&0&1
\end{bmatrix}.
$$

The Tanner graph of this matrix contains three 4-cycles with $md^\perp = 12$ and $\left\lfloor \sqrt{nm(m-1) + \frac{n^2}{4}} + \frac{n}{2} \right\rfloor = \left\lfloor \sqrt{(21)(2) + \frac{49}{4}} + \frac{7}{2} \right\rfloor = 10$, which precludes the existence of a cycle-free Tanner graph but requires that any Tanner graph for this $\mathbf{H}$ contain at least two 4-cycles. ∎

Some authors, e.g., Johnson and Weller [2003], use Steiner systems for designing LDPC codes. The following example explains the technique of such a design.

Example 15.2.3. Consider the code C^* with the parity-check matrix

$$\mathbf{H}^* = \begin{bmatrix} 1\,0\,0\,1\,0\,0\,1\,0\,0\,1\,0\,0 \\ 1\,0\,0\,0\,1\,0\,0\,1\,0\,0\,1\,0 \\ 1\,0\,0\,0\,0\,1\,0\,0\,1\,0\,0\,1 \\ 0\,1\,0\,1\,0\,0\,0\,0\,1\,0\,1\,0 \\ 0\,1\,0\,0\,1\,0\,1\,0\,0\,0\,0\,1 \\ 0\,1\,0\,0\,0\,1\,0\,1\,0\,1\,0\,0 \\ 0\,0\,1\,1\,0\,0\,0\,1\,0\,0\,0\,1 \\ 0\,0\,1\,0\,1\,0\,0\,0\,1\,1\,0\,0 \\ 0\,0\,1\,0\,0\,1\,1\,0\,0\,0\,1\,0 \end{bmatrix}.$$

This is the incidence matrix of the Steiner system $S(2,3;9)$ and the Tanner graph corresponding to this matrix meets the bound in (15.2.1), but the minimum distance for the corresponding code C^* is 2, which is not equal to the minimum row weight of $\mathbf{H}^*$, and hence C^*, in fact, does not achieve the said bound. ∎

In general, codes generated by the incidence matrices of Steiner systems do not have the minimum distance equal to the minimum row weight of corresponding parity-check matrices.

Example 15.2.4. Consider the code C' with the parity-check matrix $\mathbf{H}' = [I\ \mathbf{H}]$, where $\mathbf{H}$ is the $\binom{\nu}{2} \times \nu$ incidence matrix of the Steiner system $S(2,2;\nu)$ and I is the $\nu \times \nu$ identity matrix. Note that the code C' has length $n = \nu + \binom{\nu}{2}$, dimension $k = \binom{\nu}{2}$, and minimum dual distance $d^\perp = \nu$. The values of both the left and right side of the inequality (15.2.1) are given in Table 15.2.1 for $\nu = 2, \ldots, 10$. Note that these values are the same for $\nu = 2$ and 3, but the remaining values of ν on the two sides of the inequality are nearly equal. The corresponding parity-check matrices for $\nu = 2, 3$ are

$$\mathbf{H}_{\nu=2} = \begin{bmatrix} 1\,0\,1 \\ 0\,1\,1 \end{bmatrix}, \quad \mathbf{H}_{\nu=3} = \begin{bmatrix} 1\,0\,0\,1\,0\,1 \\ 0\,1\,0\,1\,1\,0 \\ 0\,0\,1\,0\,1\,1 \end{bmatrix}. \ \blacksquare$$

It is shown that many classical linear binary block codes do not have cycle-free Tanner graphs. Among them are the following codes:

(i) The binary Golay (23, 12,7) code $\mathcal{G}_{23}$ and the extended binary Golay (24, 12, 8) code $\mathcal{G}_{24}$ both have distance 8; however, neither of these codes satisfy the inequality (15.2.1).

(ii) The $(n = 2^m, k, d = 2^{m-\mu})$ Reed–Muller (RM) codes with $R > \frac{1}{2}$ and minimum distance $d > 2$ for $m = 3,\ldots,9$: this code has dimension $k = \sum\limits_{i=0}^{\mu} \binom{m}{i}$, and its dual has minimum distance $2^{\mu+1}$ (see MacWilliams and Sloane [1977]). For various Reed–Muller codes, the values of the two sides of the inequality (15.2.1) are given in Table 15.2.2.

(iii) The $(n = 2^m - 1, k, d)$ primitive BCH codes with $R \geq \frac{1}{2}$ for $m = 3,\ldots,8$: The computed values of both sides of inequality (15.2.1) are given in Table 15.2.3, where the lower bounds are given for the remaining codes after the first two codes.

(iv) The binary images of the $(n = 2^m - 1, k, d = n - k + 1)$ Reed–Solomon (RS) codes with $R \geq \frac{1}{2}$ for $m \geq 4$: In this case, the inequality (15.2.1) becomes

Table 15.2.1 Values of
Inequality (15.2.1)

ν	lhs (15.2.1)	rhs (15.2.1)
2	4	4
3	9	9
4	16	17
5	25	26
6	36	37
7	49	51
8	64	66
9	81	83
10	100	103

Table 15.2.2 RM Codes

RM Code	lhs (15.2.1)	rhs (15.2.1)
(8,4,4)	16	14
(16,11,4)	40	27
(32,26,4)	96	50
(32,16,8)	128	105
(64,57,4)	224	92
(64,42,8)	352	206
(128,120,4)	512	170
(128,99,8)	928	392
(128,64,16)	1024	785
(256,219,8)	2368	725
(256,136,16)	5952	1613
(256,93,32)	5216	2731
(512,502,4)	2048	562
((512,466,8)	5888	1316
(512,382,16)	8320	3197
(512,256,32)	8192	6042

Table 15.2.3 BCH Codes

BCH-Code	$d^\perp$	lhs (15.2.1)	rhs (15.2.1)
(7,4,3)	4	12	10
(15,11,3)	8	32	22
(31,26,3)	≥ 16	≥ 80	44
(31,21,5)	≥ 8	≥ 70	80
(31,16,7)	≥ 8	≥ 120	97
(63,57,3)	≥ 32	≥ 192	85
(63,51,5)	≥ 16	≥ 192	127
(63,45,7)	≥ 16	≥ 288	173
(63,39.9)	≥ 12	≥ 288	220
(63,36,11)	≥ 12	≥ 324	244
(127,120,3)	≥ 64	≥ 448	160
(127,113,5)	≥ 56	≥ 784	228
(127,106,7)	≥ 48	≥ 1008	303
(127,99,9)	≥ 40	≥ 1120	379
(127,92,11)	≥ 32	≥ 1120	457
(127,85,13)	≥ 30	≥ 1260	535
(127,78,15)	≥ 28	≥ 1372	613
(127,171,19)	≥ 22	≥ 1232	692
(127,64,21)	≥ 20	≥ 1260	770
(255,247,3)	≥ 128	≥ 1024	302
(255,239,5)	≥ 112	≥ 1792	405
(255,231,7)	≥ 96	≥ 2304	523
(255,223,9)	≥ 88	≥ 2816	646
(255,215,11)	≥ 64	≥ 2560	770
(255,207,13)	≥ 64	≥ 3072	896
(255,199,15)	≥ 60	≥ 3360	1022
(255,191,17)	≥ 42	≥ 2688	1149
(255,187,19)	≥ 42	≥ 2856	1212
(255,179,21)	≥ 40	≥ 3040	1339
(255,171,23)	≥ 32	≥ 2688	1466
(255,163,25)	≥ 32	≥ 2944	1594
(255,155,27)	≥ 32	≥ 3200	1721
(255,147,29)	≥ 28	≥ 3024	1848
(255,139,31)	≥ 26	≥ 3016	1976
(255,131,37)	≥ 22	≥ 2727	2103

$$d^\perp \leq \sqrt{mn\left(1 - \frac{1}{m(n-k)}\right) + \frac{n^2}{4(n-k)^2}} + \frac{n}{2(n-k)} \leq 1 + \sqrt{mn+1},$$

(15.2.2)

where $(n-k)/k = 1 - R \leq \frac{1}{2}$ for codes with $R \geq \frac{1}{2}$. No 4-cycle Tanner graphs can exist if $k > \sqrt{mn+1}$.

Example 15.2.5. Consider the parity-check matrix $\mathbf{H}'$ of the form

$$\mathbf{H}' = \begin{bmatrix} 1 & 0 & 0 & 0 & 0 & 0 & 1 & 1 & 1 & 1 & 0 & 0 & 0 & 0 & 0 & 0 & 0 & 0 & 0 \\ 0 & 1 & 0 & 0 & 0 & 0 & 1 & 0 & 0 & 0 & 1 & 1 & 1 & 1 & 0 & 0 & 0 & 0 & 0 \\ 0 & 0 & 1 & 0 & 0 & 0 & 0 & 1 & 0 & 0 & 0 & 1 & 0 & 0 & 0 & 1 & 1 & 1 & 0 & 0 & 0 \\ 0 & 0 & 0 & 1 & 0 & 0 & 0 & 0 & 1 & 0 & 0 & 0 & 1 & 0 & 0 & 1 & 0 & 0 & 1 & 1 & 0 \\ 0 & 0 & 0 & 0 & 1 & 0 & 0 & 0 & 0 & 1 & 0 & 0 & 0 & 1 & 0 & 0 & 1 & 0 & 1 & 0 & 1 \\ 0 & 0 & 0 & 0 & 0 & 1 & 0 & 0 & 0 & 0 & 1 & 0 & 0 & 0 & 1 & 0 & 0 & 1 & 0 & 1 & 1 \end{bmatrix},$$

which is the incidence matrix of the Steiner system $S(2,2;6)$. The above matrix $\mathbf{H}'$ can also be interpreted as the binary image of a (7,5,3) MDS (maximum distance separable) code over GF(8), and the corresponding RS code does not have the cycle-free Tanner graph (see MacWilliams and Sloane [1977: §15.5]). ∎

15.3 LDPC Codes

Low-density parity-check (LDPC) codes were invented in the 1960s by Gallager [1962]. They were forgotten for 30 years until they were rediscovered by MacKay and Neal [1996], and have now become a major area of research and application. These codes are also known as *Gallager codes*. They are decoded iteratively and have become successful in recovering the original codewords transmitted over noisy communication channels, now a major area of research and applications.

An LDPC code is a linear block code with a sparse parity-check matrix $\mathbf{H}$ that has most of its elements as 0s, and its generator matrix $\mathbf{G}$ is calculated only after the parity-check matrix $\mathbf{H}$ has been constructed. They are of two types: regular LDPC codes (§15.2) that have a constant number of 0s in each row and each column, and the irregular LDPC codes that are not regular (§15.3). It is found that the irregular LDPC codes perform better than the regular ones. The alphabet for these codes belongs to the Galois fields $GF(2^i)$, $i = 1, 2, \ldots$, and the performance of these codes is directly proportional to the size of the alphabet.

Decoding LDPC codes is an iterative process of interchanging information between the two types of nodes of the corresponding Tanner graph. As we will explain by some examples, the process is generally as follows: If at some point in the iterative process the syndrome of the estimated decoded vector (string) turns out to be the zero vector **0**, the process converges and the output is the result of the decoding. However, if the process does not converge to a solution after a predetermined number of iterations, the decoding is declared as 'failed'. The LDPC decoding algorithm is based on the belief propagation algorithm, also known as the sum-product algorithm, which is explained in Chapter 14. In practice, the polar format is generally used instead of the binary format, and computation is carried out in logarithmic format where product and division are converted into addition and subtraction, respectively, with lookup tables for logarithms to save time.

15.3.1 Channel Capacity. Shannon [1948] defines the capacity purely in terms of information theory, his research does not guarantee the existence of related schemes to achieve the capacity. Shannon also defines the notion of *code* as a finite set of vectors that is generated for transmission.

A communication channel is defined as a triple consisting of an input alphabet i, an output alphabet o, and a transition probability $\mathfrak{p}(i, o)$ for each pair (i, o) of input and output elements. The transition probability is defined as the probability that the element o is received given that i was transmitted over the channel. The importance of a code is realized in the above definition when the vector elements are correlated over the input alphabet, in the sense that all vectors have the same length, which is known as the *block length* of the code. Let there be $K = 2^k$ vectors. Then every vector can be described with k bits. Let the length of the vectors be n. Then k bits are being transmitted in n times use of the channel. Thus, the code has a rate of k/n bits per channel use (bpu).

Now suppose that a codeword over the input alphabet is transmitted, and it is received over the output alphabet. If the channel is noisy and allows errors in transmission, there is no way to verify that the output word was exactly the same codeword that was transmitted. However, we can always find the most likely codeword transmitted such that the probability that this codeword has the maximum correct transmission of the input alphabet. In fact, if we are using the maximum likelihood decoder, we proceed as follows: List all K codewords, calculate the conditional probability of each codeword, find the vector(s) that yield this maximum probability, and return the one with the highest probability. This decoder takes a lot of time and hence is not very reliable or cost-effective. One of Shannon's results states that the decoding error of the maximum likelihood decoder goes to zero exponentially fast with the block length. In other words, it establishes the existence of codes with rates close to the capacity for which the probability of error of the maximum likelihood decoder tends to zero as the block length of the code goes to infinity. This result guarantees the existence of sequences of linear codes with rates arbitrarily close to the capacity, and the probability of the maximum likelihood decoder for these codes approaches zero exponentially as the block length $n \to \infty$.

15.3.2 Types of Communication Channels. There are two types of models of communication channels used in coding theory and information theory: (i) the Binary Erasure Channel (BEC), and (ii) the Binary Symmetric Channel (BSC). They are presented graphically in Figure 15.3.1. In both channels the input alphabet is binary $\{0, 1\}$ and its elements are called *bits*. However, the output alphabet for the BEC is $\{0, 1, e\}$, where e is called the *erasure*, while for the BSC it is $\{0, 1\}$. The BEC and the BSC are described as follows.

15.3.2.1 BEC. In this model a transmitter sends a bit $(0$ or $1)^1$ and the receiver either receives the bit or it receives a message that the bit was not received, i.e., the bit was 'erased', meaning that the transmitted bit was scrambled and the receiver (decoder) has no idea what it was. A precise definition of BEC is as follows: A binary erasure channel is a channel with binary input, with erasure probability $\mathfrak{p}$. Let X be the transmitted random variable (message) with alphabet $\{0, 1\}$, and let Y be the received variable with alphabet $\{0, 1, e\}$, where e denotes the erasure (Figure 15.3.1(a)). Then the channel is characterized by the conditional probabilities

$$\mathfrak{P}(Y = 0 \,|\, X = 0) = 1 - \mathfrak{p}, \quad \mathfrak{P}(Y = e \,|\, X = 0) = \mathfrak{p},$$

$$\mathfrak{P}(Y = 1 \,|\, X = 0) = 0, \quad \mathfrak{P}(Y = 0 \,|\, X = 1) = 0,$$

$$\mathfrak{P}(Y = e \,|\, X = 1) = \mathfrak{p}, \quad \mathfrak{P}(Y = 1, \,|\, X = 1) = 1 - \mathfrak{p}.$$

The capacity of a BEC is $1 - \mathfrak{p}$. The above characterization suggests that $1 - \mathfrak{p}$ is the upper bound of the channel. When a transmitted bit is erased, the source X knows whenever a transmitted bit gets erased. The source X can do nothing about it except repeatedly transmit a bit until it is received. There is no need for X to code, because Y will simply ignore erasures, as the next successfully received bit would be the one that X intended to transmit. On average, the rate of successful transmission is $1 - \mathfrak{p}$, which is an upper bound for this transmission channel. For more details, see MacKay [2003].

15.3.2.2 BSC. In this model, a transmitter sends a bit $(0$ or $1)$ and the receiver receives a bit. It is assumed that the bit is always transmitted correctly, but during transmission it will be 'flipped' with a small probability, called the *crossover probability*. A formal definition is as follows: A binary symmetric channel with crossover probability $\mathfrak{p}$ is a channel with binary input and binary output and probability of error $\mathfrak{p}$, i.e., if X is the transmitted random variable (message) and Y the received variable (Figure 15.3.1(b)), then the channel is characterized by the conditional probabilities

$$\mathfrak{P}(Y = 0 \,|\, X = 0) = 1 - \mathfrak{p}, \quad \mathfrak{P}(Y = 0 \,|\, X = 1) = \mathfrak{p},$$

$$\mathfrak{P}(Y = 1 \,|\, X = 0) = \mathfrak{p}, \quad \mathfrak{P}(Y = 1 \,|\, X = 1) = \mathfrak{p},$$

---

[1] A nonbinary channel would be capable of transmitting more than two symbols, even possibly infinitely many symbols.

where it is assumed that $0 \leq \mathfrak{p} \leq \frac{1}{2}$. If $\mathfrak{p} > \frac{1}{2}$, the receiver can swap the output, i.e., interpret 1 when it sees 0, and conversely, and obtain an equivalent channel with crossover probability $1 = \mathfrak{p} \leq \frac{1}{2}$.

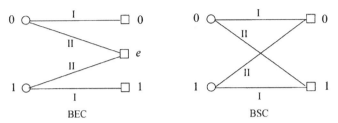

Figure 15.3.1 (a) BEC; (b) BSC
Legend: Probability $1 - \mathfrak{p}$ on edges I, and $\mathfrak{p}$ on edges II.

The capacity of this channel is $1 - \mathcal{H}(\mathfrak{p})$, where $\mathcal{H}(\mathfrak{p})$ is the binary entropy function.[2] The converse can be established using the sphere packing argument as follows: Given a codeword there are approximately $2^{n\mathcal{H}(\mathfrak{p})}$ typical output sequences. That is, there are 2^n total possible outputs, and the input selects from a code of size 2^{nR}. Thus, the receiver would choose to partition the space into 'spheres' with $2^n/2^{nR} = 2^{n(1-R)}$ potential outputs each. If $R > 1 - \mathcal{H}(\mathfrak{p})$, the spheres will be packed too tightly asymptotically and the receiver will not be able to identify the correct codeword with certainty (zero probability). The input/output alphabet in BSC is in the field F_2. It appears at first sight that the BSC is simpler than BEC, but this is not true because the capacity of BSC is $1 + \mathfrak{p} \log_2 \mathfrak{p} + (1 - \mathfrak{p}) \log_2 (1 - \mathfrak{p})$, and the maximum likelihood decoding for this channel is equivalent to finding, for a given block length n over F_2, a codeword that has the smallest Hamming distance from the output codeword (see §6.1 on bounds).

Gallager [1963] developed the LDPC codes with a sparse parity-check matrix, which is often randomly generated under certain sparsity constraints. There are two different ways to represent these codes: one as a matrix and the other as a graph. There are two kinds of LDPC codes: regular and irregular. Although the parity-check matrices are sparse, i.e., they contain many fewer 1s than the number of 0s, the examples presented in this section are of low dimensions and fail to exhibit the sparse nature of the parity-check matrix. Even the parity-check matrix in Example 15.3.1 does not come close to the sparsity expected in these codes.

Example 15.3.1. Consider a regular low-density parity-check matrix of dimension $n \times m$ for an (8,4) code follows:

---

[2] The binary entropy function is defined for a binary random variable 0 with probability $\mathfrak{p}$ and 1 with probability $1 - \mathfrak{p}$ as the function $\mathcal{H}(\mathfrak{p}) = -\mathfrak{p} \log \mathfrak{p} - (1 - \mathfrak{p}) \log(1 - \mathfrak{p})$. Note that $\mathcal{H}(0) = 0 = \mathcal{H}(1)$, with maximum at $\mathfrak{p} = \frac{1}{2}$.

$$\mathbf{H} = \begin{bmatrix} 0 & 1 & 0 & 1 & 1 & 0 & 0 & 1 \\ 1 & 1 & 1 & 0 & 0 & 1 & 0 & 0 \\ 0 & 0 & 1 & 0 & 0 & 1 & 1 & 1 \\ 1 & 0 & 0 & 1 & 1 & 0 & 1 & 0 \end{bmatrix}. \qquad (15.3.1)$$

In general, let u_c denote the number of 1s in each column and u_r of them in each row of an $n \times m$ parity-check matrix $\mathbf{H}$. Then the matrix $\mathbf{H}$ is said to be of low density if the following two conditions are satisfied: $u_c \ll n$, and $u_r \ll m$.

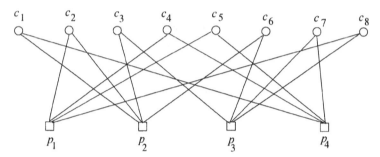

Figure 15.3.2 Parity-matrix $\mathbf{H}$ in (15.3.1).

The graphical representation of the matrix $\mathbf{H}$ defined by (15.3.1) is presented as a Tanner graph in Figure 15.3.2, which provides a simple representation of the parity-check constraints of LDPC codes and a precise description of the coding algorithms. They are bipartite graphs, in which each square represents a parity-check bit and each circle connected to a square a code bit that participates in that parity check.

Let the c-nodes in $\mathbf{H}$ represent the codeword $\mathbf{c} = \{c_1, c_2, \ldots c_n\}$ and the p-codes be $p_1, p_2, \ldots, p_m$. In Figure 15.3.2 the top line (circles) denotes the c-nodes $c_1, \ldots, c_8$, while the bottom line (squares) denotes the parity-check bits p_1, p_2, p_3, p_4. Note that the path $p_2 \to c_6 \to p_3 \to c_3 \to p_2$ in the Tanner graph in Figure 15.3.2 is an example of a 4- cycle, which is usually avoided as it does not perform very well in the decoding process for very large values of n and m. From the matrix (15.3.1) or Figure 15.3.2, the constraints at the four parity bits are given by

$$\text{At } p_1: \quad c_2 \oplus c_4 \oplus c_5 \oplus c_8 = 0,$$
$$\text{At } p_2: \quad c_1 \oplus c_2 \oplus c_3 \oplus c_5 = 0,$$
$$\text{At } p_3: \quad c_3 \oplus c_6 \oplus c_7 \oplus c_8 = 0,$$
$$\text{At } p_4: \quad c_1 \oplus c_4 \oplus c_5 \oplus c_7 = 0. \qquad (15.3.2)$$

For a regular LDPC code, the number u_c is constant for each column and

$u_r = \dfrac{n}{m} u_c$ is constant for every row. For the above matrix $\mathbf{H}$, we have $u_r = 4$ and $u_c = 2$. This is reflected in the graph in Figure 15.3.2, where we have the same number (two) of incoming edges from each c-node to p-nodes.

There are three major decoding algorithms for decoding LDPC codes: the *belief propagation algorithm* (BPA), the *message passing algorithm* (MPA), and the *sum-product algorithm* (SPA), defined and explained in Chapter 14. We now consider binary symmetric channels and describe the process of hard-decision and soft-decision decoding schemes.

15.4 Hard-Decision Decoding

We use the matrix $\mathbf{H}$ in Eq (15.3.1) and its graph in Figure 15.3.2, and assume that a message $\mathbf{c} = \{1\,0\,0\,1\,0\,1\,0\,1\}$ ($n = 8, m = 4$) is transmitted over a binary channel. We further assume that this codeword is received with one error: the bit c_3 is flipped to 1, i.e., $w_3 = 1$. The decoding scheme runs through the following steps:

STEP 1. All c-nodes c_i send a message to their p-nodes p_j that a c-node c_i has only the information that the corresponding received i-th bit of the codeword is w_i; that is, for example, node c_1 sends a message containing w_1 (which is 1) to p_2 and p_4, node c_2 sends a message containing w_2 (which is 1) to p_1 and p_2, and so on. The overall messages sent and received at this stage are presented in Table 15.4.1.

STEP 2. Every parity node p_j computes a response to every connected c-node. The response message contains the bit that p_j *believes* to be correct for the connected c-node c_i assuming that the other c-nodes connected to p_j are correct. For our example, since every p-node p_j is connected to 4 c-nodes, a p-node p_j looks at the received message from 3 c-nodes and computes the bit that the fourth c-node must be such that all the parity-check constraints (15.3.2) are satisfied.

Table 15.4.1.

p-node	Message sent/received
p_1	sent: $0 \to c_2$, $1 \to c_4$, $0 \to c_5$, $1 \to c_8$ received $w_2 \to 0$, $w_4 \to 1$, $w_5 \to 0$, $w_8 \to 1$
p_2	sent: $1 \to c_1$, $0 \to c_2$, $0 \to c_3$, $1 \to c_6$ received $w_1 \to 1$, $w_2 \to 0$, $w_3 \to 1$, $w_6 \to 1$
p_3	sent: $0 \to c_3$, $1 \to c_6$, $0 \to c_7$, $1 \to c_8$ received $w_3 \to 1$, $w_6 \to 1$, $w_7 \to 0$, $w_8 \to 1$
p_4	sent: $1 \to c_1$, $1 \to c_4$, $0 \to c_5$, $0 \to c_7$ received: $w_1 \to 1$, $w_4 \to 1$, $w_5 \to 0$, $w_7 \to 0$

STEP 3. The c-nodes receive the message from the p-nodes and use this

additional information to decide if the original message received is correct. A simple way to decide this is the *majority vote*. For our example it means that each c-node has three sources of information concerning its bit: the original bit received and two suggestions from the p-nodes.

STEP 4. Go to Step 2.

Example 15.4.1. In the case of our example of Figure 15.3.2, Step 2 terminates the decoding process since c_3 has decided for 0. This is presented in the following table, where Figure 15.3.2 has been used with both the transmitted and received code bits c_i and w_i.

c-node	w_i Received	Bit sent from p-nodes	Decision
c_1	1	$p_2 \to 1,\ p_4 \to 1$	1
c_2	0	$p_1 \to 0,\ p_2 \to 0$	0
c_3	1	$p_2 \to 0,\ p_3 \to 0$	0
c_4	1	$p_1 \to 1,\ p_4 \to 1$	1
c_5	0	$p_1 \to 0,\ p_4 \to 0$	0
c_6	1	$p_2 \to 1,\ p_3 \to 1$	1
c_7	0	$p_3 \to 0,\ p_4 \to 0$	0
c_8	1	$p_1 \to 1,\ p_3 \to 1$	1

Note that since $n = 8$ and $m = 4$, there are two p-nodes connected to each c-node. The two p-nodes in the third column are determined by the Tanner graph of Figure 15.3.2, and the related bit values of these two p-nodes are obtained from Table 15.4.1. The decision in the fourth column is by majority voting in the third column, which is straightforward in all cases. Note that when $\lfloor u_r/u_c \rfloor = \lfloor n/m \rfloor > 2$, we have more and two 0s and 1s in the third column, and the decision in the fourth column is always made by majority vote among the bit values of p-nodes in the third column.

Hence, the corrected received word is $w_1 = 1, w_2 = 0, w_3 = 0, w_4 = 1, w_5 = 0, w_6 = 1, w_7 = 0, w_8 = 1$. This is validated by verifying the constraints (15.3.2) with c_i replaced by w_i.

15.5 Soft-Decision Decoding

Unlike the soft-decision decoding of the RS codes, which uses the real-valued *soft* information available from the channel and involves the Koetter-Vardy algorithm (Koetter and Vardy [2000]), this kind of decoding for LDPC codes is based on belief propagation, and therefore, being a better decoding method for these codes, it is preferred over the hard-decision decoding scheme, although the basic idea of this decoding method is the same as for the hard-decision.

Let $\mathfrak{p}_i = \mathfrak{p}(c_i = 1 \,|\, w_i)$ be the probability that a code bit $c_i = 1$ when the received bit is w_i, and let q_{ij} be the message sent by a c-node c_i to a p-node p_j. Notice that every message always contains the pair $q_{ij}(0)$ and $q_{ij}(1)$, which is equivalent to the value of belief that w_i is a 0 or a 1. Let r_{ji} be the message sent by a p-node p_j to the c-node c_i. Again, there exists a pair $r_{ji}(0)$ and $r_{ji}(1)$ that stands for the current amount of belief that w_i is a 0 or a 1. This decoding scheme runs through the following steps:

STEP 1. Let all variable c-nodes c_i send their q_{ij} messages to a p-node p_j. Since no other information is available at this step, set $q_{ji}(1) = \mathfrak{p}_i$ and $q_{ji}(0) = 1 - \mathfrak{p}_i$.

For the next step we use the following result from Gallager [1963]: For a sequence of N of independent bits b_i with a probability $\mathfrak{p}_i$ for $b_i = 1$, the probability that the entire sequence $\{b_i\}$ contains an even number of 1s is equal to $\frac{1}{2}\left(1 + \prod_{i=1}^{N}(1 - 2\mathfrak{p}_i)\right)$.

STEP 2. The p-nodes compute their response message r_{ji} by the formulas

$$r_{ji}(0) = \frac{1}{2}\left(1 + \prod_{\nu \in p_{j\backslash i}}(1 - 2q_{\nu j}(1))\right), \quad r_{ji}(1) = 1 - r_{ji}(0), \qquad (15.5.1)$$

where $p_{j\backslash i}$ denotes the set of an even number of 1s among the p-nodes except p_i. The probability that $c_{j\backslash i}$ is true is equal to the probability $r_{ji}(0)$ that c_i is 0. Figure 15.5.1 shows how the information is used to compute the responses at this step, where $(*)$ means the bit 0 or 1.

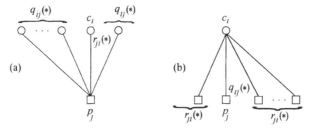

Figure 15.5.1(a) Computation of $r_{ji}(*)$ and (b) $q_{ij}(*)$ in Step 2.

STEP 3. The c-nodes update their response to the p-nodes using the following formulas:

$$q_{ij}(0) = k_{ij}(1 - \mathfrak{p}_i)\prod_{\mu \in c_{i\backslash j}} r_{\mu i}(0), \quad q_{ij}(1) = k_{ij}\mathfrak{p}_i\prod_{\mu \in c_{i\backslash j}} r_{\mu i}(1), \qquad (15.5.2)$$

where k_{ij} are constants that are chosen such that $q_{ij}(0) \oplus q_{ij}(1) = 1$. The c-nodes at this stage also update their current estimate $\hat{c}_i$ of the bit c_i by computing the probabilities for 0 and 1 and voting the larger of the two. This is done using the formulas

$$Q_i(0) = k_i(1 - \mathfrak{p}_i)\prod_{j \in c_i} r_{ji}(0), \quad Q_i(1) = k_i\mathfrak{p}_i\prod_{j \in c_i} r_{ji}(1), \qquad (15.5.3)$$

which are similar to formulas(15.5.2), so that the computed values of Q_i becomes the new message bit at every c-node, giving the updated value

$$\hat{c}_i = \begin{cases} 1 & \text{if } Q_i(1) > Q_i(0), \\ 0 & \text{otherwise.} \end{cases} \tag{15.5.4}$$

If the value of the current estimated codeword obtained from each $\hat{c}_i$ satisfies the parity-check constraints, the algorithm terminates; otherwise, termination occurs after a maximum number of iterations in the next step.

STEP 4. Go to Step 2.

The process at Steps 2 and 3 involves various multiplications of probabilities, where the result often comes very close to zero for large block lengths. This situation is avoided by using the log-domain where multiplication changes into addition, which produces a more stable form of the algorithm; and addition being less costly, it improves the performance of the algorithm. The above soft-decision decoding algorithm is suitable for BSC channels. More details are available in MacKay [1999].

An LDPC code can be implemented by a one-to-one mapping of the Tanner graph in hardware, such that each node is implemented and the connections are realized by hard wiring. The architecture of an efficient LDPC code is presented in Hamzaoui et al. [2002], but the encoding problem is not fully discussed. The resulting communication problem is, however, inferior to that of Turbo codes due to the lack of flexibility in terms of communications performance, which has always been compared with Turbo codes.

15.6 Irregular LDPC Codes

An example of LDPC $(6,3)$ code $(n = 6, k = 3)$, using the Tanner graph, is presented in Figure 15.6.1, where n variable nodes on the top line are connected to $(n - k)$ constraint nodes at the bottom of the graph.

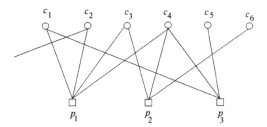

Figure 15.6.1 An irregular LDPC $(6, 3)$ code.

This type of representation is mostly used for an (n, k) LDPC code, in which the bits of the input message (codeword) when placed on top of the graph as c-nodes satisfy the graphical constraints in the following way: All edges connecting a p-node, which is denoted by a square, have the same value such that all values connecting to a p-node must sum (mod 2) to zero; that is, their sum must be an even number. All lines going out of the graph are ignored.

Example 15.6.1. The constraints for the code of Figure 15.6.1 are: $c_1 \oplus c_2 \oplus c_3 \oplus c_4 = 0$, $c_3 \oplus c_4 \oplus c_6 = 0$, $c_1 \oplus c_4 \oplus c_5 = 0$. There are 8 possible 6-bit vectors that correspond to valid codewords, namely,

$$\{000000, 001011, 010101, 011110, 100110, 101101, 110011, 111000\}$$

(see §15.1). This LDPC code represents a 3-bit message encoded as 6 bits. The parity-check matrix of this code is

$$\mathbf{H} = \begin{bmatrix} 1 & 1 & 1 & 1 & 0 & 0 \\ 0 & 0 & 1 & 1 & 0 & 1 \\ 1 & 0 & 0 & 1 & 1 & 0 \end{bmatrix}.$$

To obtain the generator matrix $\mathbf{G}$, the matrix $\mathbf{H}$ is first transformed into the form $[-P^T \mid I_{n-k}]$ by performing row and column operations. Thus,

$$\mathbf{H} = \begin{bmatrix} 1 & 1 & 1 & 1 & 0 & 0 \\ 0 & 0 & 1 & 1 & 0 & 1 \\ 1 & 0 & 0 & 1 & 1 & 0 \end{bmatrix} \overset{*}{=} \begin{bmatrix} 1 & 1 & 1 & 1 & 0 & 0 \\ 0 & 0 & 1 & 1 & 0 & 1 \\ 0 & 1 & 1 & 0 & 1 & 0 \end{bmatrix}$$

$$\overset{**}{=} \begin{bmatrix} 1 & 1 & 1 & 1 & 0 & 0 \\ 0 & 1 & 1 & 0 & 1 & 0 \\ 0 & 0 & 1 & 1 & 0 & 1 \end{bmatrix} \overset{***}{=} \begin{bmatrix} 1 & 1 & 1 & 1 & 0 & 0 \\ 0 & 1 & 1 & 0 & 1 & 0 \\ 1 & 1 & 0 & 0 & 0 & 1 \end{bmatrix},$$

where $\overset{*}{=}$ denotes the operation 'Replace R_3 by $R_1 \oplus R_3$'; $\overset{**}{=}$ the operation 'interchange C_2 and C_6, and C_5 and C_6'; and $\overset{***}{=}$ the operation 'Replace R_3 by $R_1 \oplus R_3$'. Then the generator matrix $\mathbf{G}$ is obtained from $\mathbf{H}$ as $[I_k \mid P]$.

Note that $-P^T = \begin{bmatrix} 1 & 1 & 1 \\ 0 & 1 & 1 \\ 1 & 1 & 0 \end{bmatrix} \longrightarrow P^T = \begin{bmatrix} 1 & 1 & 1 \\ 1 & 1 & 0 \\ 0 & 1 & 1 \end{bmatrix}$ by interchanging R_2

and R_3; then $P = \begin{bmatrix} 1 & 0 & 1 \\ 1 & 1 & 1 \\ 1 & 1 & 0 \end{bmatrix}$ by taking the transpose of P^T. Thus,

$$\mathbf{G} = \begin{bmatrix} 1 & 0 & 0 & 1 & 0 & 1 \\ 0 & 1 & 0 & 1 & 1 & 1 \\ 0 & 0 & 1 & 1 & 1 & 0 \end{bmatrix},$$

and after multiplying all 8 possible 3-bit vectors[3], all 8 valid codewords are obtained. For example, the codeword for the string $\{1\,1\,0\}$ is

$$\{1\,1\,0\} \cdot \begin{bmatrix} 1 & 0 & 0 & 1 & 0 & 1 \\ 0 & 1 & 0 & 1 & 1 & 1 \\ 0 & 0 & 1 & 1 & 1 & 0 \end{bmatrix} = \{1\,1\,0\,0\,1\,0\}. \blacksquare$$

---

[3] They are $\{0\,0\,0\}, \{0\,0\,1\}, \{0\,1\,0\}, \{0\,1\,1\}, \{1\,0\,0\}, \{1\,0\,1\}, \{1\,1\,0\}, \{1\,1\,1\}$.

15.6.1 Decoding. Shannon's theory (§15.2.1) has established that random linear codes in F_q approach capacity, and they can be encoded and decoded in polynomial time rather than exponential time; however, unlike encoding, the decoding process is more complicated.

Example 15.6.2. Before we derive the general theory of decoding, we will consider the above example with the codeword $\{1\,1\,0\,0\,1\,0\}$, which is transmitted across a binary erasure channel (BEC) and received with the first and fourth bits erased, i.e., the received message is $\{x\,1\,0\,x\,1\,0\}$. Since the received message has satisfied all the constraints, we create a Tanner graph for it (see Figure 15.6.2). The first bit cannot be recovered as yet because all the constraints connected with this bit have more than one unknown bit. Thus, to decode we will take into account constraints connected with only one of the erased bit, which is either the second or the third constraint. The second constraint $c_3 \oplus c_4 \oplus c_6 = 0$ gives $0 \oplus c_4 \oplus 0 = 0$, or $c_4 = 0$, so the fourth bit must be zero. Now, using this value of the fourth bit the first bit can be determined, which from the third constraint $c_1 \oplus c_4 \oplus c_5 = 0$ turns out to be 1. This decoding is completed iteratively, and the decoded message is $\mathbf{c} = \{1\,1\,0\,0\,1\,1\}$.

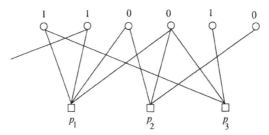

Figure 15.6.2 Tanner graph for Example 15.6.2.

This decoding can be verified by calculating the syndrome obtained by multiplying the parity-check matrix $\mathbf{H}$ by the corrected message $\mathbf{c}$:

$$\mathbf{Hc} = \begin{bmatrix} 1 & 1 & 1 & 1 & 0 & 0 \\ 0 & 0 & 1 & 1 & 0 & 1 \\ 1 & 0 & 0 & 1 & 1 & 0 \end{bmatrix} \begin{Bmatrix} 1 \\ 1 \\ 0 \\ 0 \\ 1 \\ 0 \end{Bmatrix} = \begin{Bmatrix} 0 \\ 0 \\ 0 \end{Bmatrix},$$

which is $\mathbf{0}$, thus validating the decoded message. ∎

The LDPC codes are designed along the lines of iterative coding systems. Iterative coding is also used in Turbo codes (see §16.5) but the decoding structure of LDPC codes provides better results provided that the code length is large and the parity-check matrix is sparse. In addition, the other disadvantage with LDPC codes is that their encoding schemes are more complex.

16

Special LDPC Codes

16.1 Classification of LDPC Codes

Low-density parity-check (LDPC) codes are part of many special codes, such as Gallager codes, IRA codes, Systematic codes, Turbo codes, Tornado codes, Online codes, Fountain codes, LT codes, and Raptor codes. The remaining part of the book is devoted to these codes and related topics. The following LDPC codes are discussed in this chapter:

(i) GALLAGER CODES. They encompass all codes represented with Tanner graphs (with two types of nodes: c-nodes and p-nodes) and nonsymmetric codes, and require matrix operations for encoding and decoding.

(ii) SYSTEMATIC CODES. These are the codes where each coding node has just one edge to a unique constraint.

(iii) IRA CODES. They are systematic codes where each coding node i has edges to constraints i and $i + 1$.

(iv) TURBO CODES. These codes use two or more codes, and decoding is done iteratively by feeding outputs from one decoder to the inputs of another decoder.

All these codes are based on bipartite graphs, which define parity operations for encoding and decoding. They have been proved to be asymptotically optimal. Decoding overhead is based on the number of edges in the graph (it is low density).

16.2 Gallager Codes. Let C denote an (n, k, d) code, with block length n, dimension k, and minimum Hamming distance d; and set $m = n - k$. The code C has an $m \times n$ parity-check matrix $\mathbf{H}$, and every codeword $\mathbf{c} \in \mathbf{H}$ satisfies the equation $\mathbf{c}\mathbf{H}^T = \mathbf{0}$. Suppose there exists an $\mathbf{H}$ such that $\mathbf{H}$ has some fixed number $j \in \mathbb{N}$ of 1s in each column (called the *column property*) and

some fixed number $i \in \mathbb{N}$ of 1s in each row (called the *row property*). We use
the following notation: we denote a code C by $[n, -, i]$ if an $\mathbf{H}$ for C exists
with row property, and by $[n, j, -]$ if this matrix for C exists with column
property, where '$-$' represents the corresponding parameter. Note that the
parameters i and j in this notation represent a particular check matrix for C
that need not be preserved for an equivalent parity-check matrix for the same
code. If the parameter d is not known or is unimportant, we denote such a
code as $(n, k, -)$ code. Thus, the notation $[n, j, i]$ describes the structure of a
particular $\mathbf{H}$ for the code C.

Example 16.2.1. Hamming (7,4,3) code, with $n = 7, k = 4, d = 3, m = 3$,
has one possible parity-check matrix

$$\mathbf{H} = \begin{bmatrix} 1 & 0 & 0 & 1 & 0 & 1 & 1 \\ 0 & 1 & 0 & 1 & 1 & 1 & 0 \\ 0 & 0 & 1 & 0 & 1 & 1 & 1 \end{bmatrix}.$$

It is a $(7, -, 4)$ code. ∎

We have noticed that every linear code has a certain structure. However,
there exists a *random* linear code if $\mathbf{H}$ is generated randomly, with possible
constraints. There also exist long random linear codes that achieve Shannon's capacity for many communication channels (see Reiffen [1962]). Since
the complexity of a decoding process is a practical problem, an arbitrary
random linear code may run into decoding problems that are not practically
manageable. Thus, we consider a subclass of linear codes that have effective
decoding algorithms, such as LDPC codes with a very sparse matrix $\mathbf{H}$, and
this sparcity property makes these codes very efficient for decoding. Some
additional constraints on $\mathbf{H}$ must be imposed for such codes, namely, the row
and column properties that yield (n, j, i) codes, in order to obtain the Gallager
codes.

16.2.1 Construction of Gallager Codes. We describe some constructions
of low-rate Gallager codes. According to his original construction, Gallager
[1963] constructed an (n, j, i) code in which a parity-check matrix consisted
of j blocks, with n/i rows in each block, where each column in a block has
exactly one nonzero element. This construction was modified by Gallager
[1963: 9] so that the parity-check matrix contained no short cycles.[1] The
initial approach by Gallager [op. cit.] was hampered by the existence of
such cycles in $\mathbf{H}$ since it prevented an exact error-probability analysis of his
iterative decoding process: The shorter the cycles, the shorter the breakdown period of the analysis. But this problem was resolved by the following
construction similar to that of MacKay [1999]: Generate $\mathbf{H}$ randomly with
column Hamming weight 3 and row Hamming weight as uniform as possible,

---

[1] A cycle in $\mathbf{H}$ is a sequence of distinct row-column indices $(p_1, c_1), (p_2, c_2), \ldots, (p_n, c_n)$,
n even, with $p_1 = p_2, c_2 = c_3, p_3 = p_4, \ldots$, and $c_n = c_1$, and for each index (p_i, c_i) the
corresponding entry in $\mathbf{H}$ is nonzero. See §15.1 for a 4-cycle case.

as shown in the following simple example.

Example 16.2.2. The parity-check matrix with parameters $n = 8, k = 2, m = 6$ such that there are four 1s in each row and three 1s in each column is

$$H = \begin{bmatrix} 1 & 0 & 0 & 1 & 1 & 0 & 1 & 0 \\ 0 & 0 & 0 & 1 & 1 & 1 & 0 & 1 \\ 1 & 1 & 0 & 0 & 0 & 1 & 0 & 1 \\ 1 & 0 & 1 & 0 & 0 & 1 & 1 & 0 \\ 0 & 1 & 1 & 1 & 0 & 0 & 1 & 0 \\ 0 & 1 & 1 & 0 & 1 & 0 & 0 & 1 \end{bmatrix}. \tag{16.2.1}$$

Its Tanner graph is given in Figure 16.2.1, which shows that it has a 4-cycle $(p_5 \to c_2 \to p_6 \to c_3 \to p_5)$. ∎

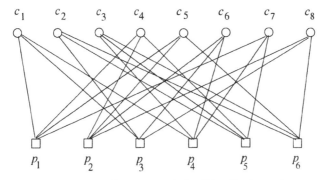

Figure 16.2.1 Tanner graph for **H** in Eq (16.2.1).

Sorokine et al. [2000] found that Gallager codes of block lengths of $n = 512$ bits and rates $R = 1/2$ and $R = 1/4$ perform similar to those designed by Gallager [1963] and MacKay [1999] with larger block lengths. Also, if the block length n is reduced to 384 bits, with $k = 192$, i.e., for a Gallager (384, 192) code, the performance decreased for $R = 1/2$. A modified construction is proposed by Sorokine et al. [op. cit.] by eliminating a 4-cycle in **H**, i.e., by requiring that no two columns of **H** have an overlap greater than 1, as in the following example.

Example 16.2.3. Consider the parity-check matrix of Example 16.2.2, where the corners of the 4-cycle are underlined, followed by its elimination:

$$H = \begin{bmatrix} 1 & 0 & 0 & 1 & 1 & 0 & 1 & 0 \\ 0 & 0 & 0 & 1 & 1 & 1 & 0 & 1 \\ 1 & 1 & 0 & 0 & 0 & 1 & 0 & 1 \\ 1 & 0 & 1 & 0 & 0 & 1 & 1 & 0 \\ 0 & \underline{1} & \underline{1} & 1 & 0 & 0 & 1 & 0 \\ 0 & \underline{1} & \underline{1} & 0 & 1 & 0 & 0 & 1 \end{bmatrix}.$$

The 4-cycle in this parity-check matrix is eliminated by the following modification:

$$H = \begin{bmatrix} 1 & 0 & 0 & 1 & 1 & 0 & 1 & 0 \\ 0 & 0 & \underline{1} & 1 & 1 & 1 & 0 & 1 \\ 1 & 1 & 0 & 0 & 0 & 1 & 0 & 1 \\ 1 & 0 & 1 & 0 & 0 & 1 & 1 & 0 \\ 0 & \underline{1} & 1 & 1 & 0 & 0 & 1 & 0 \\ 0 & \underline{1} & \underline{1} & 0 & 1 & 0 & 0 & 1 \end{bmatrix}, \qquad (16.2.2)$$

and its Tanner graph is presented in Figure 16.2.2, which shows that the 4-cycle of the previous Tanner graph is removed. ∎

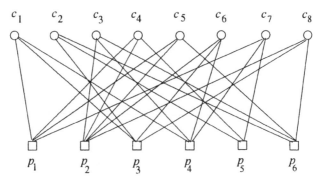

Figure 16.2.2 Tanner graph for H in (16.2.2).

Since the matrix (16.2.2) has high density (it is not sparse), all short cycles cannot be broken. Breaking all short cycles is possible for low-density matrices. Although similar to Gallager [1963: Figure 2.1], the following construction by Sorokine et al. [2000] achieves it with desired code parameters $k = 192$ at $R = 1.8$: Let $i = 3|n$ (i.e., 3 is a divisor of n). The rows of the parity-check matrix H are constructed in three parts, as follows: First row has the first i entries of 1s followed by 0s; then each of the next $(n/i) - 1$ rows is obtained by cyclic shifting of the immediately preceding row by i steps to the right, until the n-th column contains a 1. These rows form the top part of the matrix H, which is only an $(n/i) \times n$ matrix. For construction of the next (middle) part of H, let $t = \lfloor (im)/n \rfloor$; then $t + 1$ blocks of H are formed by a random permutation of the columns of the top part. The third (bottom) part of $\mathbf{H}$ is then constructed as follows: Let $s = m \bmod (n/i)$; then the bottom part of $\mathbf{H}$ is formed by permuting the columns of the first s rows of $\mathbf{H}$. This construction is derived from Gallager's construction of (n, j, i) code by deleting some rows of $\mathbf{H}$. Note that in this construction the values of n and m are restricted to the above desired code parameters.

Example 16.2.4. Parity check matrix for $n = 9, k = 2, m = 7$, where $i = 3, t = 2, s = 1$, is given by

$$
\mathbf{H} =
\left[
\begin{array}{ccccccccc}
1 & 1 & 1 & 0 & 0 & 0 & 0 & 0 & 0 \\
0 & 0 & 0 & 1 & 1 & 1 & 0 & 0 & 0 \\
0 & 0 & 0 & 0 & 0 & 0 & 1 & 1 & 1 \\
0 & 0 & 1 & 0 & 0 & 1 & 1 & 0 & 0 \\
\hline
0 & 0 & 1 & 0 & 0 & 1 & 1 & 0 & 0 \\
1 & 0 & 0 & 0 & 1 & 0 & 0 & 0 & 1 \\
0 & 1 & 0 & 1 & 0 & 0 & 0 & 1 & 0 \\
\hline
1 & 0 & 0 & 1 & 1 & 0 & 0 & 0 & 0
\end{array}
\right]
\begin{array}{l}
\\[4pt] \text{top part} \\[30pt]
\text{middle part} \\[24pt]
\text{bottom part } \blacksquare
\end{array}
$$

Notice that the upper two blocks (top and middle parts) of $\mathbf{H}$ have dimension $(n/i) \times n$ each, whereas the bottom part has dimension $\lfloor (1-R)n - 2n/i \rfloor \times n$.

16.2.2 Gallager Decoding. The decoding process uses the *message passing algorithm* (Gallager [1963]). This algorithm exchanges soft-information iteratively between variables and check nodes. The updating of nodes can be done with canonical scheduling: In the first step, all variable nodes are updated, and, in the second step all check nodes are updated. The processing of individual nodes within one step is independent, and thus, can be parallelized. The exchanged messages are assumed to be log-likelihood ratios (LLRs), defined in §14.5. Each variable node of degree i calculates an update λ_k of message k by the following formula:

$$
\lambda_k = \lambda_{\mathrm{ch}} + \sum_{l=0, l \neq k}^{i-1} \lambda_l,
\tag{16.2.3}
$$

where λ_{ch} denotes the corresponding channel LLR of the variable nodes and λ_i the LLRs of the incident edges. The check node LLR updates are calculated by formula, known as the *tanh-rule* (Chung et al. [2001]):

$$
\tanh \frac{\lambda_k}{2} = \prod_{l=0, l \neq k}^{i-1} \tanh \frac{\lambda_l}{2}.
\tag{16.2.4}
$$

Thus, the message passing algorithm provides an optimal decoding provided the Tanner graph is cycle-free. In practice, the Tanner graphs for blocks of finite length often contain cycles. In that case, once the message has completed a cycle, node updates using Eqs (16.2.3) and (16.2.4) become suboptimal. This means that an increased number of iterations obtained from longer cycles provides more gains in communication performance.

16.3 IRA Codes

Irregular-repeat-accumulate (IRA) codes belong to the class of LDPC codes and can often outperform Turbo codes (see §17.3). The IRA codes were introduced by Jin et al. [2000]. These codes have a linear-time encoding complexity with a straightforward hardware realization, and they can be presented by a Tanner graph with arbitrary connections between nodes. The decoding is an iterative process that exchanges information between the two types of nodes. The Tanner graph of an IRA code, shown in Figure 16.3.1, has n variable nodes (circles) and m parity-check nodes (squares). The variable nodes are c-nodes (CN) that can be partitioned into $k = n - m$ data (information) nodes or d-nodes (DNs) and m parity nodes or p-nodes (PNs), thus, making n CNs. The DNs are of varying degree. Let f_i denote a fraction of DNs with degree i, such that $\sum_i f_i = 1$. Each c-node (CN) is connected to a DNs, such that the edges $E_\pi = a \times m$ connections between CNs and DNs are arbitrary permutations π. The CNs are connected by a fixed zigzag pattern of edges.

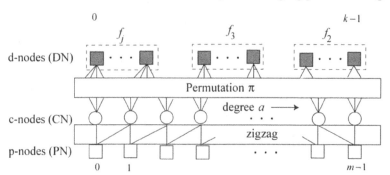

Figure 16.3.1 Tanner graph for IRA code.

The entire code ensemble can be described by the degree distribution $\mathbf{f} = (f_1, f_2, \ldots, f_j)$ and a. We obtain one code out of this ensemble for each permutation. The parameters $(\mathbf{f}, a)$ determine the convergence, while π determines the error floor. Consider a systematic code with codeword $\mathbf{c} = \{\mathbf{d}, \mathbf{p}\}$ of length $n = k + m$, where the data (message) sequence $\mathbf{d} = \{d_1, d_2, \ldots, d_k\}$ is associated with the data nodes $DN_1, \ldots, DN_k$, and each parity bit $\mathbf{p} = \{p_1, \ldots, p_m\}$ is associated with one of the parity nodes ($PN_1, \ldots, PN_m$). The code rate is

$$R = \frac{a}{a + \sum_i i f_i}.$$

ENCODING. The architecture of the encoder, presented in Figure 16.3.2, is derived from the above Tanner graph (Figure 16.3.1). The information sequence is first passed through a reception unit where the repetition pattern follows a preselected degree distribution **f**. The highest degree comes first, i.e., c_1 is repeated $n-2$ times and c_n two times. This extended data sequence is interleaved by the permutation π. Then a total of a values of the interleaved sequence are replaced by their binary sum (mod 2). The resulting sequence then passes through an accumulator, which leads to the zigzag connections between PNs and CNs. Finally, the codewords $\mathbf{c} = \{\mathbf{d}, \mathbf{p}\} = \{d_1, \ldots, d_k, p_1, \ldots, p_m\}$ are transmitted.

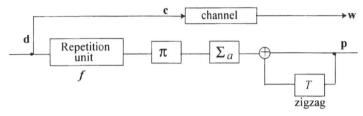

Figure 16.3.2 Encoder for IRA code.

DECODING. IRA codes can be decoded using Gallager's message passing algorithm described in §16.2.2. By selecting the permutation π for a given IRA code, the resulting Tanner graph should contain cycles that are as long as possible. The shortest cycle, called a *girth*, should be greater than 4 (see Richardson and Urbanke [2003]).

The quantity T (called *throughput*) in Figure 16.3.2 is defined by

$$T = \frac{I}{\# \text{ cycles}} f_{\text{cycle}}, \tag{16.3.1}$$

where I denotes the number of information bits to be decoded and '# cycles' refers to the number of cycles needed to decode one block, which is computed as equal to $\frac{C}{P_{\text{IO}}} + It\left(\frac{2E_\pi}{P}\right)$, where $\frac{C}{P_{\text{IO}}}$ is the number of cycles for input/output (IO) processing. Thus, from Eq (16.3.1) we obtain

$$T = \frac{I}{\dfrac{C}{P_{\text{IO}}} + It\left(\dfrac{2E_\pi}{P}\right)} f_{\text{cycle}}. \tag{16.3.2}$$

Here it is assumed that reading a new codeword C and writing the result of the preceding processed block can be done concurrently in parallel with P_{IO} (reading/writing) data. Usually, by taking $P_{\text{IO}} = 10$ and a small value of $\frac{C}{P_{\text{IO}}}$, which are reasonable assumptions, the throughput can be approximated by $T \sim \dfrac{IP}{E_\pi \, It}$, which shows that T depends on the number of iterations (It), the number processed in parallel (P), and the ratio of information bits and edges (I/E_π).

An alternative IRA code design with two degree sequences is as follows: Consider the IRA systematic code and its representation, shown in Figure 16.3.3, with two degree sequences $\lambda(d) = \sum_i \lambda_i d^{i-1}$ and $\rho(d) = \sum_j \rho_j d^{j-1}$, where λ_i and ρ_j are the fraction of degree incident on data nodes (DNs) and check nodes (PNs) with degree i, respectively. Then a received word corresponding to the codeword $\mathbf{c} = \{c_1, \ldots, c_n\}$ is given by $\mathbf{w} = \{w_1, \ldots, w_n\}$. It is also assumed that the check degrees are concentrated, i.e., $\rho(d) = d^{a-1}$ for some $a > 1$. Then for a given $\lambda(d)$ and a, the code rate is given by

$$R = \frac{\sum_i (\lambda_i/i)}{1/a + \sum_i (\lambda_i/i)}. \qquad (16.3.3)$$

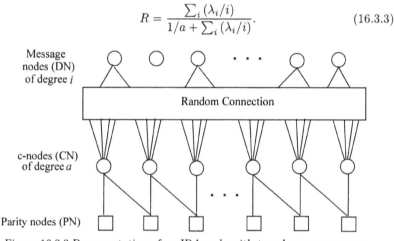

Figure 16.3.3 Representation of an IRA code with two degree sequences.

The decoding is done using an iterative message-passing algorithm, as explained above in §16.2.2, where the messages passed between nodes are extrinsic LLR. When the length n of a codeword becomes extremely large (mathematically, when $n \to \infty$), the Tanner graph is cycle-free and, thus, all messages that have passed through the decoder can be assumed as independent random variables. The probability density function (pdf) of these random variables and the probability of error are computed as follows: Assume that the all-zero codeword $\{\mathbf{0}\}$ is transmitted, and define the following notation:

(i) Let $w_{i \to c}$ and $w_{c \to i}$ denote the random variables representing the received message passed from DNs to CNs and from CNs to DNs, respectively;

(ii) Let $w_{p \to c}$ and $w_{c \to p}$ denote the random variables representing the received messages passed from PNs to CNs and from CNs to PNs, respectively;

(iii) Let w_p denote the random variables representing the LLR of PNs at the output of the channel; and

(iv) Let w_i denote the random variables representing the LLR of DNs at the output of the channel.

Then the pdf of the outgoing message from an DN or a PN is the convolution of pdfs of all incoming messages except the ones on the edge corresponding to the outgoing message. For an edge connected to a CN, the outgoing message on the edge is given by

$$\tanh \frac{w_k}{2} = \prod_{i \neq k} \tanh \frac{w_i}{2}, \tag{16.3.4}$$

similar to (16.2.4). This formula is used to compute $\text{pdf}(w_k)$ with change of measure and Fourier transform if $\text{pdf}(w_i)$, $i \neq k$, are given. Let $W_k(x)$ denote $\text{pdf}(w_k)$. Define an operator $\otimes$ as $W_k(x) = \otimes_{i \neq k} W_i(x)$, where w_k and w_i are related by (16.3.4). Then the pdf of messages passed in the Tanner graph at the $(n+1)$-st iteration can be obtained from those at the n-th iteration:

$$W_{c \to p}^{n+1}(x) = \otimes^a W_{i \to c}^n(x) \otimes W_{p \to c}^n(x),$$
$$W_{c \to i}^{n+1}(x) = \otimes^{a-1} W_{i \to c}^n(x) \otimes^2 W_{p \to c}^n(x),$$
$$W_{i \to c}^{n+1}(x) = \sum_i \lambda_i \star^{i-1} W_{c \to i}^{n+1}(x) \star W_i(x),$$
$$W_{p \to c}^{n+1}(x) = W_{c \to p}^{n+1}(x) \star W_p,$$

where $\star^i$ means applying the convolution i times (see Appendix D). When the all-zero codeword is transmitted, the probability of bit error after n iterations is given by

$$\mathfrak{p}_n(\lambda, W_i, W_p) = \int_{-\infty}^0 \sum_k \lambda_k \{ \star^k W_{c \to i}^n(x) \} \star W_i(x) \, dx.$$

Then the IRA design problem can be formulated as finding

$$\min_{\lambda \subseteq R} \mathfrak{p}_n(\lambda_1, \ldots, \lambda_d, W_{i,R}(x), W_{p,R}(x)),$$

where

$$\lambda_R \approx \left\{ \lambda_i \, \Big| \, \sum_i \lambda_i = 1, R = \frac{\sum_i (\lambda_i/i)}{1/a + \sum_i (\lambda_i/i)}, \lambda_i \geq 0 \right\},$$

and n, $W_{i,R}(x)$ and $W_{p,R}(x)$ are fixed and depend on the crossover probability or bit error rate (BER) of the BSC channel.

For decoding, assume that the transmission channel is a combination of the original BSC channel and an erasure channel. In the case of an all-zero codeword to be transmitted and BSC with cross-over probability $\mathfrak{p}$, we have $\mathrm{pdf}(w_i) = W_{\mathfrak{p},R_i}(x) = (1 - \mathfrak{p})\delta(x - t) + \mathfrak{p}\delta(x + t)$, where R_i is the desired channel code rate. The architecture of the IRA decoder with DN and PN branches operating in parallel is shown in Figure 16.3.4, e.g., for i units of DNs and PNs each with the following features: (i) the decoder is based on the Tanner graph of the code; (ii) since the permutation network Π is disjunctive, the processing of DN and PN branches is done independently, (marked by two dotted boxes in the Figure); (iii) the DN and PN units are single input and output modules, so that all incoming edges of one node can be processed sequentially; (iv) each CN unit accepts two data in parallel, one from DN and the other from PN, and calculates at most 2 data to be returned to the DN and PN branches, respectively, making sure that the symbols provided by these two branches belong to the same CN; and (v) all sequences of symbols for a node are stored in the same memory at consecutive sequences during storage. Since the number of symbols processed by this graph-decoder are given by

$$E_\pi = \begin{cases} aM, & \text{by the DN branch, } a \geq 2, \\ 2M, & \text{by the PN branch,} \end{cases}$$

the DN branch would require more cycles to process all data sequences than the PN branch. Thus, the cost to maintain this network depends on the Tanner graph of the code.

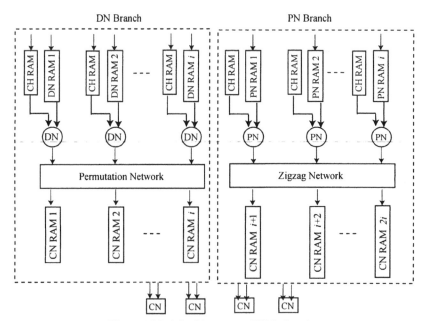

Figure 16.3.4 Architecture of IRA decoder.

According to Kienle and Wehn [2004], define $\delta(x) = \begin{cases} 0 \text{ if } x \neq 0 \\ 1 \text{ if } x = 0 \end{cases}$, and $\tau = \log \frac{1-\mathfrak{p}}{\mathfrak{p}}$. Then $W_{\mathfrak{p},R_i}(x_i) = \text{pdf}(w_i) = (1 - \mathfrak{p})\delta(x - \tau) + \mathfrak{p}\delta(x + \tau)$, which after being punctured to a desired code rate R is given by $W_{\mathfrak{p},R_i}(x_i) = (1 - \gamma)W_{i,R_i} + \gamma\delta(x)$, where γ is a fraction of parity bits being punctured. In order to require a set of rates $\{R_{\min}, R_2, \ldots, R_{\max}\}$, the optimization cost function must involve the error probability for each of these rates. To simplify with a minimal loss in performance, we minimize the sum of probabilities corresponding to $R_{\min}$ and $R_{\max}$, subject to the condition that

$$R_{\min} = \frac{\sum_i \lambda_i/i}{1/a + \sum_i \lambda_i/i}.$$

Then we design a master code that will perform well for $R_{\min}$ (with no puncturing) and for $R_{\max}$ (with maximum puncturing). Since this code is good at the two extremes, it must be good over the entire range of rates. The objective function can then be defined as

$$\mathfrak{p}_n(\lambda_1, \ldots, \lambda_d, W_{i,R_{\min}}, W_{\mathfrak{p},R_{\min}}) + \mathfrak{p}_n(\lambda_1, \ldots, \lambda_d, W_{i,R_{\max}}, W_{\mathfrak{p},R_{\max}}).$$

For example, for a nonlinear optimization routine with $n = 15$, we have $R_{\min} = 1/3$ (with $\mathfrak{p} = 0.1715$), and $R_{\max} = 5/6$ (with $\mathfrak{p} = 0.0225$), and the values of $\lambda(x)$ and $\rho(x)$ are

$$\lambda(x) = 0.24x^4 + 0.14x^5 + 0.0735x^6 + 0.026x^7 + 0.1347x^{24} + 0.3859x^{25},$$

$$\rho(x) = x^4. \tag{16.3.5}$$

For an IRA code of length 517 bytes, the results, using 4 DN branches and 4 PN branches, were compared with the Gallager code, and it was found that the former code performed better with almost no error.

16.4 Systematic Codes

In this class of codes, the input data is embedded in the encoded output. Such codes are used to add redundant information to data, and this allows errors to be detected, and possibly corrected, depending on the code used, when bits are lost or become corrupted. Examples of systematic codes are data plus checksum and Reed–Solomon codes. There are two kinds of systematic codes, one the simple systematic, an example of which is shown in Figure 16.4.1; and the other a more complex systematic code whose Tanner graph is given in Figure 16.4.2. The associated parity-check matrices $\mathbf{H}_1, \mathbf{H}_2$ are also displayed along with their graphs.

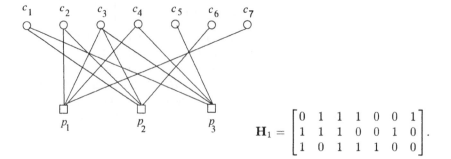

$$\mathbf{H}_1 = \begin{bmatrix} 0 & 1 & 1 & 1 & 0 & 0 & 1 \\ 1 & 1 & 1 & 0 & 0 & 1 & 0 \\ 1 & 0 & 1 & 1 & 1 & 0 & 0 \end{bmatrix}.$$

Figure 16.4.1 Simple systematic code.

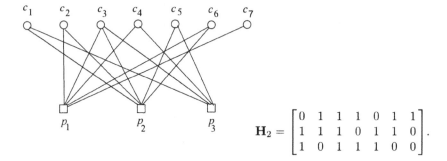

$$\mathbf{H}_2 = \begin{bmatrix} 0 & 1 & 1 & 1 & 0 & 1 & 1 \\ 1 & 1 & 1 & 0 & 1 & 1 & 0 \\ 1 & 0 & 1 & 1 & 1 & 0 & 0 \end{bmatrix}.$$

Figure 16.4.2 Complex systematic code.

The idea of combining long random codes with an iterative decoding algorithm was first introduced by Gallager [1968], who showed that these codes are 'good', meaning that they achieve an arbitrarily small probability error with maximum-likelihood decoding for rates bounded by Shannon's channel capacity. However, if the row weight of the parity-check matrix is kept fixed, then these codes cannot be 'very good'. For iterative decoding, the application of a Tanner graph is very useful and plays an important role in the decoding algorithm. We will closely look at the log-likelihood decoding of Turbo codes in §16.5. It is known that low-rate, short-frame Gallager codes perform very well in additive white Gaussian noise (AWGN) channels and Rayleigh fading channels and are used for error correction in code-division multiple-access (CDMA) systems.

Example 16.4.1. While transmitting a message with checksum, consider encoding a bit sequence $\{0\,1\,0\,1\,0\,1\,1\,0\}$ with a simple two-bit checksum. In this case, first calculate the checksum $01 + 01 + 01 + 10 = 01$; then transmit the original sequence followed by these checksum bits as $\{0\,1\,0\,1\,0\,1\,1\,0\,0\,1\}$. ∎

16.4.1 Nonsystematic Codes. A nonsystematic code is represented in Figure 16.4.3, together with its parity-check matrix $\mathbf{H}_3$.

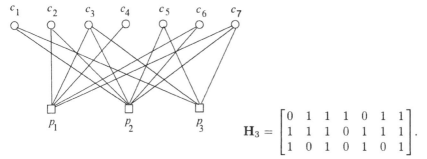

$$\mathbf{H}_3 = \begin{bmatrix} 0 & 1 & 1 & 1 & 0 & 1 & 1 \\ 1 & 1 & 1 & 0 & 1 & 1 & 1 \\ 1 & 0 & 1 & 0 & 1 & 0 & 1 \end{bmatrix}.$$

Figure 16.4.3 Nonsystematic code.

The nonsystematic LDPC-like codes were first introduced by MacKay and Neal [1996] (later to be known as MN codes). An LDPC code is *regular* if the rows and columns of $\mathbf{H}$ have uniform weight, that is, all rows have the same number of 1s (weight d_r) and all columns the same number of 1s (weight d_c). The performance of such codes is very good even though they are about 1 dB from the capacity. However, they perform worse than Turbo codes. The LDPC codes are *irregular* if the rows and columns have nonuniform weight; they outperform Turbo codes for block lengths of $n > 10^5$.

The degree distribution pair (λ, ρ) of an LDPC code is defined as

$$\lambda(x) = \sum_{i=2}^{d_v} \lambda_i x^{i-1}, \quad \rho(x) = \sum_{i=1}^{d_c} \rho_i x^{i-1},$$

where λ_i, ρ_i represent a fraction of edges emanating from p-nodes of degree i.

To construct MN codes, randomly generate an $m \times n$ matrix $\mathbf{H}$ with columns of weight d_c and rows of weight d_r, subject to some restraints.

CONSTRUCTION 1A: To avoid a 4-cycle, overlap between any two columns is no greater than 1.

CONSTRUCTION 2A: $m/2$ columns have $d_c = 2$, with no overlap between any pair of columns; remaining columns have $d_c = 3$. As in 1A, the overlap between any two columns is no greater than 1.

CONSTRUCTION 1B AND 2B: Select columns from 1A and 2A. This can result in a higher rate code.

Shamir et al. [2005] have shown that nonsystematic coding is superior to systematic provided there exists enough redundancy in the transmitted word. The redundant data bits are split (or scrambled) into coded bits. The process of splitting (or scrambling) is achieved by cascading a sparse matrix or the inverse of a sparse matrix, respectively, with an LDPC code. A nonsystematic code is one in which the encoded output data does not contain the input bits; these codes are also used to add redundant information to data bits. An example of a nonsystematic code is the Fountain code which is mostly used for online content distribution.

16.4.2 Binary Convolutional Codes. A convolutional encoder has k input sequences, n output sequences, and m delay elements arranged in a shift register. The OR gates use combinatorial logic where each of the n outputs depends on some mod-2 combination of the k current inputs and the m previous inputs in storage. The *constraint length* is the maximum number of past and present input bits that each output bit depends on. Set $m = k + 1$ and let $k = 1$. Then there are $2^m = 2^2 = 4$ total states. Since $k = 1$, two branches enter and two branches leave each state, where on each branch 'input data bit/associated output code bit' are marked on the state diagram of a finite-state machine (Figure 16.4.4).

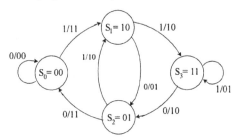

Figure 16.4.4 State diagram of a convolutional encoder.

16.4.3 Trellis Diagram. The state diagram (Figure 16.4.4) is useful in understanding the operation of the encoder. However, it does not show how the states change over time for a given input sequence. A *trellis* is an expanded state diagram that shows the passage of time, where all the possible states are shown for each instant of time; time is indicated by a movement to the right, and the input data bits and output code bits are represented by a unique path through the trellis.

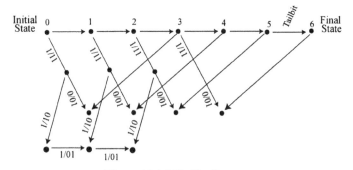

Figure 16.4.5 Trellis diagram.

A trellis diagram for the state diagram of Figure 16.4.4 is presented in Figure 16.4.5, where the states $0, 1, \ldots$, mean that 0 is the intial state, 1 is the new state after the first bit is encoded, and so on.

16.4.4 Recursive Systematic Codes. (RSC Codes) Berrou et al. [1993] have proposed a class of infinite impulse response (IIR) convolutional codes that are regarded as the building blocks for Turbo codes. Consider a binary convolutional encoder of a given rate with constraint length K, and a bit d_κ as input at time $\kappa = 1, \ldots, k$, i.e., at the time when a κ-th bit is the input, which is associated with a codeword $\mathbf{c} = \{c_i\} \equiv \{d_\kappa, p_j\}$, where

$$d_\kappa = \sum_{i=1}^{K} g_{1i} d_{\kappa-i} \pmod{2}, \quad p_j = \sum_{i=1}^{K} g_{2i} d_{j-i} \pmod{2}, \qquad (16.4.1)$$

and $\mathbf{G}_1 = \{g_{1i}\} = \{0, 1\}$ and $\mathbf{G}_2 = \{g_{2i}\} = \{0, 1\}$ are the code generators. An RSC encoder is constructed from a standard convolutional encoder by feeding back one of the outputs. This encoder is a finite IIR filter system in discrete time. The codewords $\{c_i\} = \{d_\kappa\}\{p_j\}$ in an RSC code are continuously used as feedback to the encoder inputs. The sequence $\{d_\kappa\}$ represents the data bits while $\{p_j\}$ are the parity bits. The RSC codes perform much better at higher rates for all values of E_b/N_0. These codes have a feedback loop and set either of the two outputs d_κ or p_j, where d_κ denote data bits received at the modulator. An example of this code is presented in Figure 16.4.6, where the binary a_κ is the recursively computed by the formula

$$a_\kappa = d_\kappa + \sum_{i=1}^{K} g_i^* a_{\kappa-i} \mod 2,$$

where a_κ represents the current input bit 0 or 1, and $g_i^* = \begin{cases} g_{1i} & \text{if } d_i = d_\kappa, \\ g_{2i} & \text{if } d_i = p_j. \end{cases}$

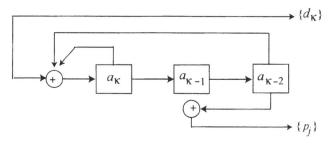

Figure 16.4.6 RSC encoder.

Example 16.4.2. Consider the input data sequence $\{d_\kappa\} = \{1010\}$ (see Figure 16.4.6). The following procedure, presented in Table 16.4.1, generates recursively the output codeword $\{c_i\}$. This table has 8 rows, the first four of which are associated with the input data bit 0 while the last four with the bit 1. The columns are numbered 1 through 7: Column (1) contains the input bit d_κ; column (2) the current bit a_κ; column (3) the starting bit $a_{\kappa-1}$; column

(4) the starting bit $a_{\kappa-2}$; column (5) the code bit pair $d_\kappa p_j$; column (6) the ending state bit a_κ; and column (7) the ending state bit $a_{\kappa-1}$.

Table 16.4.1 Data Sequence in RSC Encoder

(1)	(2)	(3)	(4)	(5)	(6)	(7)
d_κ	a_κ	$a_{\kappa-1}$	$a_{\kappa-2}$	$d_\kappa\, p_j$	a_κ	$a_{\kappa-1}$
0	0	0	0	0 0	0	0
	1	1	0	0 1	1	1
	1	0	1	0 0	1	0
	0	1	1	0 1	0	1
1	1	0	0	1 1	1	0
	0	1	0	1 0	0	1
	0	0	1	0 1	0	0
	1	1	1	1 0	1	1

This table shows the transitions, with step-by-step details as follows:

STEP 1. At any input time κ, the starting state of a transition is the entry of the two rightmost registers $a_{\kappa-1}$ and $a_{\kappa-2}$. For example, in row 1 the starting entries are $a_{\kappa-1} = 0 = a_{\kappa-2}$.

STEP 2. On any row, the entry in column (2) is obtained by computing $d_\kappa \oplus a_{\kappa-1} \oplus a_{\kappa-2}$.

STEP 3. For the code bits sequence $d_\kappa\, p_j$ in column (5), the first entry is the value of d_κ from column (1) and the value of second entry is $p_j = a_\kappa \oplus a_{\kappa-2}$.

STEP 4. The entries in columns (6) and (7) are repeated from columns (2) and (3).

The encoder constructs the codeword using Figure 16.4.6, and the procedure at times κ is shown in Table 16.4.2, where column (1) lists the times $\kappa = 1, 2, \ldots, k$; the input bits d_κ are entered in column (2); the first stage bit a_κ in column (3); the start bits $a_{\kappa-1}$ and $a_{\kappa-2}$ in columns (4) and (5), respectively; and the output $u_\kappa\, p_j$ in column (6). The entries in this table are made as follows: The entry in column (2) consists of the known data bits (1010 in this example), written one bit at one time. At any time κ, the entry in column (3) is obtained by the formula $d_\kappa \oplus a_{\kappa-1} \oplus a_{\kappa-2}$, i.e., by computing [column (2)$\oplus$ column (4) $\oplus$ column (5)]. The first entry in column (6) is d_κ from column (2) while the second entry is obtained by the encoder logic circuitry of Figure 16.4.6, which are the values of $a_{\kappa-2}$ from column (5). The row at time $\kappa = 5$ shows termination of these transitions.

Table 16.4.2 Codeword Sequence in RSC Encoder

(1)	(2)	(3)	(4)	(5)	(6)
κ	d_κ	a_κ	$a_{\kappa-1}$	$a_{\kappa-2}$	$d_\kappa\ p_j$
1	1	1	0	0	1 0
2	0	0	1	0	0 0
3	1	0	0	1	1 1
4	0	1	0	0	0 0
5			1	0	

Thus, from column (6) the data bits are 1010 (first column entry) together with the parity bits 0010 (second entry from column (5)), yielding the codeword $\mathbf{c} = \{d_\kappa, p_j\} = \{1\,0\,1\,0\,0\,0\,1\,0\}$. ∎

16.4.5 RSC Decoder. The Bahl algorithm (Bahl et al. [1974]) is used to minimize the probability of sequence errors since it generates the a *posteriori* probability or soft-decision output for each decoded bit. A modified form of this algorithm, by Berrou et al. [1993], is used for decoding RSC codes. The a *posteriori* probability of a decoded bit d_κ is derived from the joint probability $\lambda_\kappa^{i,m}$, which is defined by

$$\lambda_\kappa^{i,m} = \mathfrak{p}\{d_\kappa = i, s_\kappa = m | w_1^N\}, \qquad (16.4.2)$$

where s_κ is the decoder state at time κ, and w_1^N is a received bit sequence from time $\kappa = 1$ through time $\kappa = k$. Thus, the a *posteriori* probability for a decoded bit d_κ is given by

$$\mathfrak{p}\{d_\kappa = i | w_1^N\} = \sum_m \lambda_\kappa^{i,m}, \quad i = 0, 1, \qquad (16.4.3)$$

where the log-likelihood ratio (LLR) is the logarithm of the ratio of a *posteriori* probabilities:

$$\Lambda(d_\kappa^*) = \log\left(\frac{\sum_m \lambda_\kappa^{1,m}}{\sum_m \lambda_\kappa^{0,m}}\right), \qquad (16.4.4)$$

where d_κ^* is the decoded bit. The decoder's decision, known as the *maximum a posteriori* (MAP) rule, also known as Forney's forward-backward algorithm, is defined by

$$d_\kappa^* = \begin{cases} 1 & \text{if } \Lambda d_\kappa^* > 0, \\ 0 & \text{if } \Lambda d_\kappa^* < 0, \end{cases} \qquad (16.4.5)$$

The quantity $\Lambda(d_\kappa^*)$ is then computed using the formula (14.5.11). The computational details for computing this quantity can be found in Berrou et al. [1993], Robertson et al. [1995], Berrou and Glavieux [1996], and Pietrobon [1998].

16.5 Turbo Codes

In Turbo codes, introduced by Berrou and Glavieux [1996], the data bits drive two linear feedback shift registers that generate parity bits through the modulator. Once the codewords, which are composed of data bits and two sets of parity bits, are transmitted through the channel (mostly an AWGN) and pass through the demodulator, a sequence of received words is decoded using a stochastic algorithm known as the log-likelihood ratio method, described in the following section.

16.5.1 Log-Likelihood Ratio Method. Berrou et al. [1996] have shown that the LLR soft output $\Lambda(\hat{\theta})$ for the decoder is given by Eq (14.5.9) as $\Lambda(\hat{\theta}) = \Lambda(x|\theta) + \Lambda(\theta)$, $\hat{\theta} = \theta + x$. This definition can also be written as

$$\Lambda^*(\hat{\theta}) = \Lambda_c(x) + \Lambda(\theta), \qquad (16.5.1)$$

or

$$\Lambda(\theta) = \Lambda^*(\hat{\theta}) + \Lambda_e(\hat{\theta}), \qquad (16.5.2)$$

where $\Lambda^*(\hat{\theta})$, abbreviated for $\Lambda(\theta|x)$, is the LLR of a data bit input to the decoder and $\Lambda_e(\hat{\theta})$, called the *extrinsic* LLR, represents additional information obtained from the decoding process. From the above two equations we obtain

$$\Lambda(\hat{\theta}) = \Lambda_c(x) + \Lambda(\theta) + \Lambda_e(\hat{\theta}), \qquad (16.5.3)$$

where $\Lambda_c(x)$ refers to the LLR of a parity bit at the encoder. Now, let θ represent the data bits d_i, x as the parity bit p_i, so that $\hat{\theta} = \theta + x$ will represent the codeword c_i. Then Eq (16.5.2) is written as

$$\Lambda(d_i) = \Lambda^*(c_i) + \Lambda_e(c_i). \qquad (16.5.4)$$

Consider a systematic code in which the data bits d_i are composed of a $k_1 \times k_2$ matrix (k_1 rows and k_2 columns). Set $m_{1,2} = n_{1,2} - k_{1,2}$. The k_1 rows contain codewords with k_2 data bits and m_2 parity bits. Similarly, the k_2 columns contain codewords with k_1 data bits and m_2 parity bits. This encoding arrangement, shown in Figure 16.5.1, presents the layout of two blocks of $k_1 \times k_2$ data bits that are encoded into two codewords, one a horizontal and the other a vertical word, annexed by two sets of extrinsic LLR values, one labeled Λ_{eh} for horizontal decoding and the other Λ_{ev} for vertical decoding.

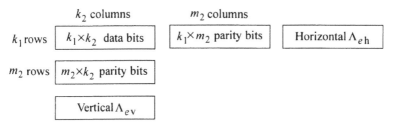

k_2 columns m_2 columns

k_1 rows $k_1 \times k_2$ data bits $k_1 \times m_2$ parity bits   Horizontal Λ_{eh}

m_2 rows $m_2 \times k_2$ parity bits

Vertical Λ_{ev}

Figure 16.5.1 Layout of systematic encoder/decoder.

In this layout the LLR values Λ_{eh} and Λ_{ev} are computed after the codewords are received. For encoding the data and parity bits are located as follows:

Data Bits				Parity Bits (Horizontal)			
d_1	d_2	$\cdots$	d_{k_2} $\mid$	p_{12}	p_{13}	$\cdots$	p_{k_2-1,k_2}
d_{k_2+1}	d_{k_2+2}	$\cdots$	d_{2k_2} $\mid$	p_{k_1+1,k_2+2}	$\cdots$	$\cdots$	$p_{2k_2-1,2k_2}$
$\cdots$	$\cdots$	$\cdots$	$\cdots$ $\mid$	$\cdots$	$\cdots$	$\cdots$	$\cdots$
$d_{(k_1-1)k_2}$	$d_{(k_1-2)k_2}$	$\cdots$	$d_{k_1 k_2}$ $\mid$	$\cdots$	$\cdots$	$\cdots$	$p_{2(k_1-1)(k_2-1),k_1 k_2}$

p_{1,k_2+1}	$p_{2,k+2}$	$\cdots$	p_{m_2,k_2}	$\mid$
$\cdots$	$\cdots$	$\cdots$	$\cdots$	$\mid \leftarrow$ Parity Bits (Vertical)
$p_{1,(m_2-1)k_2}$	$\cdots$	$\cdots$	$p_{(m_2-1)k_2,m_2 k_2}$	$\mid$

In particular, for $k_1 = 2 = k_2$, the above layout becomes

Data Bits		Parity Bits (Horizontal)	
d_1	d_2 $\mid$	p_{12}	$\mid$
d_3	d_4 $\mid$	p_{34}	$\mid$

p_{13}	$\mid$
p_{24}	$\mid$ $\leftarrow$ Parity Bits (Vertical)

It has 4 data bits and 4 parity bits, thus generating a codeword of length 8. The encoder generates the codeword

$$\mathbf{c} = \{c_i\} = \{d_1 \, d_2 \, d_3 \, d_4 \, p_{12} \, p_{34} \, p_{12} \, p_{24}\}.$$

For $k_1 = 3 = k_2$, this layout, which has 9 data bits and 18 parity bits, is

Data Bits			Parity Bits (Horizontal)			
d_1	d_2	d_3 $\mid$	p_{12}	p_{13}	p_{23}	$\mid$
d_4	d_5	d_6 $\mid$	p_{41}	p_{46}	p_{56}	$\mid$
d_7	d_8	d_9 $\mid$	p_{78}	p_{79}	p_{89}	$\mid$

$$
\begin{array}{ccc|}
p_{13} & p_{25} & p_{36} \\
p_{17} & p_{28} & p_{39} & \leftarrow \text{Parity Bits (Vertical)} \\
p_{47} & p_{58} & p_{69} \\
\end{array}
$$

In this case the encoder generates the 27-bit length codeword

$$\mathbf{c} = \{c_i\} = \{d_1\, d_2\, d_3\, d_4\, d_5\, d_6\, d_7\, d_8\, d_9\ p_{12}\, p_{13}\, p_{23}\, p_{41}\, p_{46}\, p_{56}\, p_{78}\, p_{79}\, p_{89}\, p_{13}\, p_{25}$$

$$p_{36}\, p_{17}\, p_{28}\, p_{39}\, p_{47}\, p_{58}\, p_{69}\}.$$

The parity bits are computed by the formula $p_{ij} = d_i \oplus d_j$.

After the codeword is transmitted, let the received word be $\mathbf{w} = \{w_i\} = \{u_i\} + \{q_{ij}\}$, where u_i denotes the received data bits, and q_{ij} the received parity bits. Using formula (14.5.13) we obtain $\Lambda_c^{(0)}(u_i) = 2u_i$, which is used along with formula (14.5.11) to compute $\Lambda_{eh}^{(\nu)}$ and $\Lambda_{ev}^{(\nu)}$.

The LLR decoder is based on the following algorithm, due to Berrou and Glavieux [1996], which is as follows:

STEP 1. Set the a priori LLR $\Lambda_{ev}^{(0)} = 0$.

STEP 2. Calculate the horizontal extrinsic LLR $\Lambda_{eh}^{(\nu)}$, where $\nu = 0, 1, 2, \dots$ is the iteration index, using the formula

$$\Lambda_{eh}^{(\nu+1)}(u_i) = \left[\Lambda_c^{(\nu)}(u_j) + \Lambda_{ev}^{(\nu)}(u_j) \right] \boxplus \Lambda_c^{(0)}(q_{ij}). \tag{16.5.5}$$

STEP 3. Calculate the vertical extrinsic LLR $\Lambda_{ev}^{(\nu)}$, where $\nu - 0, 1, 2, \dots$ is the iteration index, using the formula

$$\Lambda_{ev}^{(\nu+1)}(u_i) = \left[\Lambda_c^{(\nu)}(u_j) + \Lambda_{ev}^{(\nu+1)}(u_j) \right] \boxplus \Lambda_c^{(0)}(q_{ij}). \tag{16.5.6}$$

STEP 4. Continue the iteration for $\nu = 0, 1, 2, \dots$. This process may be terminated when two (or more) consecutive iterations yield the same polarity of the data bits.

STEP 5. The corrected message is the voltage value of the sum of the LLRs $\Lambda_c^{(0)}(u_i) + \Lambda_{eh}^{(\nu)} + \Lambda_{ev}^{(\nu)}$, which gives the corrected data bits 1 and 0 according to whether the voltage value is positive or negative.

Example 16.5.2. Consider the data bits $\{0\,1\,1\,0\}$ that generate the parity bits $p_{12} = d_1 \oplus d_2 = 0 \oplus 1 = 1$; $p_{34} = d_3 \oplus d_4 = 1 \oplus 01$; $p_{13} = d_1 \oplus d_3 = 0 \oplus 1 = 1$; $p_{24} = d_2 \oplus d_4 = 1 \oplus 0 = 1$. This 8-bit long codeword $\mathbf{c} = \{0\,1\,1\,0\,1\,1\,1\,1\}$ with voltage values $\{-1, +1, +1, -1, +1, +1, +1, +1\}$ is transmitted. Suppose that the received word $\mathbf{w}$ has the voltage values $\mathbf{w} = \{0.05, 0.75, 0.55, 0.1, 1.25, 2.0, 2.5, 1.1\}$. Using formula (14.5.13) we find that $\Lambda_c^{(0)}(w_i) = \{0.1, 1.5, 1.1, 0.2, 2.5, 4.0, 5.0, 2.2\}$, which gives

$$\Lambda_c^{(0)}(u_1) = 0.1,\ \Lambda_c^{(0)}(u_2) = 1.5,\ \Lambda_c^{(0)}(u_3) = 1.1,\ \Lambda_c^{(0)}(u_4) = 0.2,$$

$\Lambda_c^{(0)}(q_{12}) = 2.5$, $\Lambda_c^{(0)}(q_{34}) = 4.0$, $\Lambda_c^{(0)}(q_{13}) = 5.0$, $\Lambda_c^{(0)}(q_{24}) = 2.2$.

For the first iteration we take $\nu = 0$ in (16.5.5) and (16.5.6). Then using (14.5.11) we get

$$\Lambda_{eh}^{(1)}(u_1) = \left[\Lambda_c^{(0)}(u_2) + \Lambda_{ev}^{(0)}(u_2)\right] \boxplus \Lambda_c^{(0)}(q_{12})$$
$$= [1.5 + 0] \boxplus 2.5 = (-1)(+1)(+1)(1.5) = -1.5,$$

$$\Lambda_{eh}^{(1)}(u_2) = \left[\Lambda_c^{(0)}(u_1) + \Lambda_{ev}^{(0)}(u_1)\right] \boxplus \Lambda_c^{(0)}(q_{12})$$
$$= [0.1 + 0] \boxplus 2.5 = (-1)(+1)(+1)(0.1) = -0.1,$$

$$\Lambda_{eh}^{(1)}(u_3) = \left[\Lambda_c^{(0)}(u_4) + \Lambda_{ev}^{(0)}(u_4)\right] \boxplus \Lambda_c^{(0)}(q_{34})$$
$$= [0.2 + 0] \boxplus 4.0 = (-1)(+1)(+1)(0.2) = -0.2,$$

$$\Lambda_{eh}^{(1)}(u_4) = \left[\Lambda_c^{(0)}(u_3) + \Lambda_{ev}^{(0)}(u_3)\right] \boxplus \Lambda_c^{(0)}(q_{34})$$
$$= [1.1 + 0] \boxplus 4.0 = (-1)(+1)(+1)(1.1) = -1.1,$$

$$\Lambda_{ev}^{(1)}(u_1) = \left[\Lambda_c^{(0)}(u_3) + \Lambda_{eh}^{(1)}(u_3)\right] \boxplus \Lambda_c^{(0)}(q_{13})$$
$$= [1.1 - 0.2] \boxplus 5.0 = (-1)(+1)(+1)(0.9) = -0.9,$$

$$\Lambda_{ev}^{(1)}(u_2) = \left[\Lambda_c^{(0)}(u_4) + \Lambda_{eh}^{(1)}(u_4)\right] \boxplus \Lambda_c^{(0)}(q_{24})$$
$$= [0.2 - 1.1] \boxplus 2.2 = (-1)(-1)(+1)(0.9) = 0.9,$$

$$\Lambda_{ev}^{(1)}(u_3) = \left[\Lambda_c^{(0)}(u_1) + \Lambda_{eh}^{(1)}(u_1)\right] \boxplus \Lambda_c^{(0)}(q_{13})$$
$$= [0.1 - 1.5] \boxplus 5.0 = (-1)(-1)(+1)(1.4) = 1.4,$$

$$\Lambda_{ev}^{(1)}(u_4) = \left[\Lambda_c^{(0)}(u_2) + \Lambda_{eh}^{(1)}(u_2)\right] \boxplus \Lambda_c^{(0)}(q_{24})$$
$$= [1.5 - 0.1] \boxplus 2.2 = (-1)(+1)(+1)(1.4) = -1.4.$$

Thus, the original LLR plus both horizontal and vertical extrinsic LLRs produce the following polarity layout for the decoded data bits:

$0.1 - 1.5 - 0.9 = -2.3$	$1.5 - 0.1 + 0.9 = 2.3$
$1.1 - 0.2 + 1.4 = 2.3$	$0.2 - 1.1 - 1.4 = -2.3$

which, by the polarity rule, implies that the data bits must be $[0\,1\,1\,0]$. Now we proceed to the second iteration ($\nu = 1$), as follows:

$$\Lambda_{eh}^{(2)}(u_1) = \left[\Lambda_c^{(0)}(u_2) + \Lambda_{ev}^{(1)}(u_2)\right] \boxplus \Lambda_c^{(0)}(q_{12})$$
$$= [1.5 + 0.9] \boxplus 2.5 = (-1)(+1)(+1)(2.4) = -2.4,$$

$$\Lambda_{eh}^{(2)}(u_2) = \left[\Lambda_c^{(0)}(u_1) + \Lambda_{ev}^{(1)}(u_1)\right] \boxplus \Lambda_c^{(0)}(q_{12})$$
$$= [0.1 - 0.9] \boxplus 2.5 = (-1)(-1)(+1)(0.8) = -0.8,$$

$$\Lambda_{\text{eh}}^{(2)}(u_3) = \left[\Lambda_c^{(0)}(u_4) + \Lambda_{\text{ev}}^{(1)}(u_4)\right] \boxplus \Lambda_c^{(0)}(q_{34})$$

$$= [0.2 - 1.4] \boxplus 4.0 = (-1)(-1)(+1)(0.2) = 1.2,$$

$$\Lambda_{\text{eh}}^{(2)}(u_4) = \left[\Lambda_c^{(0)}(u_3) + \Lambda_{\text{ev}}^{(1)}(u_3)\right] \boxplus \Lambda_c^{(0)}(q_{34})$$

$$= [1.1 + 1.4] \boxplus 4.0 = (-1)(+1)(+1)(1.1) = -2.5,$$

$$\Lambda_{\text{ev}}^{(2)}(u_1) = \left[\Lambda_c^{(0)}(u_3) + \Lambda_{\text{eh}}^{(2)}(u_3)\right] \boxplus \Lambda_c^{(0)}(q_{13})$$

$$= [1.1 + 1..2] \boxplus 5.0 = (-1)(+1)(+1)(0.9) = -2.3,$$

$$\Lambda_{\text{ev}}^{(2)}(u_2) = \left[\Lambda_c^{(0)}(u_4) + \Lambda_{\text{eh}}^{(2)}(u_4)\right] \boxplus \Lambda_c^{(0)}(q_{24})$$

$$= [0.2 - 2.5] \boxplus 2.2 = (-1)(-1)(+1)(2.3) = 2.3,$$

$$\Lambda_{\text{ev}}^{(2)}(u_3) = \left[\Lambda_c^{(0)}(u_1) + \Lambda_{\text{eh}}^{(2)}(u_1)\right] \boxplus \Lambda_c^{(0)}(q_{13})$$

$$= [0.1 - 2.4] \boxplus 5.0 = (-1)(-1)(+1)(2.3) = 2.3,$$

$$\Lambda_{\text{ev}}^{(2)}(u_4) = \left[\Lambda_c^{(0)}(u_2) + \Lambda_{\text{eh}}^{(2)}(u_2)\right] \boxplus \Lambda_c^{(0)}(q_{24})$$

$$= [1.5 + 0.8] \boxplus 2.2 = (-1)(+1)(+1)(2.2) = -2.2.$$

Thus, the original LLR plus both horizontal and vertical extrinsic LLRs produce the following polarity layout for the decoded data bits:

$0.1 - 2.4 - 2.3 = -4.6$	$1.5 + 0.8 + 2.3 = 4.6$
$1.1 + 1.2 + 2.3 = 4.6$	$0.2 - 2.5 - 2.2 = -4.5$

which, by the polarity rule, implies that the data bits must be $[0\,1\,1\,0]$. Notice that the results of two consecutive iterations match. We may even go to the third iteration, which gives

$$\Lambda_{\text{eh}}^{(3)}(u_1) = -2.5; \ \Lambda_{\text{eh}}^{(3)}(u_2) = 2.2; \ \Lambda_{\text{eh}}^{(3)}(u_3) = 2.0; \ \Lambda_{\text{eh}}^{(3)}(u_4) = -3.4;$$

$$\Lambda_{\text{ev}}^{(3)}(u_1) = -3.1; \ \Lambda_{\text{ev}}^{(3)}(u_2) = 2.2; \ \Lambda_{\text{ev}}^{(3)}(u_3) = 3.3; \ \Lambda_{\text{ev}}^{(3)}(u_4) = -2.2.$$

Thus, the original LLR plus both horizontal and vertical extrinsic LLRs produce the following polarity layout for the decoded data bits:

$0.1 - 2.5 - 3.1 = -5.5$	$1.5 - +2.2 + 2.2 = 5.9$
$1.1 + 2.0 + 3.3 = 6.4$	$0.2 - 3.4 - 2.2 = -5.4$

which, by the polarity rule, again implies that the data bits must be $[0\,1\,1\,0]$. At this iterative state, we may decide to terminate the process. ∎

At this point we can reflect upon the stopping rule for a decoding stochastic process that is a martingale. It appears that such a process may be stopped if the following three conditions are met: (i) During the course of using the belief propagation algorithm, check the most probable state at each time, thus obtaining a sequence of data bits after each data bit has passed through the consecutive iterations of the LLR computations, and (ii) combine the finitely many iterations for each data bit to compute the most probable state for each data bit. This will be achieved after a finite number of iterations; and (ii) if all estimates in (i) agree with the most probable state of the data bits determined in (ii), then the decoder has arrived at a state when it can stop. However, we may encounter an occasional error, due to other causes.

Turbo codes get their name because the decoder uses feedback, like a Turbo engine. Some characteristics of Turbo codes are: (i) these codes perform extremely well at rates very close to the Shannon limit due to a low multiplicity of low weight codewords; (ii) the performance of Turbo codes becomes better the larger their block size gets, and larger block sizes are not more complex to decode. The BER floor is lower for larger frame/interleaver sizes; and (iii) the complexity of a constraint length K_T Turbo code is almost equal to a $K = K_C$ convolutional code; in fact, $-K_C \approx 2 + K_T + \log_2(\#$ of decoder iterations).

The examples given throughout this chapter and all others are very simple; they deal with codes of small length. However, in practice, the code lengths are very large. For example, we compare the performance of the maximum length UMTS Turbo code against four LDPC codes, with the following code parameters: All codes are of rate 1/2; the LDPC codes are of length $(n, k) = (15000, 5000)$, with up to 100 iterations of log-domain sum-product decoding; the Turbo code has length $(n, k) = (15354, 5114)$, with up to 16 iterations of log-MAP decoding. Both codes operate with BPSK modulation, AWGN and fully-interleaved Rayleigh fading, and there are enough trial runs to log 40 frame errors. The four LDPC codes are: (1) MacKay [1999] regular construction 2A (MacKay [1999]); (2) Richardson, Shakorallahi and Urbanke [2001] irregular construction; (3) Tian et al. [2003] Chris Jones' improved irregular construction; and (4) Yang and Ryan [2003] extended IRA code. The results for BPSK/AWGN capacity 0.5 dB for rate $R = 1/3$ are presented in Figure 16.5.2.

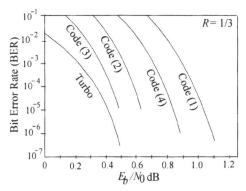

Figure 16.5.2 Performance comparison.

16.6 BP Decoding

The belief propagation (BP) decoding, published by Pearl [1988], has been used to decode LDPC codes, or other recent codes where an LDPC code is used as a pre-code. This decoding operates on a Tanner graph used as a representation of an LPDC code. The basic BP decoding described in Chapter 14 operates on a so-called *flooding schedule*, as follows:

STEP 1. Evaluate the codewords associated with all parity-check nodes.

STEP 2. 'Flood' the information obtained in Step 1 to all the variable nodes.

STEP 3. Compute all the varible nodes.

STEP 4. 'Flood' back the values found in Step 3 to the parity-check nodes.

Among other schedules, a probabilistic schedule[1] is found in Mao and Beni-hashami [2001], which uses the local girth.[2] Then a probability is assigned to each variable node[3] decoding on the local girth and then output data symbols at each variable node, except the first one, are updated with this probability while continuously updating the output data symbols. This decoding method improves the residual bit error rate (BER) for LDPC codes, although convergence of this schedule gets worse as decoding progresses. Improvement of the convergence problem has been proposed in Zhang and Fossovier [2002], which is known as the *shuffled BP decoding*, and by Sharon et al. [2004], which is known as the *dual check-shuffled BP decoding*. These improved algorithms are based on the fact that new information obtained during the updating process of nodes can be used right away in updating other nodes within the same iteration. This technique improves the convergence and the BER performance of the BP decoder. Thus, the best choice for decoding LDPC codes is a girth-based stochastic BP decoder with shuffled and check-shuffled schedules.

The original BP decoding operates as follows: The (n, k) LDPC code of length n and k data bits is associated with a parity-check matrix $\mathbf{H}$ of dimension $m \times n$, where $m = n - k$ denotes the number of parity-check bits; thus, $n = 2^m - 1$ and $k = 2^m - m - 1$ for $m \geq 3$. Let H_{jn} denote the elements in the m-th row and j-th column of the matrix $\mathbf{H}$, $j = 1, \ldots, m$, then the set $N(j) = \{l : H_{jl} \neq 0\}$ contains the symbols (codewords) that are located in the j parity-check equations. Similarly, the set $M(j) = \{j : H_{jl} \neq 0\}$ contains the check equations at the l-th symbol.

Let w_l denote the received symbol for the variable node l, and let $w_{jl}^{(0)} = w_l$ denote the received symbol at the edge connecting the variable node l with the check node j before the decoding process starts. In general, $w_{jl}^{(i)}$ will denote the received symbol at the edge connecting the variable node l with the check

---

[1] These schedules are, in fact, stochastic because of the process involved in their algorithms. However, we will keep their standard designation as 'probabilistic' schedules.

[2] The length of the shortest cycle passing through a given node is called the girth.

[3] A variable node is an input or output node that varies at each iteration.

node j at the iteration i. The BP decoder consists of the following steps:

STEP 1. Determine the check node associated with the information symbol

$$c_{jl}^{(i)} = 2 \tanh^{-1} \left(\prod_{l' \in N(j) \backslash l} \tanh \left(\tfrac{1}{2} w_{jl'}^{(i-1)} \right) \right), \qquad (16.6.1)$$

where $c_{jl}^{(i)}$ is the symbol at the edge connecting the check node j with the variable node l at the i-th iteration, and $N(j) \backslash l$ denotes the set $N(j)$ from which the l-th check equation is excluded (see §15.5 for a similar notation). The factor 2 is required in these formulas since $\tanh^{-1}(x) = \tfrac{1}{2} \ln \dfrac{1+x}{1-x}, 0 \le x < 1$. This ensures that the outputs $c_{jl}^{(i)}$ will be either 0 or 1.

STEP 2. Update the variable node symbol using the iterative relation

$$w_{jl}^{(i)} = w_l + \sum j' \in M(l) \backslash j c_{j'l}. \qquad (16.6.2)$$

STEP 3. For each variable node l, compute the hard decision (§15.4) $y_l^{(i)} = \operatorname{sgn} \left\{ w_{jl}^{(i)} + c_{jl}^{(i)} \right\}$ for $j = 1, \dots, m$.

STEP 4. Use all these values to compute the parity-check equations, and determine when the decoding can be stopped.

16.6.1 Probabilistic Shuffled Decoding. (Zang and Fossovier [2002]) Suppose that the local girth g_l of each variable node l has been computed and, after the maximum local girth $g_{\max}$ has been determined, the *updating probability* $\mathfrak{p}_l^{[g]}$ for each l is defined as the ratio of g_l and $g_{\max}$. Then all variable-to-check messages $w_{jl}^{(i)}$ are initially set equal to the value w_l, and all check-to-variable messages $c_{jl}^{(0)}$ are set equal to zero. At the beginning of the i-th iteration, all messages $c_{jl}^{(i-1)}$ and all messages $w_{jl}^{(i-1)}$ become available iteratively one-by-one. In each iteration the decoding goes through all variable nodes in a series and randomly decides using a lookup table with preassigned random numbers, whether or not to update the inputs and outputs at each node with probability $\mathfrak{p}_l^{[g]}$. (For random and pseudo-random numbers, see Appendix E.) Thus, the algorithm for the stochastic shuffled decoding is as follows:

STEP 1. Initialize $l = 1, i = 1$,

STEP 2. With probability $1 - \mathfrak{p}_l^{[g]}$, where $\mathfrak{p})l^{[g]} = g_l/g_{\max}$, set $c_{jl}^{(i)} = c_{jl}^{(i-1)}$ and $w_{jl}^{(i)} = w_{jl}^{(i-1)}$ for all $l \in M(l)$, and continue to Step 5.

STEP 3. For all inputs at the variable node l and for all $j \in M)l)$, compute

$$c_{jl}^{(i)} = 2 \tanh^{-1} \left(\prod_{\substack{l' \in N)(j) \backslash l \\ l' < l}} \tanh \left(\tfrac{1}{2} w_{jl'}^{(i)} \right) \cdot \prod_{\substack{l' \in N(j) \backslash l \\ l' > l}} \tanh \left(\tfrac{1}{2} w_{jl'}^{(i-1)} \right) \right).$$

$$(16.6.3)$$

This process is known as the *horizontal step*.

STEP 4. Compute the outputs at the variable node l for all $j \in M(l)$ from Eq (16.6.2). This is known as the *vertical step*.

STEP 5. If $l < n$, increment l and continue at Step 2.

STEP 6. If the stopping rule is satisfied (see §14.5.2), stop all decoding and computation of the output hard decision $\hat{y}_l^{(i)}$ for all l.

STEP 7. Set $i = i + 1$ and $l = 1$; start the next iteration by continuing at Step 2.

Note that in the case of 'flooding' schedule the iteration index i does not correspond to the number of iterations.

16.6.2 Probabilistic Check-Shuffled Decoding. (Sharon et al. [2004]) We use the notation of the previous section. Let $V^{(i)}$ denote the set of all variable nodes that take part in updates during the iteration i. Let all check nodes be traversed serially with probability $\mathsf{p}_l^{[g]}$, and inputs and outputs are updated only for the edges that are connected to the active variable nodes. The algorithm for the stochastic check-shuffled decoding is as follows:

STEP 1. Initialize $j = 1$ and $i = 1$.

STEP 2. Add l to the set $V^{i)}$ with probability $\mathsf{p}_l^{[g]}$ for all $l = 1, \dots, n$.

STEP 3. For all $l \in N(j) \cap V^{(i)}$, compute

$$w_{jl}^{(i)} = w_l + \sum_{\substack{j' \in M(lj) \setminus j \\ j' < j}} c_{j'l}^{(i)} + \sum_{\substack{j' \in M(lj) \setminus j \\ j' > j}} c_{j'l}^{(i-1)}. \tag{16.6.4}$$

STEP 4. Set $w_{jl}^{(i)} = w_{jl}^{(i-1)}$ for all $l \in N(j) \setminus V^{(i)}$.

STEP 5. For all $l \in N(j) \cap V^{(i)}$, compute

$$c_{jl}^{(i)} = 2 \tanh^{-1} \left(\prod_{l' \in N(j) \setminus l} \tanh \left(\tfrac{1}{2} w_{jl'}^{(i)} \right) \right). \tag{16.6.5}$$

STEP 6. For all $l \in N(j) \setminus V^{(i)}$, set $c_{jl}^{(i)} = c_{jl}^{(i-1)}$.

STEP 7. If $j < m$, increment j and continue at Step 3.

STEP 8. If the stopping rule is satisfied, stop decoding and output hard decision $\hat{y}_j^{(i)}$ for all j.

STEP 9. Set $i = i + 1$ and $j = 1$; start the next iteration by continuing at Step 2.

Although BP decoding is not optimal in terms of maximum-likelihood decoding, the latter is very complex and, therefore, suboptimal decoders have been utilized. Besides the three algorithms presented in this section there is a *tree pruning* (TP) decoding algorithm for stochastic inference on binary Markov random fields. We will not discuss this topic in this book as it demands extensive understanding of operations in these fields.

16.7 Practical Evaluation of LDPC Codes

The theoretical basis for LDPC codes can be defined by the following four steps: (i) Choose a rate $R = \dfrac{n}{n+m}$ for the code; (ii) Define probability distributions λ and ρ for cardinality of the left and right side nodes; (iii) Define f as the overhead factor of the graph. On an average, nf nodes out of a total of $(n+m)$ nodes must be downloaded in order to reconstitute the data, where $f = 1$ is the optimal case as in the RS codes; and (iv) Prove that $f = 1$ for infinite graphs where node cardinality adhere to λ and ρ.

The following questions demand close attention: What kind of overhead factor (f) can we expect for LDPC codes for large and small n? Are the above-mentioned types of codes equivalent, or do they perform differently? How do the published distributions fare in producing good codes for finite values of n? Is there a great deal of random variation in code generation for the given probability distributions? How do these codes compare with the RS codes?

A practical evaluation of LDPC codes is carried out by an experimental methodology that is based on the following steps: Choose R; choose n; calculate m from $R = \dfrac{n}{n+m}$; generate a graph in one of the three ways: (i) use a published probability distribution; (ii) use a probability distribution derived from a previously generated graph, or use a randomly generated probability distribution.; and (iii) perform a Monte-Carlo simulation of thousands of random downloads and experimentally determine the average value of the overhead factor f.

The choice of different variables is as follows: Choose $R \in \{\frac{1}{2}, \frac{1}{3}, \frac{2}{3}\}$; small $n \in \{$even numbers between 2 and 150$\}$, and large $n \in \{250, 500, 1250, 2500, 5000, 12500, 25000, 50000, 125000\}$. According to Plank [2004], there are eighty published probability distributions for all graphs and rates, so derive using the closest best graph. The best overhead factors have a maximum for $10 < n < 50$, and then descend toward 1 as $n \to \infty$. The overhead factor is the largest (about 1.24) at rate $R = 1/3$; it is about 1.15 for $R = 1/2$, and the lowest (about 1.1) at $R = 2/3$. Thus, the larger rates perform better.

Comparing the three codes, namely the systematic, Gallager, and IRA codes, at $R = 0.5$ and the overhead factor between 1 and 1.15, we find that they are different, but of all these three, the systematic code is the best for small n, whereas the IRA code is the best for large n, and Gallager codes are better than the IRA but inferior to the systematic for $n > 100$. Table

16.7.1 presents the results for these three codes, where the range refers to the overhead factor.

Table 16.7.1 Comparison of Three Codes

Code	|Product #|	Range	|1st Quartile|Median|Mean|3rd Quartile
Gallager	S99	| 1.23 – 1.29 |	1.24 | 1.25 | 1.25 | 1.26
	S99*	|1.14 – 1.297|	1.15 | 1.21 |1.555| 1.96
	RU03	| 1.15 – 1.4 |	1.15 | 1.1 |1.275| 1.4
IRA	U03	| 1.05 – 1.82 |	1.05 | 1.052 |1.095| 1.14
	R03	| 1.01 – 1.91 |	1.14 | 1.63 | 1.45 | 1.76
Systematic|	L97A	| 1.01 – 1.53 |	1.207 | 1.8 |1.265| 1.46

The question arises as to how systematic and IRA codes can outperform the Gallager code. Table 16.7.1 provides the performance variation among these three codes (Plank [2004]:42), but does not provide a clear-cut answers; some are good, some bad in different ranges of overhead factor. In the absence of research on their convergence, the question about their performance remains unanswered. On comparison with the Reed–Solomon code, LDPC codes are sometimes extremely faster for large n, fast network, and slow computation, but sometimes Reed–Solomon is much faster, especially for small n, slow network and fast computation. The difference between GF(8) and GF(16) is also very significant.

Plank's study [2004] has concluded that (i) for small n, the best codes have come out of the Monte Carlo simulation, i.e., λ and ρ are very poor metrics/constructors for finite codes; (ii) even sub-optimal LDPC codes are important alternatives to Reed–Solomon codes; and (iii) the needs for developers in wide-area storage system are not met even with these developments. More research is required in this area.

Some open questions are: Does this pattern continue for $m \geq 4$? Can we use optimal graphs for small m to construct graphs for large m? Can we generate good graphs for large n and n in an incremental manner? Do all these constructions prove anything? The answers cannot be known until more research is done. ∎

17

Discrete Distributions

The modern trend in the development of error correcting codes is to use the techniques of stochastic analysis. This outlook has lately produced the Tornado codes, Fountain codes, Luby transform (LT), and Raptor codes, as described in the following chapters. The background material for such a direction in the evolution of ECC is presented in this chapter. As James Grover Thurber (1894–1961) has so aptly said, "A pinch of probability is worth a pound of perhaps." Out of all possible distributions only the ones with direct application to the above-mentioned codes are presented.

17.1 Polynomial Interpolation

Interpolation requires estimating the values of a function $f(x)$ for arguments between $x_0, \ldots, x_n$ at which the values $y_0, \ldots, y_n$ are known, whereas prediction involves estimating values of $f(x)$ outside the interval in which the data arguments $x_0, \ldots, x_n$ fall. The central difference formulas of Stirling, Bessel, and Everett provide the basic approach to interpolation; they are used for arguments that are not very close to the beginning or end of a table. In fact, they use data from both sides of the interpolation argument x in approximately equal amounts, which seems to be a good 'common sense' practice supported by the study of errors involved in interpolation. The degree of the interpolating polynomial need not be chosen in advance: just continue to fit differences from the difference table into appropriate places in the formula used for computation. Since higher differences normally tend to zero, the terms of the formulas described below eventually diminish to negligible size.

Newton's forward formula is usually applied near the beginning of a table, whereas Newton's backward formula is chosen for interpolation near the end of a table. The Lagrange formula is also used for interpolation; it does not require prior computation of a difference table, but the disadvantage lies in the fact that the degree of the polynomial must be chosen at the beginning. Aitkin's method provides a better approach to difference formulas; it does not require the degree of the polynomial to be chosen at the outset. Finally, osculating polynomials and Taylor's polynomial are sometimes used in interpolation problems.

The simplest interpolation is linear, given by the formula: $f(i) = a + (b - a)(i - 1)$ for two given arguments a and b, where $i \geq 1$. Thus, $f(1) = a$ and $f(2) = b$. Newton's forward formula is

$$f_k = y_0 + \binom{k}{1}\Delta y_0 + \binom{k}{2}\Delta^2 y_0 + \cdots + \binom{k}{n}\Delta^n y_0, \quad \Delta y_0 = y_1 - y_0.$$

Newton's backward formula is

$$f_k = y_0 + k\Delta y_0 + \frac{k(k+1)}{2!}\Delta^2 y_0 + \cdots + \frac{k(k+1)\cdots(k+n-1)}{n!}\Delta^n y_0.$$

Everett's formula is

$$f_k = \binom{k}{1}y_1 + \binom{k+1}{3}\delta^2 y_1 + \binom{k+2}{5}\delta^4 y_1 + \cdots$$
$$- \binom{k-1}{1}y_0 + - \binom{k}{3}\delta^2 y_0 - \binom{k+1}{1}\delta^4 y_0 - \cdots.$$

The Lagrange interpolation formula is defined as follows: For $n \geq 0$, let $a_0, \ldots, a_n$ be $n + 1$ distinct elements of a field F, and let $b_0, \ldots, b_n$ be $n + 1$ arbitrary elements of F. Then there exists exactly one polynomial $f \in F[x]$ of degree $\leq n$ such that $f(a_i) = b_i$ for $i = 0, 1, \ldots, n$. This polynomial is given by

$$f(x) = \sum_{i+0}^{n} b_i \prod_{\substack{k=0 \\ k \neq i}}^{n} (a_i - a_k)^{-1}(x - a_k).$$

A polynomial f defined on a ring $R[x_1, \ldots, x_n]$ is said to be *symmetric* if $f(x_{i_1}, \ldots, x_{i_n}) = f(x_1, \ldots, x_n)$ for any permutation $i_1, \ldots, i_n$. Let $\sigma_1, \ldots, \sigma_n$ be the elementary symmetric polynomials in $x_1, \ldots, x_n$ over R, and let $s_0 = n \in F$ and $s_k = s_k(x_1, \ldots, x_n) = x_1^k + \cdots + x_n^k \in R[x_1, \ldots, x_n]$ for $k \geq 1$. Then Newton's formula is

$$s_k - s_{k-1}\sigma_1 + s_{k-2}\sigma_2 + \cdots + (-1)^{m-1}s_{k-m+1}\sigma_{m-1} + (-1)^m \frac{m}{n}s_{k-m}\sigma_m = 0$$

for $k \geq 1$, where $m = \min\{k, n\}$.

In general, given a set of n points (x_i, y_i), where no two x_i are the same, a polynomial p of degree at most n has the property

$$p(x_i) = y_i, \quad i = 0, 1, \ldots, n. \tag{17.1.1}$$

The existence and uniqueness of such a polynomial can be proved by the Vandermonde matrix, as follows: Since the polynomial interpolation defines a linear bijection $L_n : K^{n+1} \to X_n$ for the $(n + 1)$ nodes x_i, where X_n is the vector space of polynomials of degree at most n, let the polynomial be of the form

$$p(x) = a_0 + a_1 x + a_2 x^2 + \cdots + a_{n-1} x^{n-1} + a_n x^n. \qquad (17.1.2)$$

Then substituting Eq (17.1.1) into (17.1.2) we obtain a system of linear equations in the coefficients a_k defined by

$$\begin{bmatrix} 1 & x_n & x_n^2 & \cdots & x_n^{n-1} & x_n^n \\ 1 & x_{n-1} & x_{n-1}^2 & \cdots & x_{n-1}^{n-1} & x_{n-1}^n \\ \vdots & \vdots & \vdots & \vdots & \vdots & \vdots \\ 1 & x_1 & x_1^2 & \cdots & x_1^{n-1} & x_1^n \\ 1 & x_0 & x_0^2 & \cdots & x_0^{n-1} & x_0^n \end{bmatrix} \begin{Bmatrix} a_0 \\ a_1 \\ \vdots \\ a_{n-1} \\ a_n \end{Bmatrix} = \begin{Bmatrix} y_0 \\ y_1 \\ \vdots \\ y_{n-1} \\ y_n \end{Bmatrix}, \qquad (17.1.3)$$

where the matrix on the left is known as a Vandermonde matrix. The condition number of this matrix may be very large, which may cause large errors in computing a_i using the Gauss elimination method since it is of order $O(n^3)$. However, some useful algorithms to compute numerically stable solutions of order $O(n^2)$ are available in Björk and Pereyra [1970], Higham [1988], and Calvet and Reichel [1993].

The proof for uniqueness is based on the construction of the interpolant for a given Vandermonde matrix by setting up the system $\mathbf{Va} = \mathbf{y}$ and proving that $\mathbf{V}$ is nonsingular. Given $\det(\mathbf{V}) = \prod\limits_{i,j=0,\ i<j}^{n} (x_i - x_j)$ since the $(n+1)$ points x_i are distinct, the determinant cannot be zero because $x_i - x_j \neq 0$. Hence, $\mathbf{V}$ is nonsingular, and the system has a unique solution.

When interpolating a function f by a polynomial of degree n at the nodes $x_0, \ldots, x_n$, an interpolation error occurs and it is defined by

$$f(x) - p(x) = f[x_0, \ldots, x_n, x] \prod_{i=0}^{n}(x - x_i), \qquad (17.1.4)$$

where $f[x_0, \ldots, x_n, x]$ denotes the divided difference. If f is an $(n+1)$-times continuously differentiable function on the smallest interval I that contains the nodes x_i and x, we can write the error in the Lagrange form as

$$f(x) - p(x) = \frac{f^{(n+1)}(\xi)}{(n+1)!} \prod_{i=0}^{n}(x - x_i), \qquad (17.1.5)$$

for some $\xi \in I$. Thus, the remainder term in (17.1.5) is a special case of an interpolation error when all interpolation nodes x_i are identical. In the case of equally spaced interpolation nodes $x_i = x_0 + ih$, where h is the step size, the interpolation error is $O(h^n)$, but it does not tell us anything about what happens when $n \to \infty$. The answer is found when we discuss convergence, i.e., choosing the interpolation points x_i such that $|x - x_i|$ is as small as possible. A result on convergence is: For any function $f(x)$ continuous on an interval I, there exists a table of nodes for which the sequence of interpolating polynomial $p(x)$ converges uniformly to $f(x)$ on I. Moreover, for every absolutely continuous function on $[-1, 1]$, the sequence of interpolating polynomials constructed on Chebyshev nodes converges uniformly to $f(x)$.

17.2 Chernoff Bound

In probability theory, the Chernoff bound, named after Herman Chernoff [1981], provides exponentially decreasing bounds on tail distributions of sums of independent random variables. Compared to the Markov inequality, which yields only the power-law bound (related to probability distribution) on tail decay, the Chernoff bound performs better for the first or second moment based tail bounds.

Let $x_1, \ldots, x_n$ be independent Bernoulli random variables,[1] each with probability $\mathfrak{p} > 1/2$. Then the probability of simultaneous occurrence of more than $n/2$ of the events $\{x_k = 1\}$ has exact value $\mathfrak{P}$, where

$$\mathfrak{P} = \sum_{i=\lfloor \frac{n}{2} \rfloor + 1}^{n} \binom{n}{i} \mathfrak{p}^i (1 - \mathfrak{p})^i. \tag{17.2.1}$$

The Chernoff bound is the lower bound on $\mathfrak{P}$ such that

$$\mathfrak{P} \geq 1 - e^{-2n(\mathfrak{p} - 1/2)^2}. \tag{17.2.2}$$

Example 17.2.1. We determine Chernoff bounds for the success probability of majority agreement for n independent, equally likely events. Take the case of a biased coin, for which one side (say, head) is more likely to come up than the other, but it is not known in advance which side it is. Let us then flip the coin many times and then choose the side that came up most often. The question remains as to how many times the coin must be flipped to be certain that the side of your choice is correct. Let x_i denote the event that the i-th coin flip comes up heads. Suppose we want to ensure that we choose the wrong side with at most a small probability ε. Then, rearranging as above, we must have

$$n \geq \frac{1}{(\mathfrak{p} - 1/2)^2} \ln \frac{1}{\sqrt{\varepsilon}}. \tag{17.2.3}$$

If the coin is really biased, say coming up heads 60% of the time ($\mathfrak{p} = .6$), then we can guess that side with 95% accuracy ($\varepsilon = .05$) after 150 flips ($n = 150$).

---

[1] A Bernoulli random variable x is either 0 or 1, which defines success ($x = 1$) or failure ($x = 0$) such that $P_x(1) \equiv \mathfrak{P}(x = 1) = \mathfrak{p}$, $P_x(0) \equiv \mathfrak{P}(x = 0) = 1 - \mathfrak{p}$. For n independent Bernoulli trials, the number of successes $X = \sum_{i=1}^{n} \mathfrak{P}_i(x) = 1$, and $E(x) = n\mathfrak{p}$.

But if it is 90% biased, only 10 flips would suffice. If the coin is biased by a tiny amount only, like a real coin, the number of flips required for success becomes much larger. Therefore, Chernoff bound is used in a randomized algorithm to determine a bound on the number of necessary runs (flips) to obtain a value by majority agreement up to a prescribed probability. For example, if an algorithm A computes the correct value of a function f with probability $\mathfrak{p} > 1/2$, then by choosing n satisfying the inequality (17.2.3), the probability that a majority is obtained and is equal to the value is at least $1 - \varepsilon$, which for suffuciently small ε is quite reliable. Note that for a constant $\mathfrak{p}$, the value of ε decreases exponentially as n increases. If $\mathfrak{p}$ is very close to $1/2$, the necessary n can become very large. For example, if $\mathfrak{p} = 1/2 + 1/2^m$, the value of n is bounded below by an exponential function of m, i.e., $n \geq 2^{2m} \ln \dfrac{1}{\sqrt{\varepsilon}}$. The graphs for the number of flips for a 2/3 biased coin and for the probability of heads are presented in Figure 17.2.1(a) and (b). ∎

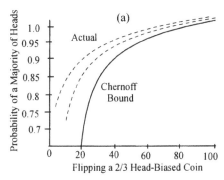
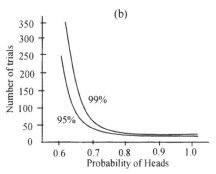

Figure 17.2.1(a) Biased coin.   Figure 17.2.1(b) Probability data.

17.3 Gaussian Distribution

We define singular distributions as limits of regular functions. As we have seen, the Dirac delta-function $\delta(x)$ can be represented as a limit of a sequence of functionals $\{T_k[\phi] = \int_{-\infty}^{\infty} f_k(x)\,\phi(x)\,dx\}$ with respect to the kernels that are regular functions. This leads to the representation of $\delta(x)$ in the limit such that $T_k[\phi] \to \delta(\phi)$ as $k \to \infty$ for each $\phi \in \mathcal{D}$. The convergence in this situation is always the weak convergence. Consider the family of Gaussian functions

$$g_\varepsilon(x) = \frac{1}{\sqrt{2\pi\varepsilon}}\, e^{-x^2/(2\varepsilon)}, \quad \varepsilon > 0, \tag{17.3.1}$$

which are normalized by the condition that $\int_{-\infty}^{\infty} g_\varepsilon(x)\,dx = 1$.

Let $f_k(x) = g_{1/k}(x)$, $k = 1, 2, \ldots$, be a weakly approximating sequence. Then $\varepsilon \to 0$ as $k \to \infty$, and the graphs of the approximating Gaussian functions $f_k(x)$ show higher and higher peaks and move more and more close toward $x = 0$, but the area under each one of them remains constant (see Figure 17.3.1). In fact, Gaussian functions are infinitely differentiable and integrable functions for all values of $x \in \mathbb{R}$, but they decay exponentially to 0 as $x \to \pm\infty$.

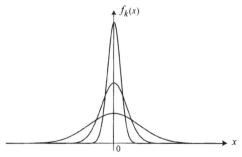

Figure 17.3.1 Graphs of the first three Gaussian functions $g_k(x)$.

We consider the Cauchy densities (or the Lorentz curves)

$$\lambda_\varepsilon(x) = \frac{1}{\pi} \frac{\varepsilon}{x^2 + \varepsilon^2} \qquad (17.3.2)$$

and define the kernels $f_k(x) = \lambda_{1/k}(x)$. These kernels, which look almost like the Gaussian functions but differ from them significantly, weakly converge to the Dirac δ-function. In fact, both the Gaussian functions and the Cauchy densities are infinitely differentiable and integrable functions for all values of $x \in \mathbb{R}$, as the integral $\displaystyle\int \frac{dx}{x^2 + 1} = \arctan x$ has finite limit at $x = -\infty$ and $x = \infty$, but at $x = 0$

$$g_\varepsilon(0) = \frac{1}{\sqrt{2\pi\varepsilon}} \quad \text{and} \quad \lambda_\varepsilon(0) = \frac{1}{\pi\varepsilon},$$

and thus both go to $+\infty$ as $\varepsilon \to 0$.

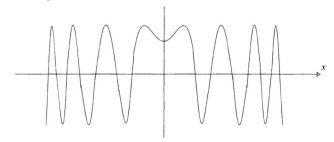

Figure 17.3.2 Graphs of $\Re\{f_\varepsilon(x)\}$.

These two functions differ from each other as follows: Gaussian functions decay exponentially to 0 as $x \to \pm\infty$, whereas the Cauchy densities, being asymptotic to $\dfrac{\varepsilon}{\pi x^2}$, decay more slowly to 0 as $x \to \pm\infty$ than the Gaussian, and the areas under them are less concentrated around $x = 0$ than in the case of the Gaussian functions. The complex-valued functions

$$f_\varepsilon(x) = \sqrt{\frac{i}{2\pi\varepsilon}}\, e^{-ix^2/(2\varepsilon)}, \quad \varepsilon > 0, \tag{17.3.3}$$

are found in quantum mechanics and quasi-optics.[2] The graph of $\Re\{f_\varepsilon(x)\}$ is given in Figure 17.3.2. Note that $|f_\varepsilon(x)| = \dfrac{1}{\sqrt{2\pi\varepsilon}} = \text{const}$ and diverges to ∞ as $\varepsilon \to 0$. However, these functions converge weakly to the δ-function as $\varepsilon \to 0$. It follows from the fact that these functions oscillate at higher and higher speed as ε becomes smaller and smaller.

17.4 Poisson Distribution

In probability theory and statistics, the Poisson distribution (or Poisson law of small numbers) is a discrete probability distribution that expresses the probability of a number of events occurring in a fixed period of time if these events occur with a known average rate and independently of the time since the last event (see Gullberg [1997]). In addition to the time interval, the Poisson distribution is also applicable to other specified intervals such as distance, area, or volume. This distribution originated with the publication of the work entitled 'Reserches sur la probabilité des jugements en matière criminelle et en matière civile' by Siméon-Denis Poisson (1781–1840), which, among other things of a civil nature, focused on the number of discrete occurrences that take place during a time interval of given length. The theory so developed states that if the expected number of occurrences in a given interval is λ, then the probability that there are exactly k occurrences, where $k = 0, 1, 2, \ldots$, is equal to

$$\mathfrak{p}(k, \lambda) = \frac{\lambda^k\, e^{-\lambda}}{k!}. \tag{17.4.1}$$

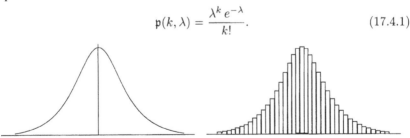

Figure 17.4.1 Continuous and discrete Poisson distribution.

For example, if the events occur on an average of four times per minute, then the probability of an event occurring k times in a 10-minute interval will

---

[2] In quasi-optics, $\Re\{f_\varepsilon(x)\}$ turn out to be Green's functions of a monochromatic wave in the Fresnel approximation (see Kythe [2011]).

be given by (17.4.1) with $\lambda = 10 \times 4 = 40$. As a function of k, the function $\mathfrak{p}(k, \lambda)$ is the probability mass function. This distribution can be applied to systems with a large number of possible events, each of which is rare. A classical example is the nuclear decay of atoms. A graphical representation of the Poisson distribution for different values of k is presented in Figure 17.4.1. An algorithm to generate Poisson-distributed random variables is given in Knuth [1969].

17.5 Degree Distribution

Consider a random bipartite graph B with $N(< n)$ nodes (c-nodes) on the left and βN nodes (p-nodes) on the right (Figure 17.5.1). Let $\{\lambda_i\}$ and $\{\rho_i\}$ be two vectors such that λ_i and ρ_i are the functions of edges of degree i on the left and right, respectively. Let a_l and a_r denote the average node degree for the d-nodes (on the left) and p-nodes (on the right), respectively.

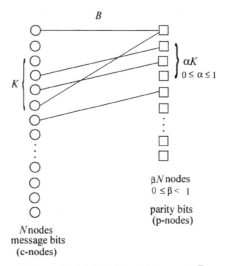

Figure 17.5.1 The bipartite graph B.

Example 17.5.1. Consider the structure of a graph B', shown in Figure 17.5.2, with given degree sequences $\{\lambda_i\}$ and $\{\rho_i\}$ for $i = 3$ and the number of edges $E = 6$. Here, $\{\lambda_1, \lambda_2, \lambda_3\} = \{\frac{1}{6}, \frac{2}{6}, \frac{3}{6}\}$ and $\{\rho_1, \rho_2\} = \{\frac{4}{6}, \frac{2}{6}\}$. ∎

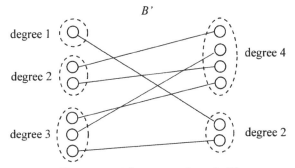

Figure 17.5.2 Structure of graph B'.

In the general case, for the d-nodes (on the left side of the graph B), we have $N = \sum_i \frac{\lambda_i E}{i}$ and $a_l = \frac{E}{N}$, where E denotes the number of edges. Thus,

$$\frac{1}{a_l} = \frac{N}{E} = \sum_i \frac{\lambda_i}{i}. \tag{17.5.1}$$

Similarly, for the p-nodes (on the right side of graph B), we have $\beta N = \sum_i \frac{\rho_i E}{i}$ and $a_r = \frac{E}{\beta N}$, $0 \le \beta < 1$. Thus,

$$\frac{1}{a_r} = \frac{\beta N}{E} = \sum_i \frac{\rho_i}{i}. \tag{17.5.2}$$

We will model this discrete construction first in the form of a difference equation in discrete time variable t, and then in the form of an ordinary differential equation in continuous time t. Let $\Delta t = 1/E$ denote the time scalar associated with one step, and let $\delta > 0$ denote the fraction of losses in the data bits. Initially, the data bits (symbols) on the left are expected to be received error-free with probability $p = 1 - \delta$ because this symbol is received error-free and therefore removed from the subgraph of B. This initial subgraph contains δN d-nodes (on the left). A successful decoding process will therefore run until $T = \delta N \Delta t = \delta N/E = \delta/a_l$.

Let $l_i(t)$ denote the fraction of edges, in terms of E, with left degree i remaining at time t, $r_i(t)$ the fraction of edges, in terms of E, with right degree i remaining at time t, and $e(t)$ the fraction of edges, in terms of E, remaining at time t. Then

$$e(t) = \sum_i l_i(t) = \sum_i r_i(t). \tag{17.5.3}$$

At each step, a random node of degree 1 on the right is chosen and the associated node on the left and all its adjacent edges are deleted. The probability that the edge adjacent to the node of degree 1 on the right has degree

i on the left is $\mathfrak{p} = \dfrac{l_i(t)}{e(t)}$. Thus, we lose i edges of degree i on the left (Figure 17.5.3(a), where $\times$ denotes the lost (or deleted) edge).

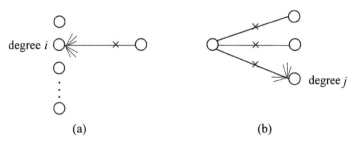

degree i

degree j

(a)   (b)

Figure 17.5.3 Deletion of edges.

Thus, the difference equation for the *expected change* in the number of edges $\mathfrak{L}_i(t)$ of degree i on the left can be written as

$$\mathfrak{L}_i(t + \Delta t) - \mathfrak{L}_i(t) = \frac{i l_i(t)}{e(t)}. \qquad (17.5.4)$$

Since $l_i(t) = \dfrac{\mathfrak{L}_i(t)}{E} = \mathfrak{L}_i(t)\,\Delta t$, we find from (17.5.4) that as $E \to \infty$,

$$\frac{dl_i(t)}{dt} = \frac{i\, l_i(t)}{e(t)}. \qquad (17.5.5)$$

Note that removal of a node of degree i on the left means deletion of not only one edge of degree i on the right but also other $i - 1$ edges adjacent to that node. Thus, the expected number of the edges removed is $a(t) - 1$, where $a(t) = \sum_i \dfrac{i l_i(t)}{e(t)}$. In this process the right endpoints of these $i - 1$ edges are randomly distributed such that one of the edges i of degree j on the right is deleted ($\times$-ed). Although the j edges of degree j are lost, yet $j - 1$ edges of degree $j - 1$ are gained (Figure 17.5.3(b)). Thus, the probability that an edge has degree j on the right is $\mathfrak{p} = \dfrac{r_j(t)}{e(t)}$. For $i > 1$ we obtain the difference equation

$$\mathfrak{R}_i(t + \Delta t) - \mathfrak{R}_i(t) = [r_{i+1}(t) - r_i(t)] \frac{i\,[a(t) - 1]}{e(t)}, \qquad (17.5.6)$$

where $\mathfrak{R}_i(t)$ is the expected change in the number of edges of degree i on the right. The associated differential equation is

$$\frac{dr_i(t)}{dt} = [r_{i+1}(t) - r_i(t)] \frac{a(t) - 1}{e(t)}. \qquad (17.5.7)$$

To solve Eqs (17.5.5) and (17.5.7), define a variable x such that $\dfrac{dx}{x} = \dfrac{dt}{e(t)}$, where $x = e^{\int_0^t du/e(u)}$. Then Eq (17.5.5) becomes $\dfrac{dl_i(x)}{dx} = \dfrac{i\, l_i(x)}{x}$, whose solution is $l_i(x) = C_i x^i$, where C_i is constant of integration. Since $x = 1$ when $t = 0$, we find that $l_i(0) = \delta \lambda_i = C_i$. Hence,

$$l_i(x) = \delta \lambda_i x^i. \tag{17.5.8}$$

The general solution of Eq (17.5.7) can be found in Luby et al. [1997]. In particular, when $i = 1$, the solution of Eq (17.5.7) for $r_1(t)$ is given by

$$r_1(t) = \delta \lambda(t) \left[t - 1 + \rho(1 - \delta)\lambda(t) \right], \quad r_1(t) > 0, \tag{17.5.9}$$

subject to the condition that $\rho\,(1 - \delta\lambda(t)) > 1 - t,\ t \in (0,1]$, where $\lambda(t) = \sum\limits_{i \geq 1} \lambda_i t^{i-1}$ and $\rho(t) = \sum\limits_{i \geq 1} \rho_i t^{i-1}$.

For any LDPC code, there is a 'worst case' channel parameter called the *threshold* such that the message distribution during belief propagation evolves in such a way that the probability of error converges to zero as the number of iterations tends to infinity.

Density distribution is used to find the degree distribution pair (λ, ρ) that maximizes the threshold. Richardson and Urbanke [2001] have provided the following steps for density evolution:

STEP 1. Fix a maximum number of iterations.

STEP 2. For an initial degree distribution, find the threshold.

STEP 3. Apply a small change to the degree distribution. If the new threshold is larger, take this as the current distribution. Repeat Steps 2 and 3 until this happens.

Richardson and Urbanke [2001] identify a rate 1/2 code with degree distribution pair that is 0.06 dB away from the Shannon capacity. Chung et al. [2001] use density evolution to design a rate 1/2 code that is 0.0045 dB away from capacity.

17.6 Probability Distributions

Some useful discrete and continuous probability distributions are as follows:

17.6.1 Discrete Probability Distributions. They are:

(a) BINOMIAL $B(n, \mathfrak{p})$: The frequency in n independent trials has a binomial distribution, where $\mathfrak{p}$ is the probability in each trial. The conditional probability is $\mathfrak{P}(X = x) = \binom{n}{x}\mathfrak{p}^x(1 - \mathfrak{p})^{n-x}$, $x = 0, 1, 2, \ldots, n$, with the expectation $\mu = n\mathfrak{p}$ and variance $\sigma^2 = n\mathfrak{p}(1 - \mathfrak{p})$.

(b) GEOMETRIC $G(\mathfrak{p})$: The number of required trials until an event with probability $\mathfrak{p}$ occurs has a geometric distribution. The data is $\mathfrak{P}(X = x) = \mathfrak{p}(1 - \mathfrak{p})^{x-1}$, expectation $\mu = 1/\mathfrak{p}$, and variance $\sigma^2 = (1 - \mathfrak{p})/\mathfrak{p}^2$.

(c) POISSON $P(\lambda)$: This is the distribution of a number of points in the random number generation process which has certain assumptions. It approximates to the binomial distribution when n is large and $\mathfrak{p}$ small, where $\lambda = n\mathfrak{p}$. The data is $\mathfrak{P}(X = x) = e^{-\lambda}\lambda^x/(x!)$, expectation $\mu = \lambda$, and variance $\sigma^2 = \lambda$.

(d) HYPERGEOMETRIC $H(N, n, \mathfrak{p})$: This distribution is used in connection with sampling without replacement from a finite population with elements of two different kinds. The data is $\mathfrak{P}(X = x) = \dfrac{\dbinom{N\mathfrak{p}}{x}\dbinom{N - N\mathfrak{p}}{n - x}}{\dbinom{N}{\mathfrak{p}}}$,

expectation $\mu = n\mathfrak{p}$, and variance $\sigma^2 = n\mathfrak{p}(1 - \mathfrak{p})\dfrac{N - n}{N - 1}$.

(e) PASCAL OR NEGATIVE BINOMIAL $NB(j, \mathfrak{p})$: This distribution is used when the number of required trials until an event with probability $\mathfrak{p}$ occurs for the j-th time. The data is $\mathfrak{P}(X = x) = \binom{x-1}{j-1}\mathfrak{p}^j(1 - \mathfrak{p})^{x-j}$, $x = j, j + 1, \ldots$, expectation $\mu = j/\mathfrak{p}$, and variance $\sigma^2 = j(1 - \mathfrak{p})/\mathfrak{p}^2$.

17.6.2 Continuous Probability Distributions. They are:

(a) UNIFORM $U(a, b)$: This is used in certain waiting times rounding off errors. The data is $f(x) = 1/(b - a)$, $a \le x \le b$, expectation $\mu = (a + b)/2$, and variance $\sigma^2 = (b - a)^2/12$.

(b) GENERAL NORMAL $N(\mu, \sigma)$: Under general conditions the sum of a large number of random variables is approximately normally distributed. The data is $f(x) = \frac{1}{\sigma}\phi\left(\dfrac{x - \mu}{\sigma}\right)$, expectation μ, and variance σ^2.

(c) NORMED NORMAL $N(0, 1)$: This X has a general normal distribution, then $(X - \mu)/\sigma$ has a normed normal distribution. The data is $f(x) = \frac{1}{\sqrt{2\pi}}e^{-x^2/2}$, expectation $\mu = 0$, and variance σ^2.

(d) GAMMA $\Gamma(n, \lambda)$: It is the distribution of the sum of n independent random variables with an exponential distribution with parameter λ. The data are: $f(x) = \dfrac{\lambda^n}{\Gamma(n)}x^{n-1}e^{-\lambda x}$; expectation $\mu = n/\lambda$; and variance $\sigma^2 = n/\lambda^2$.

(e) CHI-SQUARE $\chi^2(r)$: This defines the distribution of $a_1^2 + a_2^2 + \cdots + a_r^2$, where $a_1, a_2, \ldots, a_r$ are independent and have a normed normal distribution. The parameter r is called the number of degrees of freedom. The data is $f(x) = \dfrac{1}{2^{r/2}\Gamma(1/2)}x^{r/2-1}e^{-x/2}$, $x \ge 0$, expectation $\mu = r$, and variance $\sigma^2 = 2r$.

(f) F-DISTRIBUTION $F(r_1, r_2)$: This is the distribution of $(X_1/r_1)/(X_2/r_2)$, where X_1, X_2 are independent and have χ^2-distribution and $r_{1,2}$ are degrees of freedom, respectively. The data is $f(x) = \dfrac{a_r x^{(r_1/2)-1}}{b_r(r_2 + r_1 x)^{(r_1+r_2)/2}}, x \geq 0$, where $a_r = \Gamma\left(\dfrac{r + r_2}{2)}\right) r_1^{r_1/2} r_2^{r_2/2}$, $b_r = \Gamma\left(\dfrac{r_1}{2}\right)\Gamma\left(\dfrac{r_2}{2}\right)$, expectation $\mu = \dfrac{r_2}{r_2 - 2}, r_2 > 2$; and $\sigma^2 = \dfrac{2r_2^2(r_1 + r_2 - 2)}{r_1(r_2 - 2)^2(r_2 - 4)}, r_2 > 4$.

(g) BETA DISTRIBUTION $B(p,q)$: This is useful as a priori distribution for unknown probability in Baysian models. The data is $f(x) = a_{p,q} x^{p-1}(1 - x)^{q-1}, 0 \leq x \leq 1$, where $a_{p,q} = \dfrac{\Gamma(p + q)}{\Gamma(p)\Gamma(q)}, p > 0, q > 0$, expectation $\mu = p/(p+q)$, and $\sigma^2 = \dfrac{pq}{(p+q)^2(p+q+1)}$.

(h) RAYLEIGH DISTRIBUTION $R(\alpha)$: This is useful in communications systems and in relativity theory. The data is $f(x) = \dfrac{x}{\alpha^2} e^{-x^2/(2\alpha^2)}$, expectation $\mu = \alpha\sqrt{\pi/2}$, and variance $\sigma^2 = 2\alpha^2(1 - \pi/4)$.

Probability functions are defined in Abramowitz and Stegun [1972: 927 ff.].

17.7 Probability Computation

The computation of probability when a given encoding symbol of degree i is released while l input symbols remain unprocessed plays a very important role in Luby transform (LT) codes described in Chapter 19. We must keep in mind that the input symbols for neighbors of an encoding symbol are chosen independently of all other encoding symbols. As such the probability that this encoding symbol is released when l input symbols remain unprocessed is independent of all other encoding symbols. This observation leads to the following definitions on encoding symbol release.

An encoding symbol is said to be *released* when l input symbols remain unprocessed if it is released by the processing of the $(k - l)$-th input symbol, at which time the encoding symbol randomly covers one of the l unprocessed input symbols.

Theorem 17.7.1. (Degree release probability) *Let* $q(d, l)$ *be the probability that an encoding symbol of degree* d *is released when* l *input symbols remain unprocessed. Then*

$$q(1, k) = 1, \ and \ q(d, l) = \frac{d(d-1)l \displaystyle\prod_{j=0}^{d-3}[k - (l+1) - j]}{\displaystyle\prod_{j=0}^{d-1}(k - j)}, \tag{17.7.1}$$

for $d = 2, \ldots, k$ and for all $l = k - d + 1, \ldots, 1$, where $\mathfrak{q}(d, l) = 0$ for all other d and l.

PROOF. Note that $\mathfrak{q}(d, l)$ is the probability that $(d - 2)$ of the neighbors of the encoding symbols are among the first $k - (l + 1)$ symbols already processed, where one neighbor is processed at step $(k - l)$ and the remaining neighbor is among the l unprocessed input symbols. ∎

DEFINITION 17.7.1. (Overall release probability) Let $\mathfrak{r}(d, l)$ be the probability that an encoding symbol is chosen to be of degree d and is released when l input symbols remain unprocessed. In other words, $\mathfrak{r}(d, l) = \rho(d)\mathfrak{q}(d, l)$, where $\mathfrak{q}(d, l)$ is defined in (17.7.1) and

$$\rho(1) = 1/K,$$

$$\rho(d) = \frac{1}{d(d - 1)} = \frac{1}{d - 1} - \frac{1}{d} \text{ for all } d = 2, \ldots, K, \qquad (17.7.2)$$

such that $\sum_{i=1}^{k} \rho(d) = 1$, where $K(< k)$ denotes any transmitted symbols. Then $\mathfrak{r}(l) = \sum_{d=1}^{k} r(d, l)$ is said to be the overall release probability that an encoding symbol is released when l input symbols remain unprocessed. Note that (17.7.2) also defines the ideal soliton distribution (see §17.8.2).

17.8 Soliton Distributions

Soliton distributions are very useful in some recent codes to recover erasures. These distributions were first published by Luby [2002]. First we provide some background information about solitons before discussing these discrete distributions.

17.8.1 Solitons. The notion comes from physics and mathematics where a soliton is a self-reinforcing solitary wave or pulse that maintains its shape while it travels at constant speed. Solitons are caused by cancellation of nonlinear and dispersive effects in the medium. Thus, a soliton wave appears when dispersion balances refraction. For example, the dispersion relation for a water wave in an ocean at depth h is $\omega^2 = gh \tanh ah$, where ω is the radian frequency, g the acceleration due to gravity, and a the amplitude. Thus, $\omega = \left(gh \tanh ah\right)^{1/2} \approx ca\left(1 - \frac{1}{6}a^2h^2\right)$, where $c = \sqrt{gh}$. In 1895, Korteweg-deVries developed a nonlinear equation for long dispersive water waves in a channel of depth h as

$$\eta_t + c_0\left(1 + \frac{3}{2}\frac{\eta}{h}\right)\eta_x + \sigma\eta_{xxx} = 0, \qquad (17.8.1)$$

where $\eta(x, t)$ determines the free surface elevation of the waves, c_0 is the shallow water wave speed, and σ is a constant for fairly long waves. The dispersive term in this K-dV equation allows solitary and periodic waves that are not found in shallow water wave theory. Eq (17.8.1) is solved in the form $\eta(x,t) = hf(X), X = x - Ut$, where U is a constant wave velocity. Substituting this form into Eq (17.8.1) yields $\frac{1}{6}h^2 f''' + \frac{3}{2}ff' + \left(1 - \frac{U}{c}\right)f' = 0$, $\sigma = ch^2/6$, which after integrating twice gives

$$\frac{1}{3}h^2 f'^2 + f^3 + \left(1 - \frac{U}{c}\right)f^2 + 4Af + B = 0, \qquad (17.8.2)$$

where A and B are integration constants. Since we seek a solitary wave solution, with boundary conditions $f, f', f'' \to 0$ as $|X| \to \infty$, we have $A = B = 0$. With these values of A and B, we integrate Eq (17.8.2) and obtain

$$X = \int_0^f \frac{df}{f'} = \sqrt{\frac{h^2}{3}} \int_0^f \frac{df}{f\sqrt{\alpha - f}},$$

where $\alpha = 2\left(\frac{U}{c} - 1\right)$. The substitution $f = \alpha \operatorname{sech}^2 \theta$ reduces this result to $X - X_0 = \left(\frac{4h^2}{3\alpha}\right)^{1/2} \theta$, where X_0 is an integration constant. Hence, the solution for $f(X)$ is given by

$$f(X) = a \operatorname{sech}^2 \left[\sqrt{\frac{3\alpha}{4h^2}} (X - X_0)\right]. \qquad (17.8.3)$$

This solution increases from $f = 0$ as $X \to -\infty$ such that it attains a maximum value $f = f_{max}$ at $X = 0$, and then decreases to $f = 0$ as $X \to \infty$, as shown in Figure 17.8.1. This also implies that $X_0 = 0$ so that the solution (17.8.3) becomes $f(X) = \alpha \operatorname{sech}^2 \left[\sqrt{\frac{3\alpha}{4h^2}} X\right]$. Hence the final solution is

$$\eta(x,t) = \eta_0 \operatorname{sech}^2 \left[\sqrt{\frac{3\eta_0}{4h^2}} (x - Ut)\right], \qquad (17.8.4)$$

where $\eta_0 = \alpha h$. This is called a solitary wave solution of the K-dV equation, also known as a *soliton* (named so by Zabusky and Kruskal [1965]).

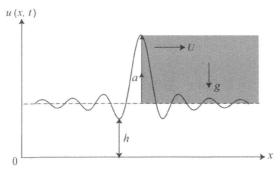

Figure 17.8.1 A soliton.

Since $\eta > 0$ for all x, the soliton is a self-reinforcing wave that propagates in the medium without change of shape with velocity $U = c\left(1 + \dfrac{\eta_0}{2h}\right)$, which is directly proportional to the amplitude η_0. The width $\sqrt{4h^3/(3\eta_0)}$ is inversely proportional to $\sqrt{\eta_0}$, which means that the soliton propagates to the right with a velocity U that is directly proportional to the amplitude. Therefore, the taller solitons travel faster and are narrower than the shorter (or slower) ones. They can overtake the shorter ones and, surprisingly, they emerge from this interaction without change of shape. These waves are stable and can travel very long distances. A sech-envelope soliton for water waves is shown in Figure 17.8.2.

Figure 17.8.2 A soliton for water waves.

Besides the K-dV equation, soliton solution can also be obtained from the nonlinear Shrödinger equation $i\psi_t + \frac{1}{2}w''\psi_{xx} = 0$, which after using a Taylor expansion for $w = w(k, a^2)$ of the form $w = w_0 + (k - k_0)\left(\dfrac{\partial^2 w}{\partial k^2}\right)_{k=k_0} + \frac{1}{2}(k - k_0)^2\left(\dfrac{\partial^2 w}{\partial k^2}\right)_{k=k_0} + \left(\dfrac{\partial w}{\partial |a|^2}\right)_{|a|^2=k_0}$, where $w_0 = w(k_0)$, reduces to

$$i\left(a_t + w_0' a_x\right) + \frac{1}{2}\, w_0'' a + xx + \gamma|a|^2 a = 0,$$

where $\gamma = -\left(\dfrac{\partial w}{\partial |a|^2}\right)_{|a|^2=k_0}$ is a constant. The amplitude satisfies the normalized nonlinear Schrödinger equation

$$i\, a_t + \frac{1}{2}\, w_0'' a_{xx} + \gamma|a|^2 a = 0. \tag{17.8.5}$$

The wave modulation is stable if $\gamma w_0'' < 0$ and unstable if $\gamma w_0'' > 0$. In this case again, writing the nonlinear Schrödinger equation in the standard form

$$i\psi_t + \psi_{xx} + \gamma|\psi|^2\psi = 0, \quad -\infty < x < \infty, \ t \geq 0, \tag{17.8.6}$$

we seek a solution of the form $\psi = f(X)e^{i(mX-at)}$, $X = x - Ut$ for some finite f and constant wave speed U, where m and n are constants. The solution of Eq (17.8.6) is

$$f(X) = \sqrt{\frac{2\alpha}{\gamma}}\, \text{sech}\left[\sqrt{\alpha}\,(x - Ut)\right], \quad \alpha > 0. \tag{17.8.7}$$

This represents a soliton that propagates with change of shape but constant velocity U. Unlike the K-dV equation, the amplitude and velocity of this wave are independent parameters. The soliton exists only for the unstable case ($\gamma > 0$), which means that small modulations of the unstable wave train lead to a series of solitons. Besides the K-dV and nonlinear Schrödinger equations, the other models include the coupled nonlinear Schrödinger equation and the sine-Gordon equation, in which, just as in the other models, dispersion and nonlinearity interact to generate permanent and localized solitary waves. These waves were noticed by John Scott Russell in 1834, who called them the 'wave of translation'. Solitons occur in fiber optics, which prompted Robin Bullough in 1973 to propose the idea of a soliton-based transmission system in optical telecommunications (see Hasegawa and Tappert [1973] and Mollenauer and Gordon [2006]). More information on solitons is available, for example, in Jefferey and Tanuiti [1964], Drazin and Johnson [1989], Johnson [1997], and Knobel [1999].

17.8.2 Ideal Soliton Distribution. Any good degree distribution must possess the property that input symbols are added to the ripple at the same rate as they are processed. The concept of a ripple is defined as follows: Let I denote the set of indices of input symbols that have already been determined and eliminated. Then the set of indices of input symbols that been determined is known as the *ripple* and denoted by R, and its size is denoted by s. This feature has led Luby [*loc. cit.*] to the concept of the soliton distribution.

The ripple effect of released encoding symbols that cover random input symbols is that they become redundant, and therefore the ripple size s must be kept small at all times. However, if the ripple vanishes, i.e., if the ripple size becomes zero before all the k input symbols are covered, then the overall process fails. Thus, the ripple size must be kept large enough to ensure that the ripple does not disappear prematurely. It is desirable that the ripple size should never be too small or too large.

The *ideal soliton distribution* has been defined by Eq (17.7.2), with the expected degree almost equal to $\ln k$. Although it provides the ideal behavior as to the expected number of encoding symbols needed to recover the data, it is not at all useful in practice because the fluctuations that appear make the decoding very difficult since there will be no degree 1 parity nodes, and a few data nodes may arrive with no neighbors at all. In this sense, this distribution is not stable.

In the definition (17.7.2) of the ideal soliton distribution, the condition $\sum\limits_{d=1}^{k} \rho(d) = 1$ is justified as follows: One way to choose a sample from this distribution is to first choose a random real number $\alpha \in (0,1)$, and then for $d = 2, \ldots, k$, let the sample value be $= \begin{cases} d \text{ if } 1/d < \alpha < 1/(d-1), \\ 1 \text{ if } 0 < \alpha \leq 1/k. \end{cases}$ Then the expected value of an encoding symbol is $\sum_{d=2}^{k} \frac{1}{d-1} = H(k) \approx \ln k$, where $H(k)$ is the harmonic sum up to k terms.[3]

17.8.3 Robust Soliton Distribution. The robust soliton distribution is successful in practice because the expected ripple size is large enough at each point in the process so that it never vanishes completely with high probability. Let δ be the allowable failure probability of the decoder to recover the data for a given number K of encoding symbols. The basic idea in the design of this distribution is to maintain the expected ripple size approximately at $\sqrt{k} \ln(k/\delta)$ throughout the process. It is based on the intuition that the probability a random walk of length k deviates from its mean by more than $\sqrt{k} \ln(k/\delta)$ is at most δ (Durrett [1996]), and this can be achieved using $K = k + O\left(\sqrt{k} \ln^2(k/\delta)\right)$ encoding symbols. The following definition shows this.

DEFINITION 17.8.3. (Robust soliton distribution) Let $s = c\sqrt{k} \ln(k/\delta)$ for some suitable constant $c > 0$. Define

$$
\tau(d) = \begin{cases}
\dfrac{s}{k}\dfrac{1}{d} & \text{for } d = 1, \ldots, (k/s) - 1, \\
\dfrac{s}{k} \ln(s/\delta) & \text{for } d = k/s, \\
0 & \text{for } d > k/s.
\end{cases} \tag{17.8.3}
$$

Let $Z = \sum_{d} [\rho(d) + \tau(d)]$. Then the mean robust soliton distribution $\mu(d)$ is given by

$$
\mu(d) = \frac{\rho(d) + \tau(d)}{Z}. \tag{17.8.4}
$$

For a given value of δ, the value of Z depends on k and c such that the value of Z must remain as close to 1 as possible.

This definition has two extra parameters, c and δ, to ensure that the expected number of degree 1 parity checks is about s rather than 1. The parameter c is of order 1, whereas the parameter δ denotes a bound on the probability that the decoding process quits only after a certain number k' of packets have been received.

Example 17.8.1. Consider the case when $k = 10000$, $c = 0.2$, and $\delta = 0.05$. Then $\sqrt{k} = 100$, $s \approx 244$, $s/\delta = 4880$, $k/s = 14400/310 = 41$, and $s/k \approx 0.0244$ (as used in MacKay [2005]). Note that $K(= k/s) = 41$ in the definition (17.7.2) for $\rho(d)$. The values of the functions $\rho(d)$ and $\tau(d)$, defined by (17.7.2) and (17.8.3) for this data and presented in Table 17.8.1,

---

[3] The harmonic sum $H(k)$ is the partial sum of the harmonic series, defined by $H(k) = \sum_{i=1}^{k} \frac{1}{i}$, and its value is given by $H(k) = \ln k + \gamma + \varepsilon_k$, where $\gamma \approx 0.577215665$ is Euler's constant and $\varepsilon_k \sim \frac{1}{2k} \to 0$ as $k \to \infty$.

are used to compute the mean $\mu(d)$ given by (17.8.4), where $\Sigma_1 = 1$, and $\Sigma_2 = \frac{1}{41}\left[\ln(40) + \gamma + \ln(4880)\right] \approx 0.311194633$, thus giving $Z = \Sigma_1 + \Sigma_2 \approx 1.3$, which is same as in MacKay [2005].

Table 17.8.1 Computed Values for a Robust Soliton Distribution

d	$\rho(d)$	$\tau(d)$	$\mu(d)$
1	$\frac{1}{41}$	$\frac{1}{41}\cdot 1$	0.0387
2	$\frac{1}{1\cdot 2} = 1 - \frac{1}{2}$	$\frac{1}{41}\cdot\frac{1}{2}$	0.3971
3	$\frac{1}{2\cdot 3} = \frac{1}{2} - \frac{1}{3}$	$\frac{1}{41}\cdot\frac{1}{3}$	0.1355
4	$\frac{1}{3\cdot 4} = \frac{1}{3} - \frac{1}{4}$	$\frac{1}{41}\cdot\frac{1}{4}$	0.0693
5	$\frac{1}{4\cdot 5} = \frac{1}{4} - \frac{1}{5}$	$\frac{1}{41}\cdot\frac{1}{5}$	0.0425
$\vdots$	$\vdots$	$\vdots$	$\vdots$
39	$\frac{1}{38\cdot 39} = \frac{1}{38} - \frac{1}{39}$	$\frac{1}{41}\cdot\frac{1}{39}$	0.0010
40	$\frac{1}{39\cdot 40} = \frac{1}{39} - \frac{1}{40}$	$\frac{1}{41}\cdot\frac{1}{40}$	0.0009
41	$\frac{1}{40\cdot 41} = \frac{1}{40} - \frac{1}{41}$	$\frac{1}{41}\ln(4880)$	0.1610
sum	Σ_1	Σ_2	

A barplot of $\mu(d)$ for $d = 1, 2, \ldots, 41$ is presented in Figure 17.8.3, which is a discrete representation of Figures 17.8.1 and 17.8.2 (both in the first quadrant). In fact, the discretization of a soliton (shaded part) in Figure 17.8.1 is the ideal soliton distribution $\rho(d)$, while that of the sech-envelope soliton of Figure 17.8.2 is the robust soliton distribution $\mu(d)$. The $\tau(d)$ distrubution is simply a correction factor.

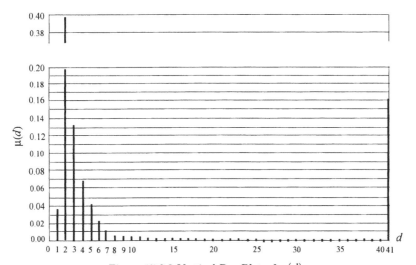

Figure 17.8.3 Vertical Bar Plot of $\mu(d)$.

Note that the value of the factor Z depends on the value of k, c, and δ. For example, with $k = 10000$ and $\delta = 0.5$ MacKay [2005] gives: (a) $Z \approx 1.01$ and $s = 1010$ for $c = 0.01$; (b) $Z \approx 1.03$ and $s = 337$ for $c = 0.03$: and (c) $Z \approx 1.1$ and $s = 99$ for $c = 0.1$. Other values of k, c and δ produce different value of Z, but Z must be as close to 1 as possible to maintain low error .

Henceforth, we use d to denote the degree distribution of classes of Fountain codes; it should not be confused with the Hamming distance d of a linear binary code.

18

Erasure Codes

There are four recently developed codes, namely the Tornado, Fountain, Luby Transform (LT), and Raptor codes. The first two are discussed in this chapter, and the last two in the subsequent two chapters, respectively.

18.1 Erasure Codes

Erasures were discussed in detail in §13.6. A more interesting situation arises in the following case. Consider a wide-area file system in which large files are usually partitioned into n blocks that are replicated among the servers. Different users (clients) download most of the n blocks, where replication is also permitted, although it is wasteful in terms of both space and performance. Suppose clients cannot access certain blocks on the server. That is where the erasure codes come in, to calculate m coding blocks, and distribute the $n + m$ blocks on the network. Then clients download the $n + a$ closest blocks, regardless of identity, and from these, recalculate the file. The erasure codes provide excellent use of space, fault-tolerance, and relief from block identity, where any $n + a$ blocks are sufficient. However, certain codes, like the Reed–Solomon (RS) code, have performance issues, whereas more recent codes, like the Luby Transform codes, have patent issues and open research questions (see Chapter 19). Hence, the realization of the success of erasure codes is not so straightforward, and some exposure to certain well-known erasure codes is desirable.

The RS code is regarded as a standard erasure code. As seen in Chapter 13, suppose we have n information devices and k coding devices. We break up each data block into words of size s such that $2^s < n + k$. Suppose there are n codewords $c_1, \ldots, c_n$, and k parity-check words $p_1, \ldots, p_k$. Then encoding and decoding processes use an $(n + k) \times n$ coding matrix M, which is defined such that $M\{c_1, \ldots, c_n\} = \{c_1, \ldots, c_n, p_1, \ldots, p_k\}$, with the additional property that all $n \times n$ matrices created by deleting k rows from M are invertible. The matrix M always exists if it is derived from the Vandermonde matrix (see below), where every row corresponds to a dataword or a codeword. To decode, suppose that n symbols are downloaded, and then create a matrix M' corresponding to these n symbols. Thus, $M^{-1} \times$ {existing words}= {codewords}.

Example 18.1.1. Consider a single channel model that has a q-ary erasure channel for $q = 8$, with an input $\{0, 1, 2, \ldots, 7\} = \{000, 001, \ldots, 111\}$ (in binary), with a probability $\mathfrak{p}$ of failure of error-free transmission (known as the *erasure probability*) and, thus, a probability $1 - \mathfrak{p}$ of transmitting the message without error. This situation is presented in Figure 18.1.1, where the probability $1 - \mathfrak{p}$ is marked by I and $\mathfrak{p}$ by II, thus showing that the probability of getting an output x (a complete failure) is definitely $\mathfrak{p}$. The recovery of the output x will be presented in the discussion of Fountain, LT, and Raptor codes. ■

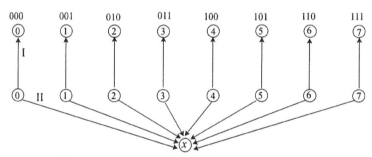

Figure 18.1.1 A channel with erasure probability $\mathfrak{p}$.

In RS coding, Galois field arithmetic is used, where addition is XOR, and multiplication or division require expensive log-lookup tables. Encoding is $O(mn)$, whereas decoding requires an $n \times n$ matrix inversion which is $O(n^3)$, followed by recalculation of codewords which is $O(n^2)$. However, with x words per block, encoding is $O(knx)$ and decoding is $O(n^3) + O(n^2)$. It is obvious that as n and k increase, the RS code becomes excessively expensive.

The common method, prior to recent codes, started theoretically in 1997 and depended on a feedback channel between the receiver and the sender for retransmission of messages that were not received. This method has the advantage that it works without worrying about the erasure probability $\mathfrak{p}$: If $\mathfrak{p}$ is large, the feedback is larger. In view of Shannon's theory, there is a 'no feedback' stage if the channel capacity $\mathfrak{C}$ of the forward channel is $1 - \mathfrak{p}$ bits. The disadvantage is that this retransmission method is wasteful.

18.1.1 Forward Error Correcting Codes. Erasure channels are discussed, where data is either transferred correctly, or where errors packets are dropped on the Internet. A file divided into n blocks is distributed. Erasure codes can be used to encode the original data, resulting in $n + m$ blocks. These blocks are then distributed such that any $n' \geq n$ blocks are sufficient to recover the original file, where the overhead factor $\gamma = n'/n \geq 1$. With different codes there is often a trade-off between computational efficiency and γ. For example, $\gamma = 1$ for RS codes, but computation is inefficient.

An erasure code is a forward error correcting (FEC) code for the binary erasure channel that transforms a message of k symbols into a longer message (codeword) with n symbols such that the original message can be recovered from a subset of the n symbols. The rate R of the code is defined by $R = k/n$, and the reception recovery is defined by the ratio K/k, where K denotes the number of symbols required for recovery.

The optimal erasure codes are those with the property that any k out of n codeword symbols are sufficient to recover the original message, thus possessing the optimal reception efficiency. Optimal codes are MDS (minimum distance separable) codes.

The parity check for erasure codes is a special case when $n = k + 1$, where a checksum is computed from a set of k values $\{w_i\}_{1 \leq i \leq k}$ and appended to the k source values $w_{k+1} = -\sum_{i=1}^{k} w_i$. The set of $(k+1)$ values $\{w_i\}_{1 \leq i \leq k+1}$ becomes consistent with regard to the checksum. If one of these values, say w_e, is *erased*, it can be easily recovered by summing the remaining variables, that is,

$$w_e = -\sum_{i=1, i \neq e}^{k+1} w_i. \tag{18.1.1}$$

Example 18.1.2. Consider the simplest case of $k = 2$, where redundancy symbols are created by sampling different points along the line between two original symbols (linear distribution). Suppose that the original message $\{3184150015\}$ (a 10-digit phone number) is sent by person X to another person Y via a faulty e-mail.[1] Because the message has ten characters, the sender X breaks it into two parts $a = 31841$ and $b = 50015$, and sends two messages: "A = 31841" and "B = 50015" to Y. Before transmission, the sender X constructs a linear distribution (interpolation) function $f(i) = a + (b-a)(i-1)$, $i = 1, 2, \ldots, 5$, where in this case $f(i) = 31841 + 18174(i-1)$, and computes (extrapolates) the values $f(3) = 68189, f(4) = 72696$, and $f(5) = 104537$, and then transmits three redundant messages: "C = 68189", "D = 86363", and "E = 104537". Now, Y knows the function $f(i)$, where $i = A, B$ are two parts of the original message. Suppose Y receives only "C" and "E". Then Y reconstructs the entire original message by computing the values of A and B from $f(4) = A + 3(B - A) = 86363$ and $f(5) = A + 4(B - A) = 104537$, just received. Since the function $f(i)$ is linear, subtracting $f(4)$ from $f(5)$ yields $B - A = 18174$, and then from $f(4)$ we get $A + 3(18174) = 86363$, which gives $A = 31841$ (same value is obtained from $f(5)$), and thus the first part A of

---

[1] This is a fictitious e-mail service in which messages longer than 5 characters are not permitted, and it is not only very expensive but also very unreliable. The receiver Y is not required to acknowledge receipt of the message. The choice of this phone number is purely random.

the message is obtained; the second part B of the message is the obtained from $B = A + 18174$, thus giving $B = 50015$. A graphical presentation of this example is given in Figure 18.1.2. Note that the erasure code in this example has a rate of 40%. ∎

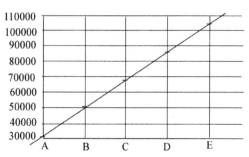

Figure 18.1.2 Faulty E-mail.

The above example is, of course, oversimplified; in practice, any true erasure code that works for any data would require a different function than $f(i)$, even some other distribution. In fact, in the general case, a polynomial distribution is used, and the arithmetic is carried out in a finite field F_{2^m}, where m is the number of bits in a symbol. The sender numbers the data symbols from 0 to k and sends them. He then constructs, for example, an equally spaced Lagrange polynomial $p(x)$ of order k such that $p(i)$ is equal to the data symbol i, and then he sends $p(k), \ldots, p(n-1)$. The receiver can now use the polynomial interpolation to recover the lost data, provided he has successfully received k symbols. If the number of bits in a symbol is less than b, multiple polynomials are used. In either case, the sender constructs the symbols k to $n-1$ 'on the fly' (i.e., distribute the workload evenly between transmission of the symbols). If the receiver wants to do the calculations 'on the fly', he must construct a new polynomial q such that

$$q(i) = \begin{cases} p(i) \text{ if the symbol } i < k \text{ is received successfully,} \\ 0 \text{ if the symbol } i < k \text{ is not received.} \end{cases}$$

Now let $r = p - q$. Then we know that

$$r(i) = \begin{cases} 0 \text{ if the symbol } i < k \text{ has been received successfully,} \\ p(i) - q(i) \text{ if the symbol } i \geq k \text{ has been received successfully.} \end{cases}$$

Then $r(i) = p(i) - q(i)$ can be computed since we have enough data points to construct r and evaluate it to find the lost packets. To complete this process, both sender and receiver require $O(n(n-k))$ operations, but only $O(n-k)$ space for operating 'on the fly'.

18.1.2 Vandermonde Matrix. The completion of the above process requires an RS code with codewords constructed over F_{2^n} using a Vandermonde matrix V which is an $n \times k$ matrix with the terms of a geometric progression in each row, that is,

$$V = \begin{bmatrix} 1 & \alpha_1 & \alpha_1^2 & \cdots & \alpha_1^{k-1} \\ 1 & \alpha_2 & \alpha_2^2 & \cdots & \alpha_2^{k-1} \\ \vdots & \vdots & \vdots & \cdots & \vdots \\ 1 & \alpha_n & \alpha_n^2 & \cdots & \alpha_n^{k-1} \end{bmatrix}. \tag{18.1.2}$$

We can write the matrix V as $V_{i,j} = \alpha_i^j$ for all $i = 1, \ldots, n$ and $j = 1, \ldots, k$. If $n = k$, the determinant of a square Vandermode matrix is given by

$$|V| = \prod_{1 \le i < j \le n} (\alpha_j - \alpha_i). \tag{18.1.3}$$

A square Vandermonde matrix is invertible iff all the α_i are distinct. By Leibniz's formula the Vandermonde determinant can be written as

$$|V| = \prod_{1 \le i < j \le n} (\alpha_j - \alpha_i) = \sum_{\pi \in S_n} \mathrm{sgn}(\pi) \alpha_1^{\pi(1)-1} \cdots \alpha_n^{\pi(n)-1}, \tag{18.1.4}$$

where S_n denotes the set of permutations of $\{1, \ldots, n\}$, and $\mathrm{sgn}(\pi)$ denotes the signature of the permutation π.[2] In fact, the determinant $|V|$ is evaluated using the sum of permutation products. Now consider the binomial $x_j - x_i$ for any $j > i$. For every product with $x_i^i x_j^j$ there is another with $x_i^j x_l^i$. Since these products are the result of permutations with opposite parity, subtract them and find the ones divisible by $x_j - x_i$, for each $j > i$. Remember that multivariable polynomials over the integers have unique factorization, and this factorization of the determinant includes $x_j - x_i$ for each $j > i$. Also, their product gives an expression of degree $n(n-1)/2$. Consider any permutation product in $|V|$, which also has degree $0 + 1 + 2 + \cdots + (n-1) = n(n-1)/2$. These factors produce an expression that divides the determinant, has the same degree, and therefore, a constant. Next, notice that the product of x_i^i

---

[2] The signature (or sign) of the permutation π is defined by $\mathrm{sgn}(\pi) = \pm 1$ according to whether π is even or odd. In fact, $\mathrm{sgn}(\pi) = (1)^{N(\pi)}$, where $N(\pi)$ is the number of inversions in π. For example, consider the permutation π of the set $\{1, 2, 3, 4, 5\}$, which changes the initial arrangement 12345 into 34521. This can be obtained by three transpositions: first exchange the places of 3 and 5, then exchange places 2 and 4, and finally exchange places 1 and 5. This permutation is odd, and $\mathrm{sgn}(\pi) = -1$. Some properties of permutation are: (i) Identity permutation is even; (ii) composition of two even permutations is even; (iii) composition of two odd permutations is even; (iv) composition of an odd and an even permutation is odd; (v) inverse of every even (odd) permutation is even (odd); (vi) a cycle is even iff its length is odd; and(vii) every permutation of odd order is even, but the converse may not be true.

along the diagonal of $|V|$ has degree 1. If we multiply the binomials together as $(x_1 - x_0)(x_2 - x_0)(x_2 - x_1)(x_3 - x_0)\cdots$ and take the first variable from each binomial, we find that the resulting expression has coefficient 1, and thus $|V|$ is the product of $x_j - x_i$ for $0 \le i < j \le n$.

The Vandermonde polynomial, multiplied with the symmetric polynomials, generates all the alternating polynomials. One of the applications of the Vandermonde matrix is the evaluation of a polynomial at a set of points. This matrix transforms the coefficients of a polynomial $a_0 + a_1 x + a_2 x^2 + \cdots + a_{n-1} x^{n-1}$ to the values the polynomials take at the points α_i. Since $|V| \ne 0$ at distinct points α_i, the mapping from the coefficients to values at the (distinct) points α_i is a one-to-one correspondence, and the resulting polynomial interpolation problem has a unique solution. Thus, for example, solving a system of linear equations $V\mathbf{u} = \mathbf{y}$ for $\mathbf{u}$ with an $n \times k$ Vanermonde matrix V is equivalent to finding the coefficients u_j of the polynomial $p(x) = \sum_{j=0}^{k-1} u_j x^j$ of degree $\le (k - 1)$ such that $p(\alpha_j) = y_i$ for $i = 1, \dots, n$.

Note that a square Vandermonde matrix can be easily inverted in terms of the Lagrange basis polynomials $p_n(x)$, such that each column is the coefficients of $p_n(x)$, with terms in increasing order going downward. The resulting solution to the interpolation polynomial is called the Lagrange polynomial $P_n(x)$. If the values of α_i range over powers of a finite field, then the determinant has application to the BCH code (see Chapter 11).

Example 18.1.3. A simple example of a square Vandermonde matrix is

$$
V_1 = \begin{bmatrix} 1 & 1 & 1 & 1 \\ 1 & 2 & 4 & 8 \\ 1 & 3 & 9 & 27 \\ 1 & 4 & 16 & 64 \end{bmatrix},
$$

in which the second column is the first column multiplied by the respective row number, and the k-th column is the second column raised to the $k - 1$ power. Now, let $\mathbf{x} = \{x_1, x_2, \dots, x_n\}$ denote the second column of the matrix (18.1.2), and assume that the elements of $\mathbf{x}$ are unknowns, yet to be determined. Then the determinant $|V|$ becomes a *Vandermonde polynomial* in n variable $x_1, \dots, x_n$. The determinant $|V_1| = (2 - 1)(3 - 1)(4 - 1)(3 - 2)(4 - 2)(4 - 3) = 12$. If we scale the second row of the matrix V_1 by 2^{17}, that row will run from 2^{17} to 2^{20}, yet the matrix will remain nonsingular. If this is done for every row of V_1, the matrix will start at exponent 17 instead of exponent 0. Thus, if we start with a Vandermonde matrix of distinct rows and scale a row or column by a nonzero constant, it will multiply the determinant $|V_1|$ by the same constant, and the matrix will remain singular. ■

To summarize, an erasure code is, in general, defined as follows: Given a signal of k blocks, recode to n blocks, $n > k$. For optimal performance, reconstruct the signal, given any k unique blocks. For suboptimal performance, reconstruct signal using $(1 + \varepsilon)k$, $\varepsilon > 0$, unique blocks. The rate is $R = k/n$, and the overhead is $\gamma = 1/R$. These suboptimal codes are also known as *near-optimal erasure codes*, which require $(1 + \varepsilon)k$ symbols to recover the message. A reduction in the value of ε increases the cost of CPU time; these codes trade correction capabilities for computational complexity. There are practical algorithms that can encode and decode in linear time.

Erasure codes are applied, for example, in (i) reliable multicast, e.g., IETF (reliable multicast working group); 3GPP (Multimedia Broadcast/Multicast Service (MBMS)); (ii) Peer-to-Peer Systems, e.g., those for solving the last block problem; and (iii) distributed storage. These codes are used for signals from deep space satellites, reliable multimedia multicasting (Digital Fountain), and reliable storage.

18.2 Tornado Codes

In 1997, Luby et al. [1997] introduced the Tornado codes, which are a class of erasure codes that belong to the family of forward error correction (FEC). They require finitely many additional redundant blocks than needed in RS codes. Compared to RS codes, they are much faster to generate and are capable of correcting erasures much faster. For example, Byers et al. [1998] showed how Tornado codes can greatly outperform the RS codes for large values of block length. Tornado codes possess a layered structure, in which all layers, except the last one, use an LDPC code for error correction, and the last layer uses an RS code (Figure 18.2.1). Note that the LDPC codes are fast but they suffer occasional failures, whereas RS codes are slower but optimal with regard to recovery from failures.

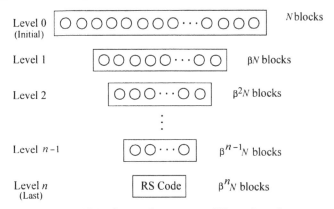

Figure 18.2.1 Layered structure of Tornado codes.

Tornado codes have a built-in algorithm that decides on the number of levels, recovery blocks and their lengths in each level, and the distribution used to generate blocks from all levels except the last one. The criterion for this algorithm is that the $\beta^n m$ blocks at the n-th level must be at most equal to the number of blocks in the RS code.

The encoder takes the data bits $\{d_i\}$, determines the parity-check bits $\{p_j\}$, and creates the input data, which is divided into blocks. These blocks are sequences of bits of the same size. The encoding process is effectively demonstrated by considering a randomly irregular bipartite graph with a carefully chosen degree sequence, like the graph B of Figure 17.5.1. As a simple example, the parity bits in the graph of Figure 18.2.2(a) yield the parity bits as $p_1 = d_1 \oplus d_2 \oplus d_3$, $p_2 = d_1 \oplus d_4$, and $p_3 = d_1 \oplus d_2 \oplus d_4$, which yields the codeword $\mathbf{c} = \{c_1, \dots, c_4\}$.

In general, we consider a bipartite graph B and a message consisting of N blocks of data bits $\{d_1, d_2, \dots, d_N\}$ and βN blocks of parity bits $\{p_1, \dots, p_{\beta N}\}$. The bipartite graph may be regarded as a mapping from the data bits to the parity bits, i.e., $B : \{d_1, d_2, \dots, d_N\} \mapsto \{p_1, p_2, \dots, p_{\beta N}\}$ (Figure 17.5.1). Each parity bit is the XOR of its neighboring data bits. Thus, encoding time is proportional to the number of edges because the XOR operation is performed for each edge in B.

Let $d(> 0)$ denote the degree of each c-nodes (on the left), and let $\beta = (1 - \varepsilon)d$, where $\varepsilon > 0$ is an arbitrarily small quantity, be as large as possible. The *unshared neighbor property* is very important. Accordingly, an unshared neighbor is a p-node that is only the neighbor of one d-node, as shown in Figure 18.2.3. Obviously, such a p-node is not shared as neighbors for multiple d-nodes. Thus, a graph has the (α, δ) unshared neighbor property if for any subset $k \leq \alpha m$ of d-nodes, at least δk have an unshared neighbor among the p-nodes. The following result holds: For a graph B that has (α, δ) unshared neighbor property and the c-nodes have degree d , then $\delta \geq \dfrac{2\beta}{d} - 1$.

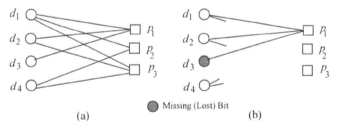

Figure 18.2.2 (a) Encoding and (b) decoding steps.

In decoding we look for a parity bit with all but one of its known neighbors (code bits). The missing code bit is the XOR of the parity bit and its known code bit (Figure 18.2.2 (b)). The decoding time is at most proportional to the number of edges in B. The construction of a sparse random bipartite graph is therefore required to ensure, with high probability, the recovery of all code bits if at most $(1 - \varepsilon)\beta N$ of the bits are lost, where $\varepsilon > 0$ is an arbitrarily small quantity.

The erasure of a block (input or recovery) is detected by other means, like a CRC check, or arrival failure of a network packet with a given sequence number.

The user decides to assign the number of recovery blocks, and the number of levels, as well as the number of blocks in each level. This number in each level is determined by a fraction β, $0 \leq \beta < 1$, such that if there are N input blocks at the initial level, then the first recovery level has βN blocks, the second $\beta^2 N$ blocks, and so on; in general, the i-th level has $\beta^i N$ blocks. Since $\beta^i \to 0$ as $i \to \infty$, the blocks in the $(n - 1)$-st recovery levels are far smaller than the initial level. For decoding, all recovery levels from the first through the $(n - 1)$-th use an LDPC code, where the bitwise XOR operates on all these input blocks. In view of Property 2 (§2.2.2), for a sequence of k bits $\{b_1, b_2, \ldots, b_k\}$, given $b_1 \oplus b_2 \oplus \cdots \oplus b_k\}$ and any $(k - 1)$ bits b_i, the missing bit can be easily determined. Using this property, the recovery blocks at each level is the XOR of some set of input blocks at level $i-1$, where $i = 1, \ldots, n-1$, and $i = 0$ is the initial set of input blocks. A random use of blocks in the XOR is allowed, except that the same blocks must not repeat. However, the number of blocks XOR-ed to complete a recovery level from each level must be chosen from a preassigned specific distribution such as the degree distribution (§17.7).

The advantage of using the XOR operation both at the input and recovery levels is twofold: It operates extremely fast, and the recovery blocks are the XOR of only a subset of the blocks. Thus, the recovery from the first to the $(n - 1)$-th level is very fast.

The last n-th level uses an RS code that is optimal in failure recovery, as we have seen in Chapter 13. Since the last level has the smallest number of blocks, the RS code, although slower than the LDPC codes, operates only on a fewer recovery blocks and recovers the amount of data equally efficiently.

During recovery, the decoding algorithm works backward, starting with the last n-th level, and going upward. It is like a Tornado, which after touchdown at the last level goes upward. Thus, the RS code recovers first, as this recovery is certainly error-free if the number of missing blocks at the $(n-1)$-th level is less than those at the n-th level. From the $(n-1)$-th level through level 1, the LDPC recovery, with the XOR operation, is used for the recovery at the $(n - i)$-th level with high probability if all the recovery blocks are present and the $(n - i + 1)$-th level is missing at least βm fewer blocks than the $(n - 1)$-th recovery level. Thus, the recovery algorithm seeks to find a recovery block from the missing $(n-i+1)$-th level, and thus, according to the above Property 2, XOR-ing the recovery block with all of the blocks that are present yields the missing block.

We will expand the above outline of the recovery algorithm to the case when no parity bits are lost but $K \leq \alpha N$ data bits are lost. This is a hypothetical situation because in practice the parity bits have the same probability to be lost (erased) as the data bits. However, a discussion of this case provides a better explanation of the recursive use of the (α, δ) unshared neighbor property to recover the lost bits. If $K \leq \alpha N$ data bits are lost, then at least δK data bits have an unshared neighbor among the parity bits (on the right). Let d_i^{lost}, $0 \leq i < \delta K$, be any lost data bit, and consider it jointly with its unshared neighbor p_i^{lost}. Suppose $\beta = 1/2$. Then if d_i^{lost} has degree d, the parity bit p_i^{lost} will have degree $2d$ (Figure 18.2.3). Moreover, the bit p_i^{lost} is the XOR of $2d - 1$ received data bits together with the lost bit d_i^{lost}. Thus, the bit d_i^{lost} can be recovered.

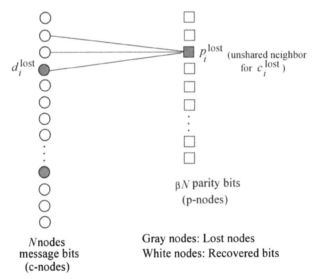

Figure 18.2.3 Unshared neighbors in the graph B.

After the recovery of the δK lost code bits that have an unshared neighbor on the right, still $K - \delta K$ lost code bits remain lost. But all of them can be recovered by recursively applying the unshared neighbor property to each one of them, as some of the $K - \delta K$ lost code bits still have their unshared neighbors that they did not have before. This procedure takes at most $2d$ XOR operations to recover a lost code bit.

The recursive structure is shown in Figure 17.5.1, where the βN parity bits are protected from erasures by the $\beta^2 N$ parity bits at the next level (Level 1), which in turn are protected by $\beta^2 N$ parity bits at Level 2, and so on. At some point when the number of parity bits reduce to $\beta^i N$ $(i > 1)$, which is at least equal to the length of the RS code (last level), the recursion of the unshared neighbor property stops. This process requires at most βN parity bits to protect N code bits from erasures. Because the decoding process is from right to left (from bottom upward in Figure 18.2.1), we may fail completely to decode the lost (erased) code bits. In the worst case, if all the lost bits are in the last level, only $O(N^{1/\beta^i})$ errors will be responsible for the code's failure. If the RS code decodes successfully due to the earlier application of the unshared neighbor property, we can correct only αN of the N code bits from erasures. If we assume that the probability of erasures being more than αN is low, then the Tornado code will recover, with high probability, all of the erasures.

Assuming that the erasures occur at random locations and are not burst errors, which is highly unlikely in practice, let the probability of each bit being erased be $\mathfrak{p}$. Let x be a random variable denoting the number of erasures that have occurred in the K code bits. The expected value of this random variable is $E(x) = K\mathfrak{p}$. Suppose x follows a binomial distribution. Then for a certain value r, the probability of x taking r values is

$$\mathfrak{P}(x = r) = \binom{K}{r}\mathfrak{p}^r(1 - \mathfrak{p})^{K-r}.$$

The probability that we will have more than r erasures is

$$\mathfrak{P}(x \geq r) = \sum_{j=r}^{K} \binom{K}{j}\mathfrak{p}^K(1 - \mathfrak{p})^{K-j} \leq \binom{K}{r}\mathfrak{p}^r$$

$$\leq \left(\frac{Ke}{r}\right)^r = \left(\frac{K\mathfrak{p}e}{r}\right)^r = \left(\frac{eE(x)}{r}\right)^r = \frac{1}{2^N},$$

where we choose the value of $r > 2eE(x)$. If $r \leq \log_2 K$, then $\mathfrak{p}(x \geq r) \leq 1/K$, which is a very good number since K is assumed to be very large. Thus, the erasures that are recovered at any level are of the order $O(K^{1/\beta^i})$. Hence, this code corrects the erasures with high probability.

Elias [1955] was the first to show that (i) the channel capacity $\mathfrak{C} = 1 - \mathfrak{p}$ for the binary erasure channel (BEC) with erasure probability $\mathfrak{p}$, and (ii) the random codes of rates arbitrarily close to $1 - \mathfrak{p}$ can be decoded on this channel with an exponentially small error probability using maximum likelihood decoding (§3.4.12.2), which reduces to solving systems of linear equations using Gaussian elimination method in F_2^M (§13.6.1), although it is slow if the code length is long. The RS codes, on the other hand, are decoded from a block with maximum possible erasures using faster algorithms based on fast polynomial arithmetic in Galois fields. Finding that the RS codes were not meeting the pressure of transmission of data on the Internet, Luby et al. [2001] constructed Tornado codes with linear time encoding and decoding algorithms that almost reach the channel capacity of the BEC. These codes are very similar to Gallager's LDPC codes, but they use a highly irregular weight distribution for their Tanner graphs because the RS codes are used prior to the Tornado codes in their decoding algorithm. The running time of encoding and decoding algorithms of Tornado codes is proportional to their block length rather than to their dimensions. The effect is that the encoding and decoding is slow for low rates, which restricts their applications in many areas as listed in Byers et al. [1998]. This disadvantage made these codes unsuitable for vast data transmission systems.

According to a study by the Fountain Group, entitled *A digital fountain approach to reliable distribution of bulk data*, at http://portal.acm.org/citat ion.cfm? id=285243,285258, software implementation of Tornado codes are almost 100 times faster than RS codes on small block lengths and about 10,000 times faster on larger block lengths. This advantage has resulted in the merger of many other similar codes, like online codes. The advantages of Tornado codes are: (i) computation involves parity (XOR) only; and (ii) each block requires a fraction of the other blocks to calculate, and thus, encoding is $O(x(n + N))$. The disadvantages are: For $\alpha > 0$ more than K blocks are needed to recalculate the file; and (ii) the theory has been developed for asymptotics and, therefore is not well understood in the finite case.

However, practical applications of these codes are not yet fully studied and implemented. These codes remain unusable for developers of wide-area storage systems. The reasons are as follows: (i) Hard-core graph theory is not well understood by systems people, since this theory dwells in asymptotics, and (ii) patent worries are not very encouraging. As a result, one cannot find a 'Tornado code' for the storage system, an therefore, the industry, including LoCI, OceanStore, and BitTorrent, still use RS codes.

A comparison of Tornado and RS codes is presented in Table 18.2.1.

Table 18.2.1. Properties of Tornado and RS Codes

	Tornado	Reed–Solomon
Decoding efficiency	$(1 + \varepsilon)$ required	1
Encoding times	$(K + l) \ln(1 + \varepsilon)P$	KlP
Decoding times	$(K + l) \ln(1 + \varepsilon)P$	KlP
Basic operation	XOR	Field operations

18.3 Rateless Codes

The LDPC codes, decoded using the belief propagation algorithm, are capable of achieving the channel capacity on the binary erasure channel (BEC). This conclusion does not necessarily mean that the problem of reliable communication on noisy channels is almost completely solved. For example, this may be satisfactory in most cases, but on the Internet which is modeled as a BEC, the probability $\mathfrak{p}$ that a given packet is dropped varies with time and depends on traffic conditions in the network. A code designed for low $\mathfrak{p}$ (good channel) would result in a complete failure on a high $\mathfrak{p}$ (bad) channel. On the other hand, a code designed for a bad channel would produce unnecessary packet transmission when used on a good channel. This problem can be solved using rateless codes.

In rateless erasure codes, unlike the block codes where the k symbols of message are encoded to a preassigned number of blocks, the transmitter encodes them into an infinite sequence of symbols and then transmits them. After the decoder receives a sufficient number of symbols, it decodes the original k symbols. Because the number of symbols required for successful decoding depends on the quality of the channel, the decoder, in case of a failure, can pick up a few more output symbols and start decoding them. If the decoder still fails, the transmitter continues to retransmit. This process is repeated until the decoding is successful; the decoder then tells the transmitter over a feedback channel to stop any further transmissions.

Luby [2002] has provided solutions in the LT codes (Chapter 19) to the following problems associated with the rateless codes: (i) The performance of rateless codes is highly sensitive to the performance of the low-rate block code, i.e., a slightly suboptimal code can result in a highly suboptimal rateless code; (ii) the rateless code has very high decoding complexity, even with low $\mathfrak{p}$ (good channel), as the receiver is decoding the same low-rate code on any channel, but with varying channel information; (iii) the complexity of rateless decoding process increases at least as $O(k/R)$, where R is the rate of the low-rate code. The LT codes are low-density generator matrix codes that are decoded using the belief propagation algorithm, which is used for decoding the LDPC codes, and they achieve capacity on every BEC. A drawback of the LT codes is that they share the error floor problem that occurs in the case of capacity achieving LDPC codes. It is shown by Shokrollahi [2002] and Shokrollahi et al. [2001] that this problem can be solved using Raptor codes (see Chapter 20), which are LT codes combined with the outer LDPC codes.

18.4 Online Codes

This is another example of rateless erasure codes that can encode a message into a number of symbols such that knowledge of any fraction of these symbols allows the recovery of the original message with a high degree of probability. These codes produce an arbitrary large number of blocks of symbols that can be broadcast until the receiver has enough symbols for the total recovery. The algorithm for an online code consists of the following phases:

(i) The message is split into n fixed-size message blocks.

(ii) The OUTER ENCODING, which is an erasure code is used to produce auxiliary blocks that are appended to the n message blocks, thus generating a composite message.

(iii) The INNER ENCODING takes the composite message and generates the check blocks that pass through a transmision channel.

(iv) After receiving a certain number of check blocks, a fraction of the composite message is covered. The gray blocks in Figure 18.4.1 represent the lost blocks.

(v) Once enough of the composite message is recovered, the outer encoding is used to recover the original message.

An outline of this algorithm is presented in Figure 18.4.1.

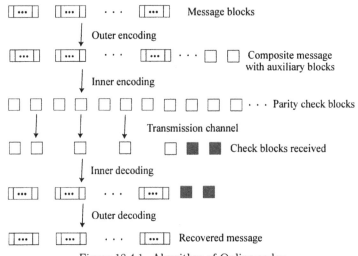

Figure 18.4.1. Algorithm of Online codes.

An online code is parameterized by the block size n and two scalars q and ε, which set a balance between the complexity and performance of encoding. Generally, we take $q = 3$ and $\varepsilon = 0.01$. A message of n blocks is recovered with high probability with $(1 + 3\varepsilon)n$ check blocks, where the probability of failure is $(\varepsilon/2)^{q-1}$. Any erasure code may be used for outer encoding, but the best results are obtained by pseudo-randomly choosing, for each message block, q auxiliary blocks from a total of $0.55qn\varepsilon$ auxiliary blocks and attaching them to each block. Then each auxiliary block is the XOR of all message blocks that have been attached to it. The inner encoding takes the composite message and generates a sequence of check blocks, where each such check block is the XOR of all the blocks from the composite message that are attached to it.

The *degree* of a check block is equal to the number of blocks that are attached to it. This degree is determined by sampling a random distribution $p = \{p_1, p_2, \ldots p_f\}$ defined by

$$p_1 = 1 - \frac{1 - 1/f}{1 + \varepsilon}. \quad p_i = \frac{(1 - p_1)f}{(f - 1)\,i(i - 1)} \quad \text{for } i = 2, \ldots, f, \qquad (18.4.1)$$

where

$$f = \left\lceil \frac{\ln(\varepsilon^2/4)}{\ln(1 - \varepsilon/2)} \right\rceil.$$

Once this degree is known, the blocks from the composite message that are attached to the check blocks are chosen uniformly.

The inner decoder holds check blocks that it cannot decode currently, as a check block can only be decoded when all but one of the blocks that is attached to it are known. The progress of the inner decoder is described as follows: Initially, after a number of check blocks of degree > 1 that are received but cannot be used increases linearly, then at a certain point some of these check blocks suddenly become usable, thus resolving more blocks which causes more check blocks to be usable. In a short time, the original message is decoded, still possibly with certain illusive blocks, which are cleared by the outer decoder, and the original message is recovered.

18.5 Fountain Codes

A Fountain code, which is a rateless erasure code, solves the problem of transmitting a fixed set of data to multiple clients over unreliable links. An earlier solution included transmitting the original data interleaved with erasure coded blocks, but it has undesirable overhead because of its slow rate. The ideal solution should be reliable (client always gets the whole file), efficient (extra work is minimized), on demand (client gets the file at their discretion), and tolerant (solution is tolerant of clients with different capabilities).

The block codes, like the Tornado, suffered disadvantages in their use for data transmission. It was realized that the model of a single erasure channel was defective and, therefore, inadequate in those cases where the data was sent concurrently from one sender to multiple users. The erasure channels from the sender to each receiver basically operated on different erasure probabilities, and adjustments to the sender's erasure probability turned into a guessing game. Matters became complicated as the number of receivers increased, or satellite and wireless transmissions were used, which introduced sudden abrupt changes in their receptions. Thus, the entire business of encoding and decoding processes started failing to keep track of the loss rates of individual receivers.

The above circumstances demanded that new 'robust and reliable transmission schemes' must be constructed. Byers et al. [1998] provided a new class of codes, known as Fountain codes, which addressed the above issues to a great extent. These codes are an ideal mechanism to transmit information on computer networks, where the server sends data to multiple receivers by generating an almost infinite stream of packets. As soon as a receiver requests data, the packets are copied and delivered to the receiver, except in the case of a broadcast transmission where there is no need to copy the data because any and all outgoing packets are received simultaneously by all receivers. In all other types of networks, the copying is done by the sender, or the network if multicast is enabled. The receiver collects all output symbols, and thereafter the transmission is terminated.

A Fountain code produces a preassigned set of k input symbols according to a given distribution on F_2^k, followed by a limitless stream of output symbols. Whereas the strength of Tornado codes lies in the fact that the value of n is kept very close to k, the decoding algorithm of a Fountain code can recover the original k symbols from any set of n output symbols. A very desirable property of Fountain codes is as follows: Each output symbol is generated independently of each other, so the receiver can collect output symbols generated from the same set of k input symbols on different devices that operate a Fountain code. This feature enables creation of the 'design of massively scalable and fault-tolerant systems over packet-based networks'.

The Fountain code is a near-optimal erasure code, also known as a *rateless erasure code*. It can transform a k symbol message into an infinite (practically as large as needed) encoded form by generating an arbitrarily large number of redundancy symbols that can be used for error correction and erasure recovery. The receiver can start the decoding process after receiving slightly more than k encoded symbols. The *repairing* is the process of rebuilding the lost encoded fragments from existing encoded fragments. This issue of regeneration arises in distributed storage systems where communication to maintain encoded redundancy is problematic.

The main goal in creating Fountain codes has been erasure correction with no feedback (in fact, almost zero feedback). The RS codes, which work well with codewords of the size $q = 2^m$, have the property that if any k of the N transmitted symbols are received, then the original k source symbols can be recovered. However, RS codes have a serious disadvantage: They work only for small k, N and q, and the cost of encoding and decoding is of the order $k(N-k)\log_2 N$ block operations. Also, for RS codes (as with any block code), the erasure probability $\mathfrak{p}$ and the code rate $R = k/N$ must be estimated prior to the transmission. The RS code fails completely if $\mathfrak{p}$ is larger than expected and the decoder receives fewer than k symbols. In fact, we seek an extension of this code 'on the fly' to create a low-rate (N, k) code. We saw an extension in the Tornado codes, and the others are in the remaining recent codes, each better than the previous one.

Fountain codes have the property that a potentially limitless sequence of encoding symbols are generated from a given set of source symbols such that the original source symbols can be recovered from any subset of the encoding symbols of size equal to or only slightly larger than the number of source symbols. There is a Fountain solution, in which the server transmits a constant stream of encoding packets, the client succeeds when a minimal number of packets are received, and it assumes fast encode/decode.

The encoder is a Fountain, a metaphor that provides an endless supply of water drops (encoded packets). To build a Fountain code, one uses Tornado erasure codes because they are fast, although they are suboptimal; reconstruction requires $(1 + \varepsilon)k$ packets. The server distributes a file containing n blocks of length m; it then encodes blocks into packets and distributes them into 2^n different packets, although infinitely many such packets can be generated. The number of blocks encoded into one packet is the *degree* of the packet. In the ideal case, the decoder needs to collect any n packets in order to decode the original content. This situation is metaphorically analogous to getting water drops from a (digital) water fountain. No retransmission of a specific packet is needed in the ideal case. In practice, erasure codes approximate this situation.

A Fountain code is *optimal* if the original k source symbols can be recovered from any k encoding symbols. There are Fountain codes known to recover the original k source symbols from any k' of the encoding symbols with high probability, where k' is just slightly larger than k. For any smaller k, say, less than 3,000, the overhead γ is defined by $k' = (1 + \gamma k)$, which amounts to a very insignificant value. However, with such a value of k, both LT and Raptor codes using the sparse bipartite graph (belief propagation algorithm) never achieve a value of $\gamma < 0.1$. The Raptor and online codes behave much better than the LT codes iff the belief propagation algorithm over the check matrix $\mathbf{H}$ can recover most of the source symbols. This is a serious drawback of Fountain codes. If a Raptor or online code is capable of recovering most of the source symbols, say more than 95%, then pre-encoding is not required, because transmission of a very few (less than 0.05%) dense encoding symbols can cover up all the source symbols with a very high probability. These dense encoding symbols also contribute to the remaining matrix $\mathbf{H}$ and provide it its full column rank, and the recovered symbols can be solved by a Gaussian elimination process over the remaining graph.

Implementation issues are as follows: (i) How to divide the data into blocks; (ii) Efficiency in terms of encoding and decoding time; (iii) Overhead factor γ must be greater than 1 for efficient coding methods, but it results in γ times more data to be sent (not desirable); (iv) security, and (v) patents issues, as LT and Raptor codes are patented.

In practical communications, the data is transmitted in blocks. Suppose a binary message of length n is transmitted in K blocks of m bits each; thus, $n = Km$. The transmitter continuously transmits blocks of m bits that are obtained by XOR-ing (i.e., by adding mod 2) subsets of the blocks, while the receiver generally collects a little bit more than the K blocks to decode and retrieve the original message.

Example 18.5.1. Consider an encoder for a block of size K packets $b_1, b_2, \ldots, b_K$, and let n denote the clock cycle. The encoder generates K random bits $\{G_{in}\}$. Let the received packet after transmission be t_n, which is the bitwise sum (mod 2) such that $\sum G_{in} = 1$. Then we have $t_n = \sum_{i=1}^{K} b_i G_{in}$, where this sum is obtained by successively XOR-ing the K packets. Each set of K random bits can be regarded as generating a new column in a finitely increasing $N \times K$ binary generator matrix $\mathbf{G}$ with K rows, as in Figure 18.5.1, where the columns marked by a $\downarrow$ indicate that the corresponding column has been erased during transmission. The receiver holds out the 'bucket' and collects N packets.

$$
\begin{array}{cccccccc}
\downarrow & \downarrow & \downarrow & \downarrow & \downarrow & \downarrow & \downarrow & \downarrow
\end{array}
$$

$$
\begin{array}{l}
1\,0\,0\,1\,1\,1\,0\,0\,0\,1\,0\,1\,1\,0\,0\,0\,0\,1\,1\,1\,0\,1\,0\,0\,1\,1\cdots \\
0\,1\,1\,1\,1\,0\,0\,1\,0\,0\,0\,0\,1\,1\,1\,0\,1\,1\,0\,1\,1\,0\,1\,1\,0\,0\cdots \\
1\,1\,1\,0\,0\,1\,1\,0\,0\,0\,1\,1\,1\,0\,0\,1\,0\,1\,0\,1\,1\,0\,0\,0\,1\,1\cdots \\
\vdots \\
0\,0\,1\,1\,1\,0\,0\,0\,0\,1\,1\,0\,1\,1\,1\,0\,0\,0\,0\,1\,0\,1\,1\,1\,1\,0\cdots \\
1\,1\,1\,0\,0\,1\,1\,1\,0\,0\,0\,0\,1\,1\,0\,1\,0\,1\,1\,1\,0\,0\,1\,0\,1\,1\cdots
\end{array}
$$

The received packets (all above columns minus those under the arrows) yield the $N \times K$ generator matrix of a random order

$$
\mathbf{G}_{\text{partial}} =
\begin{bmatrix}
1\,0\,1\,1\,0\,0\,1\,1\,1\,0\,0\,0\,1\,1\,1\,0\,1\,1 \\
0\,1\,1\,1\,0\,1\,0\,0\,1\,1\,0\,1\,0\,1\,0\,1\,0\,0 \\
1\,1\,0\,0\,1\,0\,0\,1\,1\,0\,1\,0\,0\,1\,0\,0\,1\,1 \\
\vdots \\
0\,1\,1\,1\,0\,0\,1\,0\,1\,1\,0\,0\,0\,1\,1\,1\,1\,0 \\
1\,1\,0\,0\,1\,1\,0\,0\,1\,0\,1\,0\,1\,;\,0\,1\,1\,1
\end{bmatrix}.
$$

Figure 18.5.1 Generator matrix.

It is assumed in this example that the decoder has already corrected single, double, triple, or multiple errors prior to the above process of 'catching' the generator matrix $\mathbf{G}_{\text{partial}}$. Since the erasures (columns) change at each stage and the packets keep coming continuously, the entire error-free file is eventually obtained. ∎

The simplest question at the start of the decoding process is to estimate the probability that the decoder will recover all the source packets. Assume that the decoder knows the above portion of the entire generator matrix $\mathbf{G}_{\text{partial}}$. For example, the decoder knows that this partial matrix $\mathbf{G}_{\text{partial}}$ was generated by a random or pseudo-number generator that is identically synchronized to the encoder's or equipped with a random key that determines the K bits

$\{G_{in}\}_{i=1}^{K}$. As long as the packet size is much larger than the key size of about 32 bits, the use of the key will cause a minimal cost increase.

Decoding of Fountain codes involves solving a system of equations. Let the blocks be denoted by $B_1, B_2, \ldots, B_K$. Suppose, for example, $K = 3$ and $m = 4$. Assume that the following equations hold for the received blocks: $B_1 + B_3 = 0100$, $B_2 + B_3 = 1110$, $B_1 + B_2 + B_3 = 0000$. This system is solved by adding the first and third equations, which gives $B_1 = 1110$, and then $B_3 = 1010$ and $B_2 = 0100$.

Every packet that is received generally has a header that is used by the Fountain code for its key. The blocks in a Fountain code are formed by random combination (sum) of packets. The indices of the packets involved must be known to the receiver. That is where the key comes in. After N blocks have been received, the probability that the decoder will get the entire source file error-free depends on the following three cases:

(i) If $N < K$, decoding is not possible because the decoder does not have sufficient information to recover the source file.

(ii) If $N = K$, the decoder can recover the source file with probability 0.289 for $K > 10$, which is realized if the matrix $\mathbf{G}$ is invertible (mod 2), i.e., if the decoder can compute $\mathbf{G}^{-1}$ by the Gaussian elimination method (see §13.6.1) to recover $b_i = \sum_{n=1}^{N} t_n G_{ni}^{-1}$. The probability that a $K \times K$ matrix is invertible is equivalent to the product of K probabilities, each of which being the probability that a new column of $\mathbf{G}$ is linearly independent of the preceding column. Thus, the factors in this product in succession from first to the last are $\left(1 - 2^{-K}\right), \left(1 - 2^{-(K-1)}\right), \ldots, \left(1 - 2^{(j-1)}\right)$, $\ldots, \left(1 - 2^{-3}\right), \left(1 - 2^{-2}\right), \left(1 - 2^{-1}\right)$, where the j-th factor is the probability that the j-th column (packet) of $\mathbf{G}$ is not all zeros, $j = 1, 2, \ldots, K$. The product of all these K factors is approximately equal to 0.289 for $K > 10$.

(iii) if N is slightly larger than K, i.e., if $N = K + \varpi$, decoding is possible with probability $\geq 1 - 2^{-\varpi}$. In this case, let the probability that a $K \times K$ binary matrix is invertible be $1 - \mathfrak{p}$, where $\mathfrak{p}$ is the probability that the decoder will fail to decode the source file when ϖ extra packets arrive. Thus, $\mathfrak{p}(\varpi) \leq 2^{-\varpi}$ for any K.

19

Luby Transform Codes

19.1 Transmission Methods

A large file sent over the Internet to many receivers splits into smaller packets. The packets may become corrupted or lost (erased) during transmission. Varying bandwidth may cause different receivers to receive these packets at different times. Large transmissions can be solved by the following three methods: (i) use acknowledgments for each received packet to signal which packets are to be retransmitted, as in TCP/IP, a very complex and inefficient process; (ii) use erasure correction codes in which redundancy is added by encoding a k-packet file as n packets $(n > k)$, but not without issues when different receivers have unknown and varying loss rates; and (iii) produce an indefinitely long sequence of packets such that any collection of $k(1 + \varepsilon)$ packets is sufficient to recover the entire file with high probability, where $\varepsilon > 0$ is an arbitrarily small number that measures the tradeoff between the loss recovery property of the code and the encoding and decoding processes. This process, which is independent of loss rate of the receiver, is known as a 'Fountain code', where the metaphor is analogous to filling a bucket under a fountain spraying water drops. Recall that coding theory is used to either remove redundancy from messages (source coding) or to add redundancy in order to make error correction possible (error correcting codes).

Traditional schemes for transmitting data across an erasure channel assume a continuous two-way communication between the sender encoding and the receiver decoding the received message and sending an acknowledgment back to the sender, until all the packets in the message are received successfully. However, certain networks, e.g. cellular wireless transmission, do not have a feedback channel. To ensure reliability on these networks, Fountain and LT codes adopt a one-way communication protocol, where the sender encodes and transmits the packets of the message, while the receiver evaluates each packet as it is received. If there is an error, the erroneous packet is discarded; otherwise, the packet is saved as a part of the message. Eventually the receiver has enough valid packets to reconstruct the message and can signal completion to the sender.

The transmission channel suitable for hard-decision decoding is the BSC channel, both due to Gallager [1963]. The other BEC transmission channel is not used for the the first and second classes of modern *universal Fountain codes*, which contain the LT codes and the Raptor codes. This is simply because the erasure problem is resolved using the RS codes or a pre-code, like an LDPC code, before these classes of modern codes operate. Gallager has provided two BSC algorithms, algorithm A and B, of which algorithm B is more powerful than A because A is a special case of B. Another hard-decision decoder, called the erasure decoder, is due to Robertson and Urbanke [2001]. A universal Fountain code possesses a fast encoder and decoder, and the decoder is capable of recovering the original symbols from any set of output symbols whose size is close to the optimal size with high probability. The first class of such universal Fountain codes is the Luby Transform (LT) code, developed by Luby [2001/2002; 2002].

19.2 Luby Transform (LT) Codes

The central theme of LT codes is the distribution used for their generation. As soon as an output symbol is generated in an LT code, a weight distribution is sampled that returns an integer d, $1 \leq d \leq k$, where k is the number of input (data) symbols. Then d random distinct input symbols are chosen, and their value is added to yield the value of the output symbols, and it happens with every output symbol.

Luby Transform (LT) codes, invented in 1998 and published by Luby [2002], exhibit a better performance than any random linear Fountain code, but do so by simplifying the encoding and decoding processes. In these codes, only a few blocks have random combinations (sums), where the number of blocks in the sums is given by an optimized distribution function, and decoding is done by solving a system of equations involving block files, as in the Fountain codes. The LT codes depend on sparse bipartite graphs to trade reception overhead for encoding and decoding speed. They behave like a rateless sparse random Fountain code and use a decoding algorithm with very small overhead. Their design does not depend on the erasure probability of the channel. This feature alone makes these codes capable of efficiently serving a channel with multiple receivers.

The LT codes use a simple algorithm based on the XOR operation to encode and decode the message. Although the encoding algorithm can in principle produce infinitely many message packets, the percentage of packets received for decoding the message can be arbitrarily small. In this sense, these codes are rateless. They are rateless erasure correcting codes because they can be used to transmit digital data reliably on an erasure channel, and they are very efficient as the data length increases.

19.2.1 LT Encoding. To fully understand the encoding and decoding process, along with the decoding algorithm, some new definitions are as follows: When the decoding process starts, all packets are *uncovered*. At the first stage, all degree 1 packets get *released* to cover their respective unique neighbor. All sets of covered packets that have not yet been processed form a *ripple*. At each subsequent stage, one packet from the ripple is selected randomly and *processed*, and then removed as a neighbor of all encoded packets. Any encoded packet that has degree 1 is now released and its neighbor covered. If the neighbor is not in the ripple, it is added to the ripple. The decoding process ends when the ripple is empty. The decoding fails if at least one packet remains uncovered. This decoding algorithm suggests that the decoding process is virtually similar to that of a Tornado code.

The encoding of LT codes generates distinct encoding packets of symbols of source file 'on the fly' by the server as need arises. The source file $\{x_\kappa\} = \{x_1, x_2, \ldots, x_k\}$ produces each encoded packet P_j by first choosing randomly the degree d of the packet from a degree distribution $\rho(d)$ (§17.5), where the choice of ρ depends on the size n of the source file, and second, by choosing uniformly at random P_j distinct input packets and setting P_j equal to the XOR (mod 2) of the neighboring $P_\nu, \nu \neq j$, packets. The function $\tau(d)$ (§17.8.3) is provided in the robust soliton distribution as a modification factor to maintain a reasonable size of the ripple from start to end. Consider the stage when an input symbol is being processed while j input symbols still remain to be processed. Because the ripple size decreases by one each time an input symbol is processed, the ripple should, therefore, be increased by one each time an input symbol is processed. The factor $\tau(d)$ plays this role of keeping the ripple size uniform on an average.

The encoding (as well as the decoding) process is described in Luby [2002]. Let a symbol denote a string of n bits, and let the input file be a sequence of k input symbols $\{x_\kappa\} = \{x_1, \ldots, x_k\}$. The encoding process will produce a potentially unlimited sequence of output symbols $c_1, c_2, \ldots, c_n$. It begins by dividing the uncoded message into n blocks of almost equal length. The encoded packets are produced using a pseudorandom number generator, which is determined by a probability distribution over the random variable, called the degree distribution defined in §17.5, as follows:

(i) The degree d, $d = 1, \ldots, n$, of the next packet is chosen at random.

(ii) Exactly d blocks of the message are randomly chosen.

(iii) If M_i is the i-th block of the message, the data part of the next packet is computed by $M_{i_1} \oplus M_{i_2} \oplus \cdots \oplus M_{i_d}$, where $\{i_1, i_2, \ldots, i_d\}$ are the randomly chosen indices for the d blocks included in the packet.

(iv) A prefix is attached to the encoded packet that defines how many blocks n are in the message, how many d have been XOR-ed into the data part of this packet, and the list of indices $\{i_1, i_2, \ldots, i_d\}$.

(v) Finally, a reliable error-detecting code, like a CRC check (see §2.2.14), is applied to the packet, and the packet is transmitted.

(vi) This entire process continues until the receiver signals that the transmission is complete.

Notice that the length j of the encoding symbols is chosen as desired. Although a larger value of j provides an overall more efficient encoding and decoding because of overheads with bookkeeping operations, the value of j is immaterial for the development of the theory. However, choosing j close to the length of the packet is desirable. The encoder works as follows: The data of length N is partitioned into $k = N/l$ input symbols, i.e., each input symbol is of length j. Each encoding symbol is generated as defined above.

The encoding process can be explained by Figure 19.2.1; namely, how a packet is chosen at random from degree distribution of file blocks by XOR-ing selected blocks.

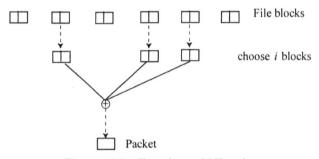

Figure 19.2.1 Encoding of LT codes.

We present a simple method for encoding and decoding of LT codes, as described in Hamzaoui et al. [2006].

Example 19.2.1. (Encoding) Given k source symbols $x_1, \ldots, x_k$ and a suitable degree distribution on d, a sequence of encoded symbols c_i, $i \geq 1$, is generated for each i by randomly selecting a degree $d \in \{1, \ldots, k\}$ according to the degree distribution, and by uniformly selecting at random d distinct source symbols and setting c_i equal to their mod 2 bitwise sum (i.e., by XOR-ing them). Consider LT encoding of $k = 3$ source symbols (here bits) $x_1 = 0, x_2 = 0, x_3 = 1$ into 5 encoded symbols $c_1, \ldots, c_5$. Figure 19.2.2 shows each step in the encoding process:

   (a) $d = 1, c_1 = x_1 \oplus x_3$,

   (b) $d = 1, c_2 = x_2$,

   (c) $d = 3, c_3 = x_1 \oplus x_2 \oplus x_3$,

   (d) $d = 2, c_4 = x_1 \oplus x_3$,

(e) $d = 2, c_5 = x_1 \oplus x_2$.

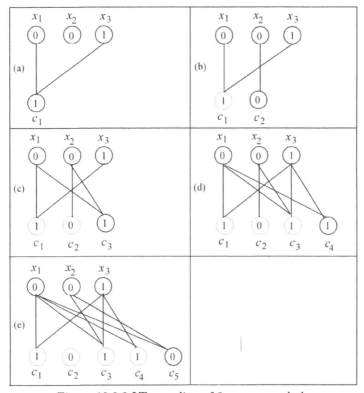

Figure 19.2.2 LT encoding of 3 source symbols.

Once the set of encoding symbols and some representation of their associated degrees and set of neighbors are obtained, the decoder repeatedly
recovers input symbols using the rule described in §19.2.2 and §19.2.3 as long
as it applies.

19.2.2 Decoding of LT Codes. This is similar to that of LDPC codes over
the erasure channel. The success or failure of the decoding algorithm depends
on the output degree distribution. Since n is kept close to k, the output
symbols of the LT codes must have an average weight of at least $ck \log(k)/n$,
where $c > 0$ is a constant defined in Definition 17.8.3. Luby [2002] has shown
that it is possible to construct a weight distribution that is almost close to
this lower bound by using a fast decoder.

The decoder can reconstruct all P_j packets with probability $1 - \mathfrak{p}$ when any
$n + O\left(\sqrt{n}\log(n/\delta)\right)$ packets have been received. While the time for encoding
each packet is of the order $O\left(\log(n/\delta)\right)$, the time for decoding each packet

is of the order $O\left(n\log(n/\delta)\right)$. Because the encoded packets are generated on the fly, how does the decoder know which message nodes are neighbors of a particular encoded packet? Luby has provided two suggestions: (i) explicitly include the information in the packet as an additional overhead, or (ii) replicate the pseudo-random process at the decoder by supplying it with suitable keys.

Let $\mathbf{b} = \{b_1, \ldots, b_k\}$ denote the set of received source files and $\mathbf{P} = \{P_1, \ldots, P_m\}$ the set of parity-check nodes. Analytically, the decoder recovers $\mathbf{b}$ by solving the matrix equation $\mathbf{P} = \mathbf{b}\,\mathbf{G}$, where $\mathbf{G}$ is the matrix associated with a Tanner graph, as in the case of a linear Fountain code (§18.5). In view of the above two suggestions, it is assumed that the decoder has prior knowledge of the pseudo-random matrix $\mathbf{G}$. The recovery of $\mathbf{b}$ is achieved by the message passing algorithm. If we regard the decoding algorithm as the sum-product algorithm (Chapter 14), we can consider that all message packets are either completely certain or completely uncertain. An uncertain message packet implies that any of the received source files b_i, $i = 1, 2, \ldots, n$, could have any value with equal probability, whereas a certain message packet means that b_i definitely (i.e., with probability 1) has a particular specific value.

In general, the encoded packets P_j will be called *parity-check nodes*. Then the decoding process goes through the following steps:

STEP 1. Find a parity-check node P_j that is connected (on a Tanner graph) to only one source packet f_i. If no such node exists, then the decoding algorithm stops and fails in the recovery of all source packets. If a node P_j exists, then

(a) Set $b_i = P_j$.

(b) Add b_i to all nodes $P_{j'}$ that correspond to b_i, where $P_{j'} := P_j + b_i$ for all j' such that $G_{j'j} = 1$.

(c) Remove all edges connected to the source packet b_i.

STEP 2. Repeat Step 1 until all b_i, $i = 1, \ldots, n$, are found and processed.

Example 19.2.2. To illustrate the above decoding process, we consider the source file $\{b_1, b_2, b_3\} = \{111\}$ and the parity as $\{P_1, P_2, P_3, P_4\} = \{1011\}$, and the Tanner graphs shown in Figure 19.2.3.

At the start, there are three source packets and four parity nodes. In the first iteration, the parity node P_1 is connected to b_1 (Panel a). Set b_1 equal to P_4; thus $b_1 = 1$, and discard the node P_1 (Panel b). The second iteration starts in Panel c, where P_4 is connected to b_2. Set b_2 equal to P_4 (i.e., $b_2 = 1$); discard P_2 (Panel d). Add b_3 to P_4 (Panel e). Finally, connect b_3 to P_4 (Panel e); then P_3 is connected to b_3 (Panel e); thus, $b_3 = 1$ (Panel f). The results can also be verified by solving the set of equations: $P_1 \oplus P_2 \oplus P_4 = 0$; $P_2 \oplus P_3 \oplus P_4 = 0$; $P_2 \oplus P_3 = 1$, which upon solving give

$\{P_j\} = \{1011\}.$ ∎

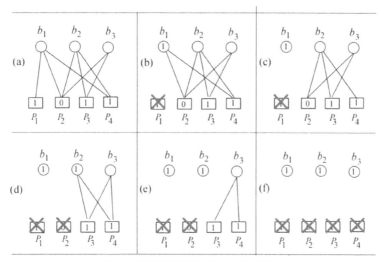

Figure 19.2.3 LT decoding algorithm.

19.2.3 Decoder Recovery Rule. There is a one-to-one correspondence between the LT process and the decoder. An encoding symbol represents an input symbol iff the encoding symbol can recover the input symbol. This means that the success of the LT process depends on a successful recovery of all data symbols.

If there is at least one encoding symbol that has exactly one neighbor, then the neighbor can be recovered immediately since it is a copy of the encoding symbol. The rule is that the recovered input symbol is XOR-ed into any remaining encoding symbols which also have that input symbol as a neighbor; then the recovered input symbol is removed as a neighbor from each of these encoding symbols, and the degree of each such encoding symbol is decreased by one to reflect this removal (Luby [2002]).

The decoding process uses XOR operations to retrieve the encoded message, as follows:

STEP 1. If the current packet is not clean, or if it replicates a packet already processed, the packet is discarded.

STEP 2. If the current clean packet is of degree $d > 1$, it is first processed against all completely decoded blocks in the message queuing area (see details in Step 3), then stored in a buffer area if its reduced degree is greater than 1.

STEP 3. After a new, clean packet of degree $d = 1$ (blocks M_i) is received, or if the degree of the current packet is reduced to 1 by Step 2, it is moved to the message queuing area, and then matched against all the packets of degree > 1 still in the buffer. It is XOR-ed into the data part of any buffered packet that was encoded using M_i; then the degree of that matching packet is determined, and the list of indices for the packet is adjusted to reflect the application of M_i.

STEP 4. After Step 3 unlocks a block of degree $= 2$ in the buffer, the block is reduced to degree 1 and is in its turn moved to the message queuing area, and then processed against the packets remaining in the buffer.

STEP 5. After all the n blocks of the message have been processed and moved to the queuing area, the receiver signals the completion of the transmission.

This decoding process works because of formula (2.2.1): $\mathbf{s} \oplus \mathbf{s} = \mathbf{0}$ for any bit string or symbol $\mathbf{s}$. Thus, after decoding $d - 1$ distinct blocks using XOR operation in a packet of degree d, the only thing left is the original uncoded content of the unmatched block, that is,

$$\left(M_{i_1} \oplus M_{i_2} \oplus \cdots \oplus M_{i_d}\right) \oplus \left(M_{i_1} \oplus \cdots \oplus M_{i_{k-1}} \oplus M_{i_{k+1}} \oplus \cdots \oplus M_{i_d}\right) = M_{i_k}.$$

This process illustrates the *Luby transform* in LT codes, performed in decoding and established using formula (2.2.1).

Figure 19.2.4 explains the process where the original file blocks (input symbols) are at the top and the encoded packets are at the bottom.

The packet of degree 1 starts the decoding. If the packet degree is 1, then the neighbor block becomes the packet. The value of the recently recovered block is XOR-ed with its neighbors, and edges are removed. This process continues iteratively until done or no more degree 1 packets remain.

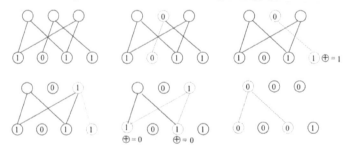

Figure 19.2.4 Decoding of LT codes.

The algorithm based on the degree distribution (§17.5) can be described by the following example. With a good degree distribution, the overhead factor can be made arbitrarily small. The ideal soliton distribution works the best. In fact, from this distribution Luby derived the robust solition distribution. For $n \approx 10,000$ Figure 17.8.2 shows that this distribution does improve upon the ideal soliton distribution, although it exhibits a spike.

Example 19.2.3. Suppose that each symbol is a 1-bit packet, and let the input consist of 5 symbols. Then 5 output symbols are generated by solving the following six equations: (1) $x_2 \oplus x_3 \oplus x_4 \oplus x_5 = 1$, (2) $x_1 \oplus x_3 \oplus x_4 = 0$, (3) $x_2 \oplus x_3 \oplus x_5 = 0$, (4) $x_3 \oplus x_4 = 1$, (5) $x_3 = 0$, and (6) $x_2 \oplus x_3 = 0$. Starting with equation (5) and solving all six equations we find that the input symbol is $\{x_1, x_2, x_3, x_4, x_5\} = \{10010\}$. ∎

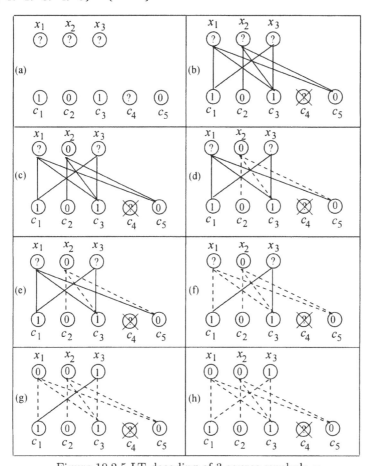

Figure 19.2.5 LT decoding of 3 source symbols. ∎

Example 19.2.4. (Decoding) Suppose that n decoded symbols have been received. The decoder finds an encoded symbol c_i that is connected to only one source symbol x_κ. If there is no such symbol, i.e., the encoded system has degree 1, the decoder stops and waits until more encoded symbols are received before proceeding to decoding. At this stage, set $x_\kappa = c_i$; set $c_j = c_j \oplus x_\kappa$, for all indices $j \neq i$, such that c_j is connected to x_κ; then remove all edges connected to x_κ. This process is repeated until all k source symbols are recovered. Consider this decoding process for $k = 3$ source symbols (here bits) x_1, x_2, x_3 from the 4 received encoded symbols c_1, c_2, c_3, c_5, while c_4 is lost. This example is a continuation of Figure 19.2.2 (Example 19.2.3), and the results of decoding are presented in Figure 19.2.5. Each step in Figure 19.2.5 is explained as follows:

Panel (a): Three unknown source symbols (bits) x_1, x_2, x_3 and 4 encoded bits c_1, c_2, c_3, c_5, where c_4 is lost.

Panel (b): Determine the degree and the source symbols associated to each received encoded symbol.

Panel (c): Find the encoded symbol with degree 1; the only one is c_2; set $x_2 = c_2 = 0$. Since x_2 is also connected to c_3 and c_5, set $c_3 = c_3 \oplus x_2$ and $c_5 = c_5 \oplus x_2$.

Panel (d): Remove all edges connected to x_2.

Panel (e): The next encoded symbol with degree 1 is c_3, which is connected to x_1, so set $x_1 = c_5$. Since x_1 is also connected to c_1 and c_3, set $c_1 = c_1 \oplus x_1$ and $c_3 = c_3 \oplus x_1$.

Panel (f): Remove all edges connected to x_1.

Panel (g): There are two symbols with degree 1: c_1 and c_3, and both are connected to x_3. Thus, x_3 can be decoded with either of them. Set x_3 equal to either c_1 or c_3.

Panel (h): All source symbols have been recovered and the decoding stops. ■

In general, let $\mathbb{N}$ denote the set of indices of the input symbols $x_i, i \in \mathbb{N}$, which have already been determined and eliminated. Let r be the ripple, which is the set of indices of input symbols that have been determined but not eliminated. Let the reduced set of symbols in x_i be denoted by $\tilde{x}_i$ and the reduced degree by $|\tilde{x}_i|$. Then the belief propagation algorithm (Chapter 14) can be modified for the LT decoding as follows:

```
N ← ∅
for i = 1, 2, . . . , j do
x̃ᵢ ← xᵢ
while N ≠ {1, 2, . . . , k} do
r ← ∪ᵢ:|x̃ᵢ|=1 x̃ᵢ
if r = ∅ then
return failure
choose an element k ∈ r; solve for xₖ
N ← N ∪ {k}
x̃ᵢ ← x̃ᵢ − {k} for i = 1, 2, . . . , j
return success  // all variables determined
```

In order to assure the success of this algorithm, we must choose the number

j of the output symbols and determine the degree of an input symbol (i.e., the distribution of $\mathbf{x} = \{x_\kappa\}$, $\kappa = 1, \ldots, k$) so that the ripple is never empty when some variables are yet to be determined. However, the value of $|\bar{x}_i|$ should not be 1 for too many equations at the same time, otherwise several equations might determine the same variables, thus presenting a wasteful situation. In other words, we should decode $x_1, \ldots, x_k$ after we have collected $c_1, \ldots, c_j$, where j is a bit greater than k, because if $j < k$, then the algorithm will always fail since in that case the equations will not have a unique solution. Since we need a set of k linearly independent equations, we must have $j \geq k$. This situation is similar to the coupon-collector problem, defined as follows: Suppose there are k coupons in a set, and at each trial a coupon is drawn at random. If we wish to collect all of the coupons in a set, the expected number of trials must be at least $k \ln k$. Now, returning to the distribution of $\mathbf{x}$, if the degree were always 1, we would have to collect approximately $k \ln k$ output symbols. Recall that receiving an output symbol of degree i is like receiving i coupons. Thus, the sum of degrees of all received symbols must be of order $O(k \ln k)$. Since the number of symbols being transmitted is k, the average degree must be approximately $\ln k$.

An output symbol contributes one of its input symbols to the ripple at the stage when all but one of its variables have been eliminated. Thus, for an output symbol of degree i, that stage is a random variable centered around $k \dfrac{i-1}{i+1}$. In view of the definition of ideal soliton distribution (§17.8.1), the expected number of output symbols of degree 1 is $k \frac{1}{k} = 1$, while for $i = 2, \ldots, k$, the expected number of output symbols of degree i is $\dfrac{k}{i(i-1)}$. Suppose that the number of output symbols of degree 1 is one, which is the expected number. Then one variable is determined and eliminated. The computation of the distribution of reduced degree is carried out as follows:

Expected number of output symbols of reduced degree 1 is

$$= k\rho(i)\left(\frac{2}{k}\right) = k\left(\frac{1}{1\cdot 2}\right)\left(\frac{2}{k}\right) = 1.$$

Expected number of output symbols of reduced degree i is

$$= k\rho(i)\left(1-\frac{i}{k}\right) + k\rho(i+1)\left(\frac{i+1}{k}\right) = \frac{k}{i(i-1)} + \frac{k}{(i+1)i}\left(\frac{i+1}{k}\right) = \frac{k-1}{i(i-1)}.$$

Thus, if at each step the actual number of input symbols joining the ripple is equal to the expected number, then the ideal soliton distribution exactly reproduces itself, and the size of the ripple is 1 at all times. However, in practice, this distribution fails because the actual number of additions to the ripple fluctuates around the expected number. The robust soliton distribution discussed in §17.8.3 was obtained by perturbing the ideal soliton distribution so that the expected size of the ripple is of order $O\left(\sqrt{k} \ln k\right)$ at all stages, and is such that $j = k + O\left(\sqrt{k}\,\ln^2(k/\delta)\right)$ with the average degree of an output symbol $\leq H(k) + O\left(\sqrt{k}\,\ln^2(k/\delta)\right)$, where $\delta > 0$ is small (failure probability) and $H(k)$ denotes the the harmonic sum up to k (see §17.8.2).

19.3 Performance

The operation of the LT encoder is described as follows: From the k message symbols it generates an infinite sequence of encoded bits, where each such bit is generated first by choosing a degree d at random according to the distribution used, and then choosing uniformly at random d distinct input bits, such that the encoded bit's value is the XOR-sum of these d bit values. Then the encoded bit is transmitted over a noisy channel.

Suppose the decoder receives a corrupted value of this bit. It is assumed that encoder and decoder are completely synchronized and share a common random number generator, which guarantees that the decoder knows which d bits are used to generate any given encoded bit, but not their values. Note that this kind of synchronization is easily achieved on the Internet because every packet has an uncorrupted packet number. Thus, the decoder can reconstruct the LT code's Tanner graph without error.

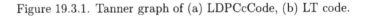

Figure 19.3.1. Tanner graph of (a) LDPCcCode, (b) LT code.

The decoder now runs a belief propagation algorithm on this Tanner graph. The message passing rules are similar to those of an LDPC decoder except for one important difference, The Tanner graph of an LDPC code contains only one kind of variable node (Figure 19.3.1(a)), but that of an LT code contains *two* kinds of variable nodes (Figure 19.3.1(b)): the message bit variable nodes that are not transmitted and as such have no channel evidence, and the encoded bit variable nodes that are transmitted over the channel. Hence, for large block lengths, the performance of such a system depends mostly on the degree distribution.

19.3.1. Error Floor Problem. In an LT code, the decoding starts after receiving a finite number n symbols from the sequence of infinitely many symbols sent out by the transmitter. At this stage, the decoder is trying to decode an (n, k) code with the rate $R = n/k \neq 0$. However, the decoding complexity depends on the rate R: it increases as R decreases. This means that the probability of decoding errors decreases. Palanki and Yedidia [2003] have studied the variation of bit error rate and word error rate against the code rate R. They have found that

(i) For the LT codes on a BSC with 11% bit flip probability ($p = 0.11$) the decoder buffers up k/R bits before it starts decoding using belief propagation, and the Shannon limit is $R^{-1} = 2$, which means that a little over $2k$ bits should suffice for reliable decoding in the limit for large k.

(ii) An LT code with $k = 10000$ using the RS(10000, 0.1, 0.5) distribution can achieve a word error rate of 10^{-2} at $R^{-1} = 2.5$ (or $n = 25000$). This data is not very impressive for most applications, although it may work out for certain applications.

(iii) Both the word error rate and the bit error rate bottom out into an error floor; thus, it is hardly possible to achieve a very small order error rate without very large overheads.

This error floor problem is found not only in LT codes that are generated using a robust soliton distribution, but also in codes generated using distributions optimized for the BEC (see Shokroallhi [2002]). Note that these optimized distributions were not designed for a direct application in LT codes. Although an advantage of such distributions is that the average number of edges per node remain constant with increasing k, resulting in the growth of decoding complexity of only $O(k)$, there is, however, a small fraction of message bit nodes that remains disconnected to any check node. Thus, even if k increases arbitrarily large, the bit error rate does not go to zero, as it theoretically should, and thus, the word error rate always remains 1.

Example 19.3.1. Palanki and Yedidia [*op. cit.*] have studied the performance of the following distribution on a BSC with $p = 0.11$:

$$\mu(x) = 0.007969x + 0.493570x^2 + 1.66220x^3 + 0.072646x^4 + 0.082558x^5$$
$$+ 0.056058x^8 + 0.037229x^9 + 0.055590x^{19}$$
$$+ 0.025023x^{63} + 0.0003125x^{66}. \tag{19.8}$$

They found that this distribution exhibits remarkably bad error floors even in the limiting case of infinite block length. However, note that these error floors exist not because of the presence of message bit variable nodes that are not connected to any check nodes; they are in part due to the fact that there exist variable nodes that are connected to a relatively small number of output nodes and, hence, they are unreliable. ∎

Luby [2002] used the robust soliton distribution, which is a modified form of the ideal soliton distribution, both defined in §17.8.2. Luby's analysis and simulation show that this distribution performs very well on the erasure channel, although the decoding complexity grows as $O(k \ln k)$. But it is shown by Shokrollahi [2002] that this growth in complexity is needed to achieve channel capacity, and the Raptor code, which is a slightly suboptimal code, is designed with decoding complexity of $O(k)$ (see Chapter 20).[1]

19.4 Comparison of LT Codes with Other Codes

The basic idea behind the full development of LT codes is the so-called 'digital fountain approach', introduced by Byers et al. [1998; 2002], which is similar to a universal erasure code. As seen before, the erasure codes are simply block codes with a fixed rate such that if k input symbols are used to generate $(n - k)$ redundant symbols for an encoding system of n symbols, then the code rate is $R = k/n$. For example, in the RS codes any k of the n encoding symbols is sufficient to recover the original k input symbols (see MacWilliams and Sloane [1977]), but more than k of the n encoding symbols are needed to recover the original k input symbols in the case of Tornado codes (see Luby [2001]). As we have seen before, the RS codes work efficiently only for small values of k and n, as only $k(n - k)|A|/2$ symbol operations, where $|A|$ is the size of the finite field used, are needed in the implementation of these codes to produce the n encoding as well as decoding systems, although the quadratic time implementations for encoding and decoding are faster in practice for given values of k and n.

Tornado codes are block codes with linear n encoding and decoding times (Byers et al. [2002]). They are similar to LT codes in the restricted sense that they both use a similar rule to recover the data and both have almost similar degree distribution on input symbols, although degree distribution used for Tornado code is not applicable to LT codes. The difference lies in the kind of distribution in these two kind of codes: For Tornado codes, this distribution is like the soliton distribution and that on the first layer of redundant symbols approximates the Poisson distribution. The soliton distribution on input symbols cannot be applied to LT codes because such a distribution cannot be the resulting degree distribution on input symbols when encoding symbols are generated independently irrespective of the choice of distribution on neighbors of encoding symbols. Alternatively, the distributions on Tornado codes could be reversed so that the Poisson distribution is used on the input symbols and the soliton distribution is used on the redundant

---

[1] Raptor codes are currently being used by Digital Fountain, a Silicon Valley-based company, to provide fast and reliable transfer of large files over the Internet.

symbols, as is done in LT codes. However, the distribution so induced on the redundant symbols by the missing input symbols in the Tornado code graph does not match with the soliton distribution, and so any application of the recovery rule to the induced distributions does not produce any advantage. However, LT codes possess more application advantages over Tornado codes. Let $c = n/k$ denote the constant stretch factor of the Tornado code design. Then for fixed n and k, the Tornado code will produce n encoding symbols and will fail to produce any more such symbols even if the need arises. On the other hand, the encoder can generate on demand as many encoding symbols as needed.

The analysis of LT codes is also different from that of Tornado codes as described in Luby et al. [1998; 2001]. In fact, where Tornado codes are applicable to graphs with constant maximum degree, the LT codes use graphs of logarithmic density, and hence, Tornado code processes do not apply to LT codes. The process and analysis of LT codes simulate throwing balls into bins, and follow the first principles of probability to control and capture the behavior of the data recovery process.

Tornado codes use a cascading sequence of bipartite graphs between several layers of symbols, such that the input symbols are at the first layer and the redundant symbols are at each subsequent layer. This cascading process requires that either the same graph structure be constructed in advance at both the encoder and decoder, or a prior graph structure be constructed at the encoder, which is then communicated to the decoder. In either construction the process is very cumbersome because the graph structure size is proportional to $n = ck$. In contrast, in LT codes the degree distributions based on the data length are easy to compute, and no other prior processing is needed for the encoder and the decoder.

The decoding for Tornado codes depends on having a sufficient number of distinct encoding symbols from the set of n encoding symbols. In many transmission situations, the total number of encoding symbols transmitted from an encoder to the decoder is several times larger than the number of encoding symbols that can be generated by the Tornado encoder. In such cases, a good policy is to randomly select an encoding symbol each time a symbol is transmitted. Then it will be easy to see that the average number of encoding symbols that a decoder needs to receive the required k distinct encoding symbols out of the $n = ck$ encoding symbols is $ck \ln \dfrac{c}{c-1}$.

Example 19.4.1. For $c = 2$, the decoder needs to receive $2k \ln 2 \approx 1.386k$ encoding symbols on average in order to receive k distinct encoding symbols. However, we must have $c > 10$ in order to receive $1.053k$ encoding symbols, and $c > 20$ for receiving $1.025k$ encoding symbols. ∎

This example shows that such values of c make the Tornado codes much less useful because the encoder and decoder memory usage and the decoding time is proportional to c times the data length. On the other hand, the encoder and decoder memory usage for LT codes is proportional to the data length and the decoding time depends only on the data length irrespective of the number of encoding symbols (small or large) generated and transmitted by the encoder.

Both RS and Tornado codes are systematic, that is, all the input (data) symbols are directly included among the encoding symbols, but LT codes are not systematic. Since the LT codes belong to the first class of universal Fountain codes, they are based on bipartite graphs, that is, graphs with two disjoint sets of nodes such that two nodes in the same set are not connected by an edge.

20

Raptor Codes

20.1 Evolution of Raptor Codes

The second class of Fountain codes is the Raptor code ('raptor' derived from *rapid tornado*), which was necessitated due to the fact that in many applications a universal Fountain code must be constructed that has a constant average weight of an output symbol and operates on a fast decoding algorithm. The basic idea behind Raptor codes is a pre-coding of the input symbols prior to the application of an appropriate LT code. We will design a class of universal Raptor codes with linear time encoder and decoder for which the probability of decoding failure converges to zero polynomially fast as the number of input symbols increases. With proper bounds on the error probability, a finite length Raptor code can be constructed with low error probability.

LT and Raptor codes are not systematic, which means that the input symbols are not nessarily reproduced among the output symbols. This is a disadvantage for these codes, since the idea of transmitting the input symbols prior to the output symbols are produced by the code seems defective as it does not guarantee a foolproof decoding from any subset of received output symbols. Thus, it is very desirable to develop an efficient version of systematic Raptor codes, which is presented below in §20.1.7. In subsequent analysis the input symbols $\mathbf{x} = \{x_i\}$ are finitely many sequences composed of codewords $\{c_i\}$. If a code is not systematic, the input symbols are not always reproduced by the encoder, and the performance of the code remains inferior.

Consider a linear code of length n and dimension k. A Raptor code is specified by the parameters $(k, \mathcal{C}, \Omega(x))$, where $\mathcal{C}$ is the (n, k) erasure correcting block code, called the *pre-code*, and $\Omega(x)$ is the generator polynomial of the degree distribution of the LT code such that $\Omega(x) = \sum_{i=1}^{k} \Omega_i x^i$, where Ω_i is the probability that the degree of the output node is 1. The pre-code $\mathcal{C}$ need not be systematic, although systematic codes are generally used. Such a pre-code with k input symbols is used to construct the codewords in $\mathcal{C}$ consisting of n intermediate symbols. The ouput symbols are those generated by the LT code from the n intermediate symbols, as shown in Figure 20.1.1, in which the input symbols are appended by redundant symbols (marked RN) in the case of systematic pre-code $\mathcal{C}$.

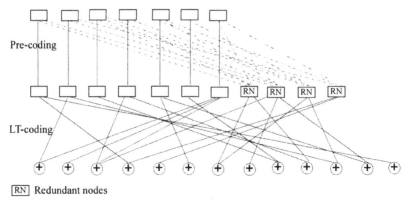

RN Redundant nodes

Figure 20.1.1 Pre-coding of a Raptor code.

As noticed before, LT codes are the first example of a Raptor code with parameters $\left(k, \mathbb{F}_2^k, \Omega(x)\right)$, where $\mathbb{F}_2^k$ is a code of block length k, and the distribution polynomial $\Omega(x)$ is a special output distribution with an overhead of order $O(\log^2(k)/\sqrt{k})$. LT codes have no pre-coding, and instead use a very elaborate degree distribution.

20.1.1 Pre-Code Only Raptor Codes. If we look at the spectrum of Raptor codes, on one end there are the LT codes and on the other end there are *pre-code only* (PCO) Raptor codes. A Raptor code with parameters $(k, \mathcal{C}, x)$ is called a PCO Raptor code with pre-code $\mathcal{C}$. In this code, the k input symbols are encoded through $\mathcal{C}$ and generate the n intermediate symbols with a fixed linear output distribution $\Omega(x) = x$. The value of every output symbol is that of a randomly and uniformly chosen input symbol. The decoding algorithm is trivial: it collects a predetermined number m of output symbols, which in turn determine the value of, say, l intermediate symbols. Finally, the decoding algorithm for the pre-code is applied to these recovered intermediate symbols, thus recovering the values of the input symbols.

20.1.2 Shokrollahi's Condition. An alternative method to construct an effective distribution, subject to the requirement that the expected size of the ripple after some step j must be at least $c\sqrt{k-j}$, where $c > 0$ is a constant, has been proposed by Shokrollahi [2002]. The basic concept is the observation that random fluctuations of additions to the ripple are not likely to produce a ripple of size zero when the expectation is so large. Define a generating function Ω by

$$\Omega(x) = \sum_{d=1}^{k} \rho(d)x^d, \quad x \in [0, 1-\delta]. \tag{20.1.1}$$

In fact, $\Omega(x)$ is the degree distribution. The probability of a given input variable is determined after j symbols have been decoded and eliminated. This is the probability that, for some output symbols, the input variable is among the elements of the input symbols such that all the other input symbols have already been determined at step j. Obviously, given the number k of input symbols, this probability for an output symbol of degree d is equal to $d\frac{1}{k}\left(\frac{j}{k}\right)^{d-1}$. If d is not known, then the average over the possible degrees of expectations is

$$\sum_{d=1}^{k} d\rho(d)\frac{1}{k}\left(\frac{j}{k}\right)^{d-1} = \frac{1}{k}\Psi\left(\frac{j}{k}\right), \tag{20.1.2}$$

where $\Psi(x) = \Omega'(x) = \sum_{d=1}^{k} d\rho(d)x^{d-1}$ is the derivative of $\Omega(x)$. Then the probability that the input variable has not been determined by the step j is $\left(1 - \frac{1}{k}\Psi\left(\frac{j}{k}\right)\right)^{(1+\varepsilon)k} \approx e^{-(1+\varepsilon)\Psi(j/k)}$, where $(1+\varepsilon)$ is the overhead. Thus, the expected number of input variables at the step j is $k\left(1 - e^{-(1+\varepsilon)\Psi(j/k)}\right)$ and the expected size of the ripple is $j = k\left(1 - e^{-(1+\varepsilon)\Psi(j/k)}\right) - j$. Since $k\left(1 - e^{-(1+\varepsilon)\Psi(j/k)}\right) - j \geq \sqrt{k-j}$, we set $x = j/k$ and get $1 - x - e^{-(1+\varepsilon)\Psi(x)} \geq c\sqrt{\frac{1-x}{k}}$. If this condition is satisfied, then the expected size of the input ripple $k\left(1 - x - e^{(1+\varepsilon)\Psi(x)}\right) \geq c\sqrt{(1-x)k}$, or $e^{-(1+\varepsilon)\Psi(x)} \leq 1 - x - c\sqrt{\frac{1-x}{k}}$, which gives the condition

$$\Psi(x) \geq -\frac{1}{1+\varepsilon}\ln\left(1 - x - c\sqrt{\frac{1-x}{k}}\right). \tag{20.1.3}$$

The inequality in (20.1.3) can only hold if $1 - x - c\sqrt{\frac{1-x}{k}} > 0$, or if

$$x = \frac{j}{k} < 1 - \frac{c^2}{k}. \tag{20.1.4}$$

For any $x \in [0, 1 - c^2/k]$, Shokrollahi's condition (20.1.3) is a linear inequality for the quantities $\rho(d)$, and it can be used to set up linear programming to minimize $\sum_{d=1}^{k} d\rho(d)$, which is the expected degree of an output symbol, such that the solution of this linear program for specified values of k, j and c will give us an efficient degree distribution on the input symbols.

Example 20.1.1. Since the output node is of degree d, the probability that ν is a neighbor of that output node is $\sum_d \Omega_d\left(1 - \frac{d}{k}\right) = 1 - a/k$, where

$a = \Psi(1)$ is the average degree of an output node, and k is number of input symbols. Then the average degree of an input symbol α is defined as $\alpha = an/k$. Shokrollahi [2004] has provided the values of ε and α for different values of k, as presented in Table 20.1.1.

Table 20.1.1 Values of ε and α for Some Values of k

k	ε	α	k	ε	α
65536	0.038	5.87	100000	0.028	5.85
80000	0.035	5.91	120000	0.020	5.83

Thus, for example, for $k = 100000$, the encoding introduces a redundancy (overhead) of $1 + \varepsilon = 1.028$, and the average degree of an input symbol $\alpha = 5.88$. This is much less than $\ln k \approx 11.5129$, which would be the average degree needed by the coupon collector's algorithm, which is discussed in §20.3. Thus, the decoder is very likely to stall, leaving about 500 symbols undetermined. To decode the remaining symbols we require some additional redundancy with an LDPC code that will be able to recover the missing packets. ∎

A listplot of the condition (20.1.3) for $\varepsilon = 0.28$, $c = 0.2$, $k = 100000$, and $x = 0.0(0.02)0.98$ is shown in Figure 20.1.2.

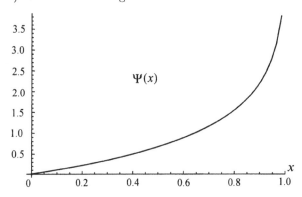

Figure 20.1.2 Listplot of Shokrollahi's condition (20.1.3).

Hence, we are now on the verge of Raptor codes that are a combination of LT and LDPC codes. The function $\Psi(x)$ is defined on and above the plot points in Figure 20/1/2. In simple terms, Shokrollahi's condition states that first, we transmit a file of k symbols by first pre-coding that file into $\hat{k} \approx k/(1 - \hat{R})$ symbols with a robust outer error-correcting code, where $\hat{R}$ is the erasure rate defined by $\hat{R} = e^{-d}$ (d being the Hamming distance), so that the original file is slightly enlarged; for example, we have $\hat{R} \approx 5\%$ for $d = 3$, and $\hat{R} \approx 1.8\%$ for $d = 4$; and second, we transmit this enlarged file by using a weak LT code. Once this enlarged file containing a little over k symbols has been received, we can recover $(1 - \hat{R})\hat{k}$ of the pre-coded symbols, approximately k symbols. Finally, we use the outer code to recover the original file, as in Figure 20.1.3.

$k = 2^4 = 16$ symbols (packets)

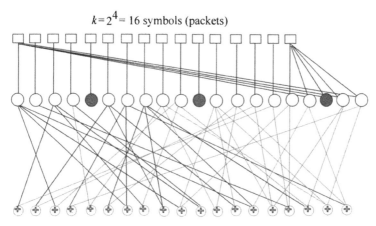

Figure 20.1.3 Pre-coding of a Raptor code.

MacKay [2005] has shown that while the LT code recovers, say, $k = 10000$ symbols within a number of received symbols $N = 110000$, the weak LT code usually recovers 8000 symbols within a received number of 9250. The lost symbols are gray-colored in Figure 20.1.3.

The performance of a PCO Raptor code depends on the performance of its pre-code. In fact, Shokrollahi [2006] has shown that for a block code $\mathcal{C}$ such that if $k\eta$ arithmetic operations ($\eta > 0$) are sufficient for encoding k-length input symbols (which are vectors) and $k\gamma$ arithmetic operations ($\gamma > 0$) suffice for decoding $\mathcal{C}$ with high probability over a binary erasure channel (BEC) with erasure probability $1 - R(1 + \varepsilon)$ for some $\varepsilon > 0$, where $R = k/n$ is the rate of the pre-code $\mathcal{C}$, then the PCO Raptor code with pre-code $\mathcal{C}$ has space consumption $1/R$, overhead $[-\ln(1 - R(1 + \varepsilon))/R - 1]$, encoding cost η, and the decoding cost γ for its decoding algorithm. More information on irregular LDPC codes and fast encoding algorithms for them is available in Richardson et al. [2001a; 2001b; 2001c].

20.1.3 Asymptotically Good Raptor Codes. We have noticed that for LT codes the overhead is close to 0 and space consumption close to 1, but the decoding cost grows with k. For PCO Raptor codes the decoding cost may become almost constant, yet its overhead becomes greater than 0 and space comsumption away from 1; thus, convergence to 0 amounts to an increase in cost as k increases. Shokroallahi [2004; 2006] has designed Raptor codes between these two extremes, with encoding and decoding algorithms of constant cost and overhead arbitrarily close to 0 and space consumption close to 1. The design of such a code is based on an appropriate output distribution $\Omega(x)$ and an appropriate pre-code $\mathcal{C}$.

First, we define the output distribution: Let $\varepsilon > 0$ be a real number, set $D = \lceil 4(1 + \varepsilon)/\varepsilon \rceil$, and define

$$\Omega_D(x) = \frac{1}{\mu + 1}\left(\mu x + \sum_{i=2}^{D} \frac{x^i}{(i-1)i} + \frac{x^{D+1}}{D}\right), \qquad (20.1.5)$$

where $\mu = (\varepsilon/2) + (\varepsilon/2)^2$. Then there exists a real number $c > 0$ that depends on ε such that, with an error probability of $\leq e^{-cn}$, any set of $(1 + \varepsilon/2)n + 1$ output symbols of the LT code with parameters $(n, \Omega_D(x))$ is sufficient to recover at least $(1 - \delta)n$ input symbols using belief propagation (BP) decoding, described in §16.6, where $\delta = (\varepsilon/4)/(1 + \varepsilon)$. Proof of this result is provided below in §20.1.4.

The above result is the basis for the construction of asymptotically good Raptor codes. Such codes are constructed using LT codes and suitable pre-codes described in the above result. Assume that for every n there is a linear code $\mathcal{C}_n$ of block length n such that (i) the rate R of $\mathcal{C}_n$ is $(1 + \varepsilon/2)/(1 + \varepsilon)$, and (ii) the BP decoder can decode $\mathcal{C}_n$ on a BEC with erasure probabiil- ity $\delta = (\varepsilon/4)/(1 + \varepsilon) = (1 - R)/2$, which involves arithmetic operations of the order $O(n \log(1 + \varepsilon))$, which can be weakened to $O(n)$ because it is not necessary for the code to achieve capacity. Examples of such codes are Tor- nado codes (§18.2), right-regular codes (Shokrollahi [1999]), certain types of Repeat-Accumulate codes, and other capacity-achieving codes (see Oswald and Shokrollahi [2002]).

The asymptotically good Raptor code with parameters $(k, \mathcal{C}_n, \Omega_D(x))$ has space consumption $1/R$, overhead ε, and a cost of $O(\log(1 + \varepsilon))$ under the BP decoding of both the pre-code and the LT code. An outline of the proof is based on edge-degree distribution and AND-OR tree evaluation, which is described in Luby et al. [1998]. A simple analysis of the probability that a tree consisting of alternating layers of AND and OR gates is equal to 1 yields an intuitive way to obtain the criterion on edge distributions that are required for decoding all lost or missing symbols when a fixed fraction of them is erased independently at random.

BP decoding is used at different steps as follows: At each step, all released output nodes, which are connected to one unrecovered input node, recover their incident input nodes, and then these recovered input nodes update their incident output nodes and then are removed. For example, consider an edge (a, b) chosen uniformly at random from the original Tanner graph. Let $\omega(x)$ denote the generator polynomial of the output edge distribution with the probability that the output node b is released at some step where the edge (a, b) has already been decoded and a portion x input nodes has been recov- ered. Similarly, let $\iota(x)$ denote the generator polynomial of the input edge distribution with the probability that the input node a is recovered at some step where edge (a, b) has not yet been deleted and a portion $1 - x$ of output nodes has already been released.

Let $\mathfrak{p}_m$ denote the probability that a randomly chosen edge has carried a symbol from its incident output symbol at the step m. Assuming that none of the input nodes are initially known, this probability is recursively defined for the adjacent step $m + 1$ by

$$\mathfrak{p}_{m+1} = \omega\big(1 - \iota(1 - \mathfrak{p}_m)\big). \tag{20.1.6}$$

Similarly, the recursive formula for the probability $\mathfrak{r}_m$ for a randomly chosen edge that has not carried a symbol from its incident input symbol at the step m is

$$\mathfrak{r}_{m+1} = \iota\big(1 - \omega(1 - \mathfrak{r}_m)\big). \tag{20.1.7}$$

This formula suggests that Shokorallahi's condition,

$$\iota\big(1 - \omega(1 - x)\big) < x, \; x \in [\delta, 1], \tag{20.1.8}$$

is necessary for reliable recovery of a fraction $(1 - \delta)n$ of the intermediate symbols.

20.1.4 Proof of Asymptotic Performance. The output degree distribution $\Omega(x)$, defined by (20.1.5), is a modification of the ideal soliton distribution (Definition 17.7.1, and §17.8.2). To show that the Raptor code with parameters $(k, \mathcal{C}_n, \Omega_D(x))$ has space consumption $1/R$, overhead $1 + \varepsilon$, and encoding and decoding costs of the order $O(\log(1/\varepsilon))$, notice that any set of $(1 + \varepsilon/2)/n + 1$ output symbols of the LT code with parameters $(n, \Omega_D(x))$ is sufficient to recover reliably at least $1 - \delta$ intermediate symbols using the BP decoding. This is accomplished by showing that the input edge distribution $\iota(x)$ and the output edge distribution $\omega(x)$ induced by $\Omega_D(x)$ satisfy the condition (20.1.8), and thus recover $(1 - \delta)n$ of the intermediate symbols. As soon as this recovery is achieved, the decoder for $\mathcal{C}_n$ is able to recover the k input symbols. Thus, the overhead is $n(1 + \varepsilon/2)/k = 1 + \varepsilon$. Since the encoding and decoding costs of the LT code are proportional to the average degree of the distribution $\Omega_D(x)$, and since this average degree is $\Omega'_D(x) = 1 + H(D)/(1 + \mu) = \ln(1/\varepsilon) + \alpha + O(\varepsilon)$, where $H(D)$ is the finite harmonic sum up to D, defined in Footnote 3, §17.8.2, and $0 < \alpha < 1 + \gamma + \ln(9)$, and γ is Euler's constant, the encoding and decoding cost of $\mathcal{C}_n$ is $O(\log(1/\varepsilon))$. This proves the result.

20.1.5 Finite-Length Raptor Codes. The asymptotic analysis presented in the previous section is not very useful in practice. The bounds on error probability establish that they are not sharp for finite-length Raptor codes and do not perform well in practical applications, as they rely on a martingale type argument of a fair play. The same is valid for the pre-code. Hence, we must develop a different kind of error analysis for Raptor codes of finite lengths with BP decoding. According to Luby et al. [2002], consider an input symbol at time T if at least one neighbor of this input symbol becomes of reduced degree 1 after T input symbols have been recovered. Define the *input ripple* at time T as the set of all input symbols that have been released at time T. For the finite-length Raptor codes it is necessary "to keep the input ripple large during as large a fraction of the decoding process as possible."

This leads to the development of an LT code component by linear programming on a heuristically obtained design problem, as follows: Consider running the BP decoding algorithm on $k(1 + \varepsilon)$ output symbols. Using Eq (20.1.6), we can obtain a recursion on the probability u_m that an input symbol is unrecovered at step m:

$$u_{m+1} = e^{(1+\varepsilon)\Omega'(u_m)}. \tag{20.1.9}$$

This recursion shows that if an expected x-fraction of input symbols has been already recovered at some step of the algorithm, then in the next step the expected fraction of the input ripple will be $1 - e^{(1+\varepsilon)\Omega'(x)}$.

The degree distribution $\Omega(x)$ is chosen to minimize the *objective function* $\Psi(1) = \Omega'(1)$, subject to the constraint that the expected number of input symbols recovered during each step is larger by a constant factor $c > 0$ than $\sqrt{(1-x)k}$, which is the square root of the number of unrecovered input symbols. Thus, the design problem reduces to the following: Given ε and δ, and given the number k of input symbols, find a degree distribution $\Omega(x)$ such that (see §20.2)

$$1 - x - e^{(1+\varepsilon)\Omega'(x)} \geq c\sqrt{\frac{1-x}{k}} \quad \text{for } x \in [0, 1-\delta]. \tag{20.1.10}$$

This inequality can be solved provided $\delta > c/\sqrt{k}$. Thus, we discretize the interval $[0, 1 - \delta]$ and make sure that the inequality (20.1.10) holds at the discretization points. This will yield linear inequalities in the unknown coefficients of $\Omega(x)$. The objective function $\Omega'(1)$ is also linear in the unknown coefficients of $\Omega(x)$. The final outcome of this procedure will be a degree distribution with the minimum possible mean degree. A table of values Ω_d for some optimized degree distribution for $\delta = 0.01$ and four different values of $k = 65536, 80000, 100000$, and 120000 is available in Shokrollahi [2004; 2006: Table I], where it is noted that the values of $\Omega_d \approx \dfrac{1}{d(d-1)}$ for small values of $d > 1$, which is the same as the soliton distribution $\rho(d)$, $d \geq 2$ (§17.8.3).

20.1.6 Design and Error Analysis of Pre-Code. The pre-coding operation with a linear code is used to transfer data reliably. A pre-code, which is sometimes called an outer-code, may be a concatenation of multiple codes; for example, it may be a Hamming code with an LDPC code, or a fixed-rate erasure code that has a usually high rate. The choice of a Tornado code as a pre-code for Raptor codes seems very appealing as it helps prove theoretical

results, but in practice it turns out to be a poor choice. A special class of
LDPC codes as a choice for pre-code is constructed as follows: Let a bipartite graph have n data (message) nodes and r parity-check nodes, and let the
linear code associated with this graph be a linear block code of length n. The
n nodes define the coordinate positions of a codeword as $c_1, c_2, \ldots, c_n$. Thus,
the codewords are vectors of length n over the base field such that for every
check node the sum of its neighbors among the message nodes is 0 (Figure
20.1.4). The inner code, which is a form of LT codes, takes the result of pre-
coding operation and generates a sequence of encoding symbols. In decoding
raptor codes, in the concatenated case the inner code is decoded using the
belief propagation algorithm, but in the combined case the pre-code and the
inner code are treated as one set of simultaneous equations, usually solved by
the Gauss elimination method.

As Luby et al. [2001] have shown, the BP decoding of LDPC codes over
an erasure channel is very similar to the BP decoding of LT codes. According
to Di et al. [2002], this decoding algorithm is successful iff the graph induced
by the erased message locations does not contain a stopping set.[2] Thus, in
the case of erasure decoding for LDPC codes, it is necessary to compute the
probability that the graph generated by the erased locations has a stopping
set of size s for each value of s.

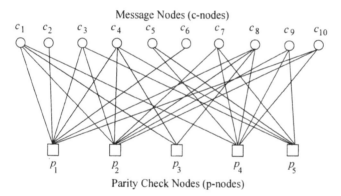

Figure 20.1.4 An LDPC code for the pre-code.

The LDPC codes as suitable candidates for pre-codes are constructed from
a node degree distribution $\Lambda(x) = \sum_d \Lambda_d x^d$, as follows: For each message
node, a degree d is chosen independently at random from the distribution,
and then d random check nodes are chosen to be the neighbors of the message

---

[2] A stopping set is a set of message nodes such that their induced graph satisfies the
condition that all the check nodes have degree greater than 1. In Figure 20.1.4 the message
nodes 1, 2, 4, 5 generate a stopping set of size 4. Another property of stopping sets is that
the union of two stopping sets is again a stopping set. Thus, a bipartite graph contains a
unique maximal stopping set, which may be the empty set.

node. This defines an ensemble of graphs, which is denoted by $\mathbf{P}\left(\Lambda(x), n, r\right)$. Let $r \in \mathbb{N}$, and $z, \omega \in \mathbb{Z}$, with $d \geq 1$. Define $A_n(z, \omega)$ recursively by

$$A_0(r, 0) = 1, \quad A_0(z, \omega) = 0 \quad \text{for } (z, \omega) \neq (r, 0),$$

$$A_{n+1}(z, \omega) = \sum_{l, k} A_n(l, k) \times$$

$$\times \sum_d \Lambda_d \frac{\binom{l}{l - z}\binom{k}{k + l - z - \omega}\binom{r - l - k}{d - k - 2l + 2z + \omega}}{\binom{r}{d}} \quad \text{for } n \geq 0.$$

$$(20.1.11)$$

Then the probability that a bipartite graph in the ensemble $\mathbf{P}\left(\Lambda(x), n, r\right)$ has a maximal stopping set of size s is at most

$$\binom{n}{s} \sum_{s=0}^{r} A_s(z, \omega) \left(1 - \sum_d \Lambda_d \frac{\binom{r - z}{d}}{\binom{r}{d}}\right)^{n - z}. \qquad (20.1.12)$$

A proof of this result is given in Shokrollahi [2002].

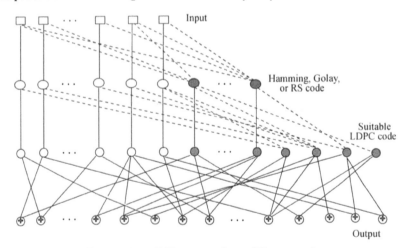

Figure 20.1.5 Different versions of Raptor codes.

The decoding error probability of a Raptor code with parameters $(k, \mathcal{C}, \Omega(x))$ is estimated by the above finite-length analysis of the corresponding LT code and of the pre-code $\mathcal{C}$ using any code $\mathcal{C}$ for which the decoding error probability is already known. For example, $\mathcal{C}$ can be chosen from $\mathbf{P}\left(\Lambda(x), n, r\right)$. Let $\mathfrak{p}_l^{\mathrm{LT}}$ denote the probability that the LT decoder fails after recovering l intermediate symbols, and let $\mathfrak{p}_j^{\mathcal{C}}$ denote the probability that the pre-code $\mathcal{C}$ fails to decode j randomly chosen erasures. Then the probability that the k input symbols cannot be recovered from the $k(1+\varepsilon)$ output symbols is $\sum_{l=0}^{n} \mathfrak{p}_l^{\mathrm{LT}} \mathfrak{p}_{n-j}^{\mathcal{C}}$.

A presentation of different versions of Raptor codes is given in Figure 20.1.5, in which the pre-coding is accomplished in multiple stages: the first stage is Hamming, Golay, or RS coding, and the second a suitable LDPC coding. The choice of intermediate codes is always dependent on the stochastic process and achievement of desired error probabilities associated with different codes. Shokrollahi [2003] uses irregular LDPC code as a suitable choice.

20.1.7 Systematic Raptor Codes. Assuming that we have a Raptor code with parameters $(k, \mathcal{C}, \Omega(x))$ that has a reliable decoding algorithm with overhead ε, let n denote the block length of the pre-code $\mathcal{C}$. A systematic version of such a Raptor code is as follows: We design an encoding algorithm that takes k input symbols $x_1, x_2, \ldots, x_k$ and produces a set $\{i_1, i_2, \ldots, i_k\}$ with distinct k indices between 1 and $k(1+\varepsilon)$, and an unbounded string $z_1, z_2, \ldots$ of output symbols such that $z_{i_1} = x_1, \ldots, z_{i_k} = x_k$, and such that the output symbols can be computed efficiently. Hence, the set $\{i_1, i_2, \ldots, i_k\}$ will be said to designate the *systematic locations*, and the associated output symbols as *systematic output symbols*, while the other remaining output symbols as *nonsystematic output symbols*. The algorithm is as follows:

(i) Compute the systematic locations $i_1, i_2, \ldots, i_k$. This will also yield an invertible binary $k \times k$ matrix $\mathbf{R}$. These data are computed by sampling $k(1+\varepsilon)$ times the distribution $\Omega(x)$ independently to obtain vectors $v_1, v_2, \ldots, v_{k(1+\varepsilon)}$ and applying a modification of the decoding algorithm to these vectors.

(ii) The matrix $\mathbf{R}$ is the product of the matrix $\mathbf{A}$, whose rows are $v_1, \ldots, v_{k(1+\varepsilon)}$, and a generator matrix $\mathbf{G}$ of the pre-code $\mathcal{C}$. The first $k(1+\varepsilon)$ output symbols of the systematic encoder are determined by these sampled vectors.

(iii) Next, to encode the input symbols $x_1, x_2, \ldots, x_k$, use $\mathbf{R}^{-1}$ to transform these into intermediate symbols $y_1, y_2, \ldots, y_k$.

(iv) Apply the Raptor code with parameters $(k, \mathcal{C}, \Omega(x))$ to the intermediate symbols. This will yield the first $k(1+\varepsilon)$ symbols using the previously sampled vectors $v_1, v_2, \ldots, v_{k(1+\varepsilon)}$. In this process the output symbols associated with the systematic locations must coincide with the input symbols.

(v) For decoding the systematic Raptor code, first apply a decoding step for the original Raptor code to obtain the intermediate symbols $y_1, y_2, \ldots, y_k$, and then use the matrix $\mathbf{R}$ to transform these intermediate symbols back to the input symbols $x_1, x_2, \ldots, x_k$.

In terms of matrix algebra, the generator matrix $\mathbf{G}$ is an $n \times k$ matrix. Let $\mathbf{x}$ denote the vector $\{x_1, x_2, \ldots, x_k\}$. Then the pre-coding step of the Raptor code corresponds to the multiplication $\mathbf{u}^T = \mathbf{G} \cdot \mathbf{x}^T$. Because each output symbol of the Raptor code is obtained by independently sampling from the distribution $\Omega(x)$ to obtain a row vector $\mathbf{v} \in F_2^n$, the output symbols are computed as the dot product $\mathbf{v} \cdot \mathbf{u}^T$, where $\mathbf{v}$ is the vector associated with the output symbols. For any given set of N output symbols, there corresponds a binary $N \times n$ matrix $\mathbf{S}$ whose rows consist of the vectors associated with the output symbols; thus,

$$\mathbf{S} \cdot \mathbf{G} \cdot \mathbf{z}^T = \mathbf{z}^T, \tag{20.1.13}$$

where $\mathbf{z} = \{z_1, z_2, \dots, z_k\}$ is the column vector consisting of the output symbols. By solving Eq (20.1.13) for the vector $\mathbf{x}$, we complete the decoding of the Raptor code.

20.2 Importance Sampling

Importance sampling (IS) is used to estimate the average number of packets (Tirronen [2006]). It is defined as the expectation of a function $h(\mathbf{x})$ by

$$E\left[h(\mathbf{x})\right] = \int h(\mathbf{x}) p(\mathbf{x}) \, d\mathbf{x} = \int h(\mathbf{x}) \frac{p(\mathbf{x})}{g(\mathbf{x})} g(\mathbf{x}) \, d\mathbf{x}, \tag{20.2.1}$$

where $p(\mathbf{x})$ and $g(\mathbf{x})$ are probability distributions. An estimate $\tilde{h}$ for expectation h can be calculated by drawing samples $\tilde{\mathbf{x}}^{(i)}$ from $g(\mathbf{x})$ so that

$$\tilde{h} = \frac{1}{K} \sum_{i=1}^{K} h(\tilde{\mathbf{x}}^{(i)} = \sum_{i=1}^{K} w(\tilde{\mathbf{x}})^{(i)} h(\tilde{\mathbf{x}})^{(i)}, \tag{20.2.2}$$

where $w(\mathbf{x}) = p(\mathbf{x})/g(\mathbf{x})$ defines the importance ratios.

IS-based optimization of LT codes is possible because it allows the calculation of an expectation using a different distribution instead of the original one. Tirronen [2006] has proposed an optimization method for LT codes using IS, where the estimate for an average number of packets needed for decoding is defined by

$$\widehat{R}(\mathbf{q}) = \frac{1}{m} \sum_{k=1}^{m} R_k \prod_i \left(\frac{q_i}{p_i}\right)^{n_i^{(k)}}, \tag{20.2.3}$$

where $\mathbf{p}$ and $\mathbf{q}$ denote degree distributions, R_k is the number of packets needed for decoding, and $n_i^{(k)}$ denotes the number of packets of degree i. Tirronen's iterative algorithm optimizes the estimate $\tilde{h}$ using the method of steepest descent, also known as the gradient method, where the accuracy of the estimate is controlled through the gradient by generating samples until the variance of the gradient is small enough. This algorithm takes some degree distribution as input and generates a better distribution, if possible. The implementation of this algorithm has been tested for ≤ 100 to keep the simulation and optimization times within reasonable limits. The results are shown in Figure 20.2.1, where point distributions (or probabilities for each degree) are optimized for $n = 1, 2, \dots, 10$, and the plot shows average overhead percentages from 100,000 simulations with the optimized distributions (Tirronen [2006]).

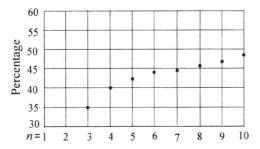

Figure 20.2.1 IS-based optimization of LT codes.

The results with $n = 100$ and $n = 1000$ for robust soliton distribution, presented in Table 20.2.1 and 20.2.2, respectively, show the probabilities for degree 1, 2 optimized; 1, 2 with spike at 50 (or 100); ideal soliton; and robust soliton distributions.

Table 20.2.1 Optimized Distributions for $n = 100$

Distribution	Avg	Std
Degree 1 and 2 optimized	125	13
Degree 1 and 2 with spike at 50	124	10
Ideal soliton	170	70
Robust soliton	130	13

Table 20.2.2 Optimized Distributions for $n = 1000$

Distribution	Avg	Std
Degree 1 and 2 optimized	1121	37
Degree 1 and 2 with spike at 100	1130	84
Robust soliton	1124	57

It is obvious that the optimized results for $n = 100$ perform equally well, although with a note of caution that the applications should be carefully designed to avoid inefficiencies. However, the unanswered question remains as to what the best form of degree distribution is, and whether the spike is really needed.

20.3 Coupon Collector's Algorithm

The coupon collector's problem is a popular topic in discrete mathematics and probability theory. The classical version of this problem is as follows:

Suppose that there are n different types of coupons, and that the collector gets every coupon with probability $1/n$ each day. What is the number of days that he needs to collect in order to have at least one of each type?

For $i = 1, 2, \ldots, n$, let x_i^n be the number of waiting days until another new type is obtained after $i - 1$ different types have been collected. Let us define $y^n = x_1^n + \cdots + x_n^n$ as the number of days the collector needs to collect in order to have at least one of the n types. The expected number of days collecting all n coupons is given by (see Durrett [1996]; Ross [2002])

$$E[y^n] = \sum_{i=1}^n E[x_i^n] = \sum_{i=1}^n \frac{n}{n_i + 1} = n \sum_{i=1}^n \frac{1}{n} \sim n \log n. \qquad (20.3.1)$$

The sum in (20.3.1) is also given by $n \sum_{i=1}^n \frac{1}{n} = nH_n$, where H_n is the harmonic sum (see §17.8.2, footnote 3). Moreover, some limit theorems are also known. For example, $y^n/n \log n$ converges to 1 in probability, i.e., for an arbitrary $\varepsilon > 0$ we have (see Durrett [1996: 38])

$$\lim_{n \to \infty} \mathfrak{p}\{(1 - \varepsilon)n \log n \le t^n \le (1 + \varepsilon)n \log n\} = 1. \qquad (20.3.2)$$

More precisely, for any real number u we have (see Durrett [1996: 144])

$$\lim_{n \to \infty} \mathfrak{P}\{y^n \le n(\log n + u)\} = e^{-e^{-u}},$$

$$\lim_{n \to \infty} \mathfrak{P}\{y^n \ge n(\log n - u)\} = 1 - e^{-e^{-u}}.$$

The right side of these two formulas is known as the *double exponential* distribution. A solution of the coupon collector's problem can be found at the websites *http://books.google.com/books?id=a_2vslx4FQMC&pg=PA80*, and *http://demonstrations.wolfram.com/CouponCollectorProblem/*.

The coupon collector's algorithm is a stochastic process involving the law of large numbers defined by $\lim_{n \to \infty} y^n/(\log n) = 1$, which is equivalent to (20.3.2). An algorithm is available at *http://books.google.com/books?id=QKVY4mDivBEC&pg=PA57*.

20.4 Open Problems

As we have noticed, all LDPC codes are based on bipartite Tanner graphs, which define parity operations for encoding and decoding. Decoding is based on the numbers of edges in graph (low density). All these codes are proven to be asymptotically optimal. Yet the following questions remain open: (i) What kind of overhead factors can we expect for LDPC codes for large and small n?; (ii) Are the three types of codes (Gallager, systematic, and IRA) equivalent, or do they perform differently?; (iii) How do the published distributions fare in producing good codes for finite values of n?; (iv) Is there a great deal of random variation in code generation for given probability distributions?; and (v) How do the codes compare to RS codes? Some of these questions have been partially answered. For example, it is shown above that the three codes (Gallager, systematic, and IRA) are different; For $R = 1/2$: systematic is best for small n; IRA is best for large n; Gallager is the worst up to about $n = 100$, and thereafter it becomes better than IRA but almost similar to IRA until $R \approx 700$, after which IRS performs better than it (see Figure 16.5.2 and Table 16.7.1). Thus, these three codes perform very poorly for small n, as well as for large n, except in certain cases as listed above. The following question still needs an answer: Although these codes converge to the overhead-factor $f = 1$ as $n \to \infty$, how do we minimize f for small n, and what about other rates?

Luby [1991] is a landmark publication, and Shokrollahi [2006] is a brilliant research article that shows deep perception in creating the Raptor codes. However, this is not the end. In 1998, Byers et al. showed how the Tornado codes can greatly outperform RS codes for large values of n. Luby et al. formed Digital Fountain soon after 1998, and patented the codes. Many others have published studies on similar LDPC codes with asymptotically optimal properties. Yet, no one has studied the practical applications of these LDPC codes. Thus, these codes still remain unusable for developers of wide-area storage systems, and RS codes still dominate the market.

Raptor codes are successfully used in commercial products marketed by Digital Fountain. As reported in Shokrollahi [2006], Raptor codes are capable of processing data with speeds of several gigabits per second on a 2.4-GHz Intel Xeon processor, with extremely reliable control of zero error probability of the decoder. A version of the Raptor codes, tailored to work for small lengths and requiring very little processing power, has been selected as Global Standard for Reliable Broadcast of Multimedia to 3G Mobile Devices; details are available in a Technical Report of the Multiple Broadcast/Multicast Services (MBMS [2005]). Some articles of the recent research deal with asynchronous reliable multicast (Byers et al. [2002]), video streaming over MBMS (Afzal et al. [2005]), and advanced receiver techniques for MBMS broadcast services (Gasiba et al. [2006]). Other open problems can be obtained by studying recent research publications. Finally, the advancements in graph theory are yet to be used for further developments in coding theory.

A

Table A.1 ASCII Table

Dec	Hex	Oct	Binary	char	Description
0	0	000	00000000	NUL	null
1	1	001	00000001	SOH	start of heading
2	2	002	00000010	STX	start of text
3	3	003	00000011	ETX	end of text
4	4	004	00000100	EOT	end of transmission
5	5	005	00000101	ENQ	enquiry
6	6	006	00000110	ACK	acknowledge
7	7	007	00000111	BEL	bell
8	8	010	00001000	BS	backspace
9	9	011	00001001	TAB	horizontal tab
10	A	012	00001010	LF/NL	line feed/new line
11	B	013	00001011	CR	carriage return
12	C	014	00001100	FF/NP	form feed/new page
13	D	015	00001101	CR	carriage return
14	E	016	00001110	SO	shift out
15	F	017	00001111	SI	shift in
16	10	020	00010000	DLE	data link escape
17	11	021	00010001	DC1	device control 1
18	12	022	00010010	DC2	device control 2
19	13	023	00010011	DC3	device control 3
20	13	024	00010100	DC4	device control 4
21	15	025	00010101	NAK	negative acknowledge
22	16	026	00010110	SYN	synchronous idle
23	17	027	00010111	ETB	end of trans. block
24	18	030	00011000	CAN	cancel
25	19	031	00011001	EM	end of medium
26	1A	032	00011010	SUB	substitute
27	1B	033	00011011	ESC	escape (␣)
28	1C	034	00011100	FS	file separator
29	1D	035	00011101	GS	group separator
30	1E	036	00011110	RS	record separator

Table A.1 ASCII Table, Contd.

Dec	Hex	Oct	Binary	char	Description
31	1F	037	00011111	US	unit separator
32	20	040	00100000	SP	space (␣)
33	21	041	00100001	!	exclamation mark
34	22	042	00100010	"	double quote
35	23	043	00100011	#	number sign
36	24	044	00100100	$	dollar sign
37	25	045	00100101	%	percent
38	26	046	00100110	&	ampersand
39	27	047	00100111	'	single quote
40	28	050	00101000	(	left/open parenthesis
41	29	051	00101001	)	right/close parenthesis
42	2A	052	00101010	*	asterisk
43	2B	053	00101011	+	plus
44	2C	054	00101100	,	comma
45	2D	055	00101101	-	minus or dash
46	2E	056	00101110	.	dot
47	2F	057	00101111	/	forward slash
48	30	060	00110000	0	
49	31	061	00110001	1	
50	32	062	00110010	2	
51	33	063	00110011	3	
52	34	064	00110100	4	
53	35	065	00110101	5	
54	36	066	00110110	6	
55	37	067	00110111	7	
56	38	070	00111000	8	
57	39	071	00111001	9	
58	3A	072	00111010	:	colon
59	3B	073	00111011	;	semicolon
60	3C	074	00111100	<	less than
61	3D	075	00111101	=	equal sign
62	3E	076	00111110	>	greater than
63	3F	077	00111111	?	question mark
64	40	100	01000000	@	at symbol
65	41	101	01000001	A	
66	42	102	01000010	B	
67	43	103	01000011	C	
68	44	104	01000100	D	

Table A.1 ASCII Table, Contd.

Dec	Hex	Oct	Binary	char	Description
69	45	105	01000101	E	
70	46	106	01000110	F	
71	47	107	01000111	G	
72	48	110	01001000	H	
73	49	111	01001001	I	
74	4A	112	01001010	J	
75	4B	113	01001011	K	
76	4C	114	01001100	L	
77	4D	115	01001101	M	
78	4E	116	01001110	N	
79	4F	117	01001111	O	
80	50	120	01010000	P	
81	51	121	01010001	Q	
82	52	122	01010010	R	
83	53	123	01010011	S	
84	54	124	01010100	T	
85	55	125	01010101	U	
86	56	126	01010110	V	
87	57	127	01010111	W	
88	58	130	01011000	X	
89	59	131	01011001	Y	
90	5A	132	01011010	Z	
91	5B	133	01011011	[	left bracket
92	5C	134	01011100	\	back/opening slash
93	5D	135	01011101	]	right/closing bracket
94	5E	136	01011110	^	caret/circumflex
95	5F	137	01011111	_	underscore
96	60	140	01100000	`	grave accent
97	61	141	01100001	a	
98	62	142	01100010	b	
99	63	143	01100011	c	
100	64	144	01100100	d	
101	65	145	01100101	e	
102	66	146	01100110	f	
103	67	147	01100111	g	
104	68	150	01101000	h	
105	69	151	01101001	i	
106	6A	152	01101010	j	

Table A.1 ASCII Table, Contd.

Dec	Hex	Oct	Binary	char	Description
107	6B	153	01101011	k	
108	6C	154	01101100	l	
109	6D	155	01101101	m	
110	6E	156	01101110	n	
111	6F	157	01101111	o	
112	70	160	01110000	p	
113	71	161	01110001	q	
114	72	162	01110010	r	
115	73	163	01110011	s	
116	74	164	01110100	t	
117	75	165	01110101	u	
118	76	166	01110110	v	
119	77	167	01110111	w	
120	78	170	01111000	x	
121	79	171	01111001	y	
122	7A	171	01111010	z	
123	7B	172	01111011	{	left/opening brace
124	7C	173	01111100	|	vertical bar
125	7D	174	01111101	}	right/closing brace
126	7E	175	01111110	~	tilde
127	7F	176	01111111	DEL	delete

In addition, there are ANSII codes that use decimal digits 128–257 to represent other international symbols that have become part of the English language. These codes are available on the Internet (e.g., www.LookupTables.com).

B

Some Useful Groups

B.1 Permutation Groups

A group whose elements are permutations on a given finite set of objects (symbols) is called a *permutation group* on these objects. Let $a_1, a_2, \ldots, a_n$ denote n distinct objects, and let $b_1, b_2, \ldots, b_n$ be any arrangement of the same n objects. The operation of replacing each a_i by b_i, $1 \leq i \leq n$, is called a *permutation* performed on the n objects. It is denoted by

$$S : \begin{pmatrix} a_1 \ a_2 \ \cdots \ a_n \\ b_1 \ b_2 \ \cdots \ b_n \end{pmatrix} \tag{B.1}$$

and is called a *permutation of degree* n. If the symbols on both lines in (B.1) are the same, the permutation is called the *identical* permutation and is denoted by I. Let

$$T : \begin{pmatrix} b_1 \ b_2 \ \cdots \ b_n \\ c_1 \ c_2 \ \cdots \ c_n \end{pmatrix}, \quad U : \begin{pmatrix} a_1 \ a_2 \ \cdots \ a_n \\ c_1 \ c_2 \ \cdots \ c_n \end{pmatrix},$$

then U is the *product* of S and T, i.e., $U = ST$, since

$$U = ST = \begin{pmatrix} a_1 \ a_2 \ \cdots \ a_n \\ b_1 \ b_2 \ \cdots \ b_n \end{pmatrix} \begin{pmatrix} b_1 \ b_2 \ \cdots \ b_n \\ c_1 \ c_2 \ \cdots \ c_n \end{pmatrix} = \begin{pmatrix} a_1 \ a_2 \ \cdots \ a_n \\ c_1 \ c_2 \ \cdots \ c_n \end{pmatrix}.$$

Multiplication of permutations is associative: $(S_1 S_2)S_3 = S_1(S_2 S_3) = S_1 S_2 S_3$, but multiplication of two permutations is not always commutative.

Example B.1.1. $\begin{pmatrix} a\ b\ c \\ b\ c\ a \end{pmatrix} \begin{pmatrix} a\ b\ c \\ b\ a\ c \end{pmatrix} = \begin{pmatrix} a\ b\ c \\ b\ c\ a \end{pmatrix} \begin{pmatrix} b\ c\ a \\ a\ c\ b \end{pmatrix} = \begin{pmatrix} a\ b\ c \\ a\ c\ b \end{pmatrix}$, but

$\begin{pmatrix} a\ b\ c \\ b\ a\ c \end{pmatrix} \begin{pmatrix} a\ b\ c \\ b\ c\ a \end{pmatrix} = \begin{pmatrix} a\ b\ c \\ b\ a\ c \end{pmatrix} \begin{pmatrix} b\ a\ c \\ c\ b\ a \end{pmatrix} = \begin{pmatrix} a\ b\ c \\ c\ b\ a \end{pmatrix}$ ∎.

If $ST = I$, then S (or T) is said to be the *inverse* of T (or S). The product of a permutation and its inverse is independent of the order in which the multiplication is performed.

431

B.2 Cyclic Permutation

A cyclic permutation (or circular permutation) is of the form

$$\begin{pmatrix} a_1 \, a_2 \, \cdots \, a_{n-1} \, a_n \\ a_2 \, a_3 \, \cdots \, a_n \, a_1 \end{pmatrix},$$

and for brevity it is denoted by $(a_1 \, a_2 \, \cdots \, a_n)$. Thus, $(a_1 \, a_2 \, \cdots \, a_n) = (a_2 \, a_3 \, \cdots \, a_n \, a_1) = (a_3 \, a_4 \, \cdots \, a_n \, a_1 \, a_2) = \cdots$. The cyclic permutation of two objects, $(a \, b)$, is called a *transportation* and (a) denotes the operation of replacing a by a.

Example B.2.1. Product of two cyclic permutations: $(a \, b \, c \, d \, e)(b \, c \, e \, d) = (a \, c \, b \, e)(d)$. ∎

Theorem B.2.1. *Any given permutation is a product of cyclic permutations, no two of which have an object in common.*

Example B.2.2. (i) $\begin{pmatrix} 1 \, 2 \, 3 \, 4 \, 5 \, 6 \, 7 \, 8 \\ 2 \, 4 \, 3 \, 8 \, 6 \, 5 \, 7 \, 1 \end{pmatrix} = (1 \, 2 \, 4 \, 8)(3)(5 \, 6)(7) = (1 \, 2 \, 4 \, 8)(5 \, 6).$

(ii) $\begin{pmatrix} 1 \, 2 \, 3 \, 4 \, 5 \, 6 \, 7 \, 8 \, 9 \\ 2 \, 3 \, 1 \, 5 \, 6 \, 4 \, 8 \, 9 \, 7 \end{pmatrix} = (1 \, 2 \, 3)(4 \, 5 \, 6)(7 \, 8 \, 9).$ ∎

Theorem B.2.2. *Any given permutation can be expressed as a product of transportations.*

Corollary B.2.1. *Any given permutation on the objects $a_1, a_2, \ldots, a_n$ can be expressed as a product in terms of the transportations $(a_1 \, a_2), \ldots, (a_1 \, a_n)$.*

Theorem B.2.3. *In any expression of a given permutation as product of transportations on objects, the number of transportations is always odd or always even.*

Example B.2.3. Consider the determinant

$$D = \begin{vmatrix} 1 & 1 & \cdots & 1 \\ a_1 & a_2 & \cdots & a_n \\ a_1^2 & a_2^2 & \cdots & a_n^2 \\ \cdots & \cdots & \cdots & \cdots \\ a_1^{n-1} & a_2^{n-1} & \cdots & a_n^{n-1} \end{vmatrix}. \tag{B.2}$$

Transportation of two objects in (B.2) changes D to $-D$. Thus, a permutation expressed as a product of an odd number of transportations changes D to $-D$, whereas a product of an even number of transportations leaves D unaltered. ∎

Theorem B.2.4. *Any even permutation may be expressed as a product of cyclic permutations, each involving just three objects.*

Corollary B.2.2. *An even permutation on objects $a_1, a_2, \ldots, a_n$ can be expressed as a product of permutation, each of which belongs to the set $(a_1 \, a_2 \, a_3), (a_1 \, a_2 \, a_4), \cdots, (a_1 \, a_2 \, a_n)$.*

A product formed by taking r factors, each of which is a given permutation S, is denoted by S^r and is called the r-th *power* of S. The associative law holds: $S^\mu S^\nu = S^{\mu+\nu}$. Let $\rho = \max\{\mu, \nu\}$, where $\mu > \nu$. Then $S^\rho = S^\mu$, $S^{\rho-\mu} = S^\mu$, $S^{\rho-\mu}S^\mu S^{-\mu} = S^\mu S_{-\mu} = I$, where I is the identical permutation.

Given a sequence of powers of S, let S^m be the first one that is equal to I. Then m is called the *order* of S. Obviously, $S^{m-1} = IS^{-1} = S^{-1}$ is the *inverse* of S. If S is of order m $(m > 2)$, then no two of the permutations $S, S^2, \ldots, S^{m-1}$ are equal. If $k > 0$ is an integer and $S^k = I$, then k is a *multiple* of order m of S. We have $S^{\mu+\nu} = S^\mu S^\nu$ for μ and ν positive integers; but if it is assumed to hold for the exponents $\mu, \nu \in \mathbb{Z}$, then S^0 is the identical permutation and $S^{-\nu}$ is the inverse of S^ν, where $S^0 = I$.

Theorem B.2.5. *If a given permutation S is written as a product of cyclic permutations, no two of which have an object in common, then the order of S is the lcm of the degrees of the cyclic permutations that compose it.*

A cyclic permutation is equal to its degree. When a given permutation S is written as in Theorem B.2.5, these component cyclic permutations are called the *cycles* of S, and S is said to be written in *standard form*. If all cycles of S are of the same degree, S is said to be *regular*. If the permutations have the same number of cycles and the cycles can be made to correspond uniquely, those of one permutation to those of the other, so that the two corresponding cycles have the same degree, then these permutations are said to be *similar*.

Example B.2.4. (i) $(a\,c\,d\,e)(f\,g)(h\,i\,j)$ and $(1\,2)(3\,4\,5)(6\,7\,8\,9\,0)$ are similar.

(ii) The third power of the cyclic permutation $(1\,2\,3\,4\,5\,6\,7\,8\,9)$ is the regular permutation $(1\,4\,7)(2\,5\,8)(3\,6\,9)$. ∎

A cyclic permutation is even or odd according to whether its degree is even or odd. Thus, a permutation is even or odd according to whether the difference between its degree and the number of its cycles is even or odd.

Given any two permutations S and T, the permutation $T^{-1}ST$ is called the *transform* of S by T, and S is said to be transformed by T, since

$$(TU)^{-1}S(TU) = (U^{-1}T^{-1})S(TU) = U^{-1} \cdot T^{-1}ST \cdot U.$$

i.e., the transform of S by TU is equal to the transform by U of the transform of S by T.

Theorem B.2.6. *The transform of S by T is found by performing the permutation T on the cycles of S.*

Let $S = (a\,b\,c\,d\,\cdots)(l\,m\,n\,o\,\cdots)$, and $T = \begin{pmatrix} a\,b\,c\,d\,\cdots\,l\,m\,n\,o\,\cdots \\ \alpha\,\beta\,\gamma\,\delta\,\cdots\,\lambda\,\mu\,\nu\,\rho\,\cdots \end{pmatrix}$. Then

$$
T^{-1}ST = \begin{pmatrix} \alpha\,\beta\,\gamma\,\delta\,\cdots\,\lambda\,\mu\,\nu\,\rho\,\cdots \\ a\,b\,c\,d\,\cdots\,l\,m\,n\,o\,\cdots \end{pmatrix} \{(a\,b\,c\,\cdots)(l\,m\,n\,\cdots)\}
$$

$$
\begin{pmatrix} a\,b\,c\,d\,\cdots\,l\,m\,n\,o\,\cdots \\ \alpha\,\beta\,\gamma\,\delta\,\cdots\,\lambda\,\mu\,\nu\,\rho\,\cdots \end{pmatrix}
$$

$$
= \begin{pmatrix} \alpha\,\beta\,\gamma\,\delta\,\cdots\,\lambda\,\mu\,\nu\,\rho\,\cdots \\ b\,c\,d\,\cdots\,m\,n\,o\,\cdots \end{pmatrix} \begin{pmatrix} a\,b\,c\,d\,\cdots\,l\,m\,n\,o\,\cdots \\ \alpha\,\beta\,\gamma\,\delta\,\cdots\,\lambda\,\mu\,\nu\,\rho\,\cdots \end{pmatrix}
$$

$$
= \begin{pmatrix} \alpha\,\beta\,\gamma\,\cdots\,\lambda\,\mu\,\nu\,\cdots \\ \beta\,\gamma\,\delta\,\cdots\,\mu\,\nu\,\rho\,\cdots \end{pmatrix} = (\alpha\,\beta\,\gamma\,\cdots)(\lambda\,\mu\,\nu\,\cdots)\,\cdots.
$$

The last expression is of the form stated in Theorem B.2.6. Thus, the permutation $T^{-1}ST$ is similar to the permutation S. In particular, if $T^{-1}ST = S$, then $TT^{-1}ST = TS$, or $ST = TS$. In this case, S and T are said to be *commutative* or *permutable*. Thus, two cyclic disjoint permutations (i.e., they have no object in common) are commutative. A necessary and sufficient condition that S and T shall be commutative is that the transform of S by T must be equal to S. Moreover, if S is permutable with both T and U, it is also permutable with TU, since $(TU)S = TUS = STU = S(TU)$. When S and T are commutable, we have $ST = TS$, whence $T^{-1}ST = S$ and $S^{-1}T^{-1}ST = I$. In general, if S and T are any two permutations, then $S^{-1}T^{-1}ST$ is called the *commutator* of S and T. Thus, the commutator of T and S is $T^{-1}S^{-1}TS$. Since the product of these two commutators is I, each of them is the inverse of the other. A necessary and sufficient condition that two permutations shall be commutative is that their commutator (in either order) must be the identical permutation I. Further, a permutation and its transform are of the same order. In particular, S_1S_2 and S_2S_1 are of the same order.

B.3 Mathieu Groups

For geometrical construction of different Golay codes we need to review the Mathieu groups and Steiner systems. We encountered a Steiner system in §7.1.4. We will also explain the relationship between the Mathieu groups and Steiner systems. The following finite simple groups are needed in this book: Alternating groups A_n; Cyclic groups Z_n; Dihedral groups D_n; Mathieu groups M_{11}, M_{12}, M_{22}, M_{23}, and M_{24}; and symmetric groups S_n.

Mathieu groups, named after the French mathematician Émile Léonard Mathieu [1861; 1873], evolved out of multiply transitive permutation groups, and are defined as follows: For a number $k \in \mathbb{N}$, a permutation group G acting on n objects (points) is *k-transitive* if, given two sets of points $\{a_1, a_2, \ldots, a_n\}$ and $\{b_1, b_2, \ldots, b_n\}$, where all a_i and all b_i are distinct, there is a group element $g \in G$ that maps a_i to b_i for $1 \leq i \leq k$. Such a group G is called *sharply k-transitive* if the element $g \in G$ is unique, i.e., the action on k-tuples $\{a_i\}$ and $\{b_i\}$ is regular rather than transitive.

For $k \leq 4$ there are only 4-transtive groups S_k that are symmetric. For $k \leq 6$ there are alternating groups A_k and the Mathieu groups $M_{11}, M_{12}, M_{22}, M_{23}$, and M_{24}. A classical theorem by Jordan (see Carmichael [1956]) states that the symmetrical and alternating groups of degrees k and $k - 2$, respectively, and M_{11} and M_{12} are the only sharply k-transitive permutation groups for $k \leq 4$. The order and transitivity of the five Mathieu groups are presented in Table B.3.1.

Table B.3.1 Order and Transitivity of Mathieu Groups

Group	Order	Factorized Order	Transitivity
M_{11}	7920	$2^4 \cdot 3^2 \cdot 5 \cdot 11$	sharply 4-transitive
M_{12}	95040	$2^6 \cdot 3^3 \cdot 5 \cdot 11$	sharply 5-transitive
M_{22}	443520	$2^7 \cdot 3^2 \cdot 5 \cdot 7 \cdot 11$	3-transitive
M_{23}	10200960	$2^7 \cdot 3^2 \cdot 5 \cdot 7 \cdot 11 \cdot 23$	4-transitive
M_{24}	244823040	$2^{10} \cdot 3^3 \cdot 5 \cdot 7 \cdot 11 \cdot 23$	5-transitive

Permutation groups are used to construct the Mathieu groups. We discuss only the M_{12} and M_{24} as they are related to the Golay codes $\mathcal{G}_{12}$ and $\mathcal{G}_{24}$.

(i) M_{12} has a maximal simple subgroup of order 660, which can be represented as a linear fractional group on the field F_{11} of 11 elements: If we write -1 as a and ∞ as b, then among three generators the two standard generators are $(0\,1\,2\,3\,4\,5\,6\,7\,8\,9\,a)$ and $(0b)(1a)(25)(37)(48)(69)$; and the third generator that yields M_{12} sends an element $x \in F_{11}$ to $4x^2 - 3x^7$, which is a permutation $(2\,6\,a\,7)(3\,9\,4\,5)$.

(ii) M_{24} has a maximal simple subgroup of order 6072; it can be represented as a linear fractional group on the field F_{23} of 23 elements. Of the three generators, one generator adds 1 to each element:

$(0\,1\,2\,3\,4\,5\,6\,7\,8\,9\,A\,B\,C\,D\,E\,F\,G\,H\,i\,J\,K\,L\,M\,N)$; the other is: $(0\,N)(1\,M)(2\,B)$ $(3\,F)(4\,H)(6\,J)(7\,D)(8\,K)(A\,G)(C\,L)(E\,I)$; and the third gives M_{24}, which sends an element $x \in F_{23}$ to $4x^4 - 3x^{15}$, and this turns out to be the permutation: $(2\,G\,9\,6\,8)(3\,C\,D\,I\,4)(7\,H\,A\,B\,M)(E\,J\,L\,K\,F)$. For details, see Carmichael [1956].

B.4 Steiner Systems

These systems, named after Jacob Steiner [1853], occur as block designs in combinatorics. A Steiner system with parameters l, m, n, written as $S(l, m, n)$, is an n-element set together with a set of m-element subsets of S, called *blocks*, with the property that each l-element subset of S is contained in exactly one block. Some examples are:

(i) $S(2, 3, n)$ is called a *Steiner triple system* and its blocks are called *triplets*, which are $n(n-1)/6$ in number. Multiplication on this Steiner system is defined by setting $aa = a$ for all $a \in S$, and $ab = c$ if $\{a, b, c\}$ is a triple.

(ii) $S(3, 4, n)$, which is called a *Steiner quadruple system*. Steiner system for higher values of m do not have names.

(iii) $S(2, m, n)$, where m is a prime power greater than 1. For this system $n = 1$ or m (mod $m(m-1)$). In particular, a Steiner system $S(2, 3, n)$ must have $n = 6k + 1$ or $6k + 3$, $k \in \mathbb{N}$, such that for each $k \in \mathbb{N}$ the systems $S(2, 3, 6k + 1)$ and $S(2, 3, 6k + 3)$ exist. For example, for $k = 1$, a Steiner system $S(2, 3, 7)$ is shown in Figure B.4.1. It is called the Fano plane, which has blocks as the 7 lines, each containing 3 points, and every pair of points belongs to a unique line.

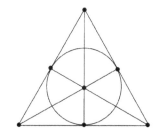

Figure B.4.1. Steiner system $S(2, 3, 7)$.

The following Steiner systems are related to respective Mathieu groups:

M_{11} is the automorphism group of a $S(4, 5, 11)$ Steiner system.

M_{12} is the automorphism group of a $S(5, 6, 12)$ Steiner system.

M_{22} is the automorphism group of a $S(3, 6, 22)$ Steiner system.

M_{23} is the automorphism group of a $S(4, 7, 23)$ Steiner system.

M_{24} is the automorphism group of a $S(3, 8, 24)$ Steiner system.

We now consider construction methods for Steiner systems $S(5, 6, 12)$ and $S(3, 8, 24)$ in detail as they are closely related to the Golay codes $\mathcal{G}_{12}$ and $\mathcal{G}_{24}$.

(a) $S(5, 6, 12)$ STEINER SYSTEM. To construct this system, we take a 12-point set and regard it as the projective line over F_{11}, since it contains the integers mod 11 together with the point at infinity. Among the integers mod 11, there are six perfect sequences, namely, $\{0, 1, 3, 4, 5, 9\}$. We will call this set a block, which will yield other blocks using the bilinear fractional transformation $z \mapsto \dfrac{az + b}{cz + d}$, $ad - bc \neq 0$. All resulting blocks form a Steiner

$S(5, 6, 12)$ system, where the automorphism group of the system is M_{12}.

(b) $S(5, 8, 24)$ STEINER SYSTEM. Also known as the *Witt system*, this system can be constructed simply using a sequence of 24 bits, which is written in lexicographic order, dropping any bit that differs from some earlier ones in fewer than 8 locations. The result is the following list, which contains $2^{12} (= 4096)$ 24-bit items:

```
0000 0000 0000 0000 0000 0000
0000 0000 0000 0000 1111 1111
0000 0000 0000 1111 0000 1111
0000 0000 0000 1111 1111 0000
0000 0000 0011 0011 0011 0011
0000 0000 0011 0011 1100 1100
0000 0000 0011 1100 0011 1100
0000 0000 0011 1100 1100 0011
0000 0000 0101 0101 0101 0101
0000 0000 0101 0101 1010 1010
```

   .
   ·   (next 4083 omitted)
   .

```
1111 1111 1111 0000 1111 0000
1111 1111 1111 1111 0000 0000
1111 1111 1111 1111 1111 1111
```

Each item in the above list is a codeword of the Golay code $\mathcal{G}_{24}$. All these 4096 codewords form a group under the XOR operation. Out of all these codewords, one has zero 1-bits, 759 of them have eight 1-bits, 2576 of them have twelve 1-bits, 759 of them have sixteen 1-bits, and one has twenty-four 1-bits. The 759 8-element blocks, called *octads*, are given by the patterns of 1's in the codewords with eight 1-bits. A more direct method is to run through all these 759 octads that are 8-element subsets of a 24-element set and skip any subset that differs from some octad already found in fewer than four locations. This leads to the following list:

```
01 02 03 04 05 06 07 08
01 02 03 04 09 10 11 12
01 02 03 04 13 14 15 16
```

   .
   ·   (next 753 octads omitted)
   .

```
13 14 15 16 17 18 19 20
13 14 15 16 21 22 23 24
17 18 19 20 21 22 23 24
```

In this list, the breakdown of different types of elements is as follows: Each single element occurs 263 time somewhere in some octad, each pair 77 times, each triple 21 times, each quadruple (tetrad) 5 time, each quintuple (pentad) once, and there is no hexad, heptad or octad.

B.5 Dihedral Groups

Let S denote a regular n-gon, $n > 2$. Then in any isometric transformations the vertices are taken to the vertices. The following results are assumed to be known; otherwise, consult a book on modern algebra.

Theorem B.5.1. *Every regular n-gon can be circumscribed by one and only one circle.*

Thus, the center of the circumscribing circle of a regular n-gon is its center.

Let I_S denote the symmetry group of S. Then an element of I_S is characterized by the mapping elements of S, and only elements of S, into S.

Theorem B.5.2. *The center of a regular n-gon S is take onto itself by any element of I_S.*

Theorem B.5.3. *If S is a regular n-gon and $\sigma \in I_S$, then the vertices of S are taken onto vertices of S by σ.*

The symmetry group of the regular n-gon S is called the *dihedral group of degree* n. Let the vertices of S with center O be denoted by $A_1, \dots, A_n$ (in clockwise order). Let σ_j, $j = 1, \dots, n$, rotate S about O in a clockwise direction through an angle $\dfrac{2(j-1)\pi}{n}$ radians so that $A_1\sigma_1 = A_j$. For example, the effect of σ_3 on the regular pentagon is shown in Figure B.5.1(a).

Let τ denote the reflection about the line through A_1 and O, so that $A_1\tau = A_1, A_2\tau = A_n$. The effect of this reflection is shown in Figure B.5.1(b). Finally the effect of the reflection τ followed by the rotation σ_3 is shown in Figure B.5.1(c). Notice that the elements $\sigma_1, \dots, \sigma_n, \tau\sigma_1, \dots, \tau\sigma_n$ are all distinct, and $\sigma_j \neq \sigma_k$ and $A_1\sigma_j \neq A_1\sigma_k$ for $k \neq k$. If $\tau\sigma_j = \sigma_k$, then $A_1\tau\sigma_j = A_1\tau\sigma_k$. Thus, $\tau\sigma_j = \sigma_k$ implies that $j = k$. But $\tau\sigma_j = \sigma_j$ implies $\tau = \sigma_1$, which is an identity, contrary to the assumption. Finally, $\tau\sigma_j = \tau\sigma_k$ implies $\sigma_j = \sigma_k$. Thus, we see that there are at least $2n$ possible elements of the dihedral group of degree n, and it can be easily shown that there are no more than $2n$ elements.

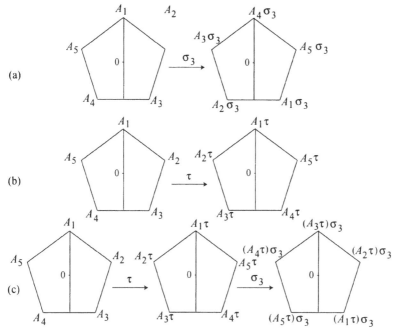

Figure B.5. (a) Mapping σ_3, (b) reflection, and (c) rotation.

Some examples of the dihedral groups are given below.

Example B.5.1. The multiplication table for the dihedral group D_3, represented by the symmetries of the regular pentagon for combining the integers from 0 to 5 is given in Table B.5.1.

Table B.5.1. Multiplication Table for the Dihedral Group D_3

$d(j,k)$	0	1	2	3	4	5
0	0	1	2	3	4	5
1	1	2	0	**5**	**3**	**4**
2	2	0	1	**4**	**5**	**3**
3	3	4	5	0	1	2
4	4	5	3	**2**	0	1
5	5	3	4	**1**	**2**	0

Note that j and k $(j,k = 0,\dots,5)$ represent the rows and columns respectively, and the entries in bold show the results that are noncommutative.

Example B.5.2. The multiplication table for the dihedral group D_4, represented by the symmetries of a square, for combining integers from 0 to 7 are given in Table B.5.2.

Table B.5.2. Multiplication Table for the Dihedral Group D_4

$d(j,k)$	0	1	2	3	4	5	6	7
0	0	1	2	3	4	5	6	7
1	1	2	3	0	**7**	**4**	**5**	**6**
2	2	3	0	1	**6**	**7**	**4**	**5**
3	3	0	1	2	**5**	**6**	**7**	**4**
4	**4**	**5**	**6**	**7**	0	1	2	3
5	**5**	**6**	**7**	**4**	1	0	1	2
6	**6**	**7**	**4**	**5**	2	3	0	1
7	**7**	**4**	**5**	**6**	1	2	3	0

As before, the entries in bold show the results that are non-commutative.

B.5 Null Space. The null space, also known as the kernel, of an $m \times n$ matrix $\mathbf{A}$, denoted by Null($\mathbf{A}$), is the set of all vectors $\mathbf{x} \in \mathbb{R}^n$ such that $\mathbf{Ax} = \mathbf{0}$. If the matrix $\mathbf{A}$ has n columns, then its null space is a linear subspace of $\mathbb{R}^n$ (see Lay [2005]). The dimension of the null space of $\mathbf{A}$ is called the *nullity* of $\mathbf{A}$. If the equation $\mathbf{Ax} = \mathbf{0}$ is written as a system of m linear equations, then the null space of $\mathbf{A}$ is the same as the solution set of this homogeneous system. The properties of Null($\mathbf{A}$) are as follows: (i) Null($\mathbf{A}$) always contains the zero vector; (ii) if $\mathbf{x} \in$ Null($\mathbf{A}$) and $\mathbf{y} \in$ Null($\mathbf{A}$), then $\mathbf{x} + \mathbf{y} \in$ Null($\mathbf{A}$); and (iii) if $\mathbf{x} \in$ Null($\mathbf{A}$) and c is a scalar, then $c\mathbf{x} \in$ Null($\mathbf{A}$). Note that the null space of a matrix is not affected by elementary row operations. The rank-nullity theorem states that rank $\mathbf{A}$+nullity $\mathbf{A} = n$.

C

Tables in Finite Fields

C.1 Table of Irreducible Polynomials in $F_2[x]$

The following definitions and explanations are necessary for understanding the usefulness of Table C.1.

1. Given a polynomial $p(x) = \sum_{k=0}^{n} a_k x^k \in F_2[x]$, the *reciprocal polynomial* $p^*(x)$ is defined by $p^*(x) = \sum_{k=0}^{n} a_k x^{n-k}$. For example, $p(x) = a + bx + cx^2 \leftrightarrow (a, b, c) \Rightarrow p^*(x) \leftrightarrow (a, b, c)^* = (c, b, a) \leftrightarrow c + bx + ax^2$.

2. If $p(x)$ is irreducible (primitive) and has a zero α, then $p^*(x)$ is irreducible (primitive) and has zero α^{-1}. It means that in Table C.1 only *one* out of the pair $p(x)$ and $p^*(x)$ is listed; that is, the sequence of coefficients may be read either from the left or from the right to obtain an irreducible polynomial.

3. For each degree, let α be a zero of the first listed polynomial. The entry following the boldface **j** is the minimum polynomial of α^j. To find the minimum polynomial not listed, use properties 2 and 7 of §8.2 and $\alpha^{2^r - 1} = 1$.

   Example C.1. Let α denote a zero of the polynomial $(100101) \leftrightarrow 1 + x^3 + x^5$ of degree 5; Table C.1 shows that α^3 is a zero of (111101) and α^5 a zero of (110111). Then the minimum polynomial of (recalling that $\alpha^{31} = 1$)

   $\alpha, \alpha^2, \alpha^4, \alpha^8, \alpha^{16}, \alpha^{32} = \alpha$ is $p_1(x) = (100101)$;

   $\alpha^3, \alpha^6, \alpha^{12}, \alpha^{24}, (\alpha^{48} = \alpha^{48-31} =)\alpha^{17}$ is $p_3(x) = (111101)$;

   $\alpha^5, \alpha^{10}, \alpha^{20}, (\alpha^{40} =)\alpha^9, \alpha^{18}$ is $p_5(x) = (110111)$.

Using the zeros of the reciprocal polynomials, the minimum polynomial of

$(\alpha^{-1} =)\alpha^{30}, (\alpha^{-2} =)\alpha^{29}, (\alpha^{-4} =)\alpha^{27}, (\alpha^{-8} =)\alpha^{23}, (\alpha^{-16} =)\alpha^{15}$ is $p_1^*(x) = (101001)$;

$(\alpha^{-3} =)\alpha^{28}, (\alpha^{-6} =)\alpha^{25}, (\alpha^{-12} =)\alpha^{19}, (\alpha^{38} =)\alpha^7, \alpha^{14}$ is $p_3^*(x) = (101111)$;

441

$(\alpha^{-5} =)\alpha^{26}, (\alpha^{-10} =)\alpha^{21}, (\alpha^{-20} =)\alpha^{11}, \alpha^{22}, (\alpha^{44} =)\alpha^{13}$ is $p_5^*(x) = (111011)$. ∎

4. The exponent e of the irreducible polynomial $p(x)$ of degree m can be calculated by the formula: $e = \dfrac{2^m - 1}{\gcd(2^m - 1, \mathbf{j})}$, where in order to factorize $2^m - 1$ we can use Table C.2 of factorization of Mersenne numbers.

Example C.2. The exponent of (1001001) is $e = \dfrac{2^6 - 1}{\gcd(2^6 - 1, 7)} = \dfrac{63}{\gcd(63, 7)} = 9$. ∎

C.2 Mersenne Numbers

Numbers of the form $M_n = 2^n - 1$ are called *Mersenne numbers*. Prime factorization of the first 32 Mersenne numbers are given in Table C.2.

C.3 Other Useful Tables

$\oplus$	0	1
0	0	1
1	1	0

Bitwise operation XOR.

Addition and Multiplication Tables in GF(8).

+	1	α	α^2	α^3	α^4	α^5	α^6
1	0	α^3	α^6	α	α^5	α^4	α^2
α	α^3	0	α^4	1	α^2	α^6	α^5
α^2	α^6	α^4	0	α^5	α	α^3	1
α^3	α	1	α^5	0	α^6	α^2	α^4
α^4	α^5	α^2	α	α^6	0	1	α^3
α^5	α^4	α^6	α^3	α^2	1	0	α
α^6	α^2	α^5	1	α^4	α^3	α	0

$\times$	1	α	α^2	α^3	α^4	α^5	α^6
1	1	α	α^2	α^3	α^4	α^5	α^6
α	α	α^2	α^3	α^4	α^5	α^6	1
α^2	α^2	α^3	α^4	α^5	α^6	1	α
α^3	α^3	α^4	α^5	α^6	1	α	α^2
α^4	α^4	α^5	α^6	1	α	α^2	α^3
α^5	α^5	α^6	1	α	α^2	α^3	α^4
α^6	α^6	1	α	α^2	α^3	α^4	α^5

Table 8.6.3 Addition.   Table 8.6.4 Multiplication.

These tables are created using formulas (8.6.2).

TABLE C.1.   443

Table C.1 All Irreducible Polynomials in $F_2[x]$ of Degree ≤ 10

Legend: (P): primitive; (NP): nonprimitive

Degree 2 ($\alpha^3 = 1$)
1 111 (P)

Degree 3 ($\alpha^7 = 1$)
1 1011 (P) †

Degree 4 ($\alpha^{15} = 1$)

1 10011 (P)	**3** 11111 (NP)	**5** 111

Degree 5 ($\alpha^{31} = 1$)

1 100 101 (P)	**3** 111 101 (P)	**5** 110 111 (P)

Degree 6 ($\alpha^{63} = 1$)

1 100 0011 (P)	**3** 101 0111 (NP)	**5** 110 0111 (P)
7 100 1011 (NP)	**9** 1101	**11** 10 1101 (P)
21 111		

Degree 7 ($\alpha^{127} = 1$)

1 1000 1001 (P)	**3** 1000 1111 (P)	**5** 1001 1101 (P)
7 1111 0111 (P)	**9** 1011 1111 (P)	**11** 1101 0101 (P)
13 1000 0011 (P)	**19** 1100 1011 (P)	**21** 1110 0101 (P)

Degree 8 ($\alpha^{255} = 1$)

1 1000 11101 (P)	**3** 1011 10111 (NP)	**5** 1111 10011 (NP)
7 1011 01001 (P)	**9** 1101 11101 (NP)	**11** 1111 00111 (P)
13 1101 01011 (P)	**15** 1110 10111 (NP)	**17** 10011
19 1011 00101 (P)	**21** 1100 01011 (NP)	**23** 1011 00011 (P)
25 1000 11011 (NP)	**27** 1101 11111 (NP)	**37** 1010 11111 (P)
43 1110 00011 (P)	**45** 1001 11001 (NP)	**51** 11111
85 111		

Degree 9 ($\alpha^{511} = 1$)

1 10000 10001 (P)	**3** 10010 11001 (P)	**5** 11001 10001 (P)
7 10100 11001 (NP)	**9** 11000 10011 (P)	**11** 10001 01101 (P)
13 10011 10111 (P)	**15** 11011 00001 (P)	**17** 10110 11011 (P)
19 11100 00101 (P)	**21** 10000 10111 (NP)	**23** 11111 01001 (P)
25 11111 00011 (P)	**27** 11100 01111 (P)	**29** 11011 01011 (P)
35 11000 00001 (NP)	**37** 10011 01111 (P)	**39** 11110 01101 (P)
41 11011 10011 (P)	**43** 11110 01011 (P)	**45** 10011 11101 (P)
51 11110 10101 (P)	**53** 10100 10101 (P)	**55** 10101 11101 (P)
73 1011	**75** 11111 11011 (P)	**77** 11010 01001 (NP)
83 11000 10101 (P)	**85** 10101 10111 (P)	

Table C.1 All Irreducible Polynomials in $F_2[x]$ $\cdots$ (*continued*)

Legend: (P): primitive; (NP): nonprimitive

Degree **10** ($\alpha^{1023} = 1$)

1 10000 001001 (P)	**3** 10000 001111 (NP)	**5** 10100 001101 (*P*)
7 11111 111001 (P)	**9** 10010 101111 (NP)	**11** 10000 110101 (NP)
13 10001 101111 (P)	**15** 10110 101011 (NP)	**17** 11101 001101 (P)
19 10111 111011 (P)	**21** 11111 101011 (P)	**23** 10000 011011 (P)
25 10100 100011 (P)	**27** 11101 111011 (NP)	**29** 10100 110001 (P)
31 11000 100011 (NP)	**33** 111101	**35** 11000 010011 (P)
37 11101 100011 (P)	**39** 10001 000111 (NP)	**41** 10111 100101 (P)
43 10100 011001 (P)	**45** 11000 110001 (NP)	**47** 11001 111111 (P)
49 11101 010101 (P)	**51** 10101 100111 (NP)	**53** 10110 001111 (P)
55 11100 101011 (NP)	**57** 11001 010001 (NP)	**59** 11100 111001 (P)
69 10111 000001 (NP)	**71** 11011 010011 (P)	**73** 11101 000111 (P)
75 10100 011111 (NP)	**77** 10100 001011 (NP)	**83** 111110 010011 (P)
85 10111 000111 (P)	**87** 10011 001001 (NP)	**89** 10011 010111 (P)
91 11010 110101 (P)	**93** 11111 111111 (NP)	**99** 110111
101 10000 101101 (P)	**103** 11101 111101 (P)	**105** 11110 000111 (NP)
107 11001 111001 (P)	**109** 10000 100111 (P)	**147** 10011 101101 (NP)
149 11000 010101 (P)	**155** 10010 101001 (NP)	**165** 101001
171 11011 001101 (NP)	**173** 11011 011111 (P)	**179** 11010 001001 (P)
341 111		

† Note that $(1011)^* = 1101$ (P) is also irreducible.

TABLE C.2. 445

Table C.2 lists primitive polynomials over F_2 for each degree $n \leq 30$. In this table only the separate terms in the polynomial are given; thus, 6 1 0 means $x^6 + x + 1$.

Table C.2 Primitive Polynomials over F_2

1	0			
2	1	0		
3	1	0		
4	1	0		
5	2	0		
6	1	0		
7	1	0		
8	4	3	2	0
9	4	0		
10	3	0		
11	2	0		
12	6	4	1	0
13	4	3	1	0
14	5	3	1	0
15	1	0		
16	5	3	2	0
17	3	0		
18	5	2	1	0
19	5	2	1	0
20	3	0		
21	2	0		
22	1	0		
23	5	0		
24	4	3	1	0
25	3	0		
26	6	2	1	0
27	5	2	1	0
28	3	0		
29	2	0		
30	6	4	1	0

An extended table for ≤ 100 can be found in Lidl and Niederreiter [1997: 563].

Table C.3 Prime Factors of First 32 Mersenne Numbers

n	Prime Factors of M_n	n	Prime Factors of M_n
2	3	17	131071
3	7	18	$3^3 \cdot 7 \cdot 19 \cdot 73$
4	$3 \cdot 5$	19	524287
5	31	20	$3 \cdot 5^2 \cdot 11 \cdot 31 \cdot 41$
6	$3^2 \cdot 7$	21	$7^2 \cdot 127 \cdot 337$
7	127	22	$3 \cdot 23 \cdot 89 \cdot 683$
8	$3 \cdot 5 \cdot 17$	23	$47 \cdot 178481$
9	$7 \cdot 73$	24	$3^2 \cdot 5 \cdot 7 \cdot 13 \cdot 17 \cdot 241$
10	$3 \cdot 11 \cdot 31$	25	$31 \cdot 601 \cdot 1801$
11	$23 \cdot 89$	26	$3 \cdot 2731 \cdot 8191$
12	$3^2 \cdot 5 \cdot 7 \cdot 13$	27	$7 \cdot 73 \cdot 262657$
13	8191	28	$3 \cdot 5 \cdot 29 \cdot 43 \cdot 113 \cdot 127$
14	$3 \cdot 43 \cdot 127$	29	$233 \cdot 1103 \cdot 2089$
15	$7 \cdot 31 \cdot 151$	30	$3^2 \cdot 7 \cdot 11 \cdot 31 \cdot 151 \cdot 331$
16	$3 \cdot 5 \cdot 17 \cdot 257$	31	2147483647

D

Discrete Fourier Transforms

D.1 Discrete Fourier Transform

The discrete Fourier transform (DFT) that transforms one function into another is called the *frequency domain representation*; it requires a discrete input function whose nonzero real or complex values are finite. Unlike the discrete-time Fourier transform (DIFT), the DFT only evaluates enough frequency components to reconstruct the finite segment that was analyzed. The DFT is used in processing information (data) stored in computers, and in signal processing and related fields where the frequencies in a signal are analyzed. An efficient computation of the DFT is provided by a fast Fourier transform (FFT) algorithm. Although DFT refers to a mathematical transformation while FFT refers to a specific family of algorithms for computing DFTs, the FFT algorithms commonly used to compute DFTs often mean DFT in common terminology, which has now become confusing by taking FFT as synonymous with DFT. For more details, see Kythe and Schäferkotter [2005: 299 ff]. The DFT is defined as follows: Let a sequence of n nonzero complex numbers $\{x_0, \ldots, x_{n-1}\}$ be transformed into a sequence of m complex numbers $\{X_0, \ldots, X_{m-1}\}$ by the formula

$$X_j = \sum_{k=0}^{n-1} x_k\, e^{-2\pi jk\, i/m} \quad \text{for } 0 \leq j \leq m-1;\ i = \sqrt{-1}, \tag{D.1}$$

where

$$x_k = \frac{1}{m} \sum_{j=0}^{m-1} X_j\, e^{2\pi kj\, i/m} \quad \text{for } 0 \leq k \leq n-1. \tag{D.2}$$

Formula (D.1) is known as the DFT analysis equation and its inverse (D.2) the DFT synthesis equation, or the *inverse discrete Fourier transform* (IDFT). The DFT pair can be represented as

$$x_k \xrightarrow{DFT} X_j \quad \text{and} \quad X_j \xrightarrow{IDFT} x_k. \tag{D.3}$$

447

Example D.1. Let $n = 4$, and define a nonperiodic sequence x_k by

$$x_k = \begin{cases} 2 & k = 0, \\ 3 & k = 1, \\ -1 & k = 2, \\ 1 & k = 3. \end{cases}$$

By (D.1), the 4-point DFT of x_k is

$$X_j = \sum_{k=0}^{3} x_k\, e^{-2\pi k j\, i/4} = 2 + 3\, e^{-\pi j\, i/2} - e^{-\pi j\, i} + e^{-3\pi j\, i/2}, \quad 0 \le j \le 3.$$

Thus, $X_0 = 2+3-1+1 = 5$; $X_1 = 2-3\,i+1+i = 3-2\,i$; $X_2 = 2-3-1-1 = -3$; $X_3 = 2 - 3\,i + 1 - i = 3 + 2\,i$. The vertical bar plots of the function x_k and its magnitude and phase are presented in Figure D.1. ∎

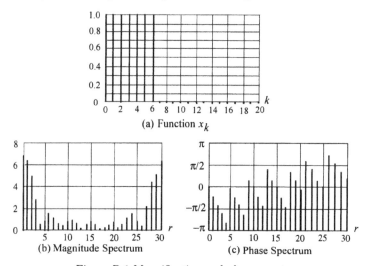

(a) Function x_k

(b) Magnitude Spectrum

(c) Phase Spectrum

Figure D.1 Magnification and phase spectra.

Example D.2. The n-point DFT of the nonperiodic sequence x_k of length n is defined by $x_k = \begin{cases} 1, & 0 \le k \le (n_1 - 1), \\ 0, & n_1 \le k \le n. \end{cases}$ Then

$$X_j = \sum_{k=0}^{n_1-1} e^{2jk\pi\, i/n} = \begin{cases} n_1, & j = 0, \\ \dfrac{1 - e^{2n_1 j\pi\, i/n}}{1 - e^{2j\pi\, i/n}}, & j \ne 0, \end{cases}$$

$$= \begin{cases} n_1, & j = 0, \\ e^{-(n_1-1)\pi j/n}\, \dfrac{\sin(n_1 j\pi/n)}{\sin(j\pi/n)}, & j \ne 0. \end{cases}$$

The magnitude $|X_j| = \begin{cases} n_1 & \text{for } j = 0, \\ \dfrac{\sin(n_1 j\pi/n)}{\sin(j\pi/n)} & \text{for } j \neq 0. \end{cases}$

The phase is given by $\phi(X_j) = \begin{cases} 0 & \text{for } j = 0, \\ \sin(n_1 j\pi/n) & \text{for } j \neq 0. \end{cases}$

D.2 DFT in Matrix Form

Expand (D.1) in terms of the time and frequency indices (k, j), and we get for $n = m$:

$$X_0 = x_0 + x_1 + x_2 \cdots + x_{n-1},$$
$$X_1 = x_0 + x_1\, e^{-2\pi i/n} + x_2\, e^{-4\pi i/n} + \cdots + x_{n-1}\, e^{-2(n-1)\pi i/n},$$
$$X_2 = x_0 + x_1\, e^{-4\pi i/n} + x_2\, e^{-8\pi i/n} + \cdots + x_{n-1}\, e^{-4(n-1)\pi i/n},$$
$$\vdots$$
$$X_{n-1} = x_0 + x_1\, e^{-2(n-1)\pi i/n} + x_2\, e^{-4(n-1)\pi i/n} + \cdots + x_{n-1}\, e^{-2(n-1)^2\pi i/n},$$

$$\tag{D.4}$$

which can be written in matrix form as

$$\begin{Bmatrix} X_0 \\ X_1 \\ X_2 \\ \vdots \\ X_{n-1} \end{Bmatrix} = \begin{bmatrix} 1 & 1 & 1 & \cdots & 1 \\ 1 & e^{-2\pi i/n} & e^{-4\pi i/n} & \cdots & e^{-2(n-1)\pi i/n} \\ 1 & e^{-4\pi i/n} & e^{-8\pi i/n} & \cdots & e^{-2(n-1)\pi i/n} \\ \vdots & \vdots & \vdots & & \vdots \\ 1 & e^{-2(n-1)\pi i/n} & e^{-4(n-1)\pi i/n} & \cdots & e^{-2(n-1)^2\pi i/n} \end{bmatrix} \begin{Bmatrix} x_0 \\ x_1 \\ x_2 \\ \vdots \\ x_{n-1} \end{Bmatrix}.$$

$$\tag{D.5}$$

Similarly, for IDFT in matrix form we have

$$\begin{Bmatrix} x_0 \\ x_1 \\ x_2 \\ \vdots \\ x_{n-1} \end{Bmatrix} = \frac{1}{n} \begin{bmatrix} 1 & 1 & 1 & \cdots & 1 \\ 1 & e^{2\pi i/n} & e^{4\pi i/n} & \cdots & e^{2(n-1)\pi i/n} \\ 1 & e^{4\pi i/n} & e^{8\pi i/n} & \cdots & e^{2(n-1)\pi i/n} \\ \vdots & \vdots & \vdots & & \vdots \\ 1 & e^{2(n-1)\pi i/n} & e^{4(n-1)\pi i/n} & \cdots & e^{2(n-1)^2\pi i/n} \end{bmatrix} \begin{Bmatrix} X_0 \\ X_1 \\ X_2 \\ \vdots \\ X_{n-1} \end{Bmatrix}.$$

$$\tag{D.6}$$

Example D.3. In Example D.2, we have $X = [5\ 3 - 2i\ -3\ 3 + 2i]$. Then

$$\begin{Bmatrix} x_0 \\ x_1 \\ x_2 \\ \vdots \\ x_{n-1} \end{Bmatrix} = \frac{1}{4} \begin{bmatrix} 1 & 1 & 1 & \cdots & 1 \\ 1 & e^{\pi i/2} & e^{\pi i} & e^{3\pi i/2} & \\ 1 & e^{\pi i} & e^{3\pi i/2} & e^{3\pi i} & \\ 1 & e^{3\pi i/2} & e^{3\pi i} & \cdots & e^{9\pi i/2} \end{bmatrix} \begin{Bmatrix} 5 \\ 3 - 2i \\ -3 \\ \vdots \\ 3 + 2i \end{Bmatrix} = \begin{Bmatrix} 2 \\ 3 \\ -1 \\ 1 \end{Bmatrix}. \ \blacksquare$$

D.3 DFT Basis Functions

If we express (D.6) in the form

$$
\begin{Bmatrix} x_0 \\ x_1 \\ x_2 \\ \vdots \\ x_{n-1} \end{Bmatrix} = \frac{1}{n} X_0 \begin{Bmatrix} 1 \\ 1 \\ 1 \\ \vdots \\ 1 \end{Bmatrix} + \frac{1}{n} X_1 \begin{Bmatrix} 1 \\ e^{2\pi i/n} \\ e^{4\pi i/n} \\ \vdots \\ e^{2(n-1)\pi i/n} \end{Bmatrix}
$$

$$
+ \frac{1}{n} X_2 \begin{Bmatrix} 1 \\ e^{4\pi i/n} \\ e^{8\pi i/n} \\ \vdots \\ e^{4(n-1)\pi i/n} \end{Bmatrix} + \cdots + \frac{1}{n} X_{n-1} \begin{Bmatrix} 1 \\ e^{2(n-1)\pi i/n} \\ e^{4(n-1)\pi i/n} \\ \vdots \\ e^{2(n-1)^2 \pi i/n} \end{Bmatrix},
\tag{D.7}
$$

then it is obvious that the DFT basis functions f_j are the columns of the right-hand square matrix B, i.e.,

$$
f_j = \frac{1}{n} \begin{bmatrix} 1 & e^{2\pi j i/n} & e^{4\pi j i/n} & \cdots & e^{2(n-1)j\pi i/n} \end{bmatrix}^T, \quad \text{for } 0 \leq j \leq n - 1.
$$

Eq (D.7) represents a DT sequence as a linear combination of complex exponentials, which are weighted by the corresponding DFT coefficients. Such a representation can be used to analyze linear time-invariant systems.

D.4 Properties of DFT

The properties of the m-point DFT are as follows:

(i) PERIODICITY. $X_j = X_{j+pm}$ for $0 \leq j \leq m - 1$, where $p \in \mathbb{R}^+$. In other words, the m-point DFT of an aperiodic sequence of length n, $n \leq m$, is periodic with period m.

(ii) LINEARITY. If x_k and y_k are two DT sequence with the m-point DFT pairs: $x_k \xrightarrow{DFT} X_j$ and $y_k \xrightarrow{DFT} Y_j$, then for any arbitrary constants a and b (which may be complex)

$$
ax_k + by_k \xrightarrow{DFT} a_1 X_j + bY_j.
\tag{D.8}
$$

(iii) ORTHOGONALITY. The column vectors f_j of the DFT matrix, defined in (D.7), form the basis vectors of the DFT and are orthogonal with respect to each other such that

$$
f_j^h \cdot f_l = \begin{cases} m & \text{for } j = l, \\ 0 & \text{for } j \neq l, \end{cases}
$$

where $\cdot$ denotes the dot product and h the Hermitian operation.

(iv) HERMITIAN SYMMETRY. This Hermitian symmetry implies that for the m-point DFT X_j of a real-valued aperiodic sequence x_k

$$X_j = \overline{X}_{m-j}, \tag{D.9}$$

where $\overline{X}_j$ is the complex conjugate of X_j. In other words, X_j is conjugate-symmetric about $j = m/2$. The magnitude $|X_{m-j}| + X_j|$, and the phase $\phi(X_{m-j}) = -\phi(X_j)$, i.e., the phase of the spectrum is odd.

(v) TIME SHIFTING. If $x_k \xrightarrow{DFT} X_j$, then for an m-point DFT and an arbitrary integer k_0

$$x_{k-k_0} \xrightarrow{DFT} e^{2k_0 j\pi\, i/m} X_j. \tag{D.10}$$

(vi) CIRCULAR CONVOLUTION. For two DT sequences x_k and y_k with the m-point DFT pairs: $x_k \xrightarrow{DFT} X_j$ and $y_k \xrightarrow{DFT} Y_j$, the circular convolution is defined by

$$x_k \otimes y_k \xrightarrow{DFT} X_j Y_j, \tag{D.11}$$

and

$$x_k\, y_k \xrightarrow{DFT} \frac{1}{m} [X_j \otimes Y_j], \tag{D.12}$$

where $\otimes$ denotes the circular convolution operation. In this operation the two sequences must be of equal length.

PARSEVAL'S THEOREM. If $x_k \xrightarrow{DFT} X_j$, then the energy E_x of an aperiodic sequence x_k of length n can be written in terms of its m-point DFT as

$$E_x = \sum_{k=0}^{n-1} |x_k|^2 = \frac{1}{m} \sum_{k=0}^{m-1} |X_j|^2. \tag{D.13}$$

In other words, the DFT energy preserves the energy of the signal within a scale factor of m.

Example D.4. Notation DFT index $= j$; DTFT frequency: $\omega_j = 2\pi j/n$; DFT coefficients X_j; DTFT coefficients X_ω. Using the DFT, calculate the DTFT of the DT decaying exponential sequence $x_k = 0.6^k u + k$. For time limitation, apply a rectangular window of length $n = 10$; then the truncated sequence is given by

$$x_k^{\mathrm{w}} = \begin{cases} 0.6^k & \text{for } 0 \le k \le 9, \\ 0 & \text{elsewhere.} \end{cases}$$

The DFT coefficients are compared with the IDFT coefficients in Table D.1.

Table D.1 Comparison of DFT Coefficients with IDFT Coefficients

j	ω_j	X_j	X_ω
-5	$-\pi$	0.6212	0.6250
-4	-0.8π	$0.6334 + 0.1504\,i$	$0.6337 + 0.3297\,i$
-3	-0.6π	$0.6807 + 0.3277\,i$	$0.6337 + 0.1513\,i$
-2	-0.4π	$0.8185 + 0.5734\,i$	$0.8235 + 0.5769\,i$
-1	-0.2π	$1.3142 + 0.9007\,i$	$1.322 + 0.9062\,i$
0	0	2.4848	2.5000
1	0.2π	$1.3142 - 0.9007\,i$	$1.322 - +0.9062\,i$
2	0.4π	$0.8185 - 0.5734\,i$	$0.8235 - 0.5769\,i$
3	0.6π	$0.6807 - 0.3277\,i$	$0.6849 - 0.3297\,i$
4	0.8π	$0.6334 - 0.1504\,i$	$0.6373 - 0.1513\,i$

Note that the above DFT coefficients can be computed from the correspondence

$$0.6^k u_k \overset{CTFT}{\longrightarrow} \frac{1}{1 - 0.6^{-\omega i}},$$

where CTFT refers to continuous-time Fourier transform. ∎

Example D.5. The DTFT of the aperiodic sequence $x_k = [2, 1, 0, 1]$ for $0 \le k \le 3$ is given by $X_j = [4, 2, 0, 2]$ for $0 \le j \le 3$. By mapping in the DTFT domain, the corresponding DTFT coefficients are given by $X_{\omega_j} = [4, 2, 0, 2]$ for $\omega_j = [0, 0.5\pi, \pi, 1.5\pi]$ rads/sec. On the other hand, if the DTFT is computed in the range $-\pi \le \omega \le \pi$, then the DTFT coefficients are given by $X_{\omega_j} = [4, 2, 0, 2]$ for $\omega_j = [\pi, -0.5\pi, 0, 0.5\pi]$ rads/sec. ∎

D.5 Zero Padding

To improve the resolution of the frequency axis ω in the DFT domain, a common practice is to append additional zero-valued samples to the DT sequences. This process, known as *zero padding*, is defined for an aperiodic sequence x_k of length n by

$$x_k^{\text{zp}} = \begin{cases} x_k & \text{for } 0 \le k \le n - 1, \\ 0 & \text{for } n \le k \le m - 1. \end{cases} \tag{D.14}$$

Thus, the zero-padded sequence x_k^{zp} has an increased length of m. This improves the frequency resolution $\Delta\omega$ of the zero-padded sequence from $2\pi/n$ to $2\pi/m$.

D.6 Properties of m-Point DFT

Let the length of the DT sequence be $n \leq m$, and let the DT sequence be zero-padded with $m - n$ zero-valued samples. Then the properties of periodicity, linearity, orthogonality, Hermitian symmetry, and Parseval's theorem are the same as those given in §D.4.

Example D.6. (Circular convolution) Consider two aperiodic sequences $x_k = [0, 1, 2, 3]$ and $y_k = [5, 5, 0, 0]$ defined over $0 \leq k \leq 3$. We will use the property (D.11) to compute circular convolution as follows: Since $X_j = [6, -2 + 2i, -2, -2 - 2i]$ and $Y_j = [10, 5 - i, 0, 5 + 5i]$ for $0 \leq j \leq 3$, we have

$$x_k \otimes y_k \xrightarrow{DFT} [60, 20i, 0, -20i],$$

which after taking the inverse DFT yields

$$x_k \otimes y_k = [15, 5, 15, 25]. \blacksquare$$

D.7 Linear Convolution versus Circular Convolution

The linear convolution $x_k \star y_k$ between two time-limited DT sequences x_k and y_k of lengths n_1 and n_2, respectively, can be expressed in terms of the circular convolution $x_k \otimes y_k$ by zero-padding both x_k and y_k such that each sequence has length $N \geq (n_1 + n_2 - 1)$. It is known that the circular convolution of the zero-padded sequences is the same as that of the linear convolution. The algorithm for implementing the linear convolution of two sequences x_k and y_k is as follows:

STEP 1. Compute the N-point DFTs X_j and Y_j of the two time-limited sequences x_k and y_k, where the value of $N \geq n_1 + n_2 - 1$.

STEP 2. Compute the product $Z_j = X_j Y_j$ for $0 \leq j \leq N - 1$.

STEP 3. Compute the sequence z_k as the IDFT of Z_j. The resulting sequence z_k is the result of the linear convolution between x_k and y_k.

Example D.7. Consider the DT sequences

$$x_k = \begin{cases} 2 & \text{for } k = 0, \\ -1 & \text{for } |k| = 1, \\ 0 & \text{otherwise,} \end{cases} \quad \text{and} \quad \begin{cases} 2 & \text{for } k = 0, \\ 3 & \text{for } |k| = 1, \\ -1 & \text{for } |k| = 2, \\ 0 & \text{otherwise.} \end{cases}$$

STEP 1. Since the sequences x_k and y_k have lengths $n_1 = 3$ and $n_2 = 5$, the value of $N \geq 5 + 3 - 1 = 7$; so we set $N = 7$. Then, zero-padding in x_k is $N - n_1 = 4$ additional zeros, which gives $x'_k = [-1, 2, -1, 0, 0, 0, 0]$; similarly, zero-padding in y_k is $N - n_2 = 2$ additional zeros, which gives

$y'_k = [-1, 3, 2, 3, -1, 0, 0]$. The values of the DFT of x'_k and of y'_k are given in Table D.2.

Table D.2 Values of X'_j, Y'_j, Z_j for $0 \le j \le 6$

j	X'_j	Y'_j	Z_j
0	0	6	0
1	$0.470 - 0.589\,i$	$-1.377 - 6.031\,i$	$-4.199 - 2.024\,i$
2	$-5.440 - 2.384\,i$	$-2.223 + 1.070\,i$	$3.760 + 4.178\,i$
3	$-3.425 - 1.650\,i$	$-2.901 - 3.638\,i$	$3.933 + 17.247\,i$
4	$-3.425 + 1.650\,i$	$-2.901 + 3.638\,i$	$3.933 - 17.247\,i$
5	$-0.544 + 2.384\,i$	$-2.223 - 31.070\,i$	$3.760 - 4.178\,i$
6	$0.470 + 0.589\,i$	$-1.377 + 6.031\,i$	$-4.199 + 2.024\,i$

STEP 2. The value of $Z_j = X'_j Y'_j$ is shown in the fourth column of Table D.2.

STE 3. Taking the IDFT of Z_j gives

$$z_k = [0.998, -5, 5.001, -1.999, 5, -5.002, 1.001]. \ \blacksquare$$

D.8 Radix-2 Algorithm for FFT

Theorem D.1. *For even values of N, the N-point DFT of a real-valued sequence x_k of length $m \le N$ can be computed from the DFT coefficients of two subsequences: (i) x_{2k}, which contains the even-valued samples of x_k, and (ii) x_{2k+1}, which contains the odd-valued samples of x_k.*

This theorem leads to the following algorithm to determine the N-point DFT:

STEP 1. Determine the $(N/2)$-point DFT G_j for $0 \le j \le N/2 - 1$ of the even-numbered samples of x_k.

STEP 2. Determine the $(N/2)$-point DFT H_j for $0 \le j \le N/2 - 1$ of the odd-numbered samples of x_k.

STEP 3. The N-point DFT coefficients X_j for $0 \le j \le k-1$ of x_k are obtained by combining the $(N/2)$ DFT coefficients of G_j and H_j using the formula $X_j = G_j + W_k^j H_j$, where $W_k^j = e^{-2\pi i/N}$ is known as the *twiddle factor*. Note that although the index $j = o, \dots, N - 1$, we only compute G_j and H_j over $0 \le j \le (N/2-1)$, and any outside value can be determined using the periodicity properties of G_j and H_j, which are defined by $G_j = G_{j+N/2}$ and $H_j = H_{j+N/2}$.

The flow graph for the above method for $N = 8$-point DFT is shown in Figure D.2.

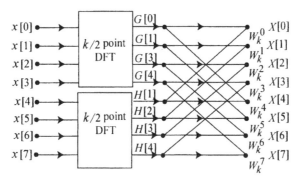

Figure D.2 Flow graph for 8-point DFT.

In general, this figure computes two $(N/2)$-point DFTs along with N complex additions and N complex multiplications. Thus, $(n/2)^2 + N$ complex additions and $(N/2)^2 + N$ complex multiplications are needed. Since $(N/2)^2 + N < N^2$ for $N > 2$, the above algorithm is considerably cost effective.

In the case when N is a power of 2, the $(N/2)$-point DFTs G_j and H_j can be computed by the formula

$$G_j = \sum_{nu=0,1,2,...}^{N/4-1} g_{2\nu}\, e^{-2\nu j\pi\, i/(N/4)} + W_{N/2}^j \sum_{\nu=0,1,2,...}^{N/4-1} g_{2\nu+1}\, e^{-2\nu j\pi\, i/(N/4)}, \text{(D.15)}$$

where the first summation will be denoted by G_j' and the second by G_j''. Similarly,

$$H_j = \sum_{nu=0,1,2,...}^{N/4-1} h_{2\nu}\, e^{-2\nu j\pi\, i/(N/4)} + W_{N/2}^j \sum_{\nu=0,1,2,...}^{N/4-1} h_{2\nu+1}\, e^{-2\nu j\pi\, i/(N/4)},$$

$$\text{(D.16)}$$

where the first summation will be denoted by H_j' and the second by H_j''. Formula (D.15) represents the $(N/2)$-point DFT G_j in terms of two $(N/4)$-point DFTs of the even- and odd-numbered samples of g_k; similarly, formula (D.16) represents the $(N/2)$-point DFT H_j in terms of two $(N/4)$-point DFTs of the even- and odd-numbered samples of h_k. The flow graphs of these cases are presented in Figure D.3.

Thus, for example, the 2-point DFTs G_0' and G_1' can be expressed as

$$G_0' = x_0\, e^{-2\nu j\pi i/2}\Big|_{\nu=0,j=0} + x_4\, e^{-\nu j\pi i/2}\Big|_{\nu=1,j=1} = x_0 + x_4,$$

and

$$G_1' = x_0\, e^{-2\nu j\pi i/2}\Big|_{\nu=0,j=1} + x_4\, e^{-\nu j\pi i/2}\Big|_{\nu=1,j=1} = x_0 - x_4.$$

Also note that in the flow graphs of Figure D.3, the twiddle factor for an 8-point DFT is

$$W_{N/2}^j = e^{-\nu j\pi i/(N/2)} = e^{-4\nu j\pi i/N} = W_N^{2j}.$$

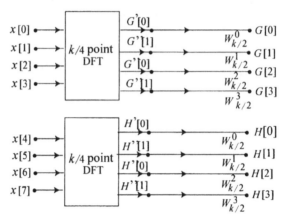

Figure D.3 Flow graph for $(N/2)$-point DFTs using $(N/4)$-point DFTs.

E

Software Resources

The following resources provide some basic software in the public domain useful to the readers, but without any warranty and without any implied warranty of merchantability or applicability for a specific purpose. The URLs were tested as being available at the time of publication, but we make no guarantees that the websites will remain available.

1. BITWISE OPERATIONS. A large number of C programs on the following topics in Boolean and bitwise operations (and more) is available (in the public domain) at *http:graphics.stanford.edu/ ˜ seander/bithacks.html*, entitled 'Bit Twiddling Hacks': (i) operation counting methodology; (ii) compute the sign of an operator; (iii) check if two integers have opposite signs; (iv) compute the integer absolute value (abs) without branching; (v) compute the min or max of two integers without branching; (vi) determine if an integer is a power of 2; (vii) sign extending; (viii) counting bit sets; (ix) compute parity; (x) swap values; (xi) reverse bit sequences; (xii) modulus division (compute remainders); (xiii) find integer $\log_2$ of an integer (position of the highest bit set); (xiv) find integer $\log_2$ of an integer with MSB n set in $O(n)$ operations; (xv) find integer $\log_2$ of an integer with 64-bit IEEE float; (xvi) find integer $\log_2$ of an integer with a look-up table; (xvii) find integer $\log_2$ of an n-bit integer in $O(\log n)$ operations; (xviii) find integer $\log_2$ of an n-bit integer in $O(\log n)$ operations with multiply and look-up; (xix) find integer $\log_{10}$ of an integer; (xx) find integer $\log_2$ of a 32-bit IEEE float; (xxi) find integer $\log_2$ of the power $(2, r)$-root of a 32-bit IEEE float (for unsigned r); (xxi) counting consecutive trailing zero bits (i.e., bit indices); (xxii) round up to the next highest power of 2; (xxiii) interleaving bits (i.e., compute Morton numbers); (xxiv) testing for ranges of bytes in a word and counting their occurrences; and (xxv) compute lexicographically next bit permutation. The details on ring counters can be found at *http;//www.allaboutcircuits. com/vol_4/chpt_12/6.html*.

2. SPARSE MODULO-2 MATRIX ROUTINES. These C subroutines are available at *http://www.cs.toronto.edu/ ˜ radford/ftp/LDPC-2006-02-08/imod2sparse. html*. See §17.8.

3. GOLAY CODES. A couple of notable websites for determining the lower bounds on Hamming and Golay codes are: *http://www.eng.tau.ac.il/~litsyn/tableand/index.html*, and *http://www.win. tue.nl/~aeb/voorlincod.html*.

The construction of the mask of a uniform dodecahedron and encoding and decoding of the Golay $\mathcal{G}_{12}$ is described in detail at the websites: *http://www.math.uic.edu/ ~fields/DecodingGolayHTML/introduction.html*, */encoding. html, decoding.html*, and */example.html* at the same website (check *google.com* also).

Details to establish a relationship between a uniform dodecadodecahedron and the extended binary Golay code $\mathcal{G}_{24}$ are available at the website: *http://homepages. wmich.edu/~drichter/golay.htm*. Also check Wolfram Math World for articles on the Golay codes (*http://mathworld.wolfram.com*).

HADAMARD MATRICES. Check Neil Sloane's library of Hadamard matrices (*http://www2.research.att.com/~njas/hadamard/*).

4. LOW-DENSITY PARITY-CHECK CODES. Software for LDPC codes is available at www.cs.toronto.edu/~radford/ftp/LDPC-2006-02-08/install.html, and www.cs.toronto.edu/~radford/ftp/LDPC-2004-05-04/pchk.html.

C programs for sparse mod-2 matrix routines are available at the website: *http://www.cs.utoronto.edu/~radford/ ftp/LDPC-2001-05-04/pchk.html*, and its 2006 version. This software deals only with linear block codes for binary (mod-2, GF(2)) vectors.

Also check the following websites: *http://www.csee.wvu.edu/wcrl/ldpc.htm (containing a list of research papers on the topic and provides other links to sites on LDPC codes), and http://www.inference.phy.cam.ac.uk/mackay/ CodesFiles. html (the homepage of MacKay, which contains graphical demonstration of various iterative decoding.) The website http://www.iterativeso lutions.com provides software for LDPC and turbo codes.*

5. FINITE FIELD ARITHMETIC. *Software for a number of topics in this area is available in books on modern algebra or abstract algebra.*

6. RANDOM AND PSEUDONUMBER GENERATORS. *Besides the Monte Carlo to generate random numbers, von Neumann [1951] generated an iterative pseuorandom number generator (PRNG). Like the Monte Carlo method, this algorithm also yields PRNs that eventually repeat themselves. Hence, these two algorithms are not satisfactory. Mersenne Twister Algorithm was invented in 1997 by Matsumoto and Nishimura to avoid many problems in PRNGs. It has a period of $2^{19937} - 1$ iterations ($> 4.3 \times 10^{6001}$), which is equidistributed in 623 dimensions for 32-bit random numbers. Details can be found at http:// www.math.sci.hiroshima-u.ac.jp/ ~m.mat/MT/SFMT/speed.htm. A PRNG suitable for cryptographic applications is called a cryptographically secure pseudorandom number generator (CSPRNG). The Mersenne Twister is not suitable for CSPRNG, but its version for cryptographic ciphering can be*

found as a pdf file at http://eprint.iacr.org/2005/165.pdf. A reliable source for CSPRNG is Luby [1996]. A PRNG can be generated from a nonuniform probability distribution, e.g., the inverse of cumulative Gaussian distribution $\mathrm{erf}^{-1}(x)$ with an ideal uniform PRNG in the domain $x \in (0,1)$ as input that produces a sequence of positive values of the Gaussian distribution. For faster computation the Ziggurat algorithm is used (see Devroye [1986]. Another dependable publication on pseudorandom numbers and their application to cryptography is by Luby [1996]. The GNU Scientific Library http://www.gnu.org/software/gsl/ is a free (GPL) C library that contains, among a lot of other useful software, a number of PRNG algorithms. Tina's Random Number Generator Library http://trng.berlios.de/ is a free BSD license C++ random number generator for sequential and Monte Carlo simulations. DieHarde (http://www.phy.duke.edu/˜rgb/General/rand_rate.php) is a free GPL C random number test suite. For linux, check the random number generator (http://eprint.icar.org/2006/-86.

Note that websites sometimes have short life spans. If any of the above websites are not found, use http://www.google.com to check the required topic.

Bibliography

M. Abramowitz and I. A. Stegun (Eds.). 1972. *Handbook of Mathematical Functions.* New York: Dover.

N. M. Abramson. 1961. Error-correcting codes from linear sequential circuits. *Proc. 4th London Symposium on Information Theory* (C. Cherry, Ed.), 26–40. London: Butterworths.

———. 1963. *Information Theory and Coding.* New York: McGraw-Hill.

J. Afzal, T. Stockhammer, T. Gasiba, and W. Xu. 2006. Video streaming over MBMS: A system design approach. *Journal of Multimedia.* 1(5): 25–35.

J. D. Alamen and D. E. Knuth. 1964. Tables of finite fields. *Sankhyā, Ser. A.* 26: 305–328.

C. A. Allen. 1966. Design of digital memories that tolerate all classes of defects. *SEL TR 4662-1.* Stanford University.

S. Amari and H. Nagaoka. 2000. *Methods of Information Geometry.* New York: Oxford University Press.

G. E. Andrews. 1976. *The Theory of Partitions* in *Encyclopedia of Mathematics and its Applications.* Vol. 2. Readding, MA: Addison-Wesley.

P. J. Antsaklis and A. N. Michel. 2006. *Linear Systems.* Boston: Birkhäuser.

S. Arimoto. 1972. An algorithm for computing the capacity of arbitrary memoryless channels. *IEEE Transactions on Information Theory.* IT-18 : 14–20.

E. F. Assmus, Jr. and J. D. Key. 1994. *8 Steiner Systems: Designs and Their Codes.* Cambridge University Press, 295–316.

L. R. Bahl, J. Cocke, F. Jelinek, and J. Raviv. 1974. Optimal decoding of linear codes for minimizing symbol error rate. *IEEE Transactions on Information Theory.* IT-20: 248–287.

A. V. Balakrishnan. 1968. *Communication Theory.* New York: McGraw-Hill.

L. Baumert, S. W. Golomb, and M. Hall, Jr. 1962. Discovery of an Hadamard matrix of order 92. *Bull. Am. Math. Soc.* 68: 237–238.

——— and M. Hall, Jr. 1965. A new construction for Hadamard matrices. *Bull. Am. Math. Soc.* 71: 169–170.

T. Bayes. 1764. An Essay Toward Solving a Problem in the Doctrine of Chances, *Philosophical Transactions of the Royal Society of London.* 53: 370-418. (Facsimile available online: the original essay with an introduction by his friend Richard Price); Bayes' Original Essay (in PDF) (UCLA Statistics Department/History of Statistics) A Short Biography of Thomas Bayes (University of St. Andrews, MacTutor History of Mathematics Archive).

J. T. B. Beard, Jr. and K. I. West. 1974. Factorization tables for $x^n - 1$ over $GF(q)$. *Math. Comp.* 28: 1167–1168.

M. Beermann, L. Schmalen, and P. Vary. 2011. Improved decoding of binary and non-binary LDPC codes by probabilistic shuffled belief propagation. *Proc. ICC, IEEE*, 231–235.

H. Behairy and S. Chang. 2000. Parallel concatenated Gallager codes. *Electonic Letters.* 36: 2025–2026.

E. R. Berlekamp. 1965. On decoding binary Bose-Chaudhuri-Hocquenghem codes. *IEEE Transactions on Information Theory.* IT-11: 577–579.

———. 1968. *Algebraic Coding Theory.* New York: McGraw-Hill.

———. 1970. Factoring polynomials over large fields. *Math. Comp.* 24: 713–735.

———. 1972. A survey of coding theory. *J. Royal Statist. Soc. Ser. A.* 135: 44–73.

——— R. E. Peile, and S. P. Pope. 1987. The application of error control to communications. *IEEE Communications Magazine.* 25: 44–47.

C. Berrou. 2003. The ten-year-old Turbo codes are entering into service. *IEEE Communications Magazine.* 41: 110–116.

——— A. Glavieux, and P. Thitimajshima. 1993. Near Shannon limit error-correcting coding and decoding: Turbo-codes. *IEEE Proceedings of the International Conf. on Communications*, Geneva, Switzerland, May 1993. ICC '93 : 1064–1070.

——— and A. Glavieux. 1996. Near optimum error correcting coding and decoding: Turbo-codes. *IEEE Trans. Commun.* 44: 1261–1271.

T. A. Berson. 1992. Differential cryptanalysis mod 2^{32} with applications to MD5. *Eurocrypt.* 71–80.

Jürgen Bienbauer. 2004. *Introduction to Coding Theory.* Boca Raton, FL: Chapman&Hall/CRC.

G. Birkoff and T. C. Bartee. 1970. *Modern Applied Algebra.* New York: McGraw-Hill.

——— and S. McLane. 1977. *A Survey of Modern Algebra.* New York: McGraw-Hill.

Birnbaum A. 1962. On the Foundations of Statistical Inference, *Journal of the American Statistical Association.* 53: 259-326.

C. M. Bishop. 2006. *Pattern Recognition and Machine Learning.* New York: Springer-Verlag. Also available at http://research.microsoft.com/enus/um/people/cmbishop/PRML/index.htm

Å. Björk and V. Pereyra. 1970. Solution of Vandermonde system of equations. *Math. Comput.* 24: 893–903.

R. E. Blahut. 1972. Computation of channel capacity and rate distortion functions. *IEEE Transactions on Information Theory.* IT-18 : 460–473.

———. 1979. Algebraic codes in the frequency domain, in *Algebraic Coding*

Theory and Applications (G. Longo, Ed.), CISM Courses and Lectures, Vol. 258: 447–494. Vienna: Springer-Verlag.

———. 1983. *Theory and Practice of Error Control Codes*. Reading, MA: Addison-Wesley.

I. F. Blake. 1973. *Algebraic Coding Theory: History and Development*. Stroudsburg, PA: Dowden-Hutchinson-Ross.

———. 1979. Codes and designs. *Math. Mag.* 52: 81–95.

——— and R. C. Mullin. 1975. *The Mathematical Theory of Coding*. New York: Academic Press.

R. W. Blockett. 1970. *Finite Dimensional Linear Systems*. New York: Wiley.

Bert den Boer and Antoon Bosselaers. 1993. *Collisions for the Compression Functions of MD5*. Berlin: Springer-Verlag.

R. C. Bose and D. K. Ray-Chaudhuri. 1960. On a class of error correcting binary group codes. *Information and Control*. 3: 68–79.

———. 1960. Further results on error correcting binary group codes. *Information and Control*. 3: 279–290.

——— and S. S. Shrikhande. 1959. A note on a result in the theory of code construction. *Information and Control*. 2: 183–194.

A. Brunton, C. Shu, and G. Roth. 2006. Belief propagation on the GPU for stereo vision. *3rd Canadian Conference on Computer and Robot Vision*.

J. W. Byers, M. Luby, M. Mitzenmacher, and A. Rege. 1998. A digital fountain approach to reliable distribution of bulk data. *Proc. ACM SIGCOMM '98*, Vancouver, Sept. 1998: 56–67.

———, M. Luby, and M. Mitzenmacher. 2002. A digital fountain approach to asynchronous reliable multicast. *IEEE Journal on Selected Areas in Communications*. 20(8): 1528–1540.

———, J. Considine, M. Mitzenmacher, and S. Rossi. 2002. Informed content delivery across adaptive overlay networks. *Proc. ACM SIGCOMM 2002*, Pittsburgh, PA, August 19-23.

G. Caire, G. Taricco and E. Bigliri. 1996. Bit-interleaved coded modulation. *IEEE Transactions on Information Theory*. 44: 927–945.

D. Calvetti and L. Reichel. 1993. Fast inversion of Vandermonde-like matrices involving orthogonal polynomials. *BIT*. 33: 473–484.

P. J. Cameron and J. H. van Lint. 1980. *Graphs, Codes and Designs*. Lecture Notes Series #43, London Mathematical Society.

M. Capalbo, O. Reingold, S. Vadhan, and A. Wigderson. 2002. Randomness conductors and constant-degree lossless expanders. *STOC '02*, May 19-21, 2002, Montreal, Quebec, Canada.

Mario Cardullo et al. 1973. Transponder apparatus and system. *U.S. Patent # 3,713,148* (*http://www.google.com/patemnts?vid=3713148*).

———, K.-M. Cheung, and R. J. McFliece. 1987. Further Results on Finite-State Codes. *TDA Progress Report 42-92*, Oct.–Dec. 1987. Jet Propulsion Laboratory, Padsadena, CA. 56–62.

R. D. Carmichael. 1937 (Reprint 1956). *Introduction to the Theory of Groups of Finite Order.* New York: Dover.

Carnap, R. 1962. *Logical Foundations of Probability,* 2nd edition. Chicago: University of Chicago Press.

G. Castagnoli, S. Bräuer, and M. Herrmann. 1993. Optimization of cyclic redundancy-check codes with 24 and 32 parity bits. *IEEE Trans. Commun.* 41: 883–892.

C. L. Chen. 1981. High-speed decoding of BCH codes. *IEEE Transactions on Information Theory.* IT-27: 254-256.

H. Chernoff. 1952. A measure of asymptotic efficiency for tests of a hypothesis based on the sum of observations. *Annals of Math. Statistics.* 23: 493–507.

———.1981. A note on an equality involving the normal distribution. *Annals of Probability.* 9:533.

R. T. Chien. 1964. Cyclic decoding procedures for Bose-Chaudhuri-Hocquenghem codes. *IEEE Transactions on Information Theory.* IT-10: 357–363.

T. Chihara. 1978. *An Introduction to Orthogonal Polynomials.* New York: Gordon and Breach.

———. 1987. Some problems for Bayesian confirmation theory. *British Journal for the Philosophy of Science.* 38: 551–560.

———and D. Stanton. 1987. Zeros of generalized Krawtchouk polynomials. *IMA Preprint Series # 361, Oct. 1987.* Minneapolis, MN: Institute for Mathematics and its Applications.

Christensen, D. 1999. Measuring Evidence. *Journal of Philosophy.* 96: 437–461.

S. Chung, T. Richardson, and R. Urbanke. 2001. Analysis of sum-product decoding of low-density parity-check codes using a Gaussian approximation. *IEEE Transactions on Information Theory.* 47: 657–670.

———, T. Richardson, and R. Urbanke. 2001. On the design of low-density parity-check codes within 0.0045 dB of the Shannon limit. *IEEE Communications Letters.* Feb. 2001.

R. Church. 1935. Tables of irreducible polynomials for the first four prime moduli. *Ann. of Math. (2).* 36: 198–209.

J. H. Conway. 1968. A tabulation of some information concerning finite fields, in *Computers in Mathematical Research* (R. F. Churchhouse and J.-C. Herz, Eds.) 17–50. Amsterdam: North-Holland.

———and Neil J. A. Sloane. 1990. A new upper bound on the minimal distance of self-dual codes. *IEEE Transactions on Information Theory.* 36: 1319–1333.

———, and N. J. A. Sloane. 1999. *Sphere Packing, lattices and Groups,* 3rd ed. New York: Springer-Verlag.

———. 1993. Self-dual codes over the integers modulo 4. *J. Comb. Theory, Ser. A.* 62: 234–254.

J. Coughlan. 2009. *Belief Propagation: A Tutorial.* The Smith-Kettlewell Eye Research Institute.

_____ and H. Shen. 2007. Dynamic quantization for belief propagation in sparse spaces. *Computer Vision and Image Understanding (CVIU)*. Special issue on Generative-Model Based Vision. 106: 47–58.

T. M. Cover and J. A Thomas. 2001. *Elements of Information Theory*. New York: Wiley.

M. F. Cowlinshaw. 2003. Decimal Floating-Point: Algorism for Computer. *Proceedings of the 16th IEEE Symposium on Computer Arithmetic*. 104–111.

H. S. M. Coxeter. 1973 *Regular Polytopes*. 3rd ed. New York: Dover.

J. A. Croswell. 2000. *A Model for Analysis of the Effects of Redundancy and Error Correction on DRAM Memory Yield and Reliability*. ME thesis, MIT.

R. T. Curtis. 2008. A new combinatorial approach to M_{24}. *Proceedings of the Cambridge Philosophical Society*. 79: 25–42.

_____. 1977. The maximal subgroups of M_{24}. *Math. Proc. Camb. Phil. Soc.* 81: 185–192.

_____. 1990. Geometric interpretations of the 'natural' generators of the Mathieu groups. *Math. Proc. Camb. Phil. Soc.* 107: 19–26.

Dale, A. I. 1989. Thomas Bayes: A Memorial. *The Mathematical Intelligencer.* 11: 18-19.

P. Delsarte. 1969. A geometric approach to a class of cyclic codes. *J. Combinatorial Theory.* 6: 340–359.

_____. 1970. On cyclic codes that are invariant under the general linear group. *IEEE Transactions on Information Theory.* IT-16: 760–769.

_____. 1973. An algebraic approach to coding theory. *Philips Research Reports Supplements.* 10.

_____. 1976. Properties and applications of the recurrence $F(i+1, k+1, n+1) = q^{k+1}F(i, k+1, n) = q^k F(i, k, n)$. *SIAM J. Appl. Math.* 31: 262–270.

_____. 1978. Bilinear forms over a finite field, with applications to coding theory. *J. Combinatorial Theory, Ser. A.* 25: 226–241.

J. Dènes and A. D. Keedwell. 1974. *Latin Squares and Their Applications.* New York: Academic Press.

L. Devroye. 1986. *Non-uniform Random Variate Generation.* New York: Springer-Verlag. (*http://cg.scs.carleton.ca/ ˜luc/rnbookindex.html*).

C. Di, D. Proietti, I. E. Telatar, T. Richardson, and R. Urbanke. 2002. Finite-length analysis of low-density parity-check codes on the binary erasure channel. *IEEE Transactions on Information Theory.* IT-48: 1570–1579.

P. Diaconis and R. L. Graham. 1985. The Radon transform on Z_2^n. *Pacific J. Math.* 118: 323–345.

L. E. Dickson. 1958. *Linear Groups with an Exposition of the Galois Field Theory.* New York: Dover.

N. Dimitrov and C. Plaxton. 2005. Optimal cover time for graph-based coupon collector process. *Proc. ICALP, 2005, LNCS 3580.* 702–716.

R. L. Dobrushin. 1972. Survey of Soviet research in coding theory. *IEEE*

Transactions on Information Theory. IT-18: 703–724.

D. Ž. Doković. 2008. Hadamard matrices of order 764 exist. *Combinatorica*. 28: 487–489.

J. L. Doob. 1953. *Stochastic Processes*. New York: Wiley.

Dragomir Ž. Doković. 2008. Hadamard matrices of order 764 exist, *Combinatorica*. 28: 487–489.

P. G. Drazin and R. S. Johnson. 1989. *Solitons: An Introduction*. 2nd ed. Cambridge. UK : Cambridge University Press.

R. Durrett. 1995. *Probability: Theory and Examples*. Belmont, CA: Duxbury Press.

———. 1996. *Stochastic Calculus*. Boca Raton, FL: CRC Press.

A. W. F. Edwards. 1972. *Likelihood*. Cambridge: Cambridge University Press.

Peter Elias. 1954. Error-free coding. *Technical Report 285, September 22, 1954*. Research Laboratory of Electronics, MIT, Cambridge, MA

———. 1955. Coding for two noisy channels. *Information Theory*. Third London Symposium, Sept. 1955.

B. Elspas and R. A. Short. 1962. A note on optimum burst-error-correcting codes. *IEEE Transactions on Information Theory*. IT-8: 39–42.

Tuvi Etzion, Ari Trachtenberg, and Alexander Vardy. 1999, Which codes have cycle-free Tanner graphs? *IEEE Transactions on Information Theory*. 45: 2173–2189.

A. Feinstein. 1954. A new basic theorem of information theory, *Technical Report No. 282*. MIT: Research Laboratory of Electronics.

P. F. Feltzenszwalb and D. P. Huttenlocher. 2006. Efficient belief propagation for early vision. *Int. J. Compu. Vision*. 70: 41–54.

Joe Fields. 2009. Decoding the Golay code by hand. Chicago: University of Illinois at Chicago. *http://www.math.uic.edu/~fields/DecodingGolay HTML/introduc-tion.html*

John G. Fletcher. 1982. An arithmetic checksum for serial transmissions. *IEEE Trans. on Commun*. COM-30: 247–252.

Ivan Flores. 1963. *The Logic of Computer Arithmetic*. Englewood Cliffs, NJ: Prentice-Hall.

G. D. Forney, Jr. 1965. On decoding BCH codes. *IEEE Transactions on Information Theory*. IT-11: 549–557.

———. 1966. *Concatenated Codes*. Cambridge, MA: MIT Press.

R. G. Gallager. 1962. Low-density parity-check codes. *IRE Trans. Information Theory*. 8: 21–28.

———. 1963. *Low-Density Parity-Check Codes*. Cambridge, MA: MIT Press.

———. 1968. *Information Theory and Reliable Communication*. New York: Wiley.

T. Gasiba, T. Stockhammer, J. Afzal, and W. Xu. 2006. System design and advanced receiver techniques for MBMS broadcast services. *Proc. IEEE*

International Conf. on Communications. Istanbul, Turkey, June 2006.

S. I. Gel'fand. 1977. Capacity of one broadcast channel. *Problemy Peredachi Informatsii (Problems in Information Transmission)*. 13: 106–108.

W. J. Gilbert and W. H. Nicholson. 2004. *Modern Algebra and Applications*, 2nd edition. New York: Wiley.

Clark Glymour. 1980. *Theory and Evidence*. Princeton, NJ: Princeton University Press.

J.-M. Goethals. 1976. Nonlinear codes defined by quadratic forms over GF(2), *Information and Control*. 31: 43–74.

———— and S. L. Snover. 1972. Nearly perfect binary codes, *Discrete Math*. 3: 65–88.

Marcel J. E. Golay. 1949. Notes on digital coding, *Proceedings IRE*. 37: 657.

————. 1977. Sieves for low autocorrelation binary sequences, *IEEE Transactions on Information Theory*. IT-23: 43–51.

————. 1982. The merit factor of long low autocorrelation binary sequences, *IEEE Transactions on Information Theory*. IT-28: 343–549.

S. W. Golomb and L. D. Baumert. 1963. The search for Hadamard matrices. *Am. Math. Monthly*. 70: 12–17.

W. C. Gore and A. B. Cooper. 1970. Comments on polynomial codes. *IEEE Transactions on Information Theory*. IT-16: 635–638.

D. C. Gorenstein and N. Zierler. 1961. A class of error-correcting codes in p^m symbols. *J. Soc. Industr. Appl. Math*. 9: 207–214.

R. Gray. 1990. *Entropy and Information Theory*. New York : Springer-Verlag. Electonic version available at *http://ee.stanford.edu/~gray/it.pdf*

D. H. Green and I. S. Taylor. 1974. Modular representation of multiple-valued logic systems. *Proc. IEEE*. 121: 409–418.

J. H. Green, Jr. and R. L. San Souchie. 1958. An error-correcting encoder and decoder of high efficiency. *Proc. IRE*. 46: 1741–1744.

J. H. Griesmer. 1960. A bound for error-correcting codes. *IBM Journal of Research and Development*. 4: 532–542.

Robert L. Griess. 1998. *Twelve Sporadic Groups*. New York: Springer-Verlag.

Jan Gullberg. 1997. *Mathematics from the Birth of Numbers*. New York: W. W. Norton.

J. Hadamard. 1893. Résolution d'une question relative aux déterminants, *Bulletin des Sciences Mathématiques*. 17: 240–246.

J. Hagenauer. 1996. Iterative decoding of binary block and convolutional codes. *IEEE Transactions on Information Theory*. IT-42: 429–445.

M. Hall, Jr. 1962. Note on the Mathieu group M_{12}, *Arch. Math*. 13: 334-340.

————. 1967. *Combinatorial Theory*. Waltham, MA: Blaisdell.

P. R. Halmos. 1958. *Finite Dimensional Vector Spaces*. Princeton, NJ: Van Nostrand.

Richard W. Hamming. 1950. Error detecting and error correction codes. *Bell System Technical Journal*. 26: 147–160.

R. Hamzaoui, V. Stankovic and Z. Xiong. 2002. Rate-based versus distortion-based optimal joint source-channel coding. *Proc. DCC '02*, Snowbird, UT, April 2002.

————, S. Ahmad, and M. Al-Akaidi. 2006. Reliable wireless videao streaming with digital fountain codes. *http://citeseerx.ist.psu.edu/viewdoc/down load?doi=10.1.188.2676&rep=rep1&type=pdf*. 1–18.

D. R. Hankerson, G. Hoffman, D. A. Leonard and C. C. Lidner. 1991. *Coding Theory and Cryptography: The Essentials*. New York: Marcel Dekker. Year 1991; 2nd edition. 2000. Boca Raton, FL: Chapman Hall/CRC.

F. J. Harris. 1978. On the use of windows for harmonic analysis with the discrete Fourier transforms. *Proc. IEEE*. 66: 51–83.

C. R. P. Hartmann and K. K. Tzeng. 1972. Generalizations of the BCH bound. *Information and Control*. 20: 489–498.

A. Hasegawa and P. Tappert. 1973. Transmission of stationary nonlinear optical pulses in dielectric fibers. I. Anomalous dispersion. *Appl. Phys. Lett.* 23: 142–144.

J. R. Higgins. 1985. Five short stories about the cardinal series, *Bull. Am. Math. Soc.* 12: 45–89.

N. J. Higham. 1988. Fast solutions of Vandermonde-like systems involving orthogonal polynomials. *J. Numer. Anal.* 8: 473–486.

Raymond Hill. 1988. *A First Course in Coding Theory*. New York: Oxford University Press.

A. Hocquenghem. 1959. Codes correcteurs d'erreurs. *Chiffres*. 2: 147–156.

P. G. Hoel, S. C. Port, and C. J. Stone. 1971. Testing Hypotheses. Chapter 3 in *Introduction to Statistical Theory*. New York: Houghton Mifflin. 56–67.

M. Y. Hsiao. 1970. A class of optimal minimum odd-weight-column SEC-DED codes. *IBM Journal of Research and Development*. 14: 395–401.

W. C. Huffman and V. Pless. 2003. *Fundamentals of Error-Correcting Codes*. Cambridge: Cambridge University Press.

D. R. Hughes and F. C. Piper. 1985. *Design Theory*. Cambridge, UK: Cambridge Univ. Press. 173–176.

Georges Ifrah. 2000. *From Prehistory to the Invention of the Computer*, English translation from the French book *Histoire universalle des chiffers* by Georges Ifrah and Robert Laffont, 1994; translated by David Bellos, E. F. Harding, Sophie Wood, and Ian Monk. New York: John Wiley.

H. Imai. 1990. *Essentials of Error-Control Coding Techniques*. San Diego, CA: Academic Press.

A. Jeffrey and T. Taniuti. 1964. *Nonlinear Wave Propagation*. New York: Academic Press.

H. Jin, A. Khandekar, and R. McEliece. 2000. Irregular Repeat-Accumulate Codes. *Proc. 2nd International Symposium on Turbo Codes & Related Topics*. Brest, France, Sept. 2000. 1–8.

R. S. Johnson, 1997. *A Modern Introduction to the Mathematical Theory of*

Water Waves. Cambridge, UK: Cambridge University Press.

S. Johnson, 1962. A new upper bound for error correcting codes. *IEEE Transactions on Information Theory*. 8: 203–207.

S. J. Johnson and S. R. Weller. 2003. Resolvable 2-designs for regular low-density parity-check codes. *IEEE Trans. Communication*. 51: 1413–1419.

_____ and S. R. Steven. 2006. Constrining LDPC degree distribution for improved error floor performance. *IEEE Comm. Letters*. 10 : 103–105.

M. Kanemasu. 1990. *Golay Codes*. Cambridge, MA: MIT Press.

M. Karlin. 1969. New binary coding results by circulants. *IEEE Transactions on Information Theory*. IT-15: 81–92.

T. Kasami, S. Lin, and W. W. Peterson. 1968. Polynomial codes. *IEEE Transactions on Information Theory*. IT-14: 807–814.

W. H. Kautz and K. N. Levitt. 1969. A survey of progress in coding theory in the Soviet Union. *IEEE Transactions on Information Theory*. IT-15: 197–245.

A. M. Kerdock. 1972. A class of low-rate non-linear binary codes. *Information and Control*. 20: 182–187.

H. Kharaghani and B. Tayfeh-Rezaie. 2005. A Hadamard matrix of order 428. *J. Combinatorial Designs*. 13: 435–440.

F. Kienle and N. Wehn. 2004. Design methodology for IRA codes. *Proc. 2004 Asia South Pacific Design Automation Conference* (ASP-DAC '04), Yokohama, Japan. January 2004.

R. Knobel. 1999. *An Introduction to the Mathematical Theory of Waves*. Providence, RI: American Mathematical Society.

Donald E. Knuth. 1969. *Seminumerical Algorithms*. *The Art of Computer Programming*. Vol. 2. Reading, MA: Addison-Wesley.

_____. 1997. *The Art of Computer Programming*. Vol. 2. *Seminumerical Algorithms*. 3rd ed. Reading, MA: Addison-Wesley.

_____. 2003. *The Art of Computer Programming*. Vol. 4., Fascicle 1: Bitwise tricks and techniques; binary designs. Reading, MA: Addison-Wesley. Available at *http://www.cs-faculty.stanford.edu/~knuth/fascicle1b.ps.gz*

R. Koetter and A. Vardy. 2000. Algebraic soft-decision decoding of Reed-Solomon codes. *Proc. IEEE International Symposium on Information Theory*. 61–64.

Israel Koren. 2002. *Computer Arithmetic Algorithms*. Natrick, MA: A. K. Peters.

Y. Kou. 1998. Low-density parity-check codes based on finite geometries: A rediscovery and new results. *IEEE Transactions on Information Theory*. IT-44: 729–734.

_____ , S. Lin. and M. Fossorier. 2001. Low-density parity-check codes based on finite geometries: A rediscovery and new results, *IEEE Transactions on Information Theory*. 47: 2711–2736.

_____ S. Lin, and M. P. Fossorier. 2001. Low-density parity-check codes based on finite geometries: A rediscovery and new results. *IEEE Transac-*

tions on Information Theory. 47: 2711–2736.

Leon G. Kraft. 1949. *A Device for Quantizing, Grouping, and Coding Amplitude Modulated Pulses.* MS thesis, Electrical Engineering Department. Cambridge, MA: MIT Press.

H. P. Kramer. 1957. A generalized sampling theorem, *J. Math. Phys.* 63: 1432–1436.

F. R. Kschischang, B. J. Frey, and H.-A. Loeliger. 2001. Factor graphs and the sum-product algorithm, *IEEE Transactions on Information Theory.* 47: 498–519.

P. K. Kythe. 1996. *Fundamental Solutions for Differential Operators and Applications.* Boston: Birkhäuser.

———. 2011. *Green's Functions and Linear Differential Equations: Theory, Applications, and Computation.* Boca Raton, FL: CRC Press.

——— and M. R. Scäferkotter. 2005. *Handbook of Computational Methods for Integration.* Boca Raton FL: Chapman & Hall/CRC.

David C. Lay. 2005. *Linear Algebra and Its Applications.* 3rd ed. Reading, MA: Addison-Wesley.

C. Y. Lee. 1958. Some properties of nonbinary error-correcting codes. *IRE Transactions on Information Theory.* 4: 77–82.

F. Lehmann and G. M. Maggio. 2003. Analysis of the iterative decoding of LDPC and product codes using the Gaussian approximation. *IEE Transactions on Information Theory.* IT-49: 2993–3000.

V. Levenshtein. 1961. Application of Hadamard matrices to one problem of coding theory. *Problemy Kibernetiki.* 5: 123–136.

Paul P. Lévy. 1925. *Calcul de Probabiltés.* Paris: Gauthier-Villars.

R. Lidl and H. Niederreiter. 1997. *Finite Fields.* 2nd ed. Cambridge, UK: Cambridge University Press.

David J. Lilja and Sachin S. Sapatnekar. 2005. *Designing Digital Computer Systems with Verilog.* Cambridge: Cambridge University Press.

S. Lin and D. Costello. 1983. *Error Control Coding: Fundamentals and Applications.* Englewood-Cliffs, NJ: Prentice-Hall.

——— and G. Pilz. 2004. *Applied Abstract Algebra,* 2nd edition. New York: Wiley.

D. Lind and B. Marcus. 1995. *An Introduction to Symbolic Dynamics and Coding.* Cambridge: Cambridge University Press.

San Ling and Chaoping Xing. 2004. *Coding Theory: A First Course.* Cambridge: Cambridge University Press.

D. B. Lloyd. 1964. Factorization of the general polynomial by means of its homomorphic congruent functions. *Am. Math. Monthly.* 71: 863–870.

——— and H. Remmers. 1967. Polynomial factor tables over finite field. *Math. Algorithms.* 2: 85–99.

M. Luby. 1996. *Pseudorandomness and Cryptographic Applications.* Princeton: NJ: Princeton University Press.

———. 2001/2002. Information Additive Code Generator and Decoder for

Communication Systems. *U.S. Patent No. 6,307,487*, Oct. 23, 2001; *U.S. Patent No. 6,373,406*. April 16, 2002.

———. 2002. LT Codes. *43rd Annual IEEE Symposium on Foundations of Computer Science.* http://ieeexplore.ieee.org/xpl/freeabs_all.jsp?arnumber =1181950

———. 2002. LT Codes. *Proc. IEEE Symposium on the Foundations of Computer Science.* (FOCS): 271–280.

———. 2004. FEC Architecture for Streaming Services Including Symbol-Based Operations and Packet Tagging. *U.S. Patent No. 7660245.* (http://v3. espacenet. com/ textdoc?DB=EPODOC&IDX=US7660245). Assigned to Qualcomm, Inc.

———, M. Mitzenmacher and A. Shokorallahi. 1998. Analysis of random processes via And-Or tree evaluation. *Proc. 9th Annual ACM-SIAM Symposium on Discrete Algorithms.* San Francisco, CA, January 25–27, 1998. 2002.

———, M. Mitzenmacher, A. Shokorallahi and D. Spielman. 2001. Efficient erasure correction codes. *IEEE Transactions on Information Theory, Special Issues on Codes and Graphs and Iterative Algorithms.* 47(2): 569–585.

David J. C. MacKay. 1999. Good error correcting codes based on very sparse matrices. *Transactions on Information Theory.* 45: 399–431.

———, S. T. Wilson, and M. C. Davey. 1999. Comparison of constructions of irregular Gallager codes. *IEEE Trans. on Communications.* 47: 1449–1454.

———. 2003. *Information Theory, Inference, and Learning.* Cambridge, UK: Cambridge University Press. Available at *http://www.inference.phy.cam.ac.uk mac kay/itila/*

———. 2005. Fountain codes. *IEE Proc.-Commun.* 152: 1062–1068.

——— and R. M. Neal. 1996. Near Shannon limit performance of low-density parity-check codes. *Electronics Letters.* 32: 1645–1646; Reprinted with errors corrected in *Electronics Letters.* 33: 457–458.

Florence J. MacWilliams and Neil J. A. Sloane. 1977. *The Theory of Error-Correcting Codes.* Amsterdam: North-Holland Publishing Company.

M. Mansour and N. Shanbhag. 2003. Architecture-Aware Low-Density-Parity-Check Codes. *Proc. 2003 IEEE International Symposium on Circuits and Systems* (ISACS '03), Bangkok, Thailand, May 2003.

Y. Mao and A. H. Banihashami. 2001. Decoding low-density parity-check codes with probabilistic scheduling. *IEEE Commun. Lett.* 5: 414–416.

James L. Massey. 1965. Step-by-step decoding of Bose-Chaudhuri-Hocquenghem codes. *IEEE Transactions on Information Theory.* IT-11: 580–585.

———. 1969. Shift-register synthesis and BCH decoding. *IEEE Transactions on Information Theory.* IT-15: 122–127.

———. 1990. Marcel J. E. Golay (1902–1989), *Information Society Newsletter.*

———. 1992. Deep Space Communications and Coding: A Match Made in Heaven, in *Advanced Methods for Satellite and Deep Space Communications*, J. Hagenauer (Ed.), Lecture Notes in Control and Information Sciences, Volume 182. Berlin: Springer-Verlag.

E. Mathieu. 1861. Mémoire sur l'étude des fonctions de plusieurs quanttés, sur la manière de les former et sur les substitutions qui les laissent invariables. *J. Math. Pures Appl.* (Liouville) (2): 241–323.

――――. 1873. Sur la fonction cinq fois transitive de 24 quantités. *Liouville J.* (2) XVIII: 25–47.

H. F. Mattson, Jr. and G. Solomon. 1961. A new treatment of Bose-Chaudhuri codes. *J. Soc. Indust. Appl. Math.* 9: 654–669.

Theresa Maximo. 2006. Revisiting Fletcher and Adler checksums. *DNS Student Forum.* (PDF file): *http//:www.zib.net/maxino06_fletcher-adler.pdf*

MBMS. 2005. Technical Specification Group Services and Systems Aspects: Multimedia Broadcast/Multicast Services (MBMS): Protocols and Codes (Release 6). *3rd Generation Partnership Project (3GPP)*, Tech. Report 3GPP TS 26 346 V6.3.0, 3GPP.

R. J. McEliece. 1977. *The Theory of Information and Coding.* Encyclopedia of Math. and Its Applications, Vol. 3. Reading, MA: Addison-Wesley.

――――. 1987. *Finite Fields for Computer Scientists and Engineers.* Boston: Kluwer Academic.

――――, E. R., Rodemich, H. C. Rumsey, and L. R. Welch. 1977. New upper bounds on the rate of a code via the Delsarte-MacWilliams inequalities. *IEEE Transactions on Information Theory.* IT-23: 157–166.

――――, D. J. C. MacKay, and J.-F. Cheng. 1998. Turbo decoding as an instance of Pearl's 'belief propagation' algorithm. *IEEE Journal on Selected Areas of Communications.* 16(2): 140–152.

Brockway McMillan. 1956. Two inequalities implied by unique decipherability, *IEEE Transactions on Information Theory.* 2: 115–116.

T. Meltzer, C. Yanover, and Y. Weiss. 2005. Globally optimal solutions for energy minimization in stereo vision using reweighted belief propagation. *Int. Conf. on Computer Vision 2005.*

L. F. Mollenauer and J. P. Gordon. 2006. *Solitons in Optical Fibers.* San Diego, CA: Elsevier Academic Press.

Todd K. Moon. 2005. *Error Correction Coding.* Newark, NJ: John Wiley.

D. E. Muller. 1954. Application of Boolean algebra to switching circuit design and error detection. *IRE Trans. Electron. Comp.* EC-3: 6–12.

H. Murakami and I. S. Reed. 1977. Recursive realization of the finite impulse filters using finite field arithmetic. *IEEE Transactions on Information Theory.* IT-23: 232–242.

K. P. Murphy, Y. Weiss, and M. I. Jordan. 1999. Loopy belief propagation for approximate inference: An empirical study. *Uncertainty in AI.*

R. M. Neal. 1999. Sparse matrix methods and probabilistic inference algorithm. Part I: Faster encoding for LDPC codes using sparse matrix methods. *IMA Program on Codes, Systems and Graphical Models. http://www*

.cs.utorento.ca/ ~radford/ radford@stat.utorento.ca

J. Neyman and E. Pearson. 1933. On the problem of the most efficient tests of statistical hypotheses. *Phil. Trans. Royal London Society, Ser. A.* 231: 289–337.

H. Niederreiter. 1977. Weights of cyclic codes. *Information and Control.* 34: 130–140.

A. F. Nikiforov, S. K. Suslov and V. B. Uranov. 1991. *Classical Orthogonal Polynomials of a Discrete Variable.* Berlin: Springer-Verlag.

A. W. Nordstrom and J. P. Robinson. 1967. An optimal nonlinear code. *Inform. Control.* 11: 613–616.

O. Novac, St. Vari-Kakas and O. Poszet. 2007. Aspects regarding the use of error detecting and error correcting codes in cache memories. *EMES'07.* University of Oradea.

J. P. Odenwalder. 1976. *Error Control Coding Handbook.* San Diego: Linkabit Corporation, July 15, 1976.

J. K. Ord. 1972. *Families of Frequency Distributions.* New York: Griffin.

A. Orlitsky, K. Viswanathan, and J. Zhang. 2005. Stopping set distribution of LDPC code ensembles. *IEEE Trans. Inform. Theory.* IT-51:929–953.

P. Oswald and A. Shokorallahi. 2002. Capacity-achieving sequences for the erasure channel. *IEEE Transactions on Information Theory.* IT-48: 3017–3028.

T. Ott and R. Stoop. 2006. The neurodynamics of belief propagation on binary Markov random fields. *NIPS.*

C. Pal, C. Sutton, and A. McCallum. 2006. Sparse forward-backward using minimum divergence beams for fast training of conditional random fields. *Proc. Intl. Conf. on Acoustics, Speech, and Signal Processing.* 5: 581–584.

R. Palanki and J. S. Yedidia. 2003. Rateless Codes on Noisy Channels. *TR2003-124.* October 2003. Cambridge, MA: Mitsubishi Electric Research Laboratories. (*http://www.merl.com*).

R. E. A. C. Paley. 1933. On orthogonal matrices. *J. Math. Phys.* 12: 311–320.

J. Pearl. 1982. Reverend Bayes on inference engines: A distributed hierarchical approach. *AAAI-82:* Pittsburgh, PA, *2nd National Conference on Artificial Intelligence,* Menlo Park, CA: AAAI Press. 133–136.

————. 1988. *Reasoning in Intelligent Systems: Networks of Plausible Inference.* San Mateo, CA: Morgan Kaufmann.

W. W. Peterson. 1960. Encoding and error-correction procedures for the Bose-Chaudhuri codes. *IEEE Transactions on Information Theory.* IT-6: 459–470.

————. 1961. *Error Correcting Codes,* 2nd edition. Cambridge, MA: MIT Press.

————and E. J. Weldon. 1972. *Error Correcting Codes,* 2nd edition. Cambridge, MA: MIT Press.

————. 1969. Some new results on finite fields and their application to the theory of BCH codes. *Proc. Conf. Math. and Its Appls.* Chapel Hill, NC, 1967. pp. 329–334. Chapel Hill, NC: University of North Carolina.

———— and D. T. Brown. Cyclic codes for error detection. *Proc. IRE.* 49: 228–235.

———— and E. J. Weldon, Jr. 1972. *Error-Correcting Codes.* Cambridge, MA: M.I.T. Press.

S. Pietrobon. 1998. Implementation and performance of a Turbo/MAP decoder. *International Journal Satellite Communication.* 16: 23–46.

James S. Plank. 2004. All About Erasure Codes — Reed-Solomon Coding — LDPC Coding. *ICL, August 20, 2004.* Logistical Computing and Internetworking Laboratory, Department of Computer Science, University of Tennessee.

Vera Pless. 1982. *Introduction to the Theory of Error-Correcting Codes.* New York: John Wiley.

M. Plotkin. 1960. Binary codes with specified minimum distance. *IRE Transactions on Information Theory.* 6: 445–450.

A. Poli and L. Huguet. 1992. *Error Correcting Codes: Theory and Applications.* Englewood Cliffs, NJ: Prentice-Hall.

F. Pollara, R. J. McEliece, and K. Abdel-Ghaffar. 1987. Construction of Finite-State Codes, *TDA Progress Report 42-90,* April-June 1987. Jet Propulsion Laboratory, Padsadena, CA. 42–49.

B. Potentz and T. S. Lee. 2008. Efficient belief propagation for higher-order cliques using linear constraint nodes. *Comput. Vis. Image Understanding.* 112: 39–54.

E. Prange. 1957. *Cyclic error-correcting codes in two symbols.* AFCRC-TN-57. 103: Sept 1957.

F. P. Preparata. 1968. A class of optimum nonlinear double-error-correcting codes. *Information and Control.* 13: 378–400.

W. V. Quine. 1982. *Methods of Logic,* 4th ed. Cambridge, MA: Harvard University Press.

L. Rabiner. 1989. A tutorial on hidden Markov models and selected applications in speech recognition. *Proc. IEEE.* 77: 257–286.

Lennart Råde and Bertil Westergren. 1995. *Mathematics Handbook for Science and Engineering.* Boston: Birkhäuser.

C. R. Rao. 1947. Factorial experiments derivable from combinatorial arrangements of arrays. *J. Royal Statist. Soc. Suppl.* 29: 128–139.

T. R. N. Rao and E. Fujiwara. 1989. *Error-Control Coding for Computer Systems.* Englewood Cliffs, NJ: Prentice-Hall International.

I. S. Reed. 1954. A class of multiple-error-correcting codes and the decoding scheme. *IRE Trans. Information Theory.* PGIT-4: 38–49.

———— and G. Solomon. 1960. Polynomial codes over certain finite fields. *J. Society Indust. Appl. Math.* 8: 300–304.

———— R. A. Scholtz, T. K.. Truong, and L. R. Welch. 1978. The fast decoding of Reed-Solomom codes using Fermat theoretic transforms and continued fractions. *IEEE Transactions on Information Theory.* IT-24: 100–106.

———— T. K. Truong, and R. L. Miller. 1979. Decoding of B.C.H. and R.S. codes with errors and erasures using continued fractions. *Electron Letters.* 15: 493–544.

———— T. K. Truong and L. R. Welch. 1978. The fast decoding of Reed-Solomon codes using Fermat transforms. *IEEE Transactions on Information Theory.* IT-24: 497–499.

B. Reiffen. 1962. Sequential decoding for discrete input memoryless channels. *Proc. IRE.* IT-8: 208–220.

E. M. Reingold, J. Nivergeld and N. Deo. 1977. *Combinatorial Algorithm. Theory and Practice.* Englewood Cliffs, NJ: Prentice-Hall.

M. Y. Rhee. 1989. *Error-Correcting Coding Theory.* New York: McGraw-Hill.

T. J. Richardson and R. L. Urbanke. 2001a. The capacity of low-density parity-check codes under message-passing decoding. *IEEE Transactions on Information Theory.* 47: 599–618.

————, A. Shokrollahi and R. Urbanke. 2001b. Design of capacity approaching irregular low-density parity-check codes. *IEEE Transactions on Information Theory.* 47: 619–637.

————, A. Shokrollahi and R. Urbanke. 2001c. Efficient encoding of low-density parity-check codes. *IEEE Transactions on Information Theory.* 47: 638–656.

———— and R. Urbanke. 2003. The renaissance of Gallager's low-density parity-check codes. *IEEE Communications Magazine.* 41: 126–131.

————, A. Shokrollahi, and R. L. Urbanke. 2001. Design of capacity-approaching irregular low-density parity-check codes. *IEEE Transactions in Information Theory.* 47: 619–637.

———— and R. Urbanke. 2008. *Modern Coding Theory.* Cambridge, UK: Cambridge University Press.

P. Robertson, E. Villeburn, and P. Hoeher. 1995. A comparison of optimal and sub-optimal MAP decoding algorithms operating in the log domain. *Proceedings of ICC 1995*, Seattle, WA, June 1995. 1009–1013.

Steven Roman. 1992. *Coding and Information Theory.* New York: Springer-Verlag.

S. Ross. 2002. *Probability Models for Computer Science.* New York: Harcourt/Academic Press.

W. Rudin. 1976. *Principles of Mathematical Analysis.* 3rd ed. New York: McGraw-Hill.

I. Sason, H. D. Pfister, and R. Urbanke. 2006. Capacity-achieving codes for the binary erasure channel with bounded complexity. *Summer Research Institute 2006, School of Computer and Communication Sciences (I&C)*, EPFL — Swiss Federal Institute of Technology, Lausanne, CH, July 18th, 2006.

C. Schlegel and L. Perez. 2004. *Trellis and Turbo Coding*. New York: Wiley Interscience.

H. Schmidt. 1974. *Decimal Computation*. New York: John Wiley. Reprinted by Robert E. Krieger Publishing Co. in 1983.

J. T. Schwarz. 1980. Fast probabilistic algorithms for verification of polynomial identities. *J. ACM*. 27: 701–717.

J. Seberry and M. Yamada. 1992. Hadamard matrices, sequences and block designs, in *Contemporary Design Theory: A Collection of Surveys*, (Ed. J. H. Dinitz and D. R. Stinson). 431–560. New York: Wiley.

―――― , B. Wysocki, and T. Wysocki. 2005. Some applications of Hadamard matrices. *Metrika*. 62: 221–239.

E. Senata. 1973. *Non-Negative Matrices*. New York : Wiley.

G. I. Shamir and J. J. Boutros. 2005. Non-systematic low-density parity-check codes for nonuniform sources. *Proc. ISIT-2005*, Adelaide, Australia, Sept. 2005. 1898–1902.

C. E. Shannon. 1948. A mathematical theory of communication, *Bell System Tech. J*. 27: 379–423.

E. Sharon, S. Lytsin, and J. Goldberger. 2004. An efficient message-passing schedule for LDPC decoding. *Proc. 23rd IEEE Convention of Electrical and Electronics Engineers in Israel*. Sept. 2004.

A. Shokrollahi. 1999. New sequences of linear time erasure codes approaching the channel capacity. *Proc. 13th Int. Symposium Applied Algebra, Algebraic Algorithms, and Error-Correcting Codes*. Lecture Notes in Computer Science. M. Fossorier, H. Imai, S. Lin, and A. Poli, Eds. Berlin: Springer-Verlag, vol. 1719, 65–76.

―――― . 2003. Raptor codes. *Tecnical Report, Laboratoire d'algorithmique, École Polytechnique Fedérale de Lausanne, Lausanne, Switzerland*. Available at *http://www.inference.phy.cam.ac.uk/mackay/ dfountain/Raptor Paper.pdf*: 1–37.

―――― . 2004. Raptor codes. *http"//citeseerx.ist.psu.edu/viewdoc/download ?doi=10.1.1.105*: 1–35.

―――― . 2006. Raptor codes. *IEEE Transactions on Information Theory*. IT-52: 2551–2567.

―――― , S. Lassen, and M. Luby. 2001. Multi-stage code generator and decoder for communication systems. *U.S. Patent Application # 20030058958*, Dec. 2001.

R. C. Singleton. Maximum distance q-ary codes. *IEEE Transactions on Information Theory*. IT-10: 116–118.

B. Skalar. 2001. *Digital Communications: Fundamentals and Applications*. 2nd ed., Upper Saddle River, NJ: Prentice-Hall.

D. Slepian. 1956. A note on binary signaling alphabets. *Bell Systems Tech. J*. 35: 203–234.

―――― . 1960. Some further theory of group codes. *Bell Systems Tech. J*. 39: 1219–1252.

————. 1974. *Key Papers in the Development of Information Theory.* New York: IEEE Press.

N. I. A. Sloane. 1975. *A short Course on Error Correcting Codes.* New York: Springer-Verlag.

I. N. Sneddon. 1957. *Partial Differential Equations.* New York: McGraw-Hill.

————. 1958. *Fourier Transforms and Their Applications.* Berlin: Springer-Verlag.

G. Solomon and J. J. Stiffler. 1965. Algebraically punctured cyclic codes. *Information and Control.* 8:170–179.

D. M. Y Sommerville. 1958. *An Introduction to the Geometry of n Dimensions.* New York : Dover.

H . Y. Song and S. W. Golomb. 1994. Some new constructions for simplex codes. *IEEE Transactions on Information Theory.* 40: 504–507.

V. Sorkine, F. R. Kschlischang, and S. Pasupathy. 2000. Gallager codes for CDMA applications — Part I: Generalizations, Constructions, and Performance bounds. *IEEE Trans. on Communications.* 48: 1660–1668.

D. A. Spielman. 1996. Linear-time encodable and decodable codes. *IEEE Transactions on Information Theory.* 42: 1723–1731.

J. Steiner. 1853. Combinatorische Aufgabe. *Journal für die reine und angewandte Mathematik.* 45: 181–182.

I. Stoica, R. Morris, D. Liben-Nowell, D. R. Karger, M. F. Kaashoek, F. Dabek, and H. Balakrishnan. 2003. Chord: a scalable peer-to-peer lookup protocol for internet applications. *IEEE/ACM Trans. Network* 11: 17–23.

E. Sudderth, A. Ihier, W. Freeman, and A. Willsky. 2003. Nonparametric belief propagation. *Conference on Computer Vision and Pattern Recognition* (CVOR).

J. J. Sylvester. 1867. Thoughts on inverse orthogonal matrices, simultaneous sign successions, and tessellated pavements in two or more colours, with applications to Newton's rule, ornamental tile work, and the theory of numbers, *Philosophical Magazine.* 34: 461–475.

R. Szeliski, R. Zabih, D. Scharstein, O.Veksler, V. Kolmogorov, A. Agarwala, M. Tappen, and C. Rother. 2008. A comparative study of energy minimization methods for Markov random fields with smoothness-based priors. *IEEE Trans. on Pattern Analysis and Machine Intelligence.* 30: 1068–1080.

Simon Tam. 2006. Single Error Correction and Double Error Detection, *XAPP645.2.2*; Aug 9, 2006, available at *http://www.xilinx.com*

A. S. Tanenbaum. 2003. *Computer Networks.* 4th ed. Englewood Cliffs, NJ: Prentice-Hall.

M. R. Tanner. 1981. A recursive approach to low complexity codes. *IEEE Transactions on Information Theory.* 27: 533–547.

M. F. Tappen and W. T. Freeman. Comparison of graph cuts with belief propagation for stereo, using identical MRF parameters. *International Conference on Computer Vision* (ICCV).

T. M. Thompson. 1983. *From Error Correcting Codes through Sphere Packing to Simple Groups*, Carus Mathematical monographs. Mathematical Association of America.

T. Tian, C. Jones, J. D. Villasenor and R. D. Wesel. 2003. Construction of irregular LDPC codes with low error floors. *Proceddings IEEE International Conference on Communications, Anchorage, AK, May 2003.* 5: 3125–3129.

A. Tiertäväinen. 1974. On the nonexistence of perfect codes over finite fields. *SIAM J. Applied Math.* 24: 88–96.

_____. 1974. A short proof of the nonexistence unknown perfect codes over GF(4), $q > 2$. *Ann. Acad. Sci. Fenn. Ser. AI.* 580: 6 pp.

Tuomas Tirronen. 2006. *Optimizing the Degree Distribution of LT Codes.* Master's thesis. March 16, 2006. Helsinki University of Technology.

_____ T. K. Truong, and R. L. Miller. 1979. Decoding of B.C.H. and R.S. codes with errors and erasures using continued fractions. *Electron Lett.* 15: 542–544.

_____ and L. R. Welch. 1978. The fast decoding of Reed-Solomon codes using Fermat transforms. *IEEE Transactions on Information Theory.* IT-24: 497–499.

M. A. Tsfasman, S, G, Vlaut, and T. Zink. 1982. Modular curves, Shimura curves and Goppa codes which are better than the Varshamov–Gilbert Bound. *Mathematische Nachrichten.* 109: 21–28.

R. J. Turyn. 1974. Hadamard matrices, Baumert-Hall units, four-symbol sequences, pulse compression, and surface wave encodings. *Journal Combin. Theory. Ser. A* 16: 313–333.

N. G. Ushakov. 2001. Density of a probability distribution, in M. Hazewinkel, *Encyclopaedia of Mathematics.* New York: Springer-Verlag.

J. H. van Lint. 1971. *Coding Theory.* Lecture Notes in Math., Vol. 201. Berlin: Springer-Verlag.

_____. 1975. A survey of perfect codes, *Rockey Mountain Math. Journal.* 5: 199–224.

_____. 1983. Kerdock Codes and Preparata Codes, *Congressus Numeratium.* 39: 25–41.

_____. 1992. *Indtroduction to Coding Theory*, Graduate Text in Mathematics # 86. Berlin: Springer-Verlag.

_____ and R. M. Wilson. 1992. *A Course in Combinatorics.* Cambridge, UK: Cambridge University Press.

R. R. Varshamov. 1957. Estimate of the number of signals in error correcting codes (Russian). *Dokladi Akad. Nauk SSSR.* 117: 739–741.

J. Verhoeff. 1969. Error Detecting Decimal Codes. *Mathematical Centre Tract.* 29. Amsterdam: The Mathematical Centre.

John von Neumann. 1951. Various techniques used in connection with random digits, in *Monte Carlo Methods*, A. S. Householder, G. E. Forsythe and H. H. Germond (Eds.) Washington, D.C.: U.S. Government Printing

Office: 36–38.

P. Wagner. 1960. A technique for couting ones in a binary computer. *Communications of the ACM*. 3: 322.

M. Wainwright, T. Jaakkola, and A. Willsky. 2005. Map function via agreement on trees: Message-passing and linear programming. *IEEE Transactions on Information Theory*. 51: 3697–3717.

John F. Wakerly. 2000. *Digital Design Principles and Practices*, 3rd ed. Englewood Cliffs, NJ: Prentice-Hall.

E. J. Watson. 1962. Primitive polynomials (mod 2). *Math. Comp.* 16: 368–369.

H. Weatherspoon and J. D. Kubiatowicz. 2002. Erasure coding vs. replication: A quantative comparison. *Peer-to-Peer Systems: First International Workshop*, IPTPS, LNCS 2429: 328–337.

C. L. Weber. 1987. *Elements of Detection and Signal Designs*. New York: Springer-Verlag.

R. L. Welch. 1997. *The Original View of Reed-Solomon Codes*. http://csi.usc. edu/ PDF/RSoriginal.pdf

D. B. West. 1996. *Introduction to Graph Theory*. Englewood Cliffs, NJ: Prentice-Hall.

S. B. Wicker. 1994. *Error Control Systems for Digital Communication and Storage*. Englewood Cliffs, NJ: Prentice-Hall.

———— and V. K. Bhargava (Eds.). 1999. *An Introduction to Reed-Solomon Codes*. New York: Institute of Electrical and Electronics Engineers.

S. S. Wilks. 1938. The large-sample distribution of the likelihood ratio for testing composite hypotheses. *The Annals of Mathematical Statistics*. 9: 60–62.

J. Williamson. 1944. Hadamard's determinant theorem and the sum of four squares. *Duke Math. J.* 11: 65–81.

———— and R. M. Wilson. 1992. *A Course in Combinatorics*. Cambridge: Cambridge University Press.

R. R. Yale. 1958. Error correcting codes and linear recurring sequences. *Report MIT Lincoln Laboratory, Lexington, MA*. 34–77.

M. Yang and W. E. Ryan. 2003. Lowering the error-rate floors of moderate-length high-rate irregular LDPC codes. *ISIT*.

C. Yanover and Y. Weiss. 2003. Finding the most probable configurations using loopy belief propagation. *NIPS*.

J. Yedidia, W. T. Freeman, and Y. Weiss. 2001. Characterizing belief propagation and its generalizations. TR2001-015. Available online at *http://www. merl.com/ reports/TR2001-15/index.html*

————, W. T. Freeman, and Y. Weiss. 2002. Understanding Belief Propagation and its Generalizations. Mitsubishi Electric Research Laboratories. *http://www.merl. com*

————, W. T. Freeman, and Y. Weiss. 2001. Bethe free energy, Kikuchi approximations, and belief propagation algorithms. TR2001-016. Available online at *http://www.merl.com/reports/TR2001-16/index.html*

N. J. Zabusky and M. D. Kruskal. 2002. Interaction of solitons in a collisionless plasma and recurrence of initial states. *Phys. Rev. Letters.* 15: 240–243.

Ahmed I. Zayed. 1993. *Advances in Shannon's Sampling Theory.* Boca Raton, FL: CRC Press.

L.-H. Zetterberg. 1962. Cyclic codes from irreducible polynomials for correction of multiple errors. *IEEE Transactions on Information Theory.* IT-8: 13–21.

J. Zhang and M. P. C. Fossorier. 2002. Stuffed belief propagation decoding. *Proc. Anu. Asilomar Conf.*, Nov. 2002.

N. Zierler. 1968. Linear recurring sequences and error-correcting codes, in *Error Correcting Codes* (H. B. Mann, Ed.), 47–59. New York: Wiley.

V. A. Zinov'ev and V. K. Leont'e. 1973. On non-existence of perfect codes over Galois fields (Russian). *Problemy Uprav. i Theor. Informacii*, 2: 123–132. *Problems of Control and Information Theory.* 2: 16–24.

Richard Zippel. 1979. Probabilistic algorithms for sparse polynomials. In *Symbolic and Algebraic Computation.* Springer-Verlag: LNCS 72: 216–226.

V. V. Zyablov. 1971. An estimate of the complexity of constructing binary linear cascade codes. *Probl. Peredachi Informacii*, 7(1): 5–13.

Index